63210130

INSECTS AND HYGIENE

INSECTS AND HYGIENE

The biology and control of insect pests of medical and domestic importance

James R. Busvine

Ph.D., D.Sc., F.I. Biol.

Emeritus Professor of Entomology as Applied to Hygiene, London School of Hygiene and Tropical Medicine

THIRD EDITION

1980

London New York

Chapman and Hall

150th Anniversary

First published 1951 by Methuen and Co., Ltd
Second edition 1966
Third edition published in 1980 by
Chapman and Hall Ltd, 11 New Fetter Lane, London EC4P 4EE

Published in the U.S.A.
by Chapman and Hall
in association with
Methuen, Inc. 733 Third Avenue, New York, N.Y. 10017

Printed in Great Britain by J. W. Arrowsmith Ltd, Bristol

ISBN 0 412 15910 4

British Library Cataloguing in Publication Data

Busvine, James Ronald
Insects and hygiene. – 3rd ed.
1. Insects, Injurious and beneficial
I. Title
632′.7 SB931 79–41115

ISBN 0–412–15910–4

Contents

Preface to the third edition

It was gratifying to be invited to prepare a third edition of this book, which first appeared in 1951. Preliminary discussions with the publishers, however, revealed a considerable challenge in the present high costs of printing, so that changes and some improvements were clearly necessary to justify the venture.

It was immediately apparent that the chapter on chemical control measures would have to be substantially re-written, because of the great changes in usage due to resistance and the regulations introduced to prevent environmental pollution. Also, I decided to expand the scope of the book by increased coverage of the pests of continental Europe and North America, including some new figures and keys in the Appendix.

These two undertakings resulted in considerable expansion in length, with about 370 new references and 250 additional specific names in the Index. In order to avoid too alarming an increase in price, I decided to sacrifice three chapters from the earlier editions: those dealing with the structure and classification of insects, their anatomy and physiology, and their ecology. Readers who require basic biological information on insects should buy one of the various short introductions to entomology available.

A further small saving in length has been made by eliminating the various notes on historical curiosities. Those interested may care to obtain a complete account of this bizarre but fascinating subject in my recent *Insects, Hygiene and History*, (1976), Athlone Press, London, 262 pp.

CHAPTER ONE

Insects and hygiene

1.1 NUMBERS AND VARIETY OF INSECTS AND ACARINES

If numbers and variety are taken as the criteria of successful evolution, then insects are the predominant form of life on earth. Other terrestrial arthropods are also more numerous and varied than vertebrates. Thus, there are about 50 000 kinds of spider and perhaps 30 000 different mites, compared to a mere 4500 species of mammals and 8600 birds; but the insects far exceed all these figures, with an estimated number of species near to 900 000. This enormous proliferation is a consequence of their evolutionary history, which began far back in geological time (see Fig. 1.1).

The predominance of insects over other land arthropods is almost certainly due to the innovation of wings, which conferred immense advantages to such small creatures. All arthropods have to moult at intervals to grow, however, and the possession of wings renders this difficult. The evolutionary solution was to postpone the emergence of wings until the insect was a fully-grown adult, which can take advantage of them for migration and colonization. A further advance was the evolution of a life history with radically different larvae and adults, the former specialized for feeding and growing, and the latter for mating and dispersal. This 'complete metamorphosis' is familiar to all of us from the example of the caterpillar and the butterfly, and is common to most of the more successful types of insect.

Although the actual magnitude of the insect hordes is not generally realized, most people are familiar with the typical forms of the main orders. These include the following: Coleoptera (beetles) 360 000 species; Lepidoptera (moths and butterflies) 160 000; Hymenoptera (ants, bees, wasps etc) 150 000; Diptera (flies, gnats etc) 90 000; Hemiptera (bugs, aphids, cicadas etc) 61 000; Orthoptera (cockroaches, grasshoppers etc) 30 000; Mallophaga (biting lice) 3000; Siphonaptera (fleas) 2000; Anoplura (sucking lice) 250.

Although these vast numbers of insects have exploited every conceivable niche in the terrestrial environment, only a very small proportion of species (about 0·5 per cent) have become serious pests. Most of these attack plants or plant products. Human agriculture upsets the 'balance of nature' by planting large areas of unmixed crops and storing large quantities after harvest. These great supplies of food invite certain kinds of insect to proliferate excessively and become pests. Most of them belong to the orders Coleoptera, Lepidoptera and Hemiptera, and the losses caused by them are enormous.

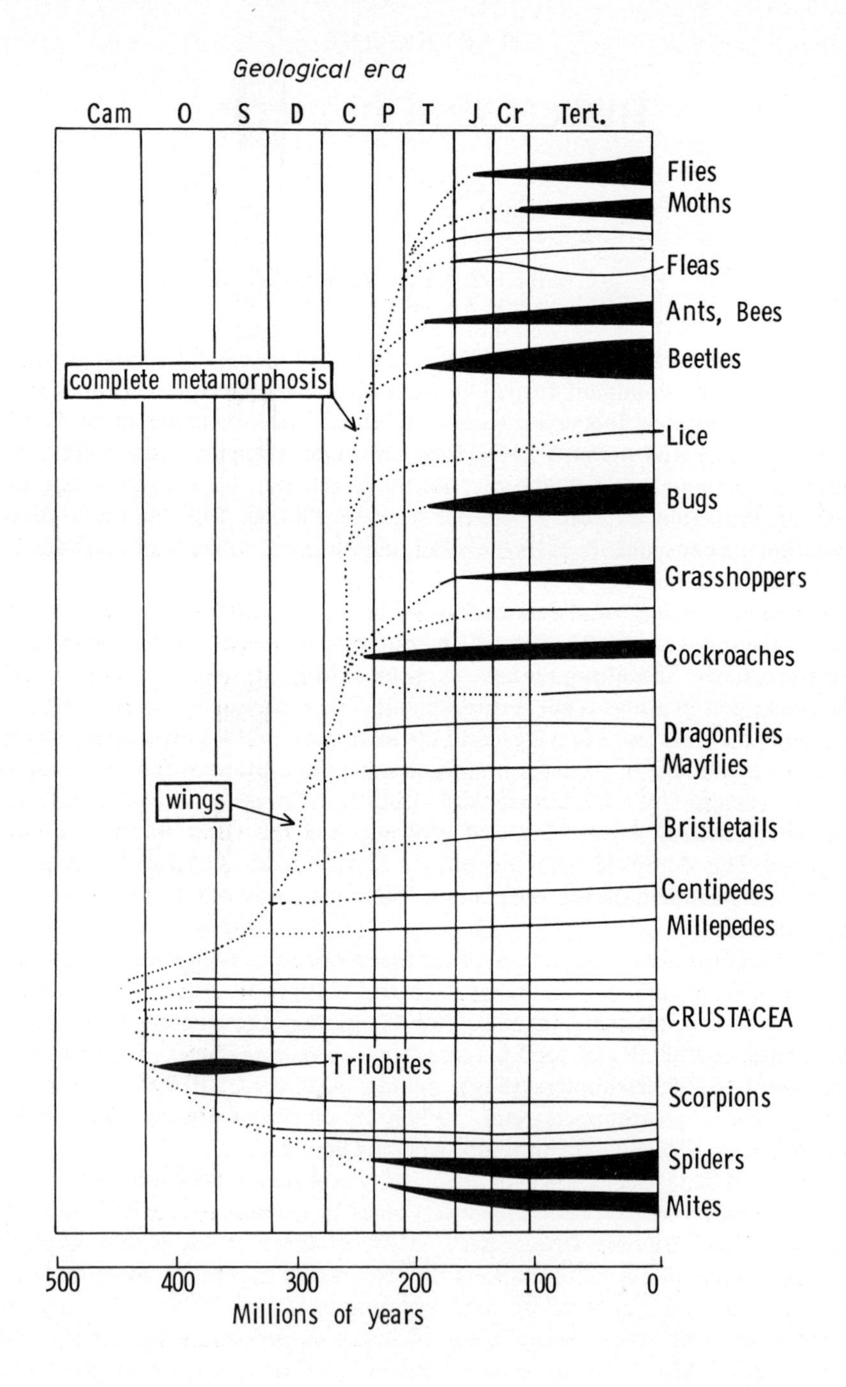

Geological era
Cam
O
S
D
C
P
T
J
Cr
Tert.
Flies
Moths
Fleas
Ants, Bees
Beetles
complete metamorphosis
Lice
Bugs
Grasshoppers
Cockroaches
Dragonflies
Mayflies
wings
Bristletails
Centipedes
Millepedes
CRUSTACEA
Trilobites
Scorpions
Spiders
Mites
500
400
300
200
100
0
Millions of years

Relatively fewer forms of insect are harmful to man and his animals because of differences in their evolutionary time scales. At the time of the emergence of the first mammals, 200 million years ago, insects had already established themselves as the most successful arthropod group on land. Certain kinds (especially in the order Diptera) took advantage of the new sources of food provided by mammals by sucking their blood or by breeding in their excrement or dead bodies. By the Oligocene (20 million years ago) the insect's main evolutionary history was over, so that specimens trapped in resin and fossilized in amber scarcely differ from their present day descendants. The past 100 000 years, which saw the emergence of man, have been too short to attract the undivided attention of more than a very few species. Nevertheless, insects which attack man as well as other mammals are vectors of some of the most serious human diseases, such as malaria, yellow fever, sleeping sickness, plague and various forms of filariasis.

Most of this book concerns insect pests; but it is convenient to include certain mites and ticks of public health importance. Most acarines are very small mites and relatively few of the larger, exclusively blood-sucking, ticks. Most are harmless to man and, because of their small size and cryptic habits, seldom noticed. However, a small number are parasitic on man or his domestic animals and some of them transmit dangerous disease organisms. Again, however, scarcely any are specific to man and the diseases they carry are brought from other animals. Examples are mite-typhus, tick-borne relapsing fever and Rocky Mountain spotted fever.

1.2 INSECTS AS DISEASE VECTORS

1.2.1 EVOLUTION OF VECTOR TRANSMISSION

The micro-organisms which are involved in vector-borne diseases vary greatly, ranging from viruses to tiny nematode worms. The *viruses* are very small, from 10 to 250 nm*, the smallest being comparable to a large protein molecule. Next in size are the *rickettsiae*, extending up to the dimensions of a small bacterium, rather less than 1 μm*. Both viruses and rickettsiae (which are probably degenerate bacteria) are exclusively parasitic and cannot be cultured in non-organic media. Bacteria are a large and variable assembly, with free-living as well

* A nanometre (nm) is one millionth part of a millimetre; a micrometre (μm) is one thousanth part of a millimetre.

Figure 1.1 Some of the main branches of the arthropod family tree. The firm lines show the geological period from which the earliest fossils of various orders occur; dotted lines show putative relationships. *Key to geological eras*: Cam., Cambrian; O, Ordovician; S, Silurian; D, Devonian; C, Carboniferous; P, Permian; T, Triassic; J, Jurassic; Cr., Cretaceous; T, Tertiary. From Busvine [94], by permission of the Athlone Press.

as parasitic forms. Two representatives of the latter are the plague bacillus ($1{\cdot}8 \times 0{\cdot}6$ μm) and the spirochaetes of relapsing fever (8–15 μm long and 0·3 μm thick). Protozoa are slightly larger and a higher form of life, with a true nucleus and chromosomes. Representatives are the malaria parasite *Plasmodium* (2–4 μm) and the larger flagellates, such as the *Leishmania* of kala azar, and the long trypanosomes of sleeping sickness, which may be 12–40 μm long. Finally, there are the parasitic nematode worms known as filaria, of which the infective stages are about 300 μm long and 6 μm thick.

The largest of these pathogens must weigh only about 10^{-5} g; this compares with 10^{-2}g as an approximation for most of the insect vectors and 10^{3}g, the weight of a man. The very small pathogens have problems of existence very different from those of larger animals, especially the parasitic forms, which have to get from one host to another. Some animal pathogens solve this by passing to a new host during close social, familial or sexual contact. Others pass out in faeces with the possibility of contaminating food or drinking water. Still others may be transferred in droplets sneezed or coughed out by the host. The most curious and complex modes of transfer, however, have evolved in a series of adaptions involving two different-sized organisms: a large animal (or plant) and a small creature such as an insect. The advantages of the alternation is that it combines a large host with a longer life, constituting a lasting reservoir with a small mobile vector, which facilitates long-range transmission. This system seems to have developed both from forms originally parasitic on large animals and from those parasitic on arthropods, though it is not always easy to guess which. Some clues can be gathered from the degree of adaption of the parasite, which usually evolves from being a pathogen to a more or less benign parasite. Many of the vector-borne diseases affecting man seem to have begun as parasites of animals, which have evolved to a more or less harmless stage. The cycle of animal and vector continues in the wild and human cases are incidental offshoots. Their malign effects on man are evidence of recent extension to a new host. Sometimes, as in louse-borne typhus, a new insect vector seems to be involved and the parasite is lethal to the vector as well.

The various mechanisms involved in the alternating system allow for speculation on the ways in which the most efficient ones were evolved. Perhaps the simplest cycle is exemplified by the fortuitous transmission of enteric diseases which occurs when flies (or cockroaches) transfer pathogens from faeces or septic matter to food, by accidental contamination due to their habits. The advent of bloodsucking forms provided a new and efficient mode of transfer of blood parasites. This probably began through traces of infected blood on the biting mouthparts of the vector, which still occurs in some cases; but because the traces are so small and may dry up and kill the pathogen, it is not efficient. A more prolonged infection of the biting vector would be the next step; and at this point, began the obligatory association of each pathogen with a particular vector. The importance of this is obvious. Thus, malaria cannot be transmitted without anopheline mosquitoes; if they are exterminated, the disease is eradicated. But enteric diseases can and do continue (though possibly somewhat abated), if all flies and other potential vectors are controlled.

The first step towards parasitism of the vector may have been a gut infection, either of the foregut or the hindgut. Infection of the next host could either be by regurgitation at the next meal, or by passing infective faeces which could enter a scratch or mucous membrane of the next host. In the most complete form of parasitism of the vector, however, the pathogen passes through the gut wall into the vector's body cavity. A crude method of escape for the pathogen was by the vertebrate host swallowing the vector, which can often happen to parasitic insects when animals groom themselves. A human example in some primitive communities is the practice of killing lice by biting them, which can result in the transmission of relapsing fever. In this arrangement, the vector can obviously only infect one new host. A more efficient method of passing on the infection is at the next blood meal. Thus, filarial worms invade the insect's proboscis and burst out of it when the insect begins to feed. In the most highly-developed system, the pathogens invade the salivary glands and are injected with saliva during feeding. This is the mode of infection of a number of serious diseases, including malaria, sleeping sickness, yellow fever, dengue, mosquito-borne encephalitides and some forms of tick-borne relapsing fever (see Table 1.1).

In these more complex processes, the pathogen requires a period of development in the vector, after which the latter may be able to pass it on to more than one new host. This depends on the life-span of the vector. There are some long-lived arthropods, such as triatomine bugs or some ticks; and in some cases, their reservoir capacity may be prolonged by passing the pathogen (via the egg) to the next generation. However, they never approach the span of human life. Very often, especially with short-lived insects like mosquitoes, survival of the insect is critical, because the insect may not live long enough for the pathogen to develop to the infective stage. Such development will be prolonged by cold and, therefore, in temperate climates, the vector often dies before it becomes infective. This is one reason why diseases like malaria and yellow fever are so much more prevalent in the tropics. Another reason is that some of these disease cycles are not highly efficient, so that many bites of a vector species may be necessary before one results in infection. This may be achieved by the more frequent attacks of the numerous insects which breed under tropical conditions.

1.2.2 PREVALENCE OF VECTOR-BORNE DISEASE, ESPECIALLY IN TEMPERATE REGIONS

For reasons mentioned earlier, vector-borne diseases have always been more prevalent in hot countries. Some of them have never extended to northern temperate countries; but others have, and though they may be rare now, they are potentially able to return. The more serious epidemics (malaria, plague, typhus) have been largely eradicated from Europe and the U.S.A., though the vectors are still present. Other vector-borne diseases are indigenous, with reservoirs in wild

Table 1.1 Pathogens and vectors of some human diseases; (sg) indicates definite involvement of the salivary glands.

Pathogen				
Type	Designation	Disease	Vector	Mode of transmission
Viruses	WEE, EEE, etc	Equine encephalitides	Mosquitoes	Bite (sg)
	YF	Yellow fever*	Mosquitoes	Bite (sg)
	RSSE, CSE etc	Spring-summer fevers	Ticks	Bite (sg)
	SFN etc	Sandfly fevers	Sandflies	Bite
Rickettsiae	*Rickettsia prowazeki*	Typhus	Lice	Faeces
	Rickettsia rickettsi	Rocky Mountain Spotted fever	Ticks	Bite (sg)
	Rickettsia tsutsugamushi	Scrub typhus*	Mites	Bite
Bacteria	*Yersinia pestis*	Plague	Fleas	Bite
	Borellia recurrentis	Relapsing fever	Lice	Crushing lice
	Borellia duttoni	Relapsing fever*	Ticks	Bite (sg)
Protozoa	*Plasmodium* spp.	Malaria	Mosquitoes	Bite (sg)
	Trypanosoma spp.	Sleeping sickness*	Tsetse	Bite (sg)
	Trypanosoma cruzi	Chagas' disease*	Triatomid bugs	Faeces
	Leishmania spp.	Kala azar and Oriental sore	Sandflies	Bite
Filarial nematodes	*Wucheria bancrofti*	Filariasis*	Mosquitoes	Bite
	Onchocerca volvula	Onchocerciasis*	Blackflies	Bite

*Tropical diseases.

animals and not readily amenable to eradication; but these are relatively uncommon and sporadic in occurrence. The present status of some of these infections will be briefly reviewed.

(a) Protozoal diseases

(i) Malaria

The enormous range of malaria can be realized from the fact that some 3000 million people live in potentially malarious areas. With the introduction of DDT house spraying about 1950, a new simple and economical method of control became available, which was outstandingly successful in temperate and subtropical regions. This encouraged the World Health Organization to formulate a programme for the total eradication of the disease, which soon achieved remarkable progress in about 60 per cent of the malarious areas, mainly on the periphery of distribution of the disease. However, little or no progress was made in the intensely malarious parts of Africa. Even in the more amenable areas, technical difficulties (such as insecticide-resistant strains of vectors) slowed the advance about 1967. Since about 1970, the situation has actually deteriorated, with severe epidemics in Ceylon, India and Pakistan etc. The goal of world eradication has been abandoned for the forseeable future and countries are reverting to the more modest concept of control. Even this is hampered by rising costs of insecticides, combined with declining use of DDT, which is cheap and safe to use.

In the temperate zones, the story is more encouraging. The northern limits of the disease once extended up to 45 or 55 degrees of latitude and, in the 19th century, there were several hundred cases a year in England. In the U.S.A., as late as the 1930s, there were 6 or 7 thousand cases a year. For the past century, however, the range of malaria has been contracting. At first there was slow improvement, due to better farming practice, with drainage and with cattle barns remote from houses to draw away the mosquitoes. Finally, in the 1950s and 1960s, the use of modern insecticides virtually eradicated malaria from temperate regions. There remains, however, a risk of re-introduction of indigenous malaria by means of travellers from the tropics, because the vectors are still present here. In surveying this risk, the W.H.O. has suggested the following malaria terminology. (1) *Indigenous* (contracted in a malarious country); (2) *imported* (contracted abroad in a malarious area); (3) *introduced* (contracted by a local vector, from an imported case); and (4) *induced* (deliberately, for malaria therapy, or by blood transfusion, accidentally) [630].

Periods of exceptional risk in Europe followed the return of soldiers from World Wars. An analysis of the British experience suggests that about half a million soldiers contracted malaria in each of the World Wars. Many thousands returned infected to Britain (over 33 000 between 1919 and 1921; and some 14 000 relapsed). Local mosquito vectors were responsible for 480 introduced cases after the First War, but only 46 after the Second [75]. The improvement was due to better surveillance, better treatment and more effective anti-mosquito measures. In the U.S.A., there was a third wave of military infections during and after the Korean War. From 1959 to 1965, both military and civilian cases ranged around 50–100 per year. With the Korean War, military cases rose to between three and four thousand, with civilian cases slightly up around 150, per year. There was a welcome decline after 1970 and by 1972 the total for both types was only about 800. Throughout the whole period, introduced cases remained at the low level of 0–5 p.a. [305].

In more recent years, the hazards of introduced malaria has shifted to civilians [75]. The monitoring of such cases in Britain is carried out by the Malaria Reference Laboratory. During the 1960s, about 100 cases were recorded annually, but the numbers shot up in 1971–3 to over 500 p.a., probably because of more efficient diagnosis with serological methods. An analysis of these last three years showed that native travellers accounted for over 30 per cent, while immigrants were responsible for only 18 per cent of verified examples. There were 26 deaths among the 548 *falciparum* cases, due largely to delayed diagnosis.

(ii) Diseases due to haemoflagellates

Certain flagellate protozoans are responsible for serious diseases which, fortunately, do not extend into the northern temperate zone. For this reason, they will be discussed only very briefly.

Sleeping sickness, which is endemic over an immense area of tropical Africa, is due to *Trypanosoma gambiense* and *T. rhodesiense*. These trypanosomes are transmitted by various species of tsetse fly, as are related forms which cause the fatal disease *nagana* affecting horses and cattle.

Chagas' disease, which occurs in many parts of South and Central America, is due to *Trypanosoma cruzi*, which is transmitted by bloodsucking bugs of the family Reduviidae.

Kala azar (visceral leishmaniasis), which, is highly dangerous if untreated, and Oriental sore are due to *Leishmania tropica* and *L. donovani*, respectively. They are spread by sandflies of the genus *Phlebotomus*, in various parts of the tropics.

(b) Arboviruses

Viruses transmitted from one vertebrate to another by insects or acarines are known as arboviruses ('arthropod-borne viruses'). In the past decade, the number of examples of these viruses has risen to over 200. Three-quarters of them are transmitted by mosquitoes, a few by sandflies and the remainder by ticks. They are generally harmless to the vectors, but in the vertebrate they can cause high fever and various unpleasant symptoms. The most dangerous types tend to be haemorrhagic (with bleeding from gums, nose, kidneys etc) or encephalitic (with nervous involvement resembling poliomyelitis). The two types are not sharply distinct, however; and there are milder viruses which merely cause severe pains in joints and bones. Man is the sole vertebrate host in very few arboviruses; various wild or domestic animals and birds constitute the natural reservoir and are often little harmed by the virus. The infection cycle is complex, in that the degree and duration of the viraemia varies in different vertebrates and so does the susceptibility to infection in different vectors. The danger to man depends on the feeding habits and prevalence of the vector. In many cases, man is not very infectious and represents a dead end of the infection chain.

(i) Mosquito-borne arboviruses; mainly tropical

Yellow fever is the most important of the tropical arboviruses, being both widespread and potentially dangerous. For nearly three centuries, it decimated

European explorers in Africa and the New World; however, it does not extend into Asia. In the Americas, epidemics have extended into the temperate regions as far north as New York and even Quebec but, since the urban vector (*Aedes aegypti*) is fairly easy to control, the disease has been confined to the tropics for some 50 years. There, the virus persists in an enormous reservoir of forest monkeys in Africa and the Neotropics, and from them, the infection can be passed to men working in such areas. Since human cases of this disease are infectious to mosquitoes, they can carry it to towns and cities, where it can be further spread by bites of the urban mosquito vector.

Because of the reservoir of 'jungle yellow fever', total eradication of the disease is not feasible. However, about 1950, the Pan American Health Organization initiated a campaign to eradicate the mosquito *Aedes aegypti* from the Western Hemisphere, using DDT etc. Good progress was made for the first 15 years but eradication was not achieved in all the countries involved, so the project was quietly abandoned. However, despite intermittent infections of forest workers, there has been no threat of an urban epidemic and, in any case, the danger of this disease is greatly mitigated by protection by a simple and effective vaccination.

(ii) Other tropical arboviruses

Various 'break-bone' fevers due to mosquito-borne arboviruses occur in different parts of the tropics. Most widely distributed is dengue, which is mainly transmitted by *Aedes aegypti* (though other species of *Aedes* are involved in the Pacific region). This is normally non-lethal but, in South-East Asia, there have been epidemics of a more dangerous disease involving a similar virus, known as haemorrhagic dengue. This is an urban infection, especially serious in Asian children, causing up to 7 per cent mortality.

(iii) Mosquito-borne arboviruses in the temperate region

Several of these occur in North America, though their nature was only recognized about 1933. Three distinct forms, which are well known, are rather similar in that wild or domestic birds are the principal reservoir and are little affected. Passage from one wild host to another is by bird-feeding mosquitoes, which sometimes bite and infect man with moderate or severe results. Horses are also sometimes afflicted; and, because the encephalitis syndrome predominates, these diseases are often called 'equine encephalitides'.

Eastern equine encephalitis (E.E.E.) is most dangerous, with fatalities up to 75 per cent, but fortunately it is comparatively rare. Its main distribution is the Atlantic seaboard and the Gulf coast of the U.S.A., but outbreaks have occurred far inland. It is associated with swamps in which breed the vectors, *Culiseta melanura* and *Aedes sollicitans*; also *Mansonia perturbans* and *Culex salinarius*.

Saint Louis encephalitis (S.L.E.) has caused larger outbreaks than E.E.E., but it is less dangerous, though mortalities up to 10–30 per cent have been recorded. Its distribution ranges from a central area around St Louis (Missouri) and over the western half of the U.S.A. However, a few outlying outbreaks have reached Florida and even Trinidad. The main vectors in the west are *Culex tarsalis* in rural districts and members of the *Culex pipiens* complex in urban areas. In Florida, *Culex nigripalpus* and *Aedes crucians* were implicated.

Western equine encephalitis (W.E.E.) is less dangerous than S.L.E. It is distributed over the U.S.A. and Canada west of the Mississippi. The main vector is *Culex tarsalis*, but the virus has been recovered from several other mosquito species.

Another American disease of this type is Californian encephalitis (C.E.). It is less well studied, but appears to be spread by species of *Aedes* and the wild reservoirs are various rodents (squirrels, ground squirrels and foxes).

In the Far East there is Japanese B encephalitis (J.E.) which extends over the tropical and temperate region of Eastern Asia. Severe outbreaks have occurred in Japan. Many vertebrates, including pigs, horses, various birds, certain lizards and possibly certain bats are infected, maintaining a continuous reservoir. The principal northern vector is *Culex tritaeniorhynchus*, which breeds in rice fields. In man the disease can cause considerable mortality (up to 30 per cent) especially among children, and it tends to have harmful after-effects.

(iv) Tick-borne arboviruses in the temperate region

Only one tick-borne arbovirus is prevalent in the U.S.A., that is, Colorado tick fever (C.T.F.). It attacks people camping in the Rocky Mountain area (Colorado, Montana) and, although it is a relatively mild disease, it may cause symptoms of encephalitis in children. The vector is the Rocky Mountain wood tick, *Dermacentor andersoni* and the reservoir, various wild rodents (e.g. ground squirrels).

In the Old World, various tick-borne viruses occur, both those causing haemorrhagic and those causing encephalitis symptoms.

Tick-borne haemorrhagic viruses include Crimean haemorrhagic fever (C.H.F.) and Omsk haemorrhagic fever (O.M.S.K.). The former is an acute infection with up to 8 per cent mortality. It occurs in late spring or early summer in the steppes of south-eastern Europe, especially the U.S.S.R. The vectors are *Hyalomma plumbeum* and *H. anatolicum*, while various small mammals serve as the wild reservoir. O.M.S.K. occurs in the steppe forests of south-west Siberia where ticks of the genus *Dermacentor* act both as vectors and reservoirs. Though a milder human disease than C.H.F., it is highly pathogenic for some wild mammals.

The tick-borne encephalitides include Russian spring–summer encephalitis (R.S.S.E.) and Central European encephalitis (C.E.E.). The former is the most dangerous, with mortality rates reaching 30 per cent. It occurs in the thickly forested regions of eastern U.S.S.R. The reservoir hosts are wild rodents and the vectors *Ixodes persulcatus* and *Haemaphysalis* spp. C.E.E. is less severe than R.S.S.E. and is rarely fatal. The wild reservoir again is wild rodents and the vectors *Ixodes ricinus* and *D. marginalis*; but goats can become infected and human cases can develop from drinking their milk. Finally, in Britain, a related virus causes louping-ill, a dangerous disease of sheep. This can sometimes be transmitted to man (by the vector, *I. ricinus*) when it presents a similar form to C.E.E.

(v) Sandfly fever

This is an example of a virus infection transmitted by diptera other than mosquitoes. The main vector is *Phlebotomus papatasii* and the distribution, in the Mediterranean and Near East corresponds with that species. In recent years, the widespread use of DDT for malaria control has largely eradicated the sandfly from many areas, and the disease is now rare. It was, in any case, of short duration and not dangerous.

(c) Rickettsial diseases

(i) Louse-borne typhus

In the past, typhus has rivalled plague in the number of victims it has claimed. It is always associated with bad hygiene and commonly accompanied many historical wars, while in peace time it fulminated in armies, navies and prisons (as ship fever, jail fever etc). The last great epidemic in Eastern Europe after the First World War, is estimated to have affected 30 million people and killed three million of them. In recent years, however, the danger has receded, with a global total of some 10 000–20 000 cases a year and a few hundred deaths, mostly in Africa [636]. A few thousand cases have been reported in the U.S.S.R. during the last decade and a few dozen in Western Europe.

The causal organism is *Rickettsia prowazeki*, transmitted by the human body louse. Although head and crab lice can transmit the pathogen experimentally, they have never been responsible for an epidemic. The rickettsiae taken up from the blood of an infected patient multiply in its gut and destroy the tissues, eventually killing it. Before it dies, however, the louse passes faeces which are highly infectious if inhaled, are scratched into the skin, or contaminate a mucous membrane.

The association of typhus with body lice restricts epidemics to communities with low standards of hygiene, which explains its past and present distribution. Body lice, however, are by no means extinct even in advanced countries. Disasters or intense warfare could destroy civilized amenities and allow them to proliferate.

(ii) Murine typhus

This is a much milder disease due to a related pathogen, *R. typhi.* It occurs in various parts of the tropics and sub-tropics and, in the New World, extends up to the southern U.S.A. It resembles plague in being primarily a disease of rodents, from which it can occasionally be transmitted to man. Transmission between the rats is by the louse *Polyplax spinulosa*, the mite *Ornithonyssus bacoti* or by various rodent fleas. Few of these are liable to attack man, except the flea *Xenopsylla cheopis*, which is the most likely cause of human infections. As with louse-borne typhus, it is the faeces of the flea which are the infectious agent.

R. typhi is not dangerous to wild rodents or their fleas, so it appears to be a well-adapted parasite. In contrast, the lethal effects of *R. prowazeki* to both man and louse, indicates a relatively new parasitic cycle, probably evolved from *R. typhi.* It has even been suggested that such an extension can occur now, if murine typhus is introduced into a louse-infested community.

(iii) Tick-borne typhus

In North America there are two varieties of spotted fever due to *R. rickettsiae*, which causes a severe illness (and a mortality, before antibiotics, of about 20 per cent). The pathogen is maintained in a variety of wild animals and transmitted by ticks, especially *Dermacentor andersoni.* All tissues of the ticks tend to become infected and transmission usually occurs during tick bite; however, this must be prolonged for several hours to ensure infection. The rickettsiae invade the ovaries and can be passed on to the next generation, so that ticks constitute an additional reservoir.

Western Rocky Mountain fever, which has been known for about 100 years, occurs in the foothills of these mountains, from Washington to New Mexico. Tick bites occasionally transfer the infection to men, usually tourists or trappers camping in the area.

Eastern Rocky Mountain fever seems to have originated farther north (British Columbia, Alberta, Saskatchewan) and has spread extensively eastward as far as Maryland. Dogs are often infected by the tick *Dermacentor variabilis*, so that the disease thus enters suburban gardens and may infect women and children.

Another tick-borne typhus is *fievre boutonneuse*, due to *R. conori.* This occurs in the Mediterranean region, the Near East and south-east Asia. Wild rodents, rabbits and dogs constitute the reservoir and various ticks, especially *Rhipicephalus sanguineus*, are the vectors. It is rarely fatal, except in aged or debilitated people.

(d) Bacterial diseases

(i) Plague

As a result of the horrifying impact of plague epidemics, many historical records of it are available. A terrible and widespread epidemic occurred in the 6th century A.D. and another in the 13th century (the Black Death) which killed a quarter of the population of Europe. Other epidemics, including the 'Great Plague' of London in 1665, flared and smouldered in Europe, gradually dying out in the 18th century. At the end of the 19th century, the greatly increased shipping carried the infection all round the world, from a fulminating epidemic in China. At first, sea ports were afflicted: but from them, rural plague foci were established, which still remain as persistent threats. With modern methods of treatment and control, much of the terror of plague is past history. In the past decade, the total number of cases reported to W.H.O. annually ranged from 1000 to 6000, with only one or two hundred deaths [634].

Plague is primarily a disease of wild rodents, among which it persists endemically, sometimes flaring into an epizootic which kills many of them. It is transmitted among the rodents by their fleas; but since these are reluctant to bite man, the wild rodent disease is seldom dangerous. Occasionally, fur trappers may skin infected animals and acquire the disease by contagion.

Plague epidemics in towns are begun by the interchange of fleas from wild rodents to domestic rats around the periphery of towns. The disease then spreads rapidly among urban rodents, to which it is highly lethal. There are two common urban rats. One of them, the black rat (*Rattus rattus*) is liable to infest wooden dwellings, though it rarely infests modern buildings. This brings it into close proximity to man; and the last step is a passage to man by a species of flea which will bite both rats and men. The tropical rat flea *Xenopsylla cheopis* is usually responsible. Human fleas seldom bite rats and human plague cases are rarely infectious to fleas (the germs are concentrated in the buboes, or plague swellings). Sometimes, however, plague bacilli accumulate in the lungs and develop a pneumonic form of plague which is highly infectious.

The present plague hazard in the northern temperate zone consists of the persistant foci in rural areas, among wild rodents, known as 'sylvatic plague'. These pose a certain danger to hunters and tourists in the vicinity, not only from rare flea bites but from skinning infected animals. With public health vigilance, however, there is little chance of disease from these foci and causing an epidemic. Furthermore, the rats common in most modern cities are Norway rats (*Rattus norwegicus*) which live in sewers and not in close contact with man.

(ii) Tularemia

This is a rural disease, occurring in Western U.S.A., Canada, Northern Europe and Japan. It is a septic, febrile disease with a mortality if untreated of about 5 per cent. The pathogen responsible is *Franciscella tularensis*, distantly related to the plague bacillus. There is a natural reservoir in various wild animals and birds, among whom it is transmitted by biting arthropods. In the U.S.A., rabbits, ground squirrels, coyotes and some birds are involved. Ticks, especially *Dermacentor andersoni*, are important vectors and the biting deer fly *Chrysops discalis* has been incriminated. In Europe, rabbits and hares (and in Sweden, lemmings) act as reservoirs. *Aedes* mosquitoes have been found infected and are known to transmit the infection to birds.

Apart from bites of vectors, human infections can arise from contact with infected carcases and water polluted with them.

(iii) Enteric diseases

These infections range from relatively mild attacks of traveller's diarrhoea to potentially dangerous conditions like typhoid. Generally speaking, they are considerably more common in hot countries, but even more important is an association with bad sanitation, especially in relation to defaecation. There is a rough analogy with malaria in that new residents in such areas are very vulnerable to infections, whereas the natives are largely immune. This immunity is due to early and repeated infections and is often purchased at the price of infant mortality. There is a Russian proverb to the effect 'fly in April, dead child in July'.

Most of these diseases are caused by members of the enterobacteriaciae. Several species of *Shigella* are responsible for bacillary dysentery, *S. dysenteriae* being the most virulent. *Salmonella typhi* which is responsible for typhoid, is restricted to man; but there are numerous other species and strains of *Salmonella* which occur in a very wide range of different animals and can be transmitted from them to man. These can be responsible for various degrees of food poisoning. *Eschericha coli* is a normal commensal inhabitant of the human intestine, but there are virulent strains which can cause dysentery. It seems likely that these are responsible for infantile summer diarrhoea and traveller's diarrhoea.

In all these diseases, pathogens are shed with the faeces. Transmission can occur if drinking water is contaminated; and also by hands, contaminated after defaecation, when preparing food. In addition to these routes, the pathogens can be transmitted by insects which visit human faeces (or *Salmonella* infected meat) and subsequently contaminate human food. The most important of such vectors is the common housefly, since it is the most difficult to exclude from foodstuffs. The germs transmitted are not those acquired during the larval stage. Although flies often breed in faeces and the maggot's guts teem with bacteria, nearly all are eliminated on pupation. It is the adult female fly which is dangerous. She will visit putrifying matter to lay her eggs and often suck up some of the liquid on the surface. On return to the kitchen or dining room, she will crawl over food and deposit drops of vomit or infected faeces on it.

It is obvious that the risk of flies picking up pathogens from human faeces is more likely under primitive conditions where unscreened privies are used. In modern towns with water-borne sewage, there is less opportunity for this. There are other insects which combine the dangerous habit of visiting septic matter and human food. These include certain blowflies and cockroaches. The former may encounter *Salmonella* infected offal in slaughter houses, while the latter tend to visit drains or liquid excrement for water. Such insects, as well as the ubiquitous housefly constitute a special danger in hospitals, unless all septic matter is protected from them.

1.3 INSECTS AND HYGIENE

1.3.1 MYIASIS

This subject will be dealt with more extensively in Chapter 6, so it will only be considered briefly here. 'Myiasis' is the invasion of the bodies of men or animals by living insects, which are almost exclusively maggots, or the larvae of diptera. It is one of the unpleasant consequences of the close association of some of the higher diptera with vertebrates; and it seems to have developed as an accidental extension of the breeding habits of these insects, making ever closer contacts with higher animals, as follows.

Most higher diptera tend to breed in putrifying organic matter. Originally, this could have been decaying vegetation; but certain species (the housefly, the stable fly) have found that contamination with vertebrate excrement greatly improves its nutritive value. Many species (housefly, lesser housefly) prefer to breed in faecal matter alone. Some species of blowfly will also breed in faeces, though the primary breeding material of this group was probably cadavers. From these bases, various types of myiasis arose. Some are *accidental* and non-productive; others are *facultative* (that is, accidental, but potentially successful); and still others *obligatory* (in species which adopt myiasis as a normal breeding method).

Accidental myiasis can result from swallowing blowfly eggs laid on food eaten without cooking (cold cooked or smoked meat etc). These may hatch in the stomach or intestine and cause vomiting or diarrhoea. It has been claimed that such eggs can complete their larval development in the stomach of a mammal, but there are some doubts about this. Coprophagous flies will readily lay eggs which soon hatch in newly-passed faeces (and some are larviparous), so that people may believe that the larvae were passed in a stool. Another type of accidental myiasis is the peri-anal or urinogenital type, which can occur in unclean or diseased people in hot climates, where few clothes are worn.

Cases of facultative myiasis can arise when the eggs of carrion-breeding blowflies are laid on wounds or sores. Though distressing to human patients, the maggots which develop are not very harmful, since they tend to feed on necrotic tissue. Indeed, for a short time after the First World War, sterile maggots were used for wound therapy.

The habit of obligatory myiasis has developed in some blowflies which invariably lay eggs in wounds and complete their larval development in living animals, or occasionally, men. Other flies with this habit do not even require injured flesh to make their entry. In some cases, these attacks can cause extensive pain and injury. The myiasis-producing flies mentioned so far have been aberrant members of families, the great majority of which perform a useful function in clearing up cadavers or faecal matter. Last to be considered, however, is a group all members of which have adopted this method of breeding. It includes the sheep nostril flies, which develop in the nasal sinuses of domestic animals, the bot flies which pass their larval life in the stomach of horses etc, and the warble flies, which migrate inside cattle and emerge in the skin of the back. Nearly all of these attack wild or domestic animals; but there are occasional human cases and there is one American species which regularly attacks man.

1.3.2 POISONING, IRRITATION AND ALLERGY

Many arthropods have developed poisoning mechanisms which they use either offensively, to paralyse their prey, or defensively against large animals such as vertebrates. The best known are the stinging bees, wasps and ants, the scorpions and certain spiders which inject venom with the bite. The poisons used may be as simple as formic acid (in certain ants); but usually more complex substances are involved, apparently proteins of low molecular weight. The symptoms caused range from slight local reaction to acute neurotoxic effects, not unlike those caused by bites of poisonous snakes. Usually, insect stings are much less serious and only lethal to small animals; however, numerous stings can be dangerous to man and especially to children. Probably the most serious are the stings of bees and the bites of spiders of the genus *Latrodectus* ('Black-widow') both of which are responsible for a small number of deaths annually in certain countries.

Another chemical defence method used by insects consists of urticating hairs, rather like those of stinging nettles, which are found on a considerable variety of moth caterpillars. These larvae cause irritating weals if they are handled and considerably more severe symptoms if the detached hairs are inhaled into the respiratory tract.

Vesicating substances are present in the blood of certain beetles, especially those of the family Meloidae ('Blister beetles'). The fluid exuded if these insects are crushed, or even roughly handled, causes painful blistering of the skin. Presumably this is a defence against being eaten by birds or other insectivorous vertebrates.

(a) Allergic effects

Insect venoms contain toxic proteins and these are liable to elicit the formation of antibodies. Therefore, repeated poisoning may cause anaphylactic shock. Bee stings in particular may be responsible for very severe effects, far worse than those due to simple poisoning.

Allergic reactions frequently occur as a result of the bites of various bloodsucking arthropods, though the effects are usually localized. The foreign proteins responsible occur in the saliva of the parasite which is injected into the wound during feeding.

Another type of allergy associated with mites and insects is a type of asthma due to inhalation of fragments of their bodies.

These different types of allergy can be observed in various stages. There are a proportion of naturally immune people; the remainder develop sensitivity to different degrees after the initial exposures; finally most of them eventually become desensitized.

Stinging and biting insects and the poisoning, irritation and allergies associated with them, are further discussed in Chapter 12.

1.3.3 DISGUST. PATHOLOGICAL CASES

Many household pests do very little harm but are regarded as highly unpleasant. Earwigs, woodlice and furniture mites are quite harmless and such pests as cockroaches and ants really consume very little food. Even the disgust caused by bed bugs or lice is not wholly alarm about the effects of their bites. The feelings of repulsion aroused by infestations of insects may be perfectly healthy instincts like the abhorrence of refuse and excrement. On such feelings are based high standards of hygiene.

Nevertheless, these feelings may exist in exaggerated pathological form associated with delusions. Nearly every advisory entomologist encounters one or two such cases annually. The afflicted person imagines his (or her) body or house to be infested with numerous invisible biting insects or mites. The sufferer experiences an anguished feeling of being unclean and persecuted. Advice is sought from chemists, doctors or (by post) from popular newspapers and all sorts of preparations are tried without avail.

One type of sufferer characteristically brings 'specimens' to the laboratory, to prove the existence of the infestation. These turn out to be pieces of scurf or dirt, fibres or flecks from garments, small harmless insects, etc. The delusions are not easy to recognize immediately, except by an experienced entomologist, and not infrequently convince members of their own family. The sufferers are often people of intelligence, who talk rationally except in regard to their unshakable idea of being infested. The following extracts, however, reveal some of the delusions.

Example 1 (from a man):

I will smear some glue on a microscope slide and hold it close to my body in bed. With luck one of the creatures may be held by the glue. . . . I would say that the creature is smaller in size than the pores of the body. I often feel their bite on the flat part of the heel. The bite is sharp and intense. When I am still, I can feel their movements in my clothes. This gives rise to a most horrid and fearsome sensation.

The creature introduced itself to me when I lay on a dirty mattress during fire-watching. . . .

I saw a number of doctors the first two years. They could no nothing for me. Some of them said there was no such insects as I described. One doctor made up a liquid preparation which a druggist informed me contained oil of sassafras.

I tried DDT without effect.

It is possible that a large number of the creatures *are* killed by some of the substances I have used. But then the eggs hatch out so the condition is no better. . . .

Example 2 (from a woman):

I found a nasty black insect that got under the skin and worried me nearly silly. I have tried everything, Lysol, kerosene, sulphur and carbolic acid. I had a blister as large as a dinner plate on my chest and cannot get rid of the horrid things in fact I have nearly *killed myself*. . . . I saw in the local Western Argus that 200 children were not fit to go to school also I had some books from the library full of nasty vermin, they were black. I wrote to the (a national newspaper) and they said it was a kind of skin trouble *lots* of people had written to them. They sent a 5% emulsion of Benzyl benzoate. I put some on my hair and find it affects my brain, I behave funny like I was drunk so I was afraid to use any more of it. . . . If you will advise me and let me know the *price* of your ointment I shall be *very very grateful*. I only wish I had never come to the dirty place (Kismet) the place should be burnt down . . .

The following points are noteworthy: (1) The sense of persecution. (2) The frequent attempts at self-medication, sometimes with unpleasant results. (3) Statements that 'other doctors' had been consulted. Such characteristics strongly suggest the existence of what American authors have termed 'delusory parasitosis', though the possible existence of a real arthropod pest should be checked. Thus, people have been suspected of such delusions when no cause was immediately apparent; but subsequent investigations have shown that there were real grounds for complaint. (In one case due to bites of *Culicoides*, in others to infestations of mites or even a dermatitis due to a kind of paint.)

The psychological causes of delusory parasitosis are beyond the scope of this book; but it may be noted that some kind of 'insect shock' is a common precipitating cause. Sometimes a real infestation (of lice, bugs or fleas) has existed and after it is cleared up the insectophobia remains. In other cases, talks, articles or films about insect pests may be responsible.

People suffering from this neurosis cannot be assisted by an entomologist and it does not really help to offer them insecticides as a placebo. The best course is to communicate with the sufferer's physician, or the local health department if this has become involved. Presumably these cases need a psychiatrist.

Table 1.2 Data on various insect infestations from certain British cities.

	1940	1943	1947	1950	1953	1957	1960	1963	1967	1969	1972
City A											
Head lice (%) children infested: Boys, Lice	3·7	2·6	2·6	2·7	1·6	0·9	0·7	0·1	0·1	1·9 (Lice and Nits combined)	*
Head lice (%) children infested: Boys, Nits	17	6·8	5·7	6·3	6·3	3·7	4·0	2·7	3·1		
Head lice (%) children infested: Girls, Lice	15	13	9·4	6·4	5·0	3·6	2·9	0·1	0·2	4·8 (Lice and Nits combined)	*
Head lice (%) children infested: Girls, Nits	40	27	23	23	21	12	13	11	8·3		
Bugs: % houses infested	2·1	1·3	1·2	1·2	0·5	0·18	0·11	0·01	*	*	*
Scabies: no. persons treated	*	21 200	4055	353	354	1846	3451	1539	2110	3169	1034
City B											
Scabies: no. treated											
Adults	623	16 685	4910	615	423	540	691	984	4112	5104	3024
Children	316	10 715	2522	325	68	65	101	307	686	1713	1004
City C											
Bugs: no. houses treated	*	*	367	614	580	142	340	182	242	260	142
Fleas: no. houses treated	*	*	23	25	25	8	8	7	34	53	29

*Data not available or unreliable.

It may be mentioned that this type of neurosis is by no means confined to Britain; it has been noted in the U.S.A. [527], Germany [165] and Hungary [370].

1.3.4 INSECTS AS DESTRUCTIVE HOUSEHOLD PESTS

There are three ways in which insects can be destructive household pests. They may consume foodstuffs, they may damage clothing and other fabrics, or they may destroy woodwork. The pests responsible are dealt with in three subsequent chapters.

In the temperate climate of Britain, insect damage is less rapid and dramatic than in many tropical countries. But even partial damage may result in considerable loss, because food may be rendered unfit for human consumption by the presence in it of insects or their larvae or by the contamination of their excrement, silk webbing or peculiar smell. Again, the presence of a few small holes may 'ruin' a suit of clothes. Only the minor attacks of the furniture beetle are sometimes viewed with complacence, because they are supposed to guarantee antiquity!

1.3.5 INCIDENCE OF INSECT PESTS OF HYGIENIC IMPORTANCE

In areas where insect-borne diseases occur, there are very often records of the prevalence of the insect vectors. There are, however, only scattered data on the prevalence of insect pests which are merely unpleasant or of hygienic importance. For some years in Britain, the Medical Officers of local Health Departments of urban boroughs published annual reports, some of which recorded the incidence of head lice and scabies, especially among school children. The records, however, varied from place to place (e.g. some recorded numbers treated, others the percentage infested); and since the reorganization of the National Health Service and the replacement of Medical Officers by Community Physicians, current data are not strictly comparable with earlier records. Some of the latter are set out in Table 1.2. This shows the general improvement in these unpleasant pests following the Second World War, probably due to the introduction of the new synthetic insecticides. Later records, however, suggest that the improvement was not maintained; indeed there has been a resurgence in scabies in the early 1970s (also noted in other countries; see p. 280).

An indication of the wide range of insect pests submitted to public health entomologists can be gathered from the ten year records of an advisory entomologist, from 1945–1955 [88]. Other information on the prevalence of certain insect pests in Britain are available from publications of a large servicing company (Rentokil Ltd.). These relate to cockroaches [124], bed bugs and fleas [125], and wood pests [269].

In Denmark, the records of the Government Pest Control Laboratory provide interesting information on the changing pest control problems [238]. Some of these, for lice, bugs and fleas are discussed in Chapter 7.

CHAPTER TWO

Organization of preventive and control measures

The prevention and control of domestic insect pests may be the concern of several kinds of people. In the first place, architects and builders are able to assist by giving thought to the liability of infestation of different types of construction. Secondly, the ordinary citizens resident in dwelling houses and other buildings are involved as the main sufferers, and also because the development of many infestations depends very much on the thoroughness with which domestic hygiene is pursued. Furthermore, the likelihood of a request for expert assistance in case of need depends on the proper outlook of the ordinary householder. To disseminate information about pests and the right way of destroying them, hygiene propaganda is useful; various official and unofficial bodies can help here. Thirdly, the Community Physicians and the Health Inspectors of local authority Health Departments are involved in more serious infestations in their general duty of safeguarding public health. Some of their functions in this respect are legally defined. Lastly, commercial firms are interested in marketing insecticides and sometimes in providing labour to apply them. These various implications of domestic pest infestation are dealt with in this chapter under the following headings:

(1) Building construction and pest infestation.
(2) Hygiene propaganda relating to insect pests.
(3) Sources of special technical information.
(4) Official pest control in Britain.
(5) Official pest control in other countries.
(6) Commercial aspects of pest control.

2.1 BUILDING CONSTRUCTION AND PEST INFESTATION

For many years it has been recognized that insect infestation in buildings is favoured by certain constructional features and commonly occurring faults. The insects take advantage of different types of hiding place, from tiny crevices to large hollow spaces. The benefits of these hiding places to the insect should be clearly understood; they are two-fold: (a) the concealment allows quite large populations of insects to develop unnoticed and renders them difficult to find; (b) the protection afforded prevents or reduces extermination by cleansing operations or by insecticides. Consequently, infestations are more likely to occur and persist in buildings with ample harbourage than in those providing little or no shelter.

2.1.1 PREVENTION OF HARBOURAGE FOR BED BUGS

Bed bugs will lodge in very narrow crevices (a millimetre or two wide) and many small bugs can be accommodated in a single nail hole. Apart from crevices in furniture, bugs will infest cracks in plaster or woodwork, e.g. badly fitting joints around door or window frames, skirting boards or picture rails. Except with very careful workmanship, it is difficult to avoid this type of crevice developing in houses of conventional type, owing to warping or shrinkage of materials as they dry out and sometimes to settlement after building is complete. Details of building design, intended to assist in the prevention of bug infestation of new houses, are given in Section VI of the Medical Research Council Report on the Bed Bug (1942) and also in Chapter X of Functional Requirements of Buildings, British Standards Code of Practice III (1950). The importance of eliminating harbourages for bed bugs was considerably reduced, however, with the introduction of DDT and other residual insecticides, which have substantially reduced infestation in Britain and similar temperate countries. This does not mean that the matter should be entirely neglected for it is still important, so far as possible, to prevent bugs hiding from the cleaning operations of the housewife and the torch of the Health Inspector.

2.1.2 PREVENTION OF HARBOURAGE FOR COCKROACHES

Cockroaches still infest kitchens of old-fashioned houses, but they are much more common and widespread in permanently warm buildings such as centrally-heated institutions and bakeries. In particular, the species *Blattella germanica*, which appears to be the most frequent offender at present, is a lover of warmth and is usually to be found in crevices near hot-water pipes. Cockroaches are large and active insects and there are records of their travelling quite long distances inside buildings.

For these reasons, the principal structures inhabited by cockroaches are pipe runs, ducts and chases. They are also troublesome in hollow wall spaces adjacent to ovens or kitchen stoves. On the whole, they are considerably more difficult to eliminate than bed bugs so that it is especially important to deny them access to deep and inaccessible harbourages. The spaces surrounding pipes passing through walls should be properly flanged and solidly packed with steel or glass wool. Intermediate lengths of ducts and chases should be capable of opening easily for inspection. Fibreboard partitions and similar constructions in kitchens should be carefully sealed with strips of gummed tape.

2.1.3 HOUSEFLIES AND BLOWFLIES

The prevalence of flies and blowflies is not related to building design, except where it is proposed to fix screens to the windows of such rooms as kitchens, larders, canteens or lavatories (see page 40). In that case, the windows must be designed to give adequate illumination from the unscreened fixed portions, while the ventilating sections must be made to open inwards.

Liability to a fly nuisance is, of course, affected by the location of a building. The breeding of flies is especially prevalent in certain places, e.g. refuse dumps, stable yards, slaughter houses, etc. Buildings, particularly dwelling houses, should not be sited near such places.

2.1.4 OTHER INSECTS

Certain pests (e.g. silverfish, booklice, earwigs, furniture mites, woodlice) are favoured by damp conditions and the best defences against them are warmth and good ventilation. Of the pests mentioned, the earwigs and woodlice are invaders from the garden and are most likely to enter windows surrounded by creeper or bushes. Silverfish have a preference for starchy substances and may therefore attack wallpaper and other pasted or gummed materials. Booklice which exist on fungi or moulds, may be found in new buildings but these will disappear when brickwork and plaster have thoroughly dried out.

2.2 HYGIENE PROPAGANDA RELATING TO INSECT PESTS

There is very little doubt that the prevention of many insect pests in houses in Britain depends largely on the standards of hygiene of the occupants. To some extent these have been moulded by upbringing and education (or lack of it) and resistance to improvement may be augmented by obstructive prejudices about insect pests. At the lowest level, one sometimes encounters archaic ideas in which the pullulation of parasites is associated with bodily vigour. More frequently there is evidence of belief in spontaneous generation. Finally, at a slightly higher level, there is often a false sense of shame which prevents the sufferer obtaining expert advice and treatment.

In its simplest form, therefore, hygiene propaganda should seek to combat prejudice and ignorance by dissemination of the simple facts of insect pest biology and of the best control measures. This may be done by use of leaflets, posters and articles, by exhibitions and by instructional films.

The following material is available in Britain.

2.2.1 BOOKLETS AND PAMPHLETS

(1) From the Health Education Council, 78 New Oxford Street, London WC1

1AH. Two pamphlets are available written in simple clear English. *What to do about Head Lice* and *What to do about Scabies.*

(2) From the British Museum (Natural History), Cromwell Road, London SW7.

Economic series:

2a *Lice*, 24pp; **4a** *British Mosquitoes*, 28pp; **5** *The Bed Bug*, 17 pp; **11a** *Domestic Wood-boring Beetles*, 40pp; **14** *Clothes Moths and Carpet Beetles*, 15pp; **15** *Common Insect Pests of Stored Food*, 62pp.

Economic leaflets:

6 *Plaster Beetles*, 4pp; **8** *Carpet Beetles*, 4pp; **11** *Wharfborer*, 4pp; **12** *Blowflies*, 4pp; **13** *Hover Fly Larvae*, 4pp; **17** *The Flour Beetle*, 4pp; **18** *Grain Thrips or Thunderflies*.

2.2.2 FILMS

Films can usually be borrowed, either for a fee or, in some cases, they are free. If they are required regularly, it may be possible to purchase them. The following are available in Britain.

(1) From the Central Film Library, Bromyard Avenue, London W3 7JB. UK1187 *Fly about the House* (1 reel, 9 min). UK2553 *The Intruders* (Cockroaches) (2 reels, 15 min). UK2757 *Start from Scratch* (Fleas) (2 reels, 18 min). MS23 *Scabies* (3 reels, 26 min). MS24 *The Scabies Mite* (1 reel).

(2) From Peter Darvill Associates Ltd, Chesham, Bucks. HP5 2SG. *War to the Last Itch* (Head lice) (1 reel, 16 min). *Knockdown and Kill* (Insecticidal action) (28 min).

(3) From Shell Film Library, 25 The Burrows, London NW4 4AT. *The Rival World* (26 min); *Unseen Enemies* (32 min); *Malaria* (18 min).

(4) From Boulton-Hawkes Films Ltd, Hadleigh, Suffolk. *The Life-cycles of insects*: *1. Incomplete metamorphosis* (17·5 min) *2. Complete metamorphosis* (21 min); *Insects Harmful to Man* (16 min).

(5) From BASF (U.K.) Ltd, Lady Lane, Hadley, Surrey IP7 6BQ. *Ten thousand to one* (seeking insecticides for pests of food) (17 min).

(6) From The Building Research Establishment, Garston WD2 7 JR. *Testing Wood Preservatives against Furniture Beetle* (9 min).

2.2.3 EXHIBITIONS

A useful method of interesting and instructing the public about insect pests and the best way of dealing with them is to arrange suitable exhibitions. Progressive Health Departments may find it desirable to plan such exhibitions in conjunction with rehousing programmes. Living specimens and other material as well as advice can often be obtained from the biology department of a local university or technical college.

2.3 SOURCES OF SPECIAL TECHNICAL INFORMATION

2.3.1 TECHNICAL PUBLICATIONS

(a) From the Ministry of Agriculture, Fisheries and Food, U.K.

Single copies may be obtained, free of charge, from the Ministry: (Publications) Tolcarne Drive, Pinner, England HA5 2DT. The following are relevant to pests mentioned in this book. *Houseflies, Blowflies and Clusterflies*; *Ants Indoors*; *Cockroaches*; *Wasps*; *Insect Nuisances in Stores and Homes*; *Pea and Bean Weevils*; *Grain Weevils*; *Insect Pests in Stored Food*; *Insects and Mites in Farm Stored Grain*; *Insects Infesting Bacon and Hams*; *Insects and Mites in Sacks*; *Mites in Stored Commodities*; *Saw-toothed Grain Beetle*; *Flies and other Insects in Poultry Houses*; *Spider Beetles*; and *The Warehouse Moth.*

(b) From The Building Research Establishment, U.K.

Publications formerly produced by the Forest Products Research Laboratory on wood pests, are now available from the Distribution Unit, Building Research Unit, Garston, England WD2 7JR (technical note numbers given). These are: **47** *The Common Furniture Beetle*; **58** *'Woodworm' Control in Domestic Roofs by Dichlorvos Vapour*; **45** *The Death Watch Beetle*; **7** *Insecticidal Smokes for Control of Wood-boring Insects*; **55** *Defects Caused by Ambrosia Beetles*; **39** *The House Longhorn Beetle*; **60** *Lyctus Powder-post Beetles*; and **43** *The Kiln Sterilization of Lyctus Infested Timber.*

(c) The British Crop Protection Council

This organization publishes a Pesticide Manual annually, giving common and chemical names, structural formulae and other data on a wide range of pesticides. Copies (requiring payment) from Mr Billitt, Clacks Farm, Boreley, Ombersley, Droitwich, Worcs., England.

(d) The World Health Organization, Geneva, Switzerland

Vector Control in International Health, published 1972. Also, there are Reports issued periodically by the Expert Committee on Vector Biology and Control. These follow a series of Reports of the Expert Committee on Insecticides.

2.3.2 REFERENCE CENTRES

(a) In Britain

Information and advice concerning insect pests are available from various experts, free of charge in most cases, to *bona fide* inquirers.

(*i*) The British Museum (Natural History), Cromwell Road, London, SW7, will usually undertake identification of specimens of pest insects.

(*ii*) The Public Health Laboratory Service Reference Laboratory for Entomology, London School of Hygiene and Tropical Medicine, Keppel Street, Gower Street, London, WC1, is comparable to other reference centres of the P.H.L.S. and provides specialist information on insect pests of public health importance.

(*iii*) The Ministry of Agriculture, Fisheries and Food, Pest Infestation Control Laboratory, Slough, Bucks, is concerned with infestation of food stores and will usually identify the pests concerned and give advice on control measures.

(b) In the U.S.A.

In many states, the County Health Departments have an Entomological Section, which will undertake the identification of specimens.

The Department of Entomology in various universities will perform the same function; many have an Agricultural Experimental Station able to assist with advice.

2.3.4 PRESERVATION AND DESPATCH OF SPECIMENS FOR IDENTIFICATION

Very often specimens are sent through the post badly packed or improperly preserved and they arrive in a condition which renders identification difficult or impossible. A few simple precautions will prevent exasperation on the part of the entomological expert and disappointment on the part of the inquirer.

(a) General points

It is not generally realized that postal regulations prohibit the sending of live insects through the post but, in any case, it is usually more satisfactory to kill the specimens before despatch. A few drops of chloroform should be applied to the cork or stopper of the bottle or tube containing the insects. The more fragile insects (flies, mosquitoes, moths) are usually killed by five or ten min exposure to the saturated vapour; but more resistant insects (beetles, cockroaches, bugs) may require an hour or so. The insects should not be allowed to come in contact with the liquid chloroform.

The only other general point to emphasize is the seemingly obvious one of packing carefully to avoid damage in transit. Dry specimens can be sent in tins or *strong* cardboard boxes. Specimens in fluid are best sent in tubes, which can be sent in hollow wooden blocks made for this purpose, or else, well padded, in tins.

(b) Insect specimens

Adults of most insects, and any moderately robust nymphs or larvae, retain their characteristic form and colour sufficiently well if merely allowed to dry without any special precautions (except under very damp conditions, when they may go mouldy). Dry insects are most easily handled if they are properly mounted on entomological pins, in which case they can be despatched in cork-lined boxes. However, it is usually nearly as satisfactory to send such specimens unpinned and lightly (i.e. not tightly) packed between layers of tissue paper and cotton wool.

(c) Insect larvae

Almost all soft-bodied insect larvae are best killed and preserved by putting them into 70 per cent alcohol (or 70 per cent methylated spirit, 30 per cent water). This is also the method of choice for soft-bodied or small and fragile adult insects (such as lice, fleas and psocids). It is a uniformly satisfactory preservative, which does little harm to most specimens. The only exception is in regard to aquatic larvae. These can be sent in water in which they were caught, if a little formalin is added, to give a solution containing about 2 per cent formaldehyde. Aquatic specimens are useless without this preservative, as they rapidly decay in ordinary water.

(d) Mites

Mites can be preserved and despatched in 70 per cent alcohol. It is desirable to put them in a small volume of liquid in a small tube, otherwise there may be difficulty in finding them again, unless they are very numerous.

2.4 OFFICIAL PEST CONTROL IN BRITAIN

2.4.1 LEGAL ASPECTS

Sometimes domestic insect pests are sufficiently troublesome to involve legal liability; for instance where they are dangerous to health or a nuisance. Certain measures of legislative control empower local authorities and school authorities to deal with verminous conditions of either people, houses, or movable articles. 'Verminous' in this context is defined as infestation by such insects as bugs, fleas or lice and their eggs, larvae or pupae. In actual fact, 'verminous' applied to a person normally involves lice or scabies mites, whereas applied to a building it usually implies bugs or fleas.

(a) Public Health Act, 1936; Public Health Act, 1961

The provisions of these acts are slightly affected by the Local Government Act, 1972, in that local authorities are now quite free to appoint officers for various duties, each such officer being referred to as 'the appropriate officer'. The great majority of local authorities no longer have Medical Officers of Health and the relevant duties in connection with pest hygiene are now undertaken by the Environmental Health Officer.

Where there is reason to believe that verminous conditions exist, the local authority may proceed as follows.

(i) Cleaning or destruction of verminous articles [Sections 84 and (London) 122]

This is to be done at the expense of the local authority and provision is made for compensation.

(ii) Cleansing of verminous houses [Sections 83 and (London) 123]

The local authority may serve notice on the owner to have the house properly cleansed and, in default, may carry out the cleansing and recover the cost. Where, however, it is deemed necessary to employ hydrogen cyanide fumigation, the operation is carried out by trained personnel and at public expense.

(iii) Cleansing station [Sections 86 and (London) 124]

Local authorities are authorized to set up (either separately or jointly) public cleansing stations provided with the necessary apparatus and staff.

(iv) Cleansing of school children

A School Medical Officer, or a person authorized by him, has power to inspect children at County Council Schools and, if they are in a verminous condition, to serve notice on their parents or guardians requiring them to cleanse the child within 24 hours. In default of this, the child may be taken to a public cleansing station to be disinfested.

(v) Cleansing of verminous persons by order of Court (Section 85)

Where a verminous person does not consent to treatment at a public cleansing station, the local authority may apply to a petty sessional court for an order to proceed with the case. The Court may then enforce attendance at the cleansing station under penalty for disobedience.

(b) Housing Act, 1957 (Section 25)

Provision is made for local authorities to disinfest buildings scheduled for demolition.

(c) Hydrogen Cyanide (Fumigation of Buildings) Act, 1951

Fumigation of buildings with hydrogen cyanide is now seldom practised because safer and better methods of disinfestation are available, and fumigation of food stores is often done with other fumigants, such as methyl bromide. Accordingly, this act is out of date. Since 1974, the matter of revising and extending this legislation has been under consideration by an interdepartmental working party. It is likely that responsibility for fumigation operations will be transferred from the Home Office to the Health and Safety Executive.

(d) Prevention of Damage by Pests Act, 1949 (S.I. 1950, no. 417)

It imposes the duty on persons concerned in the manufacture, transport, storage and sale of food, of notifying the Ministry of Agriculture of infestations in premises or vehicles, in food or in containers likely to be used for food. An infestation is defined as 'the presence of rats, mice, insects or mites in numbers or conditions which involve immediate or potential risk of substantial loss or damage to food'.

Notification of infestations in a wide variety of imported foods is not essential in certain circumstances which are detailed in S.I. 1950, no. 416.

(e) Food and Drugs Act, 1955

Proceedings may be taken under Section 8 of this Act, which refers to food 'unfit for human consumption'. The presence of insects, their dead bodies or excreta in food, however, could be defended in some cases, as not likely to cause actual illness. Accordingly, when prosecutions are brought under the Act (e.g. by a local authority) they usually rely on Section 2 which refers to 'prejudice of the purchaser . . . not of the substance or not of the quality demanded'.

(f) Food Hygiene Regulations

These require business selling food for human consumption to keep their premises in such a condition as to prevent, so far as is reasonably practicable, infestation by insects.

(g) Insects as nuisances

Infestations of unpleasant insects may sometimes be deemed statutory nuisances as defined in the Public Health Act, 1936. Thus, verminous houses might be covered by Section 92(1)(a) 'Any premises in such a state as to be prejudicial to health or a nuisance'. Similarly, accumulations of refuse or manure in which flies or other insects were breeding extensively and invading buildings might be dealt with under Section 92(1)(c) 'Any accumulation or deposit which is prejudicial to health or a nuisance'.

Legal action under these headings is taken by local authorities through their public health officers, and entomologists may be called as expert witnesses. It may be remarked that difficult points are often raised in these cases, such as the proof of responsibility for introducing bugs or fleas into a house, before or after a change of tenancy. Fortunately such actions are comparatively rare.

(h) Sale of infested articles

The Public Health Act, 1961, makes it illegal for persons trading or dealing in household articles (furniture, bedding, clothing) to prepare for sale, sell, offer, expose for sale, or deposit with another person (e.g. an auctioneer) any such household article which to his knowledge is verminous.

(i) Hygiene at airports

The Public Health (Aircraft) Regulations, 1966, provides for examination of persons suspected of being verminous and authorizes measures to be taken for preventing danger to public health.

(j) Hygiene at ports

The Public Health (Ships) Regulations, 1970, provides for medical inspection of persons arriving at, or proposing to leave an authority's port who are verminous and where necessary for the isolation or admission to hospital of such persons.

2.4.2 FUNCTIONS OF LOCAL HEALTH DEPARTMENTS IN REGARD TO INSECT PESTS

The principal duty of a local health department is the inspection and supervision of public health. In conducting this duty, the Environmental Health Officers have the right of entry into houses which they believe to be infested.

In addition to this supervision, local authorities are required, by the Public Health Acts, to provide certain facilities for the disinfestation of persons and property. The more progressive health departments may wish to extend these services to improve the local standards of hygiene; for instance they may institute special campaigns for the eradication of bugs, head lice, scabies, etc. To carry out the operations entailed, special disinfestation staffs are required; but it is very desirable for the Environmental Health Officers who supervise the work to be familiar with the various methods involved.

(a) Disinfestation of persons

Local authorities usually set up cleansing centres for the cleansing of verminous people and their clothing. There are places where these centres are merely unused corners of the local public baths, or they are improvised in part of any building which happens to be available. It is surely not too much to expect that the disinfestation centre of a large urban or metropolitan borough should be a properly designed building with adequate equipment and an efficient and self-respecting staff.

The principal treatments to be given are for head lice, scabies and body lice. The staff required must include men and women to deal with male and female patients. They must be careful, conscientious and preferably experienced. The qualifications are put in that order because the modern treatments are not difficult but require thorough application to be effective. In connection with scabies treatments, bathing facilities are desirable; and for disinfesting clothing of people infested with body lice, a hot-air disinfestor may be required.

(b) Disinfestation of buildings

Until recently the only *reliable* method of disinfesting houses from bugs or fleas was by fumigation with hydrogen cyanide. On account of the dangerous nature of this gas, it is impossible to carry out treatments rapidly and easily. Alternative accommodation must be found for the inhabitants of the houses to be fumigated, and often for those in adjoining premises, for at least one night. Often it is necessary for local health departments to employ commercial operators at considerable expense because of lack of staff trained to the use of hydrogen cyanide.

In consequence of these difficulties, house fumigation with hydrogen cyanide was seldom done on the wide scale necessary for a radical reduction of bed bug infestation. Such alternative treatments as sulphur dioxide fumigation and the use of sprays were generally regarded as palliatives, seldom able to eradicate heavy infestations.

With the introduction of persistent synthetic insecticides, especially DDT, a simple and effective method of disinfestation is available which can be applied on a wide scale. It is beginning to be realized that every large borough should have at least one specially trained mobile disinfestation squad to carry out disinfestation by this method. The men should be provided with suitable equipment which will include protective clothing, hand-operated and power-driven sprayers and a small van. Teams of two men are adequate for room spraying. With hand-powered sprayers, they alternate in spraying and pumping up the compressed air. The labour and the likelihood of spilling insecticide are considerably reduced by the use of power-operated sprayers. Spray treatment of a small bedroom takes only about 15 min with a power-sprayer and a little longer with hand-operated sprays. In contrast to a fumigation, the room is quite safe for occupation immediately after treatment.

(c) Disinfestation of articles

The use of DDT is not the answer to all disinfestation problems. Thus, it is sometimes necessary to disinfest the personal belongings of tenants before rehousing them; in which case a complete disinfestation of such articles as upholstered furniture, bedding and clothing is required within a few hours. The most satisfactory methods of achieving this are by heat or fumigation and it is highly desirable that local authority health departments should possess efficient apparatus for the purpose.

Hot air disinfestation is suitable for bedding, clothing and other fabrics which tend to absorb gases (especially hydrogen cyanide) and require very careful airing after fumigation to remove dangerous residues. Most urban authorities possess steam *disinfestors* for destroying germs but these are not entirely satisfactory for use against insect pests. The temperatures reached are unnecessarily high and the capacity is relatively small so that the method is inefficient and slow for dealing with quantities of bedding and clothing. Suitable hot-air chambers are, however, rare.

Fumigation is the most suitable method of disinfesting large pieces of furniture which would be difficult to heat thoroughly right through; also polished wooden articles might be damaged by heat.

The treatment of batches of furniture can best be done in specially designed fumigation vans. The contents of one or two small houses are loaded into such a van which is then driven to a suitable fumigation site (in a reasonably isolated open space). After subsequent forced-draught airing, the furniture can be delivered to new premises on the same day.

Small-scale fumigation, by relatively non-toxic compounds such as ethyl formate, in boxes, bins or plastic sacks, has been revived as a convenient method of disinfesting clothing of people entering a clean environment (e.g. prison or public assistance hostel) (see page 268).

2.5 OFFICIAL PEST CONTROL IN OTHER COUNTRIES

Some summarized information concerning the organization of pest control in various European countries was obtained by correspondence with appropriate experts. These will be given in alphabetical order, first for some west European states, then for east European countries.

2.5.1 WEST EUROPEAN COUNTRIES

(a) Denmark [439]

According to the law concerning protection of the environment (*Lov om miljøbeskyttelse*) the Minister can establish rules on pest control when it is necessary to prevent unhygienic conditions or considerable nuisances. This is done particularly in establishments which manufacture, sell or serve food. In these regulations it is stated that the premises must be protected against pests and that the pests must be controlled.

In the law of tenancy, it is stated that it is the landlord's obligation to control bed bugs and other pests of hygienic importance.

The law and regulations on quarantine state that the Minister can establish rules for the disinsection of ships, aeroplanes etc which present a danger to the spread of communicable diseases.

Regulations exist about scabies and lice in schools. The control is normally carried out by private disinfestors, but the local government has its own disinfestors, which take care of special cases.

(b) France [435]

There are laws on the control of mosquitoes, some of which relate to the establishment of the control measures in the Carmargue (see page 159). There is a decree on the use of pest control agents (1974). Little further information has been obtained.

(c) Germany (Federal Republic) [332]

There is a law which deals with the prevention and control of communicable diseases (*Bundes-Seuchen-Gesetz*, 1962) which is concerned with measures against insect vectors. In practice, the paragraphs dealing with the pests (*tierische Schädlinge*) are unsatisfactory, since the 'animals' are ill-defined; and furthermore, the measures against them can only be enforced when the first cases of disease have appeared; e.g. louse control in families when typhus or relapsing fever has occurred. A new amendment is believed to be being prepared.

There are disinsecting stations in the health departments of single federal states; e.g. Hamburg, Berlin, Düsseldorf, which deal with lice infestations in individuals or groups of people, such as school classes, on a voluntary basis.

Control of bed bugs, fleas, cockroaches and occasionally flies or mosquitoes is performed by special arrangements with the government, e.g. in hospitals, barracks or social service establishments. Treatment of scabies is not done officially. One mobile disinfestation unit was operating formerly in the Rhineland, but has now given up.

(d) Italy [46]

Responsibilities of the Ministry of Health of the central government is limited to maintaining quarantine stations at the most important ports and airports, where disinfestation of incoming travellers is carried out, if necessary.

In the major cities, there are large *Uffici d'Igiene* with laboratories and mobile pest control units. Small communes also have mobile disinfestation squads (2–3 people) directed by the local *Ufficiali Sanitari*, who are the hygiene experts of the communes.

Formerly, all malarious provinces had their *Comitati provinciali antimalarici* but, now that the disease has been eradicated, only a few remain, reconstituted to cope with control of mosquitoes and other pests (Grosseto, Latina, Frosinone, Sardinia) (see page 162).

As regards environmental pollution dangers, all pesticides must be registered and used according to *Decreto Presidenziale* No. 1225 (1968) and *Decreto Ministeriale* (1967). Those pesticides to be used against disease vectors will be regulated by instructions to be issued shortly by the Ministry of Health.

(e) Netherlands [55]

The scabies mite, pubic and head lice are legally considered to be medical pests and the insecticides used against them are deemed to be drugs. Accordingly, their use is governed by the Drug Administration of the Ministry of Public Health and Environmental Hygiene. Body lice are not included in this group, curiously enough, but are considered as a household pest; but they are very rare.

In other respects, pest control is the concern only of local and not central government. Disinfesting stations exist only in Amsterdam (1) and in Rotterdam (2); these are able to deal with personal pests. However, their activities are limited, due to lack of patients. Mobile disinfestors do not exist.

In most large cities, there exists a body (either part of the local health service or the sanitary service) which deals with house pests and local nuisances. The same authority undertakes cleaning and disinfesting of hospital linen and bedding (e.g. Amsterdam), or the control of pests in stored products (e.g. Rotterdam). These units are described as Departments for control of noxious animals (*Afdeling Ongedierte-bestrijding*).

2.5.2 EAST EUROPEAN COUNTRIES

In regard to sanitary operations, the national socialist countries of eastern Europe have in common the existence of 'DDD' services; that is units for Disinfection, Disinsection and De-ratization. These are generally administered by the Ministries of Public Health, but financed by the local District Councils.

(a) Bulgaria [330]

In 1951, the Praesidium of the National Assembly passed a decree relating to the control of communicable disease and malaria. However, with the virtual extinction of malaria, the importance of this has declined. In 1968 a further decree of the Council of Ministers related to the improvement of the organization of pest control services. As indicated above, this is now the responsibility of the

DDD stations, whose staff includes epidemiologists, biologists, disinfectors and disinsecters. They carry out control measures against all pests relating to human health (lice, bugs, fleas, cockroaches, flies, mosquitoes and rats). The control of personal and domestic pests is done free of charge; but contracts involving payment are arranged with shops, restaurants, warehouses and farms. The efficiency of the operations is checked by biologists with laboratory facilities.

(b) Poland [532]

The control of public health pests such as flies and lice was regulated by the Control of Infectious Diseases Act, followed for many years, and finally formalized in 1975. An extension of this law to certain other pests (cockroaches, Pharaoh's ants, fleas, bed bugs and scabies) is currently (1978) being formulated.

The control work is done by DDD stations and this is inspected by Sanitary Epidemiological Units. In addition, the control of some domestic pests is undertaken by the Corn Flour Control Service. Pediculosis, which mainly occurs in summer, is treated by nurses in the schools. Appropriate insecticides are provided free by the Sanitary Epidemiological Stations.

Instructions and advice on disinsection problems are provided by the State Institute of Hygiene.

(c) Yugoslavia [539]

A law relating to the protection of the population from infectious diseases, including scabies and pediculosis corporis, was enacted in 1975; these conditions are now notifiable. Laws on the safe residues in foodstuff and on the general use of toxic substances, were passed in 1972 and 1973.

All public health institutes in the country (about 20) have DDD services. These are under the control of the local authorities.

2.5.3 ORGANIZATION OF PEST CONTROL IN THE U.S.A. [158, 203]

(a) Legal aspects

The organization of pest control in the U.S.A. is of considerable complexity. As in other countries, the main emphasis is on agricultural pests, but insects of public health importance are involved marginally. Thus, strict quarantine procedures are maintained at ports and airports, under the Federal Plant Pests Act, 1957 (which is a revision and codification of earlier legislation). Though specifically intended to intercept plant pests, the inspection service provides a barrier to the introduction of pests of public health importance.

Under the Federal Insecticide, Fungicide and Rodenticide Act, 1947 and the Food, Drug and Cosmetic Act, the use of pesticides is regulated by registration and licensing. Since 1970, this supervision has been done by the Environmental

Protection Agency. The protection of those handling and applying insecticides is covered by the provisions of the Occupational Safety and Health Act of 1970.

Various legal sanctions on pest control operations have been enacted by individual states. For example, in California, the business of controlling domestic insect pests is regulated by the Structural Pest Control Act, 1935.

(b) Bodies concerned with research, advice etc

(i) Federal

Some important work on insects of public health importance is done by various laboratories of the U.S. Department of Agriculture. Thus, the branch concerned with Insects Affecting Animals and Man (at Gainesville, Florida) deals with a number of domestic pests. The pests of stored food are studied at the Stored-products Insects Research Branch, with headquarters at Hyattsville, Maryland. Wood-boring insects are investigated at the Forest Products Laboratory at Madison, Wisconsin; and ticks (especially of livestock) at Kerrville, Texas.

The relevant institution of the U.S. Public Health Service is the Center for Disease Control, at Gainesville, Georgia, where the control of many arthropod disease vectors is studied. The toxicology of pesticides is investigated and the Center also gives co-operative advice to state health departments.

Among the agencies under the aegis of the National Research Council are: the Chemical–Biological Co-ordination Center, which investigates the selection of new pesticides; and the Building Research Advisory Board, which advises on protection of Buildings against wood-boring insects, especially termites.

The Department of Defence is also involved in the protection of military personnel against insects of medical importance. Accordingly, the Army Medical Corps has an Entomology Branch to study and advise on this special problem.

(ii) State agencies

Many states maintain institutes for laboratory and field research to assist in advising those directly concerned with control operations. For example, in California, the Bureau of Vector Control conducts extensive research, both directly and in conjunction with the University of California.

Actual control operations are undertaken by sanitarians attached to County Health Departments. These workers deal with a variety of domestic pests, such as flies, fleas, bed bugs and cockroaches. In addition, there are over 200 agencies organized for mosquito control in the U.S.A., now mainly because of the bite nuisance. Many of the operatives belong to the American Mosquito Control Association.

The various official employees of state and federal agencies who are engaged in research or administration of pest control may become members of the

Association of American Pest Control Officials. This is concerned with legislation, standardization and research on pest control. Other organizations of officials who are occasionally concerned with pesticides are the Association of Official Analytical Chemists and the American Conference of Governmental Industrial Hygienists.

2.6 COMMERCIAL ASPECTS OF PEST CONTROL

2.6.1 INSECTICIDES

The situation with regard to the manufacture, distribution and selling of insecticides in Britain is briefly as follows. Primary insecticidal materials are either synthesized by big chemical manufacturers or else imported as raw materials by large firms. These large concerns are usually quite reliable in the specification of their products; but it is often inconvenient for them to sell insecticides direct to the user. Instead, they sell them in bulk to 'middlemen' who compound the materials in various ways and sell them to large purchasers or retail shops at an increased price. Very often they are entirely justified in doing this because they are charging for compounding the insecticides in forms suitable for use and for packing and distributing them. There are, however, certain firms who sell preparations of undisclosed composition and charge excessively for their secret formula.

Unlike patent medicines, secret insecticides are not legally required to print a declaration of ingredients on the container. It is, however, very desirable that a large purchaser (particularly on behalf of a health department or public institution) should find out the exact composition of the insecticides he buys. He will then be in a position to judge whether the price is reasonable and to compare the composition with officially recommended formulae.

2.6.2 COMMERCIAL PEST CONTROL SERVICES IN BRITAIN

Perhaps the earliest widely-developed commercial pest control service is in the field of fumigation, which needs operators with regular experience. There are now increasing numbers of firms which undertake simpler operations for eradicating such pests as woodworm, bugs or cockroaches. Large hotels, restaurants and institutions, which are often chronic sufferers, may employ such firms to make regular visits under contract. There is much to be said for this arrangement provided that the charges made are not excessive, for the commercial operators soon obtain a good deal of experience in the best materials and methods. Once again, however, it is rather desirable for a person authorizing treatments of this type in a public institution such as a hospital to be fully aware of their nature. There is no reason why payment should not be made for a service regularly and efficiently done, rather than for some quasi-magic remedy.

All the larger firms undertaking pest control services belong to the Industrial

Pest Control Association. This organization will supply a list of firms competent to undertake various types of disinfestation.

2.6.3 COMMERCIAL ASPECTS OF PEST CONTROL IN THE U.S.A.

The pesticide industry in the U.S.A. has developed a large number of new products and, without doubt, it is the largest producing country in the world. The companies and the workers in the industry have organized several technical associations. These include National Agricultural Chemicals Association, which deals with all aspects of pesticides and the Chemical Specialities Manufacturers Association (formerly the National Association of Insecticide and Disinfectant Manufacturers). The latter is somewhat orientated towards household insecticides.

Pest control servicing companies are, as might be expected, numerous and generally efficient. Many of the operators are members of the National Pest Control Association, which upholds standards and arranges periodic meetings on technical subjects.

CHAPTER THREE

Mechanical, physical and biological control measures

3.1 MECHANICAL MEASURES

Modern synthetic insecticides have achieved excellent results in many spheres; in particular, they have revolutionized the control of insect-borne diseases, especially in the tropics. These successes are not unqualified, however, as their regular use frequently leads to the emergence of resistant strains of the pests. Furthermore, the extensive use of chemical poisons has been denigrated on account of their possible danger to man and animals. For these reasons, it may be desirable to review certain alternative measures, which may seem rather old-fashioned. It is worth remembering that certain insect-borne diseases which were formerly endemic in England were not eliminated by the skill or knowledge of doctors or entomologists, but died out as a result of the changing habits of the citizens. Thus, malaria receded as agricultural practices changed and improved; typhus died out when body lice became exceedingly rare; plague has long been extinct since better housing prevented close contact with the black rat and its fleas.

3.1.1 CLEANSING OPERATIONS

The importance of cleanliness in reducing liability to insect pests is very often stressed. There are several ways in which domestic hygiene handicaps the pests. Firstly, infestations tend to be discovered early in well-regulated houses and they can be 'nipped in the bud' before colonizing many inaccessible harbourages. Secondly, the actual cleaning operations may destroy the insects (e.g. frequent combing removes and destroys head lice; laundering clothes kills body lice; scrubbing of woodwork often kills bug eggs). Thirdly, cleanliness prevents the accumulation of rubbish which may provide food for some pests (e.g. spilt foods for stored product pests) and shelter for others (e.g. debris necessary for the development of flea larvae). Analogous to cleanliness is the maintenance of a warm, well-ventilated house which is inimical to those pests which require damp conditions to flourish.

On a larger scale, cleanliness is very desirable in warehouses and similar large food stores; and the Ministry of Agriculture Infestation Control Division have usefully employed vacuum cleaners in such buildings to remove debris. Another type of hygiene concerns the elimination of waste products (see Chapter 9). A water-borne sewage system, almost universal in Britain, protects us from the swarms of flies and blowflies which can breed in primitive privies in a hot summer. The frequent removal and disposal of domestic refuse is essential to prevent flies breeding in dustbins. The household can help by keeping the bins in good condition and the lids in place.

3.1.2 SCREENING

To cope with widely distributed pests whose breeding grounds are numerous or unknown, a logical method of protection is to exclude them by fine-mesh screening. Screening material exposed to outdoor weathering must be proof against corrosion and may be made of galvanized steel, copper, phosphor bronze or aluminium. The size necessary to exclude various insects depends on the aperture, which is determined by the mesh number (holes per linear inch) and the thickness of the wire. Wire thickness is defined by standard wire gauge (swg) numbers, the following being representative [91, 137]:

20 = 0·914 mm or 0·036 in 30 = 0·315 mm or 0·0124 in
25 = 0·508 mm or 0·020 in 35 = 0·213 mm or 0·0084 in

Some suitable specifications for excluding different insect pests are as follows:

Mesh No.	*swg*	*Aperture length*	*Excludes*
10	32	2·27 mm (0·0892 in)	Houseflies, blowflies, etc
16	31	1·3 mm (0·0510 in)	Most mosquitoes
18	33	1·15 mm (0·0455 in)	All mosquitoes
20	27	0·853 mm (0·0336 in)	Sandflies

It must be remembered that apart from reducing light, screening reduces air flow and thus cuts down ventilation, largely in proportion to the fineness of the mesh.

There are also other difficulties, and the complete screening of a large building is not generally feasible unless the windows have been specially designed for it. Thus, the screening must be restricted to ventilation windows, or it will cut off too much light. Such windows will not be able to open outwards (as usual) because of the screens. Attention must be paid to ensuring door closure with self-closing springs, since it is useless to exclude insects from the windows if they can enter elsewhere.

Simple screening measures can be adopted by the householder. Thus, food can be protected from flies and blowflies by a wire-mesh safe and, on the table, by meat or milk covers. Where outer doors are kept open in summer, hanging bead screens may give a limited protection from flies. Another type of screening, on a small scale, is the protection of clothing from clothes moths by sealing it in paper or plastic bags before storage.

3.1.3 TRAPPING

Trapping has been used to combat insects for many years, generally with little more than local and temporary relief. Owing to the extraordinary powers of rapid reproduction of many insects, their numbers are not noticeably reduced, except by the most efficient traps operating under favourable circumstances. In recent years, some improved trapping methods have resulted from a better understanding of the insect behaviour involved. In some cases, however, the theories are vague and superficial. Two aspects of trapping require consideration: attraction and retention. Since the latter is considerably simpler, it will be convenient to deal with it first.

(a) Retention

Perhaps the simplest type of trap is one which has no attractant, but relies on retaining insects which encounter it accidentally. Thus, primative traps of crumpled paper have been used to collect bed bugs from infested beds, for burning in the morning. Leaves of runner beans have been used for the same purpose in the Balkans [57], their special value being the presence of tiny hooks which can arrest bugs and other small insects. In India, a bug trap consisting of two hinged strips of wood, with grooves for bugs on the inner faces, has been used to collect specimens from a natural infestation [603]. All these traps rely on the thigmotactic response of the bugs which induces immobility when they find crevices to rest in.

A very simple method of retaining insects is by using a sticky substance to entangle them, as in the old fly-papers or the sticky bands used on orchard trees. Another method of retaining insects is to use a cage with a 'no-return' inlet. For example, the so-called 'balloon fly trap' is a round cage suspended over a bait, with a conicle inlet, which is difficult for the flies to negotiate from the inside.

A third alternative for retaining insects is to kill them. Water can be used to drown them (adding soap or emulsifier to reduce surface tension). Another method is to electrocute them, using a grid of high tension electric wires. Although this is relatively expensive, it is becoming popular for fly control in restaurants because it avoids the use of insecticides. The only disadvantages are the occasional hiss of a fly being incinerated and the slight smell produced by the cremation.

(b) Attraction

It would be convenient if insects behaved like simple cybernetic machines so that one could select a simple stimulus to attract them. Unfortunately, though their reactions are largely dependent on predetermined reflexes, there are usually several complex systems which interact in apparently straightforward behaviour, when they seek food, mates or shelter. Therefore, many attractants have been found empirically, rather than deductively.

(i) Visual attractants

From what has been stated, it is probably an over-simplification to seek certain colours which are 'attractive' to (say) houseflies. Generally speaking, dark surfaces are chosen by flies for resting sites. This will include colours which appear dark to insects, such as red which is near to the invisible end of the insect's visual spectrum. On the other hand, when they are disturbed mechanically, flies become attracted to light and, in the day, fly to the window. On this basis, the use of mercury vapour or other strip lights on commercial fly-electrocuting devices would be most likely to work if the insects are periodically disturbed.

The popular idea that light is attractive to insects probably depends on the fact that insects collect round a lamp at night. This appears to depend on the 'light compass' reaction of many flying insects, which use a distant light source (sun or moon) as a directional guide. Such an insect, flying at a constant angle to a distant source, will be confused by a single, near, bright light. A simple diagram will show that an attempt to keep the angle constant to such a focus will result in a spiral path towards it. It is certainly feasible to use light traps to collect outdoor flying insects at night. 'Black light' lamps which only emit ultra-violet radiation are often used, since they are perceived by the insects but not by animals or humans who might interfere with the traps.

For houseflies indoors, there are two visual attractants which depend on shape and orientation. It has been noticed that flies prefer to rest on vertical strips or edges and they often chose the cords of pendant lamps. The old-fashioned fly papers accidentally took advantage of this habit, which has also been exploited in insecticide impregnated strings and strips. The other visual attractant for houseflies is an aggregation of other flies; and since their visual discrimination is not acute, this can be mimicked by dark markings on a light background. This has been used to attract flies to insecticide-impregnated paper discs [610].

(ii) Odour attractants: food and oviposition lures

It is possible to make use of several types of insect behaviour governed by odour stimuli to attract them to traps. Volatile substances guide many insects to a source of food. Thus, many biting insects are attracted to their hosts by their smell. Complex mixtures are involved, which may comprise skin and blood constituents, so that single chemicals are seldom entirely effective. On the other hand, most biting insects are also attracted by carbon dioxide, which can be added to outdoor traps for mosquitoes, midges or horseflies, in the form of 'dry ice'. While such traps can have no perceptible effect in controlling the vast numbers in the field, they are useful for sampling the local population of biting insects.

In many cases, it is not easy to distinguish food attractants from oviposition attractants, unless the food source is quite different from the breeding medium, or only females are attracted to the lure. The subject is discussed further on page 85.

(iii) Odour attractants: pheromones

A pheromone is an odorous emanation from an animal which affects the behaviour of another animal of the same species. The most powerful insect pheromones are sex attractants, which are not only more specific than food lures but act at very much lower concentrations. As mentioned elsewhere (page 86), some sex attractants have been synthesized and are available commercially. So far, however, the only one available for insects of public health importance is muscalure, the sex attractant of the housefly [102]. This does not seem to be as potent an attractant as good food lures.

3.1.4 MECHANICAL DESTRUCTION

The violent destruction of insects by mechanical means (as by a 'fly-swatter') is an obviously inefficient way of destroying insect pests in any numbers. In one sphere, however, mechanical destruction on a large scale has been found practical. Stored grain or seeds of various kinds as well as milled products, which have become infested with weevils or other pests, may be disinfested by passing through a machine known as the 'Entoleter'. The cereal is spouted into the machine and flung out by centrifugal force between two flat steel discs revolving rapidly on a central shaft (about 1500 rev min^{-1} for wheat to 3000 rev min^{-1} for flour). These discs are studded with small round posts of hardened steel, set in two concentric rings. The impact of the cereal against the revolving discs and posts and against the housing of the machine, is so great that all stages of insects and mites (including the egg) are killed.

3.1.5 VACUUM

The creation of a vacuum is an unusual method to be used against insect pests, but it has been employed once or twice and is included here for the sake of completeness.

The fall in pressure consequent on evacuation of air is harmless to most insects on account of their small size and comparatively robust structure. Therefore, a vacuum merely acts as does deprivation of oxygen, the effect of which will be discussed later.

3.1.6 MISCELLANEOUS METHODS: MOSQUITO LARVAE

(1) A simple, natural means of controlling mosquito larvae on a small scale is to use the mucilaginous seeds of certain cruciferous plants. The mouth brushes of

the larvae become entangled in them and the larvae eventually die of drowning or exhaustion [499].

(2) It has been found that a very thin film of lethicin spread over the surface of water prevents mosquito pupae from breaking the surface, so that they drown [392].

3.2 PHYSICAL MEASURES

3.2.1 HEAT

(a) General remarks

At temperatures above the normal range, the feverish activity of the insects induced by warmth (involves opening spiracles!) combined with the drying powers of warm air, increases the rate of water loss to a dangerous level which is often fatal after some hours. At even higher temperatures the insect falls into a heat stupor and the effects of the temperature itself are harmful in quite short exposures. The lethal effects of heat on some varieties of domestic pests are shown in Table 3.1; it has been found that all of them are killed within five minutes at 55° C and in less than an hour at 50° C. During a short exposure to high temperature, some insects, especially large ones, are appreciably cooled by the water evaporating from their bodies. Therefore, hot dry air is less dangerous to insects than hot damp air which precludes cooling by evaporation; however, the reverse is true at somewhat lower temperatures where the effect of prolonged drying is the lethal factor.

So far as domestic and medical pest insects are concerned, by far the most important use of hot air for disinfestation is for treating clothing and bedding

Table 3.1 Lethal exposures to hot or cold air, for various pests.

Insect	Stage	Hot air	Cold air
Pediculus (louse)	Egg	0·5 h at 50° C [98]	10 h at −15° C [83]
	Adult	1 h at 46° C	2 h at −15° C
Cimex (bed bug)	Egg	1 h at 45° C [396]	—
	Adult	1 h at 44° C	2 h at −17° C [317]
Xenopsylla (flea)	Larva	1 h at 39·5° C [395]	—
	Adult	1 h at 40·5° C	—
Sarcoptes (itch mite)	Adult	0·5 h at 47·5° C [405]	—
Tineola (clothes moth)	Egg	—	24 h at −15° C [21]
	Larva	0·5 h at 43·5° C [19]	48 h at −15° C
Blatta (cockroach)	Adult	—	1 h at −8° C [398]

suspected of harbouring bugs or lice. Kiln sterilization of timber to destroy wood-boring beetles is also important; but this is more of a commercial process than a matter of domestic hygiene.

The use of heat does not call for any special experience. Such risks as do exist (scorching, fire) are obvious to the simplest workman. Though heat treatment clearly cannot have any lasting protective action like some modern insecticides, it is usually cheap and often immediately effective. There are, however, certain difficulties limiting its usefulness which will be discussed shortly.

Theoretically, there are various methods of applying heat. Radiant heat is seldom satisfactory since it is difficult to apply evenly and may cause too high superficial temperatures. On a small scale, the use of blowlamps to destroy bed bugs provides an example of this method, but it is only used in buildings where scorching of paintwork is comparatively unimportant.

A more satisfactory method is to rely on heat exchange by conduction. For applying the heat, water, steam or hot air can be used.

Water is very effective in many ways but immersion in water is often a grave disadvantage and sometimes impossible. Nevertheless, it is worth noting such a simple expedient as soaking clothing in hot water (over 60° C) for a short time will destroy all vermin, even though it results in wet garments to be dried afterwards.

Steam is employed in steam sterilizers for disinfecting bedding from fever patients, etc, in such apparatus as the Washington Lyon. Machines of this type are designed for the more difficult task of destroying bacteria and, to this end, a very high temperature is produced by steam under pressure, followed by evacuation and repeated steam pressure. The method is unnecessarily slow, cumbrous and expensive for destroying insects and furthermore, the articles are left in a damp condition, necessitating drying with hot air. Similar reasons render steam unsatisfactory for other types of disinfestation, namely the unnecessarily high temperature (sometimes causing damage, for example, to polished wooden articles and leather goods) and the dampness finally produced.

Hot air is the most satisfactory heat-disinfesting agent for destroying insect pests. Any desired temperature can be achieved and the articles are left finally in a dry condition. Hot air, however, has its own disadvantages. Air has a low thermal capacity and it is a poor conductor of heat. Furthermore, since hot air is lighter than cold air, there is usually a sharp temperature gradient in any heated chamber from the cold floor to the hot ceiling. These disadvantages have handicapped hot-air disinfestation to a considerable extent. Added to them must be the poor conduction of heat through many articles to be disinfested, for example, clothing and bedding fabrics which are specially constructed to be heat insulators. This obstacle is illustrated by Fig. 3.1, which shows the lag in penetration of heat through layers of quite poor quality blanket materials [83].

Owing to the protecting effect of the fabrics (wood or other substance in which the insects are embedded) the exposure times and temperatures necessary in practice are enormously greater than those necessary to kill the unprotected

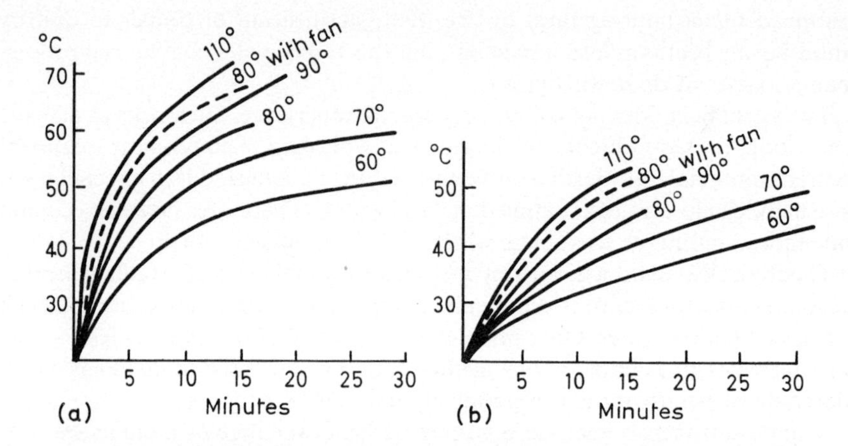

Figure 3.1 The penetration of heat through fabrics. Ordinates: temperatures recorded by thermocouples under (*a*) one and (*b*) three layers of blanket. Abscissae: time in minutes. Curves are marked with the respective external air temperatures (in ° C); in one test the air was circulated by a fan. (After Busvine [83].)

insects. Furthermore, some of the materials to be disinfested may be damp and a considerable amount of heat will be lost in drying them (latent heat of evaporation of water!).

In recent years, it has been realized that the disabilities of hot-air disinfestation can be mitigated by forced draughts induced by air circulation. Thus local cooling due to the low heat capacity of air and its poor conductivity is obviated by continual movement across the sites of heat transfer. Again, the layering of hot air and its consequent bad distribution is reduced by mixing of the air throughout the chamber. Finally, where fabrics are concerned, they are more likely to be penetrated by moving air currents than by still air.

(b) Hot-air disinfestation of clothing and bedding

Prior to the advent of DDT, the usual method of combating lousiness among troops on active service was by hot-air disinfestation of clothing and bedding. Various types of mobile disinfestors were described in the British Army Manual of Hygiene, culminating in the 'Millbank' apparatus, introduced in the Second World War, which employed the forced-draught principle [502].

Two serious disadvantages hamper hot-air disinfestation; it is slow and it does not protect the treated clothing from immediate reinfestation. Consequently, it was never able to prevent the chronic lousiness of front-line men in the First World War. Finally, the remarkable success of DDT dusting in the Naples typhus epidemic (1943) rendered hot-air disinfestors obsolete for dealing with large-scale lousiness, especially during an epidemic.

There is, however, some justification for retaining hot-air disinfestation for the routine delousing of vagrants. In some countries, regular disinfestation of infested people by modern insecticides has provoked DDT resistance in some areas, so that this insecticide might no longer prove effective if the calamity of a typhus epidemic occurred.

If it is desired to improvise a workable hot-air disinfestor for use in an urban community, it is simply necessary to adapt a small room by insulating it as well as possible and suitably lagging the door. Heat can be generated by a stove or by hot water or electric radiators. All that is necessary now is to provide hooks or rails on which to suspend the garments or bedding so that they hang separately and loosely in the upper part of the chamber *where the temperature is above* 70° C. Provided that the insect pests are not protected by more than the equivalent of two thicknesses of blanket, an exposure of one hour at this temperature should be effective.

A much more reliable and more rapid disinfestation plant is obtained if an air circulation system is installed. It is possible to improvise this arrangement in an existing room or chamber by constructing ducts containing a fan to extract air from around the top of the room and inject it again round the bottom. A specially designed disinfestor of this type was produced by the Gas Company for the Ministry of Home Security in the Second World War [7]. It was intended to disinfest bedding from air-raid shelters and had a working capacity of 380 blankets per hour.

(c) Disinfestation of foodstuffs by heat

Small quantities of food in domestic larders may develop infestations of various stored product insect pests (see page 387). It is comparatively easy to destroy these pests by heat in dry foods such as flour, rolled oats, various cereals and spices. Lightly infested food treated in this way should be quite safe to eat provided there are no squeamish objections. In any case, it should be fed to animals unless the food value is very badly deteriorated.

To allow for penetration of heat through a mass food, an exposure of 1–1·5 h to a temperature of 100° C may be necessary. The food should be heated at atmospheric pressure in closed tins in a moderate oven. Biscuit tins do very well for this purpose; their lids should be kept on during the subsequent cooling in order to conserve moisture.

(d) Kiln sterilization of wood infested with beetle grubs

Heat treatment is one effective method of destroying various types of wood-boring bettle larvae (see page 366) either in unworked wood or in furniture, plywood, etc. The exposures necessary with different temperatures and humidities have been determined for the commonly occurring powder-post bettles (*Lyctus* spp.) and these probably hold for other wood-boring grubs. The steam-heated kilns which are used industrially for this purpose, can be operated at various combinations of temperature and humidity. Moisture saturated air is more rapid in action but lower relative humidities have the advantage of preventing appreciable changes in the moisture content of the wood and surface wetting, due to moisture condensation, on cooling. The latter is sometimes detrimental to French-polished articles.

The periods for which timbers of various thicknesses must be heated are given, with other details, in a Building Research Institute leaflet. They range from 2·5 h for a 25 mm (1 in.) and 6·5 h for a 75 mm (3 in.) plank at 55° C, 8–12 h for these two planks at 50° C and about 48 h at 47° C.

Hot-air treatment of buildings to destroy the boring beetle *Hylotrupes bajulus* was formerly practised in Denmark; but the apparatus necessary is elaborate and the method has never been adopted in Britain [302].

3.2.2 COLD

Insects vary considerably in their resistance to cold. Most species, living out of doors, have to pass long periods in the winter at some stage in their life history at temperatures near the freezing point. These insects adapt themselves gradually to low temperature, apparently by eliminating much of the water in their bodies and preventing the freezing of the remainder by biophysical means. Insects adapted to warm climatic conditions are, however, unable to adapt themselves to cold and eventually die even at temperatures several degrees above zero. Among the British domestic pests there are many insects of this type, invaders from tropical regions living in the artificially warmed winter climates of dwelling houses and other buildings. Nevertheless, some insects which are adversely affected by prolonged cooling are able to endure short exposures to extreme cold (e.g. the bed bug and the body louse can survive exposures of several hours to – 15° C).

The data in Table 3.1 (page 44) show that the resistance to low temperatures varies considerably in different insects and depends to a large extent on the previous environment of the insect. These data are for exposed insects and in considering pests protected in various ways, the same principles apply as in hot-air disinfestation. Thus the 'penetration of cold' (i.e. progressive loss of heat) is retarded by the bad conduction of heat in many articles to be disinfested. Consequently it is necessary to submit them to considerably lower temperatures or for longer exposures than those necessary to kill the exposed insects. For example, all lice and their eggs are killed by an exposure to air at – 20° C for 4 h; but a period of 12 h at this temperature might be necessary to disinfest a thick fur or woollen garment. In certain parts of the world where the night temperatures during winter fall to such very low levels, it might, under some circumstances (e.g. a typhus epidemic) be advisable to take advantage of the fact to delouse clothing. But, in general, cold treatment of articles for disinfestation is clearly much more difficult and expensive than other methods. Disinfestation by chilling is clearly out of the question for rooms or buildings.

Whereas the disinfestation of articles by refrigeration is not practical, it is feasible, on a limited scale, to protect them from damage by cold storage. All insect activity is inhibited by temperatures about 5°C though some species can survive very long exposures to these conditions (e.g. clothes moth larvae have survived 12 months at such temperatures). This method of preventing insect damage is relatively expensive so that it is only likely to be employed to protect such things as valuable carpets or fur coats from clothes moths or carpet beetles, while in storage.

3.2.3 ASPHYXIANTS

The respiratory system of insects is remarkably resistant to deprivation of oxygen. It has been shown that some common insect pests require periods between 0·5–8 days to be killed by asphyxia [82]. It is therefore clearly impossible to expect to drown insects by immersing them in water for a few hours.

At the time of the First World War, it was suggested that stores of grain could be protected from attack by stored product pests by sealing the grain in air-tight silos or other containers. In theory, any insects present in the grain would be asphyxiated after the oxygen had all been used up. Comparatively little attention was paid to this method of grain storage until, shortly before the Second World War, interest was revived by some French trials. The method was put into practical use in Argentina during the war, at a time when grain could not be exported and had to be held for long periods. Subsequently, further investigations were conducted in Britain, the U.S.A. and Russia. A practical development has been the erection of eight 1000-ton storage bins in Cyprus. Work in Britain with a comparatively small container (7 tons) has shown how the method could be adopted for use on farms [297].

An alternative method of asphyxiating insects is to occlude the respiratory system by clogging the main tracheae or merely blocking the spiracles. It is even possible that water might achieve this, if, by addition of a wetting agent, it is enabled to penetrate the spiracles. This is probably the basis of the moderately effective action of soap and water sprayed on plants to control aphids. Clogging of the tracheae may also be one of the modes of action of the mineral oil emulsions used against orchard pests or the oil films applied to ponds to destroy mosquito larvae.

It would be misleading to suggest that all oil type insecticides rely on an asphyxiating action; very often the inert oil contains toxic substances, either as impurities or specially added to enhance its effect. For the reasons already stated, the asphyxiating action is not highly reliable in controlling insects and the role of oils has gradually shifted from that of principal toxic agent, to that of mildly toxic carriers of pharmacologically active substances. They will be considered from this point of view in a later section (page 115).

3.2.4 DEHYDRANTS

Owing to a relatively large surface area compared to volume, a very small animal is liable to harmful desiccation in dry air. Insects living in such conditions are protected by cuticles which prevent water loss mainly by virtue of an exceedingly thin waxy outer layer. It has been found that certain mineral dusts damage this layer and cause the death of the insect from desiccation.

The story is a rather complex one, beginning with the empirical observation that certain dusts mixed with grain protect it, to some extent, from damage by weevils and similar pests. (This is said to have been known 2000 years ago by the ancient Egyptians.) The subject attracted the attention of biologists in recent years and the principles involved have been gradually discovered. It was first shown that dusts of quite different chemical constitutions were effective provided they were fine enough; it was also found that dusts with hard sharp particles were more effective than softer substances. These facts suggested that the action was physical or mechanical rather than chemical. Secondly, the insects dying from the action of the dust were observed to lose water very rapidly, the lethal action being much more rapid in a dry than in a moist atmosphere. The inference was that the dusts were causing a harmful water loss from the insects, in some way. It was shown that the dusts acted without entering the spiracles of the respiratory system and that, although the dust was normally swallowed with the food, it was also effective in desiccating pupae which do not feed. The site of action is therefore limited to the cuticle, which loses its waterproofing powers following the disruption of the waxy epicuticular layer. Some workers attribute this action of finely divided dusts primarily or entirely to their abrasive action [325, 613]. Other investigators, particularly in more recent years, have recognized that adsorption can be the dominant factor in removal of lipoid from the epicuticle. This controversy is of more than academic interest. If the desiccating action of the dusts is due to abrasion, they are only likely to be effective against insects whose habits render them liable to this process; for example, granary beetles, which constantly rub their bodies as they crawl among the grain. If, on the other hand, the epicuticular wax can be removed by adsorption, a wide variety of insects with different habits and habitats should be susceptible to the lethal effect of the powders.

Possibly both types of desiccating action can occur in different circumstances. It is noteworthy that most of the earlier work supporting the abrasion theory was, in fact, done with grain weevils. The essential qualities of the dust required were hardness (preferably above 6·5 on Moh's scale) and fineness (mainly between 1 and 5 μm) [325].

More recent work on sorptive powders has shown, as predicted, that they can be used against a wide variety of arthropod pests [155]. The essential characters are a porous material with a large specific surface, which can be shown to absorb wax. Dusts with a pore size ranging from 2 to 30 μm and a specific surface of over 100 m^2g^{-1} are most effective. The most promising materials, which have been exploited commercially in the U.S.A., are silica aerogels (e.g. 'Silikil').

Desiccant dusts, as one might expect, are more rapidly lethal to insects in dry conditions. The more efficient types, such as the silica aerogels, will nevertheless continue to cause water loss at very high humidities. Their lethal action on insects is in some way accentuated under moist conditions if a monolayer of ammonium fluosilicate is applied to the aerogel. This modified powder kills cockroaches and fleas as rapidly at 100 per cent relative humidity as in drier air. It is suggested that the removal of the lipoid layer allows the fluoride access to the aqueous protein layers of the cuticle and thus initiates poisoning.

3.2.5 MOSQUITO REPELLING SOUND

In recent years, various electric devices have appeared on the market which, the manufacturers claim, emit sound waves repellent to mosquitoes. It is difficult to understand the rationale of this. Although there is some evidence to suggest that *male* mosquitoes are *attracted* by sounds corresponding to the wing-beat frequency of female mosquitoes, there seems no reason why such sounds should be repellent to female mosquitoes seeking a blood meal. In fact, various independent scientific tests have shown no evidence of effective repellency from these devices [333].

3.2.6 RADIOACTIVITY

The development of atomic reactors for nuclear power has resulted in the incidental production of radioactive by-products which have been put to various uses; one of these is the destruction or sterilization of insect pests. A constant and controllable radioactive source is required for this purpose, preferably one emitting γ-rays at an adequate energy level. It might perhaps be possible to separate a suitable isotope from the mixture of waste fission products; but the most convenient source, in fact, is cobalt-60, an isotope produced by the spare neutron flux available in nuclear reactors. This produces γ-rays at two energy levels and has a half-life of 5·3 years.

The effects of γ-radiation on tissues depends on the dosage, which is measured in roentgens (R) or rather similar units, radians (rad). While all cells are damaged by high doses, actively dividing cells (as in the genital system) are more sensitive to lower doses, which can cause sterility; and still smaller amounts of radiation may induce mutations (usually harmful) in the progeny of a treated animal.

For reasons which are unknown, different forms of life differ considerably in their sensitivity to radiation. Insects generally are much less sensitive than higher animals and enormous doses (over 300 000 R) may be necessary to kill them rapidly, while about half this dose would be lethal in about a week [127]. This contrasts with a mere 600 R which is lethal to man.

The large doses necessary to kill insects, together with various difficulties and precautions, render the direct destruction of pests by this means uneconomical. Accordingly, the main use of radioactivity in pest control is to produce sterile males for release to decimate the natural wild population (as described under biological control, below).

Sterility can be caused by dosages between 2000 and 20 000 R (as compared with 300 R in man). The relations between dose of radiation and effect may be represented by the normal curve of distribution in the population; a few individuals being easily affected, a few being resistant, while most cluster round the average. For this reason, it is difficult to decide the exact dosage for sterility, since a few tolerant individuals straggle up into the higher dose range. At this level, though well below the dose causing 50 per cent mortality, the lethal effect begins and it may be difficult or impossible to find a dose which will cause sterility without some curtailment of life. A compromise must be made and the doses which have been chosen, to sterilize various flies and mosquitoes for field trials, range from 2000 to 18 000 R.

3.3 BIOLOGICAL MEASURES

It should be emphasized that most of the methods discussed in this section are only in the experimental stage, at present, and none has been successfully employed in Britain. Nevertheless, the subject has attracted much attention as a possible alternative to insecticides, which have become somewhat handicapped by the growth of resistance and increasing complaints about toxic hazards. Rachel Carson, for example, has suggested that insufficient attention is paid to biological control; therefore it is useful to know something of the possibilities and limitations of such methods.

3.3.1 STERILIZATION PROCEDURES

It is convenient to group sterilization procedures together, since they involve the mating behaviour of insects, even though the sterilization may be physical (radioactivity, page 51), chemical (chemosterilants, page 81) or biological.

(a) Release of radioactively sterilized males [326]

The idea of checking populations of insects by releasing sterile males has been developed in the U.S.A. over a considerable time, beginning with a suggestion of E. F. Knipling about 1938. The method has achieved remarkable success in the eradication of the screw-worm fly (*Cochliomyia hominovorax*) from the island of Curaçao and the Florida peninsular; but it is far from applicable to all pests, since the following criteria must be satisfied:

(1) The females must mate once only or rarely more than once, so that after mating with a sterile male, they are rendered infertile.

(2) It must be possible to find a dose of radiation which will sterilize virtually all males, without much effect on their behaviour or longevity.
(3) A mass-rearing technique must be available.
(4) The insects must be easily distributed and mobile.
(5) The wild population must be small in relation to the numbers of sterile individuals which can be released.
(6) Since the aim is complete eradication, an island or an isolated infested area must be chosen.

The operations against the screw-worm, which are the only successful field campaigns to date, were enormous undertakings. The area of Florida cleared was 70 000 square miles and it needed the release of 2000 million flies over 18 months. More than 40 tons of meat were used to breed the flies and 20 aircraft employed to distribute them.

Experimental field trials have been conducted with mosquitoes (southern U.S.A.) [434,605] sheep blowflies (Scotland) [365] and houseflies (Italy) [508]; but none was very encouraging. Perhaps the main difficulty is the size of natural populations of insects, which approximate to the following numbers per square mile: screw-worm flies, 500; tsetse flies, 1000; sheep blowflies, 50 000; mosquitoes, 750 000.

From a little reflection, it is evident that the efficiency of the sterilization method is just the opposite of that of insecticides. The latter easily achieve initial reductions, but become less and less efficient in killing the last few percentages of the insect population. With sterile male release, however, the main difficulty is to make the initial reduction. If this succeeds, the last few normal females have less and less chance of escaping the growing preponderance of sterile males. For this reason, sterile male release against populous insects might perhaps be feasible, if the wild population was first reduced by other measures (e.g. insecticides) or was low because it had newly colonized new territory.

(b) Sterilization of populations by chemosterilants [543]

Chemicals capable of causing sterility in insects or other organisms are called chemosterilants. Their nature and physiological effects are discussed on pages 81–82, this section deals only with their potentialities for controlling insect populations.

Chemosterilants, however, suffer from one very great disadvantage: their danger to higher animals and man. This limits their use as much as the most toxic insecticides, though they can be used both by contact or ingestion. For stomach action, they can be incorporated in baits; this has been tried in field trials against houseflies, applying the bait to remote refuse dumps where it is not likely to do harm [335]. For other insects, attractants (possibly sex lures) may be tried, to bring them into contact with a chemosterilant source. Alternatively, chemosterilants can be used to produce sterile males for release as described in the preceding section.

(c) Release of insects carrying harmful genes [130]

Populations of most insect species (like man) carry small numbers of harmful genes, usually in the recessive form. It has been suggested that it might be possible to select strains carrying high proportions of deleterious genes which, while permitting laboratory rearing, might severely handicap insects in nature. A sustained release of such gene carriers might transmit these harmful traits to the wild population.

Alternatively, it is known that certain sub-species of several insects, while freely mating together, produce more or less infertile offspring. This represents another way of decreasing wild populations.

Both these methods are being explored in genetical laboratories, but neither has yet reached the stage of practical trial.

3.3.2 CONTROL BY PARASITES AND PREDATORS [606]

Natural populations of insects, like other animals, are to some extent limited by the depredations of predators and the regular toll of lethal parasites. Environmental factors which alter the balance, by favouring either the animal or its parasites or predators, can greatly alter the population size. The use of parasites or predators, for controlling insect pests, has been attempted on many occasions in the past few decades, with varying success. The principal successes have been among agricultural pests, especially when the pest has been able to colonize a new country without bringing along the parasites or predators which held it in check in its homeland. Unfortunately, there seem to be fewer obvious biological checks of this kind for insects of public health importance than for other pests; however, a renewed search is being made.

A wide variety of different organisms can act as parasites or predators of insects; they range from viruses to mammals. Among insects of medical importance, there is evidence of mosquito larvae being heavily parasitized by fungi (*Coelomomyces*) and Microsporidia (*Nosema* and *Thelohania*) while adults may be attacked by the fungus *Entomophthora*. Housefly eggs are apparently eaten in quantities by mites (Macrochelidae) [18] while the adults, at certain times, are attacked by *Entomophthora* (*Empusa*). While these facts are of interest and deserve further study, there are many unsolved difficulties of mass culturing and dissemination of these parasites and none has been used successfully in practice.

One of the few organisms pathogenic to insects, which has been extensively used, is the bacterium *Bacillus thuringiensis* [81]. While this has been mainly tried for agricultural insects, there have been moderately promising trials of its use as an additive to food of chickens, to prevent flies breeding in their droppings.

It has been known for over 100 years that insects suffer from virus diseases, but it was not until the development of the electron microscope that the various causal organisms were positively identified. They are classified in four arbitrary groups, of which the polyhedral virus diseases are best known. These are so-called because of the growth, in the insect tissues, of large numbers of polyhedral crystals which contain the actual virus. Some forms develop in the cytoplasm of the cells, others in the nuclei. Generally speaking larvae of Lepidoptera are most prone to infection (e.g. clothes moth grubs); but it is difficult to manipulate the disease for control purposes.

CHAPTER FOUR

Chemical control measures

4.1 INTRODUCTION

4.1.1 HISTORY OF THE CHEMICAL CONTROL OF INSECTS

Since ancient times there have been intermittent records of various substances being used to destroy insects, most of them based on folklore or on such writers as Pliny. Few of these nostrums could have been reliable, however, because they were often based on specious logic rather than on reason and experience. Thus, parts of a plant having a fancied resemblance to a noxious pest were often believed to be a specific for destroying it. During the last hundred years, however, there has been steady progress in the development of really effective insecticides. This period can be roughly divided into three phases, the first of which extended up to the Second World War. The substances which were in use at the beginning of this phase were those known from general experience to be toxic to most forms of life. They included crude inorganic chemicals such as arsenicals and mercury compounds and plant products such as nicotine and derris. Pyrethrum was a fortunate exception, which happened to be specifically active against insects. Then, to such more or less natural products, were added various by-products of the growing chemical industry; especially those from coal tar and petroleum distillation. Later, attempts were made to increase potency by chemical modifications; by introducing chlorine or thiocyanate radicles. Meanwhile, organic chemists had identified the active principles of pyrethrum and derris, so that these vegetable insecticides could be standardized; but synthesis was not then feasible.

Towards the end of this first phase, intense research in Switzerland was devoted to developing organic moth-proofing compounds; and as a by-product of this work, the remarkable powers of the quite simple chemical DDT were discovered (about 1939). It is fair to say that this opened a new era in pest control, which constitutes the second phase of development of modern insecticides. Other chlorinated hydrocarbon insecticides followed, and then a great variety of organo-phosphorus compounds and carbamates, as well as some synthetic analogues of pyrethrum. These new synthetic insecticides, being relatively cheap, stable and easy and (mostly) safe to use, were very widely employed. Production of insecticides increased about a hundred-fold between 1939 and 1961. In addition to extensive use in agriculture, the remarkable success of insecticides against insects of public health importance radically changed the methods of controlling insect-borne diseases, especially in the tropics.

These achievements, however, have been gradually eroded by two serious drawbacks: the development of resistant strains of many pests and the growing concern about toxic hazards to man, his livestock and wildlife. These two problems will be considered in some detail later (pages 119–133). It is sufficient to say here that they have caused some considerable anxiety about the future of these synthetic chemicals and great efforts are being made to find efficient alternative means of control. The mechanical, physical and biological methods described in the last chapter have been re-evaluated. Some, useful, ingenious, new techniques have been discovered and some older ones improved; but none seems to be able adequately to fill the gap if conventional insecticides fail, or are withdrawn. However, it is possible that new chemical control measures may be found which will be free of the toxic hazard and, possibly, less liable to induce pest resistance. Much of the burden of searching for such new chemicals falls on commercial firms (which have been responsible for introducing the great majority of past insecticides). Unfortunately, the enormous cost of screening new compounds for all kinds of toxic and harmful properties is deterring many chemical companies from investing in this field. Nevertheless, some promising new compounds have become available in the form of the Insect Development Inhibitors. These are specifically toxic to insects and virtually harmless to vertebrates. It is true that some conventional insecticides are considerably more toxic to insects than vertebrates, despite the fact that they attack the same biochemical targets in both organisms. Their specificity depends on more rapid penetration into insects or more rapid detoxication in vertebrates. On the other hand, the Insect Development Inhibitors attack systems absent from vertebrates; usually those concerned with moulting or metamorphosis. The best known compounds of this type are the juvenile hormone mimics (see page 76) other examples interfere with moulting, and there may also be hormone antagonists.

At one time, it was hoped that insects would not be able to develop resistance to hormone mimics, for it was supposed that this would abolish response to their own hormones. Unfortunately, it has been found that the insects are able to distinguish between their own endogenous hormones and those applied externally; so that resistance can definitely develop to such new compounds. Despite this disappointing finding, however, the possibilities of the new compounds of this type for pest control are far from exhausted; and much is to be hoped from future research in this field.

4.1.2 COMMON NAMES FOR PESTICIDES

The great multiplication of organic chemicals used as pesticides has introduced many compounds with names too complicated for general use. Accordingly, shortened forms and trade names were employed but, as several of these could be used for the same compound, confusion arose. There was clearly a need for short common names, on the lines adopted for pharmaceutical products. Britain took the lead in standardizing these names, when a Committee on Common Names for Pesticides was set up by the British Standards Institution in 1950. Since then, other nationals have followed; and there is also a list of such names recommended by the International Standards Organization in Geneva. Designing common names is less simple than might appear, since a new name must not resemble a patented name in the same field. In contrast to proprietary names, common names are written or printed without capitals (e.g. parathion). In the exceptional case where names are formed of initials, they are written in capitals without stops (e.g. DDT). Despite the convenience of common names, many commercial and other designations are still found in the literature. Some of these are given in the Chemical Appendix (page 514).

4.1.3 CLASSIFICATION OF CHEMICAL CONTROL MEASURES

Chemicals used for controlling insect pests can be *insecticides*, which are intended to kill (this includes insect development inhibitors, though their action may be delayed); *chemosterilants*, which cause sterility; *repellents*, which drive away; or *attractants*, which lure into traps or poison baits.

Insecticides can be further sub-divided into *stomach poisons* (which need to be eaten by the insects), *contact poisons* (which are lethal after contact with the cuticle) and *fumigants* (which enter by the insect's respiratory system). These categories are not rigid, for several stomach poisons have a slow toxic action on contact; moreover, most contact poisons are lethal on ingestion and some have a fumigant or a repellent action.

4.2 CHARACTERISTICS OF SUBSTANCES USED FOR PEST CONTROL

4.2.1 STOMACH POISONS

(a) Typical uses

To be effective, stomach poisons must be swallowed by the insects and so they are usually incorporated, in some way, with their food. This presents little difficulty for insects which chew solid foods or lick up exposed liquids. In agriculture and horticulture, stomach poisons are simply applied by spraying or dusting vegetation, to kill leaf-eating caterpillars or beetles. There are, however, obvious difficulties in incorporating insecticides into the food of insects which pierce plants to suck sap or pierce the skin of vertebrates to drink blood. Stomach poisons used in this way are said to have *systemic action*; this method has been quite widely developed in recent years against plant pests, but only to a very limited extent against bloodsuckers.

A certain number of pests of public health importance take solid food, which facilitates the use of stomach poisons. Thus poisons in powder form can be used against mosquito larvae, which swallow solid particles. Powder poisons have also been used against cockroaches, which are supposed to swallow them when cleaning their appendages.

In the domestic sphere, impregnation of woolen garments can be used to protect them against clothes moths or carpet beetles, which are killed when they chew the fibres. Again, wood is sometimes impregnated to protect it against boring beetles or (in the tropics) against termites.

Instead of incorporating poison with the insects' staple food, it can be offered in poison bait, which should be specially attractive or else offered in quantity, so that large numbers of the pest feed on it. Poison baits can be moist, solid or liquid. They are not suitable for use in houses against domestic insect pests. Apart from their untidy appearance, the fact that they contain poison potentially harmful to children or domestic animals usually precludes their employment. They may, however, find a limited use against nuisances such as earwigs, crickets, ants or woodlice, which sometimes invade houses from the garden. Another use for poison baits is against houseflies in stables and barns.

(b) Typical substances

(i) Inorganic compounds

The earliest stomach poison insecticides were inorganic chemicals which, as already mentioned, were toxic to higher animals as well as insects, since they are general protoplasmic poisons. They include metallic salts, such as those of mercury and arsenic, and various fluorine compounds. The heavy metals poison by forming complexes with the sulphydryl groups of glutathione and cysteine in certain enzymes, while the fluorine compounds affect cell wall permeability and inhibit enzyme action in unknown ways.

In their use as pesticides, the water solubility of these compounds was an important consideration. Insoluble compounds (e.g. lead or calcium arsenates) were extensively used in agriculture since they could be sprayed on to foliage without rapidly penetrating and damaging the plants or being too easily washed off by rain. Another insoluble compound, the aceto-arsenite of copper, known as Paris Green, was widely used to destroy mosquito larvae, just before the introduction of DDT.

The more soluble compounds include the highly toxic arsenious oxide, mercuric chloride and thallium sulphate; rather less dangerous fluorine compounds such as sodium fluoride or fluosilicate; and still less active borax and boric acid. Most of these have been used in poison baits, though they are seldom employed now, on account of the hazard and because of more effective alternatives.

Some of these substances, especially various fluorine derivatives, were used to impregnate woollen materials for protection against clothes moths. Impregnation of wood against timber pests has also been practised with sodium fluoride mercuric chloride or borax.

(ii) Organic compounds for poison baits

For circumstances where poison baits are considered desirable, a number of organic compounds, primarily developed for contact action, can be used instead of the older inorganic substances. In particular, the chlorinated cyclodienes and organo-phosphorus compounds have been used in this way.

(iii) Organic compounds for moth-proofing [598]

The inorganic compounds used to impregnate wool against clothes moth were not only toxic to man but they were not fast to washing, and in later years they were superseded by organic materials, applied in the dye-bath, and firmly fixed to the wool like colourless dyes. These all derive from the original observation of E. Meckbach about 1920, that the dye Martius Yellow imparted resistance to moth attack. A great deal of research by German and Swiss chemists in the next two decades produced some very efficient moth-proofing agents, especially the 'Eulans' of I.G. Farbenindustrie and Mitin of Geigy. Both these chemicals have a repellent effect and discourage moth larvae from eating treated wool fibres. In addition, they are stomach poisons with distinct toxic action on the grubs. Another interesting approach to moth-proofing is a method of altering the molecular structure of the wool protein, to render it indigestible to moth larvae [199] (see page 408).

(iv) Organic compounds for systemic action

Toxic chemicals can be given by mouth to animals for protection against various types of pest: bloodsuckers, flies causing cutaneous myiasis, or flies breeding in the dung of treated animals. (This last is not, perhaps, strictly systemic action.) These methods of pest control are at present in the experimental stage; they have not reached the stage where drugs given by mouth can be safely recommended for protecting human beings against biting insects.

The use of chemicals taken by mouth against bloodsuckers began with dosing rabbits with ordinary (contact) poisons, such as DDT, *gamma* HCH or pyrethrins. Limited successes were claimed in killing mosquitoes, bed bugs, lice or ticks fed on the rabbits shortly afterwards [353]. The effective doses, however, were rather close to those dangerous to the rabbits. Later research revealed certain less familiar chemicals as more suitable for this particular use; notably certain indandione derivatives and phenylbutazone. Certain organo-phosphorus compounds tested for this use showed some action at safe doses, but were less promising than phenylbutazone.

The use against flies causing skin myiasis (warble flies, cattle grubs) is restricted to veterinary uses, but may be mentioned here for interest. The insecticides are given to cattle as a single drench, or added to the diet or drinking water, injected or merely poured over the skin. In any case, they are absorbed and kill the cattle grubs without harming the beasts. Suitable materials are organo-phosphorus compounds which are detoxified in mammalian tissue faster than in insects, e.g. coumaphos, fenchlorphos and dimethoate.

4.2.2 CONTACT POISONS

(a) Typical uses

There has been an increasingly sharp dichotomy in the uses of contact insecticides, according to whether an immediate kill is required, or whether the emphasis is on long continued insecticidal action.

To accomplish a rapid 'knock-down' and kill, the insecticide is dispersed as a fine spray or, for flying insects, as an air-borne mist, aerosol or smoke. This follows in the tradition of the domestic fly spray, intended to give immediate, though perhaps only temporary, relief from the nuisance. The hand atomizing sprayers used with domestic fly sprays have been largely superseded by liquefied gas aerosol dispensers. An important public health use of these is for the disinsection of aircraft, to prevent the dissemination of mosquitoes (especially infected ones) from country to country. Large aerosol or smoke generators have been employed for the treatment of large buildings or outdoor areas, to secure temporary relief from various unpleasant or noxious insects; but this is expensive and perhaps only justified in special circumstances.

Rudiments of the alternative objective of prolonged insecticidal action are found in the long-established use of powder insecticides. In dusting a dog for fleas or blowing insecticide powder into crevices to kill cockroaches, one expected some continuing insecticidal effect from the residues. The exceptional importance of 'residual action', however, was not apparent until the introduction of DDT and other persistent synthetic contact poisons. When these are applied to walls etc of dwellings, a wide range of domestic pests can be attacked; these include any insects likely to alight on or crawl over the treated surfaces (e.g. houseflies, mosquitoes, cockroaches, bed bugs, ants).

Surface treatments or impregnation of various articles can give lasting protection from other pests: for example, the treatment of undergarments against body lice, of woollen materials against clothes moths or of furniture against timber pests.

(b) Typical substances

(i) Substances of vegetable origin

Miscellaneous substances of minor importance

Various plant products have been used as insecticides, but nearly all have been displaced by modern synthetic compounds.

Among the alkaloids, the best known is nicotine from tobacco, which has been widely used in horticulture. Anabasine is a similar alkaloid, from the plant *Anabasis aphylla* which grows wild over large parts of Asia; and this is used to some extent in the U.S.S.R. A mixture of alkaloids known as the veratrine group is obtained from the rhizomes of *Veratrum album* ('hellebore') or from the seeds of *Sabadilla officinale* ('sabadilla'), both liliaceous plants; but they are little used.

Rotenone and allied compounds

Another type of insecticide from plants comprises rotenone and allied compounds. These are the active ingredients of plants used as fish poisons by the natives of various tropical countries. They have been mainly extracted from the roots of *Derris* spp. ('derris' or 'tuba root') in Malaya or from *Lonchocarpus* spp. ('cube' or 'timbo') in South America. These plants belong to the Leguminoseae, which includes other rotenone-bearing genera, such as *Tephrosia*, *Lonchocarpus* and *Millettia*. Rotenone, like the other active compounds (toxicarol, deguelin, elliptone and sumatrol, etc), is a white crystalline material, very insoluble in water or mineral oils, but dissolving in ether, chloroform and some vegetable oils. These compounds are toxic to insects and acarines by contact (and also as stomach poisons) but the speed of action is rather slow. In oil solution, they are moderately toxic to mammals, though the dry powders are not dangerous.

Pyrethrum

Pyrethrum insecticides are obtained from the flowerheads of *Chrysanthemum cinerariifolium*. Formerly this was mainly grown in the Balkans and the insecticide derived from the powdered flowers was known as Dalmatian Insect Powder; but since 1920 it has been extensively grown in Japan and from 1933 large quantities have been exported from Kenya.

The active principle of pyrethrum is an oily liquid, which occurs to the extent of about 1 per cent by weight of the dried flowerheads. This liquid, which is insoluble in water but dissolves in various organic solvents, contains several related compounds of varying degrees of insecticidal potency. They are difficult to extract pure owing to the presence of resinous impurities which are difficult to separate from them. The active constituents are rather unstable chemically, owing to oxidation, which occurs rapidly on exposure to sunlight. For this reason, residual films do not retain their potency for more than a few days. Even in bulk, it is desirable to keep pyrethrum preparations in darkness, with a trace of anti-oxidant.

The insecticidal effects of the pyrethrum constituents are extremely rapid and they have been widely used on account of their remarkable 'knock-down' effect, which is due to a rapid paralysis induced in contaminated insects. These compounds are comparatively innocuous to vertebrates, which makes them valuable for use against insects of medical or domestic importance. A disadvantage of pyrethrum is the unreliable kill obtained with low concentrations and the relatively high cost of strong preparations. This difficulty can be largely overcome by the addition of potentiating compounds, usually known as synergists. Most of them have little or no insecticidal action alone, but when mixed with pyrethrum preparations, they greatly augment their insecticidal powers. The subject of synergists is discussed more fully on pages 73–74.

(ii) Miscellaneous synthetic substances

Partly synthetic compounds can be prepared from naturally occurring materials, chemically modified in some way. Among the simplest of these are fractions from coal tar distillate, which have been nitrated or chlorinated; e.g. dinitro-*ortho*-cresol. Most of these, however, have fairly considerable toxicity to mammals and are not used against insects of public health importance.

Plant extracts also can be modified chemically, to produce insecticides; e.g. by chlorination or addition of thiocyanate or thiocyanoacetate radicles. Some of these will be considered in their appropriate sections.

Thiocyanates and thiocyanoacetates

Organic thiocyanate molecules carry the group SCN, which is probably the essential toxic unit. The lower alkyl thiocyanates liberate the cyanide radicle in animal tissues and probably act as respiratory poisons. This has not be demonstrated with the more complex thiocyanates used as insecticides, however, and these may have a more involved mode of action.

Thiocyanate contact poisons include: 'Lethane 384' $C_4H_9O(CH_2)_2O(CH_2)_2SCN$ and lauryl thiocyanate $CH_3(CH_2)_{11}SCN$.

The organic thiocyanates, though moderately rapid in action, are handicapped by unpleasant odours and their insecticidal potency is not very good in relation to their mammalian toxicity.

Thiocyanoacetates have generally a less unpleasant smell, though they may be irritant and they are not highly insecticidal. An example is 'Thanite', which consists mainly of *iso*-bornyl thiocyanoacetate.

Both the organic thiocyanates and thiocyanoacetate contact poisons are oily liquids, insoluble in water but miscible with oils and organic solvents.

(iii) Synthetic pyrethroids [30, 167]

The development of the synthetic pyrethroids has a long history of esoteric organic chemical research. It began with the identification of the active constituents of natural pyrethrum by the Swiss chemists Staudinger and Ruzicka in the 1920s. They comprise esters of two ketonic alcohols (pyrethrolone and cinerolone) giving pyrethrin and cinerin, respectively; these are combined, either with chrysanthemic acid (to give pyrethrin I or cinerin I), or with pyrethric acid (to give pyrethrin II or cinerin II). The matter is further complicated by the fact that both the acid and the keto-alcohol can exhibit steric as well as geometric isomerism. This is important because the natural compounds, which are formed from D-*trans*-acids and D-*cis*-alcohols are more toxic to insects than the racemic mixtures formed by simple synthesis. Accordingly, in the more sophisticated synthetic compounds, the more active isomer of the acid is used. (Most of the synthetic alcohols do not have geometric isomers; but in some, the more active isomer is employed here too.) Often the more active insecticide based on an active isomer is distinguished from the mixture by the prefix bio added to the common name.

The development of synthetic forms began with variations on the alcohol components of the molecules and more recently has involved addition of halogens to the acid portion (see Fig. 4.1). This synthetic work was accompanied by extensive toxicological research, much of it done by M. Elliott and colleagues in England; and other advances were made in Japan and in France. From these toxicological studies, it is clear that pyrethroids act as nerve poisons and affect the central ganglia rather than the peripheral system. Nervous tissues of mammals are affected too; but their detoxication enzymes are so much more efficient than those of insects that the contact and oral toxicity of pyrethroids to the latter are generally a thousand fold higher than to mammals.

Insects, however, are able to detoxify different pyrethroids to varying extents (especially the natural ones). For this reason, they are often strongly potentiated by synergists (see page 73) which tend to block detoxication. An additional value of synergists is that they are relatively cheap and therefore economize in use of the expensive pyrethroids.

An additional reason for the selective contact toxicity of the pyrethroids to insects is that they have very low water–oil partition coefficients, so that they readily penetrate insect cuticle. Their rapid action is responsible for their valuable property of 'knock-down' (i.e. rapid paralysing action) on insects. This varies in different analogues and is not necessarily correlated with eventual lethal action, since insects may recover later, unless detoxication is prevented by synergists.

The first synthetic pyrethroid to be developed commercially was allethrin (1949) and this has been (and still is) quite extensively used. During the 1950s, there appeared furethrin, cyclethrin, barthrin and dimethrin. All of these were esters of chrysanthemic acid with various synthetic alcohols; but none surpassed the potency of natural pyrethrins. An advance was the introduction of tetramethrin in Japan in 1965 and a few years later, bioallethrin was introduced by French workers. In the meantime, the first compounds with activity higher than natural pyrethrins were synthesized (at Rothamstead). These incorporated a furan cyclic group in the alcohol component with the production of K-othrin, kadethrin and, above all, resmethrin. A further advance, made independently in England and Japan, was that a meta-substituted benzene linked to a phenoxy group, could improve on the furan plus benzyl side chain. This group includes phenothrin, which is cheaper, though somewhat less active than resmethrin.

The next step involved the acid component of the pyrethroid molecule, when it was discovered that the introduction of chlorine atoms greatly enhanced chemical stability, thus preventing the rapid loss of potency characteristic of

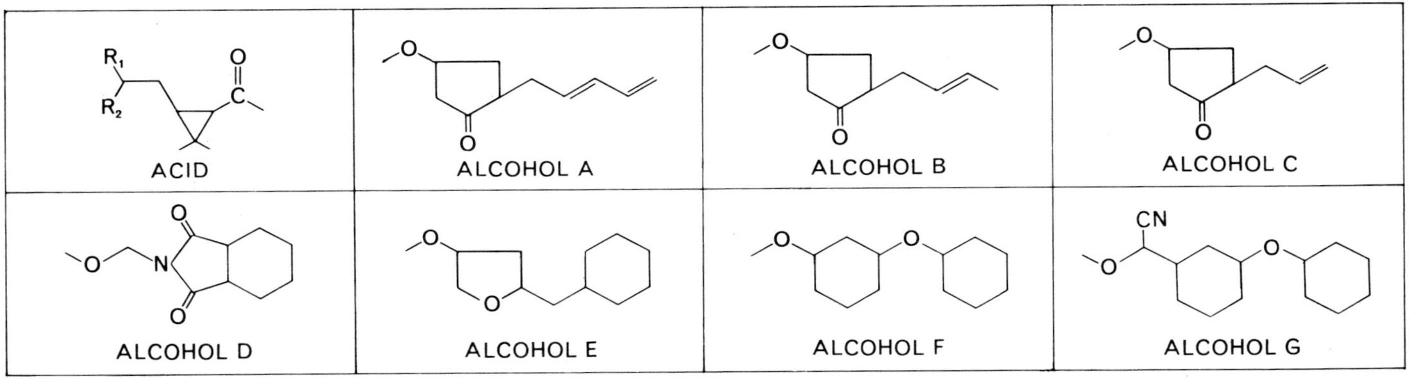

ACID | ALCOHOL A | ALCOHOL B | ALCOHOL C

ALCOHOL D | ALCOHOL E | ALCOHOL F | ALCOHOL G

NRDC no.	Name	R_1	R_2	Stereo-chemistry	Alcohol	Relative toxicity	
						Housefly	Mosquito
–	Pyrethrin I	CH_3	CH_3	(+)-trans	A	0·05	–
–	Cinerin I	CH_3	CH_3	(+)-trans	B	–	–
–	Allethrin	CH_3	CH_3	mixture	C	0·05	–
–	Tetramethrin	CH_3	CH_3	mixture	D	0·05	0·4
–	Phenothrin	CH_3	CH_3	mixture	F	0·30	0·35
104	Resmethrin	CH_3	CH_3	mixture	E	1·0	1·0
107	Bioresmethrin	CH_3	CH_3	(+)-trans	E	2·5	1·5
143	Permethrin	Cl	Cl	mixture	F	1·5	1·3
147	Biopermethrin	Cl	Cl	(+)-trans	F	2·5	2·1
149	Cypermethrin	Cl	Cl	mixture	G	–	2·1
161	Decamethrin	Br	Br	(+)-trans	G	55	40

Figure 4.1 Stereochemistry and toxicity of synthetic pyrethroids. (Toxicities based on the data of Barlow and Hadaway [30] and Elliott *et al.* [168].)

pyrethroids, especially in the light [168]. This resulted in the insecticide permethrin, which has residual insecticidal powers comparable to most organophosphorus or carbamate insecticides. Finally, the latest advance has been the addition of a CN group on the alcohol group, which increases potency; and this, combined with the use of bromine atoms on the acid (instead of the chlorine) produced decamethrin [169]. This has outstanding potency against insects, with lethal doses of the order of 0·002 mg kg^{-1}, compared to over 50 mg kg^{-1} for rats.

As with all insecticides, the relative potency of the various pyrethroids changes from one insect species to another; but, in general, they are clearly of a high order of efficiency and relatively very low hazard to vertebrates. The residual persistence of permethrin and decamethrin are particularly valuable. A word of caution must be added, however. So far, they have not been very widely used, compared to the vast quantities of organochlorine insecticides. Thus, in 1975, about 50 tons of pyrethroids were manufactured, compared with 80 000 tons of DDT which was produced in the U.S.A. alone in 1962. Nevertheless, there are already signs that these new compounds may be afflicted by the problem of resistance. Part of this may be inherited from widespread DDT-resistance, which seems to have a component which involves resistance to pyrethroids.

(iv) The organochlorine insecticides [407]

The organochlorine insecticides comprise (i) DDT and analogous compounds, (ii) *gamma* HCH and (iii) the chlorinated cyclodiene compounds and similar insecticides. All of them primarily attack nervous systems (the DDT peripherally, the other compounds centrally). Apart from this, despite extensive research and much ingenious speculation, little is certain about the mode of action of DDT and even less about the others. All of them are potentially toxic to vertebrates, though their acute toxicity to arthropods is much higher, because they penetrate into them much more readily. Because of this differential toxicity, organochlorine insecticides have been found to be safe and convenient in use. However, the presence of chlorine in their molecules has conferred on them chemical stability, which was originally hailed as a great advantage in providing residual insecticidal action. Unfortunately, this stability was later found to result in persistent residues in the environment and also in the bodies of animals and men. Actual harm to humans was rare, but recurrent cases of death among wildlife and the disquiet about the long term effects in man, led to restrictions or banning of their use in most countries.

The ingenuity of chemists has been able to overcome this particular drawback by devising biodegradable analogues of DDT (by replacing the *para*-chlorine atoms with alkyloxy groups [281, 408]) and also chlorinated cyclodiene compounds with much less persistance. Though somewhat less insecticidal than the original compounds, it is possible that these might have come into general use, except for the fact that widespread resistance to organochlorine compounds has developed in many insect pests. It is therefore fairly certain that the use of these insecticides has passed its peak and is gradually declining. Unfortunately, although alternative insecticides exist, all are more expensive than the organochlorines and almost equally prone to suffer from resistance eventually.

DDT and analogous compounds

The remarkable insecticidal powers of this relatively simple chemical were discovered in Switzerland about 1939. When supplies became generally available after the Second World War, its cheapness, efficacy and safety in use resulted in an enormous demand for it. The abbreviation DDT is taken from the initials of the generic chemical name: DichloroDiphenylTrichlorethane, the most active isomer of which is the pp′ form, shown in Fig. 4.2 (page 68). A considerable number of analogous compounds show similar insecticidal action, but none equals DDT in overall potency. Only a few have been produced commercially; these include TDE (or DDD) and methoxychlor, which have the advantage of lower mammalian toxicity.

The toxic effect of DDT depends on the steric configuration of the molecule, which somehow reduces the stability of the electric potential across the axonic nerve membranes. As a result, the temporary discharges (action potentials) which propagate along the axons, tend to replicate, so that there is constant nervous stimulation. Characteristic symptoms are tremors and inco-ordinated movements.

Gamma HCH

These letters stand for HexachloroCycloHexane. The compound was formerly known as BHC for Benzene HexaChloride; but this term is incorrect. It can exist in a number of steric isomers, differing in the orientation of the chlorine atoms in the molecule. They are designated *alpha*, *beta*, *gamma*, *delta* etc; and it was discovered in Britain about 1942 that only one of them (the *gamma* isomer) is highly insecticidal. This is remarkable, since scores of other polyhalogenated cyclohexanes have been synthesized, but none was found to approach *gamma* HCH in potency. With other organochlorine insecticides, changes in the molecule do not have such a drastic effect.

Chlorinated cyclodiene insecticides [407]

These are highly chlorinated cyclic hydrocarbons with a characteristic 'endo methylene bridge'. Alternatively, the essential nucleus of the molecule can be visualized as being formed of two bent pentagons having three points in common. There are two distinct types: the semi-synthetic mixtures produced by chlorinating pine oil and the synthetic compounds made by Diels–Alder condensation.

It has been known for some years that chlorination of the terpenes in pine oil increased their insecticidal powers. As early as 1942, the Russians were using 'Substance SK', subsequently identified as chlorinated camphene. Insecticides of this type have been developed in other countries, the best-known examples being Toxaphene (U.S.A., 1947) and Strobane (U.S.A., 1953). Both are waxy aromatic

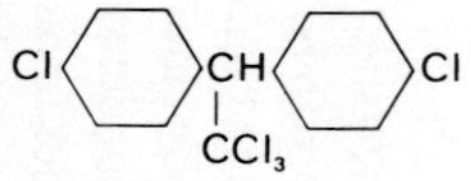

Cl, CH, CCl_3, Cl — 1	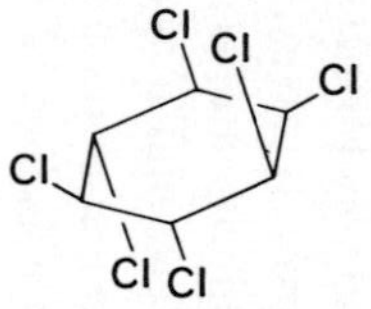 Cl, Cl, Cl, Cl, Cl, Cl — 2	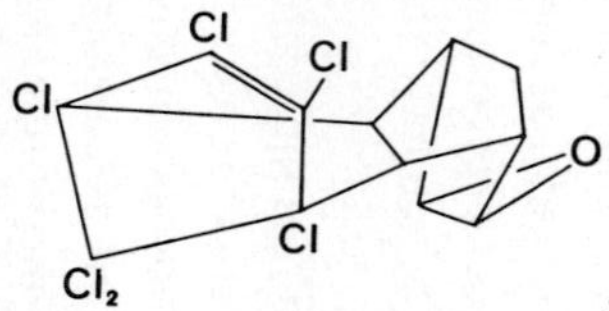 Cl, Cl, Cl, O, Cl, Cl_2 — 3	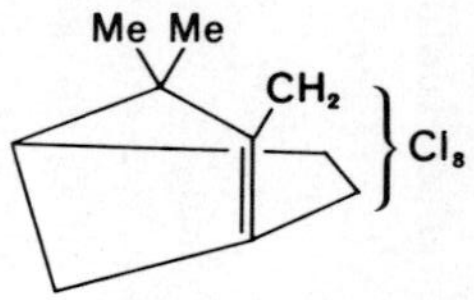 Me Me, CH_2, Cl_8 — 4
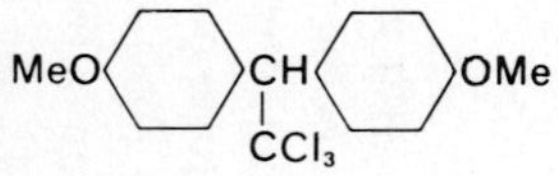 MeO, CH, CCl_3, OMe — 5	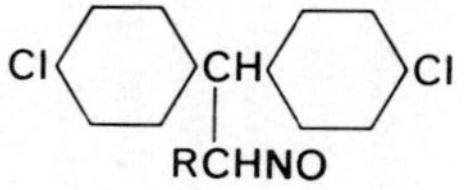 Cl, CH, RCHNO, Cl — 6	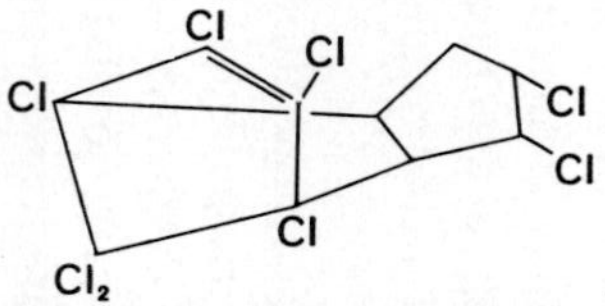 Cl, Cl, Cl, Cl, Cl, Cl, Cl_2 — 7	CCl_3, Cl, CHOCOMe, Cl — 8
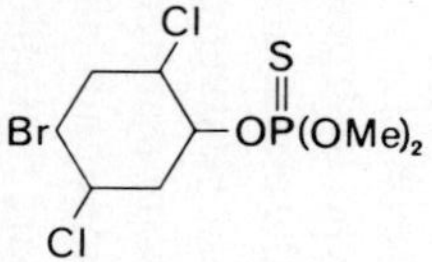 Cl, Br, S, $OP(OMe)_2$, Cl — 9	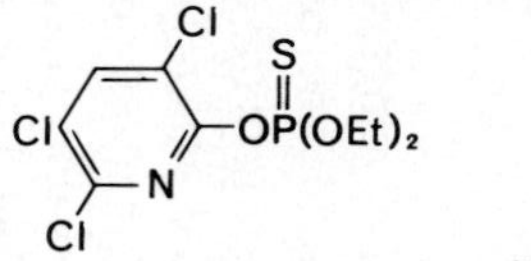 Cl, S, Cl, $OP(OEt)_2$, N, Cl — 10	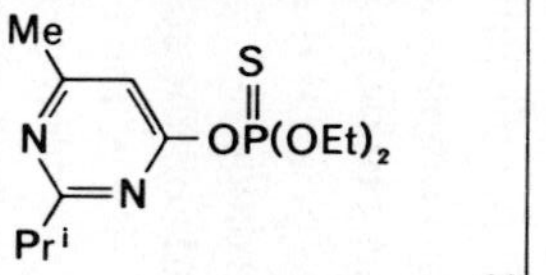 Me, S, N, $OP(OEt)_2$, N, Pr^i — 11	S, NO_2, $OP(OMe)_2$, Cl — 12
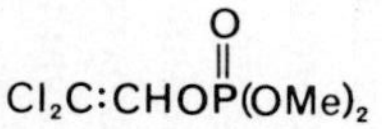 O, $Cl_2C:CHOP(OMe)_2$ — 13	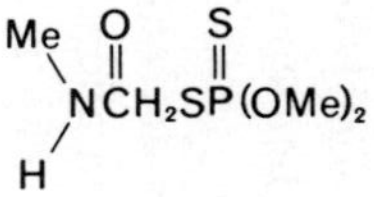 Me, O, S, $NCH_2SP(OMe)_2$, H — 14	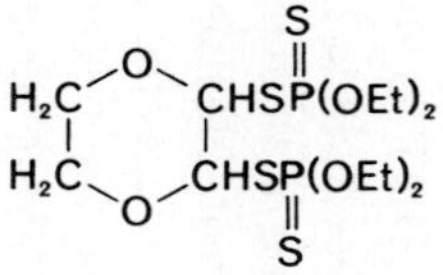 S, H_2C, O, $CHSP(OEt)_2$, H_2C, O, $CHSP(OEt)_2$, S — 15	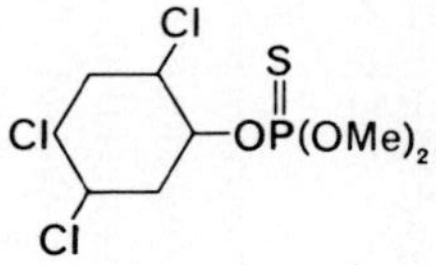 Cl, S, Cl, $OP(OMe)_2$, Cl — 16
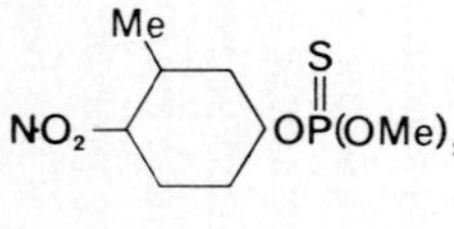 Me, S, NO_2, $OP(OMe)_2$ — 17	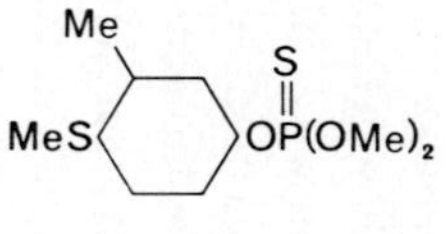 Me, S, MeS, $OP(OMe)_2$ — 18	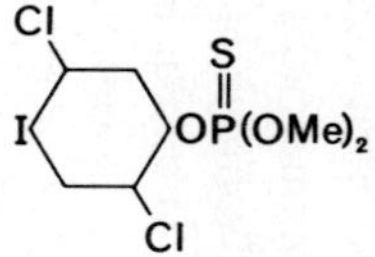 Cl, S, I, $OP(OMe)_2$, Cl — 19	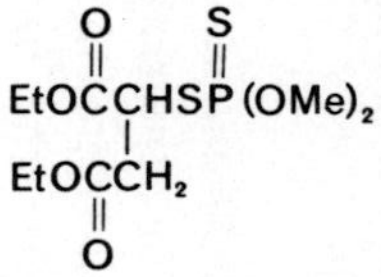 O, S, $EtOCCHSP(OMe)_2$, $EtOCCH_2$, O — 20

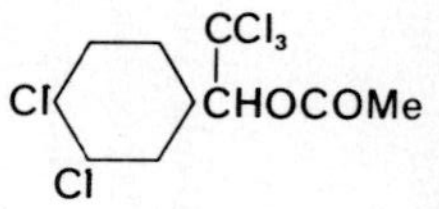

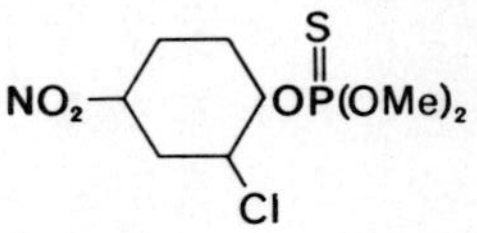

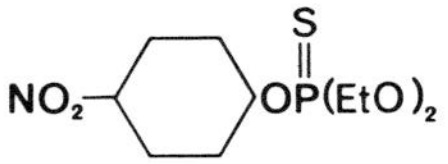

NO_2 S OP(EtO)$_2$ 21	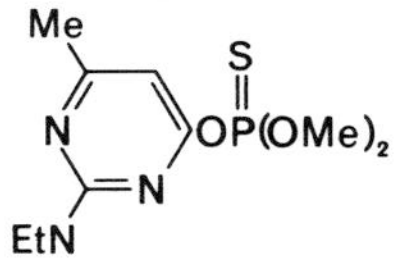 Me N S OP(OMe)$_2$ N EtN 22	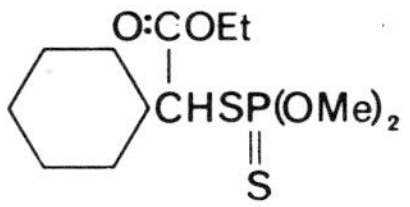 O:COEt CHSP(OMe)$_2$ S 23	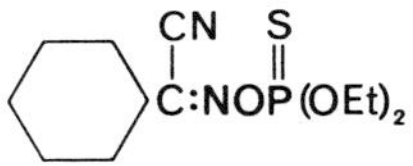 CN S C:NOP(OEt)$_2$ 24
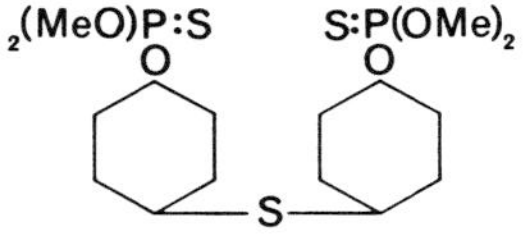 $_2$(MeO)P:S O S:P(OMe)$_2$ O S 25	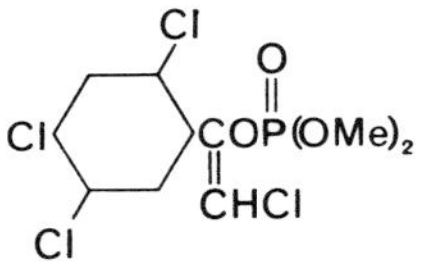 Cl Cl O COP(OMe)$_2$ CHCl Cl 26	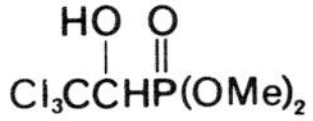 HO O Cl_3CCHP(OMe)$_2$ 27	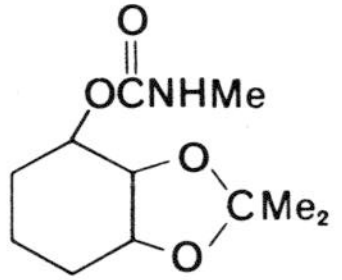 O OCNHMe O CMe$_2$ O 28
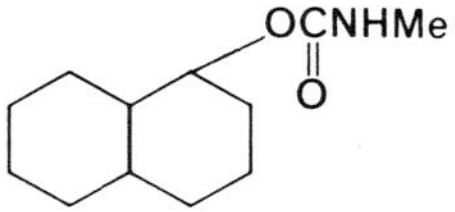 OCNHMe O 29	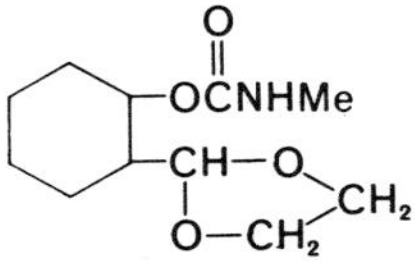 O OCNHMe CH—O CH_2 O—CH_2 30	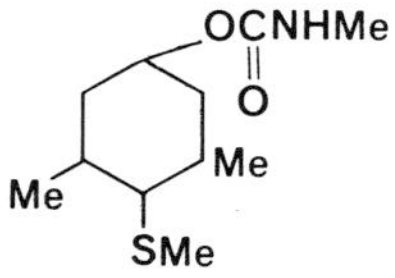 OCNHMe O Me Me SMe 31	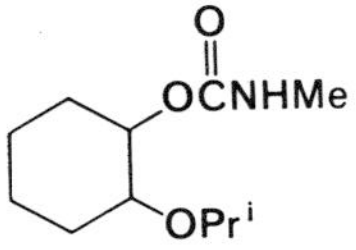 O OCNHMe OPr^i 32
MeO O OCNHMe OMe 33	Bu O OCNHMe 34	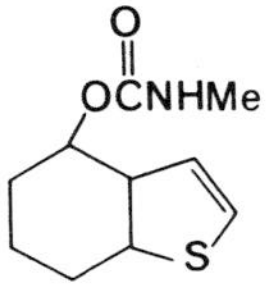 O OCNHMe S 35	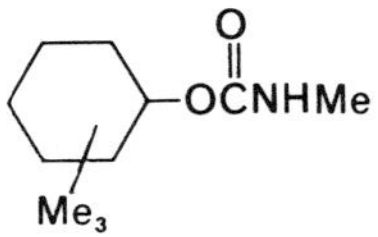 O OCNHMe Me$_3$ 36
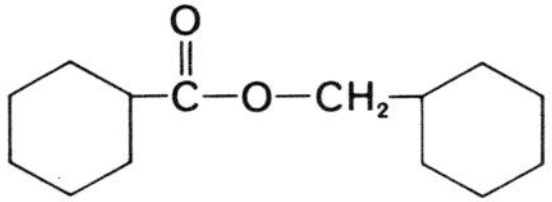 O C—O—CH_2 37	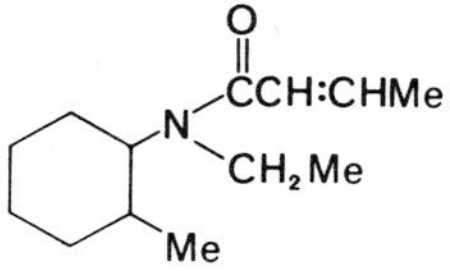 O N CCH:CHMe CH_2Me Me 38	Et S S Et N—C—S—C—N Et Et 39	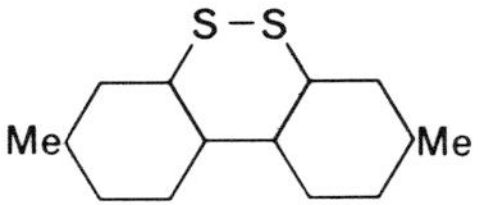 S—S Me Me 40

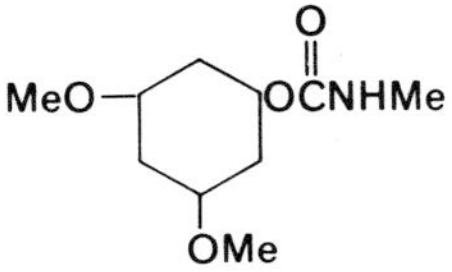

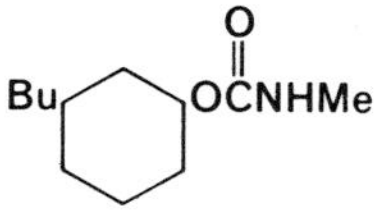

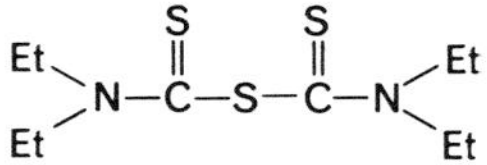

materials containing mixtures of compounds averaging 66–69 per cent Cl; virtually insoluble in water, they readily dissolve in organic solvents, especially aromatic ones. The general structure of these chlorinated compounds resembles that of the synthetic cyclodiene (see Fig. 4.2). Their mode of action appears to be similar and they are involved in the same resistance group.

The synthetic cyclodiene insecticides have been mainly developed in the U.S.A. since 1945. They are produced by Diels–Alder molecular condensation; i.e. a fusion of two five-membered rings. They include the following compounds which are all crystalline solids (m.p. 103–346° C): aldrin, chlordane, dieldrin, endosulphan, endrin, heptachlor and isobenzan. These compounds have the same general solubility pattern, being very insoluble in water (<0·1 ppm), slightly soluble (<5 per cent) in alcohol or odourless kerosene and readily soluble (25–50 per cent) in aromatic solvents. They are all widely toxic to insects, some being very potent indeed. Unfortunately, their mammalian toxicity also tends to be moderately high. Their mode of action is obscure, but is known to involve the central nervous system.

(v) The anti-cholinesterase insecticides [453]

The anti-cholinesterase insecticides, which were introduced in the decades after the Second World War, comprise the organo-phosphorus compounds and the carbamates. Many of them are considerably more volatile than the organochlorine compounds, and most are less resistant to chemical or enzymatic degradation. Therefore, they lack the advantage of long residual action but, for the same reason, they do not leave such persistent residues in the environment.

They came into prominence as resistance impaired the efficacy of the organochlorine insecticides but, as a consequence of their more extensive usage, they too have suffered from the same trouble.

Unlike the organochlorines, their mode of intoxication is fairly well understood. It affects vertebrates as well as arthropods and, among the wide range of compounds in the two groups, all degrees of toxicity occur. Selective toxicity to insects in the compounds used as insecticides depends on more efficient detoxifying enzymes in mammals rather than a difference in penetration.

Figure 4.2 (See previous pages.) Structural formulae of some synthetic insecticides. *Organochlorines*: 1, DDT; 2, *gamma* HCH; 3, dieldrin; 4, camphechlor; 5, methoxychlor; 6, Dilan; 7, chlordane; 8, perfenate. *Organophosphates*: 9, bromophos; 10, chlorpyrifos; 11, diazinon; 12, dicapthon; 13, dichlorvos; 14, dimethoate; 15, dioxathion; 16, fenchlorphos; 17, fenitrothion; 18, fenthion; 19, iodofenphos; 20, malathion; 21, parathion; 22, pirimiphos-methyl; 23, phenthoate; 24, phoxim; 25, temephos; 26, tetrachlorvinphos; 27, trichlorphon. *Carbamates*: 28, bendiocarb; 29, carbaryl; 30, dioxacarb; 31, methiocarb; 32, propoxur; 33, butacarb; 34, bufencarb; 35, Mobam; 36, Landrin. *Acaricides*: 37, benzyl benzoate; 38, crotamiton; 39, tetraethyl thiuram monosulphide; 40, dimethyl thianthrene.

Organo-phosphorus insecticides [453]

The introduction of this type of insecticide originates largely from the poineer work of G. Schrader in Germany, beginning about 1934. Following the publication of his work in 1947, thousands of different organic phosphorus compounds were synthesized and tested and a considerable number is in commercial production. The first examples in wide use (e.g. parathion, TEPP) were dangerously poisonous to higher animals, but later safer compounds were found; finally the trend of research has been to seek high insecticidal potency in combination with low mammalian toxicity.

The relevant structure of the organo-phosphorus insecticides can be understood better in relation to their mode of action, which concerns the enzyme acetylcholinesterase. In normal nervous activity, the chemical acetylcholine is rapidly secreted in small amounts at vital points in the system, especially the nerve junctions or synapses. The stimulating effect of traces of this chemical causes transmission of a nerve impulse; but too much would cause chaos and the excess is rapidly destroyed by the enzyme. This is done by temporary union of acetylcholine with the enzyme, followed by its degradation and elimination of the waste products, leaving the enzyme unchanged (as in Scheme I, below). The organic phosphorus esters, however, also can react with the enzyme but, in this case, the final break-up occurs very slowly, so that the enzyme remains blocked or poisoned and ACh accumulates (Scheme II).

Scheme I	*Scheme II*
$E + ACh \rightleftharpoons EACh$	$E + PX \rightleftharpoons EPX$
$EACh \rightarrow EA + \text{choline}$	$EPX \rightarrow EP + X$
$EA \rightarrow E + \text{acetic acid}$	$EP \rightarrow ?$

(where E = enzyme; ACh = acetylcholine; P = $(\text{Alk. O})_2P(O)–$; X = miscellaneous groups)

There are, in fact, several basic types of organo-phosphorus ester which can cause this type of poisoning; but the majority of satisfactory insecticides conform to the following pattern:

$$(YO)_2P(=X^1)–X^2–R$$

X^1 or X^2 = O or S
Y = methyl or ethyl
R = miscellaneous groups

The activity of the whole molecule is regulated by changes in X, Y or R.

The active anti-cholinesterase agents have $X^1 = O$; but in animal tissue P=S can be oxidized to P=O, thus greatly increasing the insecticidal potency. Accordingly, insecticidal compounds may have either oxygen or sulphur in this position. As regards Y, steric considerations restrict the more effective compounds to those with methyl or ethyl radicles (though in EPN, the whole alkyloxy group is replaced by phenyl).

A great variety of groups can be introduced at R. They affect the overall physical nature of the molecule; they can affect the reactivity of the phosphorus atom (through electron induction effects); finally, they may be important sterically.

This briefly summarizes the factors affecting potency. Other factors concerned with detoxication are also important, especially in regard to selective toxicity (insects as opposed to mammals). Thus, animal tissues contain a range of enzymes which may be able to detoxify organo-phosphorus compounds. Certain phosphatases can attack them at different points in the molecule, rendering it harmless. In consequence, the outcome of contamination of an insect or higher animal with a potentially toxic organo-phosphorus compound will be the outcome of various physical processes of penetration and distribution within the body and of chemical processes of toxication and detoxication. If insects differ in some of these processes from higher animals, it should be possible to take advantage of this and choose compounds destroyed in the mammal but not in the insect. Some progress in this direction has been made.

In physical properties, the organo-phosphorus insecticides are usually liquids, soluble in some organic solvents but rarely so in petroleum oils; sometimes sparingly soluble in water. Some examples with low mammalian toxicity are: bromophos, fenchlorphos, iodophenphos, malathion and pirimephos-methyl. They tend to have disagreeable smells, largely due to impurities.

Carbamate insecticides [407]

It has been known for many years, that various carbamic acid esters are toxic to animals; but these active compounds were water soluble and not suitable as insecticides. From about 1947, however, the Swiss firm Geigy began a search for suitable contact insecticides of this type, which resulted in the commercial production of one or two compounds (Dimetan, Pyrolan, Isolan). Subsequently a variety of insecticidal carbamates have been developed in various countries.

The mode of action of the carbamate poisons resembles that of the organo-phosphorus compounds, in that both attack the enzyme acetylcholinesterase. The signs of poisoning are likely to be similar in both mammals and insects, since both derive from excess of acetylcholine. There are, however, differences. The organo-phosphorus compounds actually combine with the enzyme, which becomes phosphorylated at one point. In contrast, the active carbamates possess steric resemblance to acetylcholine and compete with it for fixation at two sites on the enzyme. Once *in situ*, they are less easily hydrolysed than the natural substrate and thus block the enzyme. On the other hand, the inhibition is not so irreversible as that caused by the more potent organo-phosphorus poisons.

For steric reasons, the most active carbamates are based on the *N*-methyl or *N*, *N*-dimethyl forms of the acid:

$$-O-\overset{\overset{\displaystyle O}{\|}}{C}-N(CH_3)_2 \qquad \text{or} \qquad -O-\overset{\overset{\displaystyle O}{\|}}{C}-N(H)CH_3$$

Considerable variation in the alcoholic portion of the molecule is possible: it can consist of a phenol, pyrazolol or other heterocyclic alcohol; but steric considerations affect the nature and position of substituents to some extent. A well-known example, with low mammalian toxicity is carbaryl. Of moderate mammalian toxicity are: propoxur, bendiocarb and dioxacarb.

4.2.3 SYNERGISM AND SYNERGISTS [104]

(a) Theoretical considerations

In the context of pest control, a synergist is expected to have little of no intrinsic toxicity, but to be able to increase the potency of pesticide when added to it. In fact, however, this is a special case of synergism, which describes the situation where the joint action of two drugs or poisons is greater than would be expected from their separate effects. Sometimes, indeed, two compounds together produce *less* effect than expected, and this is described as antagonism. Either effect can occur with a mixture of two insecticides or with a non-insecticidal compound added to an insecticide. Although some of these results were discovered quite empirically, biochemical research has revealed some of the underlying causes. Synergism or antagonism depends on the inhibition of various enzyme systems. Synergism results from the suppression of a detoxifying system, which thus preserves the toxicant. In some cases, however, compounds used as insecticides need to be potentiated by enzymes in the insect tissues in order to produce their toxic effects (e.g. the conversion of P=S organo-phosphorus compounds to P=O, or the oxidation of aldrin to dieldrin). Then, the ultimate effect of a synergist depends on whether or not it inhibits the potentiating system more than the detoxifying one.

The ways in which synergists inhibit enzyme systems is not completely understood. Often the synergist itself is metabolized by the enzyme and one of its metabolites may block further action in some way. The synergist can act also as an alternative substrate for the enzyme, having more affinity for it than the pesticide. Because synergists are involved with particular enzyme systems, they tend to be specific for certain detoxication pathways and can be used as research tools to indicate the particular mechanisms responsible for tolerance or resistance. An interesting case is that of 'analogue synergists' whose molecules resemble those of the insecticide, which may explain the affinity for the enzyme. There are, for example, pairs of organo-phosphorus or carbamate insecticides which show greater than expected potency. Some organo-phosphorus compounds tend to inhibit the detoxifying enzyme carboxylesterase more strongly than their target enzyme cholinesterase; thus they facilitate the action of other cholinesterase-inhibiting insecticides. These include actual insecticides (EPN) as well as non-insecticidal compounds such as triphenyl phosphate or tricresyl phosphate. Other examples of analogue synergists are the compounds DMC and FDMC, which inhibit the detoxication of DDT by dehydrochlorination.

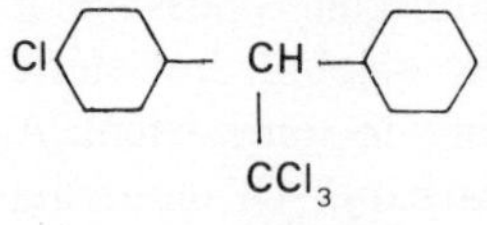

Cl⟨⬡⟩— CH —⟨⬡⟩Cl   Cl⟨⬡⟩— COH —⟨⬡⟩Cl   Cl⟨⬡⟩— COH —⟨⬡⟩Cl

CCl_3 (DDT) CH_3 (DMC) CF_3 (FDMC)

DDT 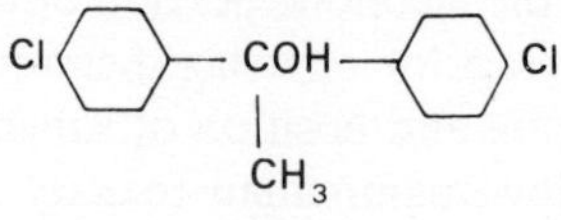 DMC 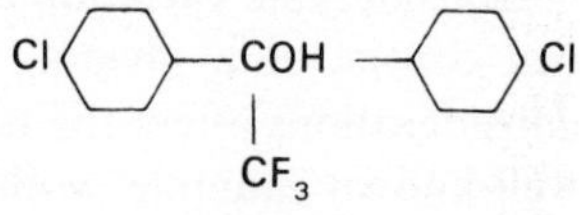 FDMC

The forgoing cases, however, represent rather special forms of synergism. By far the most important synergists are those which inhibit the microsomal mixed function oxidase (mfo) systems which are responsible for degradation of various harmful substances in both vertebrates and arthropods. Thus, it is not surprising that insects find them effective for detoxifying insecticides, including some organo-phosphorus compounds, many carbamates and above all, natural and synthetic pyrethroids. Several kinds of compound have been found to synergize these insecticides by suppressing oxidative metabolism by mfo systems. Most common are compounds with molecules incorporating a methylene dioxyphenol group (1); others include those with -N-alkyl (2), or O-(2-propynyl) groups (3).

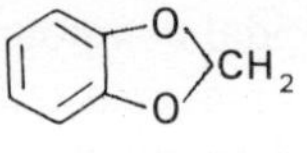

(1) benzene ring–O–CH_2–O (methylenedioxy)   (2) $—N(CH_2—CH_3)(CH_2—CH_3)$ (3) $—O—CH_2C{\equiv}CH$

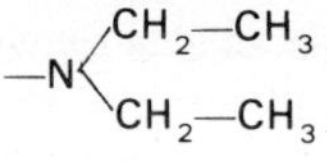

(1)	(2)	(3)
Piperonyl butoxide	SKF 525A	NIA 16824
Sulphoxide	WARF-anti-resistant	
Propyl isome		

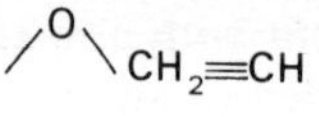

(b) Practical considerations

In pest control practice, synergists are normally of the non-toxic variety and nearly all are concerned with protection from oxidative detoxication. Although this is quite important with some carbamate insecticides, practically all the usage of synergists has been with the pyrethroids. These nearly all respond well to synergists and their active ingredients are rather expensive. One of the earliest commercially available synergists was sesame oil (about 1940), the activity of which was later (about 1946) traced to sesamin and sesamolin. About this time, other related compounds were introduced: propyl isome, sulphoxide and piperonyl butoxide. The last mentioned is perhaps the most widely used of all, even today. Piprotal (Tropital) was developed as a synergist about 1965. These are mostly amber-coloured liquids, insoluble in water, moderately soluble in water or miscible with organic solvents. They are of low mammalian toxicity and comparatively inexpensive. Their potentiation of pyrethroids generally increases with the amount of synergist, up to a point; above this, deleterious effects on the physical characteristics of the formulation and reduced penetration into the insect begin to outweigh the synergism. Most of them are used in synergist:insecticide ratios of 1:1 up to 1:10.

4.2.4 CONTACT POISONS FOR MITES AND TICKS [407]

The Acarina are sufficiently different from the Insecta to require some rather different types of poison to control them, although some poisons are effective against both.

(a) Typical uses

Various types of mite are troublesome agricultural and horticultural pests and are attacked by certain acaricides. Other mites are harmful to stored food but these are not easily attacked by chemicals other than fumigants. Comparatively few mites are pests of domestic or public health importance. Probably the most serious is the itch mite, to control which medicaments are applied to the human body. Allied to it are various mange mites of domestic animals, requiring analogous treatments. Poultry keepers may suffer from the bloodsucking poultry red mite, which also invades houses from wild bird nests. Apart from this, a few nuisance mites (e.g. the clover mite) may require the spraying of walls of infested houses. Harvest mites are sporadically annoying pests (and, in S.E. Asia, may transmit scrub typhus); but, because they are widely dispersed in pastures and scrubland, they are difficult to attack by treating their habitats. Although this has been done occasionally, repellents are generally a more practical defence. The same strategy applies to many ticks which sporadically attack man and sometimes transmit disease. Ectoparasitic ticks, which may heavily infest some domestic animals, are a veterinary problem. They are commonly attacked by dipping or spraying the affected animals.

(b) Typical substances

DDT is generally ineffective against mites and also argasid ticks, although it can be effective against larval ixodid ticks and has sometimes been used to treat infested areas used by campers and agricultural workers. There is, however, a range of chlorinated compounds, somewhat similar to DDT, which are effective against agriculturally-important mites. These all have two *para*-chlorobenzene groups joined by various atomic bridges. They include: dicofol, chlorfenethol, chlorobenzilate, chloropropylate, chlorbenside; others have additional chlorine atoms on one phenyl (chlorfensulphide, tetradifon).

Gamma-HCH is effective against most acarines, both mites and ticks and it is unfortunate that this useful organochlorine compound is becoming unavailable because of the general ban on this class of chemicals.

Various organo-phosphorus compounds are also effective against acarines. Ticks of veterinary importance can be controlled by chlordimeform and other formamidines.

The unusual habitat of the itch mite would seem to require rather special control agents. *Gamma*-HCH is highly effective and still used, although it may soon be banned in some countries. Benzyl benzoate and sulphur are two long-standing remedies. Others are dimethyl thianthrene, tetraethyl thiuram monosulphide and crotamiton (Fig. 4.2).

4.2.5 INSECT DEVELOPMENT INHIBITORS (see Fig. 4.3)

(a) Theoretical considerations

Insect development inhibitors essentially interfere with physiological processes of insects which are absent from vertebrates. So far, the most successful advances in this field have been made with endocrine systems which interfere with metamorphosis. Next most successful are compounds which interfere with moulting, perhaps by a direct attack on the physiology without involving hormones. In addition, there appear to be other systems which could be exploited for control purposes; for example, mimics of other hormone systems and anti-hormones. These, however, are currently in early stages of development.

(i) Hormones associated with moulting and metamorphosis [600]

The endocrine systems which have proved fruitful in the development of control agents are those associated with the insect brain, the *corpora allata* and the thoracic glands. Brain hormone apparently stimulates the production of the moulting hormone, *ecdysone*, from some source (possibly the abdominal oenocytes, and perhaps with involvement of the thoracic glands.) The ecdysone, or a derivative, stimulates appropriate genes, as indicated by 'puffs' visible on large chromosomes. The actual response depends on the level of production of another substance, the '*juvenile hormone*', secreted by the corpora allata. During larval or nymphal life, this hormone is freely secreted, with the result that the moulting activity initiated by ecdysone causes reduplication of juvenile tissues. But at the end of this period, the juvenile hormone secretion gradually ceases; then the ecdysone-induced moult produces pupal or adult tissues.

In early researches, it was found that the application of excess juvenile hormone at a time when its natural concentration was falling, caused abortion of metamorphosis. Excess juvenile hormone applied to female adults was found to interfere with reproduction (their eggs did not complete embryonic development; and in some cases reproductive diapause was broken). These adverse effects suggested that hormones could be used as control agents. Extraction of natural hormones was naturally very difficult and expensive but, when the chemists had worked out the structure of the hormones, it was found possible to synthesize them.

Ecdysone-type compounds, indeed, have complex steroid molecules, which makes synthesis troublesome; and, in any case, these compounds do not penetrate easily into insects, so that their practical use is limited. However, the juvenile hormone was found to be relatively simple and a considerable number of

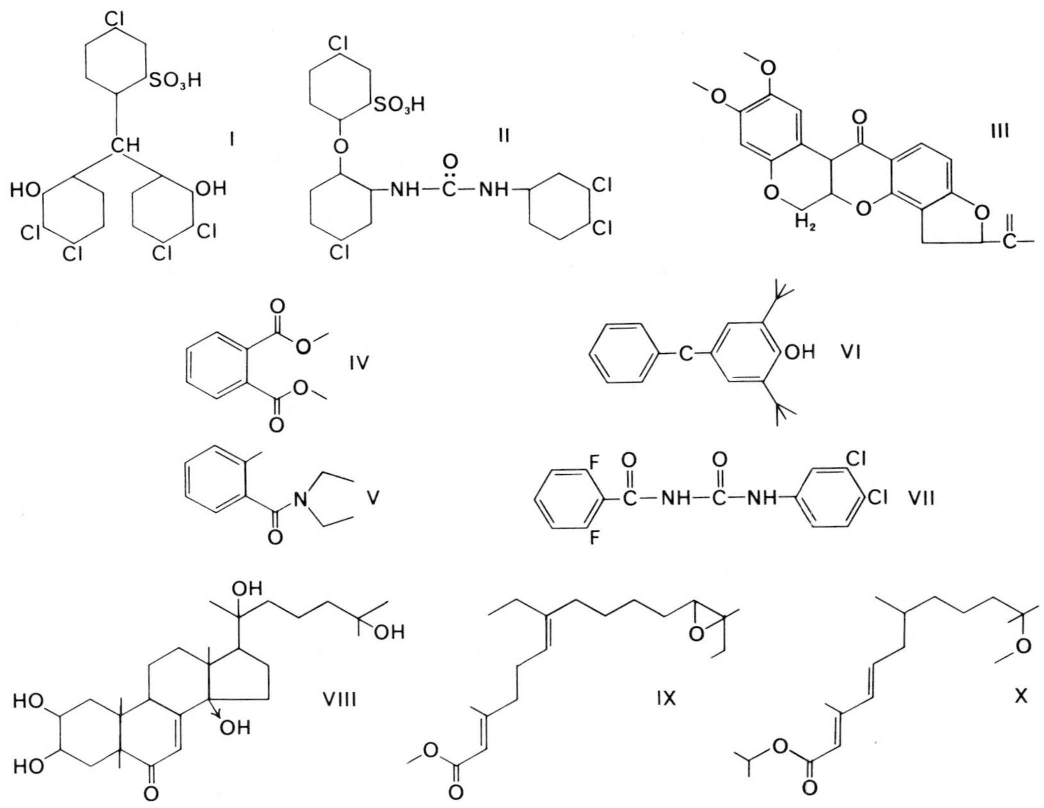

Figure 4.3 Some insect control chemicals.
I, Eulan CN; II Mitin FF; III, rotenone; IV, dimethyl phthalate; V, *N*, *N*-diethyl-*m*-toluamide; VI, Mon–0585; VII, diflubenzuron; VIII, ecdysone; IX, natural moulting hormone; X, methoprene.

analogues have been prepared. Although these juvenile hormone mimics (or juvenoids) can affect reproductive processes in adults, their practical use has been concentrated on the abortion of metamorphosis. This involves the disadvantage that the insect is only at a phase of maximum sensitivity for a relatively short time during its last juvenile stage. During this period, however, it is very sensitive indeed, though the lethal effect is not immediate but delayed. Insects which normally undergo partial metamorphosis either moult to an abnormal intermediate between a nymph and an adult or they may produce a non-reproductive additional nymphal stage. Insects with complete metamorphosis tend to die at various, often late, stages of pupation.

(ii) Disturbance of moulting

The practical use of juvenoids can be hailed as the final commercial development of basic research in insect physiology. The diflubenzuron type of compound, which interferes with moulting, appears to have been discovered empirically and the mode of action investigated subsequently. It seems that these compounds interfere with chitin synthesis and there is even evidence as to the stage in synthesis at which it is blocked.

(b) Practical considerations

Anyone concerned with pest control must welcome the introduction of insect development inhibitors. Scientists will be fascinated by their unusual modes of action, which are still not completely understood. Practical men will be glad of a pest control agent virtually free of toxic hazard. However, certain drawbacks must be recognized. One defect is their delayed action, which is inconvenient for pests such as lice, which need rapid control. Another important defect is the fact that they only act on the young stages; and, in the case of the juvenoids, for a short period in the last juvenile stage. Therefore, their most promising applications are against insects which are mainly or solely harmful in the adult stage. Alternatively, persistent residues are necessary to achieve population limitation. Finally, the most unfortunate finding is that resistance to conventional insecticides appears to extend to these new chemicals.

Nevertheless, there have been many practical trials against a wide variety of insects, with varying degrees of promise. Even by 1975, there had been the following number of tests against different orders of insect: Diptera, 47 (25 mosquitoes); Hemiptera, 47; Coleoptera, 32; Lepidoptera, 19; Hymenoptera, 10; Orthoptera, 7; Isoptera, 8; Mallophaga, 5; Anoplura, 2 [552].

The first two of these compounds to be developed commercially were the juvenoid methoprene, by Zoecon in the U.S.A. and diflubenzuron by Philips-Duphar in the Netherlands.

4.2.6 FUMIGANTS

(a) Typical uses

Fumigants can secure the most rapid eradication of pests from a limited space. They do not normally leave residues, however, so that they give no protection from reinfestation; and this seriously handicaps them (in comparison with residual contact insecticides) for most uses in public health entomology. Added to this is the fact that most of the more efficient fumigants are dangerously toxic to mammals. Accordingly, fumigation with hydrogen cyanide was regulated by an Act of Parliament in Britain in 1937 (amended 1951) which is likely to be extended to other fumigants (see page 29). However, while fumigants still offer useful service in the disinfestation of bulk food and other stored products, their use against insects of public health importance has greatly diminished. Thus hydrogen cyanide was formerly used to fumigate human dwellings against bed bugs and similar pests. This was a troublesome procedure and potentially dangerous, so that with the advent of synthetic contact insecticides, this type of fumigation has virtually ceased.

Hydrogen cyanide and methyl bromide still find a limited use in fumigating furniture and bedding against bed bugs, clothes moths etc in specially constructed steel vans [73]. On a smaller scale, various organic liquid fumigants can be used to disinfest verminous clothing in boxes, bins or plastic sacks [98].

Domestic uses of fumigants comprise the injection of woodwork against furniture beetle (*ortho*-dichlorbenzene) and the use of solid fumigants (naphthalene, etc) against clothes moths.

Perhaps the only interesting modern development is the introduction of the 'residual fumigant'. This is an insecticide of low volatility but high insecticidal potency, e.g. dichlorvos (see Fig. 4.2 and Table 4.1). Containers which gradually emit the vapour are hung up in human dwellings. With restricted ventilation, enough vapour may be present to kill sensitive insects, like mosquitoes, while being harmless to man.

(b) Typical substances [431, 460]

(i) Gaseous fumigants

Substances gaseous at normal temperatures must have low molecular weight and this limits the number available. Many true gases are inorganic compounds, some being rather inert, others very reactive and perhaps corrosive. Comparatively few successful fumigants have a boiling point below or near to ordinary room temperatures; some of them are shown in Group I of Table 4.1. Most of them can be dispensed from metal cylinders, sometimes with additional heat to assist volatilization. Hydrogen cyanide and phosphine can be generated chemically, *in situ*. In general, all this group are rather difficult and dangerous to handle without special training.

Phosphine (generated from small aluminium phosphide–ammonium carbonate tablets; 'Phostoxin') and ethylene oxide are mainly used for grain fumigation. Sulphuryl fluoride, a new fumigant, has been employed against termites in the U.S.A. Hydrogen cyanide and methyl bromide have been used in furniture fumigation (page 441). Methyl bromide has the advantage that its residues can be more easily removed by airing; but it causes an unpleasant taint with some materials [74].

Table 4.1 Properties of fumigants

Group	Name and formula	Boiling point (°C)	Vapour pressure at 20° C (mm Hg)	Saturation at 20° C (mg/l)	Vapour density (air = 1)	Acute toxicity to mammals	Inflammability
I	Phosphine PH_3	−125	> 760	—		***	***
	Sulphuryl fluoride SO_2F_2	− 55	> 760	—		*	0
	Methyl bromide CH_3Br	3·5	> 760	—	3·3	***	0
	Ethylene oxide $(CH_2)_2O$	11	> 760	—	1·5	*	***
	Hydrogen cyanide HCN	26	612	905	0·9	***	*
	Ethyl formate $H{\cdot}COOC_2H_5$	54	196	837	0·92	*	**
II	Methallyl chloride $CH_2{:}C(CH_3)CH_2Cl$	72	102	505	3·1	**	0
	Ethylene dichloride CH_2ClCH_2Cl	84	63	345	3·4	*	*
	Ethylene dibromide CH_2BrCH_2Br	131	11	112	6·5	*	0
	o-Dichlorbenzene *o*-$C_6H_4Cl_2$	179	1·2	9·3	—	0	0
III	*p*-Dichlorbenzene *p*-$C_6H_4Cl_2$	—	0·64	4·9	—	0	0
	Naphthalene $C_{10}H_8$	—	0·08	0·6	—	0	0
	Dichlorvos $C_4H_7PO_4Cl_2$	—	0·15	1·8	—	0	0

*Data not available or unreliable.

(ii) Vapour fumigants

The fumigants in Groups II and III of Table 4.1 are vapours in equilibrium with liquids or solids at room temperature. For each compound, there is a saturation concentration dependent on its vapour pressure and molecular weight and the temperature, thus:

$$C = \frac{P}{760} \times \frac{273}{T} \times \frac{M}{22{\cdot}4} \times 1000 = \frac{16\ M.P}{T}$$

C = saturation concentration (mg l^{-1}); P = vapour pressure (mm Hg); M = molecular weight (g); T = absolute temperature.

If excess of the compound is put into a closed space, the vapour will theoretically reach the saturation concentration; but, in practice, it is seldom attained because of leakage and absorption.

The compounds in Group II are relatively safe to handle on a small scale, for box or bin fumigation. Carbon disulphide (B.P. 46° C; sat. conc. 20° C, 1250 mg l^{-1}) has been omitted, because it is not only highly inflammable but stinks horribly. The other compounds are not objectionable, but all are slightly toxic to man and should not be inhaled excessively. Carbon disulphide is also highly inflammable.

The fumigants in Group III are virtually safe for normal handling.

4.2.7 CHEMOSTERILANTS

The theoretical advantages of sterilizing insects, rather than killing them, is that if the males can be sterilized without affecting their vigour, they will seek out normal females and by mating with them, discourage their union with normal males. Thus the insects are employed in their own extermination and this subject is discussed under the heading of biological control.

This section deals specifically with chemicals which induce sterility. Radioactive sterilization is dealt with elsewhere (page 52–53).

(a) Typical uses

Chemosterilants can be employed in two general ways against insects.

Firstly, they can be used under controlled conditions, to sterilize large numbers of insects from an artificial colony to produce sterile males for release. This is merely an alternative to radioactive sterilization, but it has some advantages of cheapness and convenience.

Alternatively, for insects which either cannot be reared in vast numbers or are too objectionable to release in quantity, chemosterilants can be exposed in baits to the wild insect population. Since many of these compounds are dangerous to other animals, their use in baits requires careful consideration and precautions.

So far, they have been used with some success and safety in isolated refuse tips, against houseflies.

(b) Typical substances [59, 545]

Chemosterilants work by attacking active cell nuclei, i.e. on rapidly dividing cells as in the genitalia. Similar compounds have been used to inhibit malignant growths in man; and some of the drugs used in cancer therapy provided an early source of chemosterilants for insects. Two types of compound have been tried: anti-metabolites and alkylating agents, the latter being the most powerful but also the most dangerous.

(i) Anti-metabolites

During cell proliferation, there is continual synthesis of nucleic acid. This can be inhibited by certain anti-metabolites which antagonize the synthesis of purines and pyrimidines or prevent their incorporation into nucleic acid (e.g. methotrexate, aminopterin). Insects given small doses of these are prevented from developing normal eggs or sperms.

More recently, certain plant products have been discovered with sterilizing action on insects; e.g. camtothecin [414] and β-asarone[521]. While their mode of action is not yet clear, it is possible that they may act as anti-metabolites.

(ii) Alkylating agents

These compounds attack the nucleic acid after it has been synthesized. Large doses damage the nuclei so severely that the sperms or eggs may die. Smaller doses merely disrupt the chromatic material, causing lethal mutations. The treated insects produce eggs and motile sperm, but the progeny fail to mature. This parallels the effect of moderate doses of radiation and it is described as the radio-mimetic effect.

Some of the earlier alkylating agents were the nitrogen mustards which, in water, tend to form cyclic imonium compounds. This aroused interest in compounds with a cyclic aziridine ring, which is the basis of various potent chemosterilants (e.g. tepa, tretamine, apholate).

4.2.8 REPELLENTS

Since ancient times, men have tried to keep noxious insects away from their persons or their dwellings. Primitive methods include the smoke from smouldering fires, odorous substances applied to garments or hung up in the dwelling, and various plant or mineral preparations rubbed over the body. In the early decades of this century, essential oils were commonly used and recommended. From the 1930s onward, however, many thousands of synthetic chemicals were tested as repellents and some of them have now virtually displaced the natural products. The advantages of these synthetic repellents are as follows. They are almost odourless and therefore acceptable in use; they are more persistent; they are more consistent than natural substances, which are mixtures of components varying in effectiveness; and lastly, they are relatively cheap.

The use of repellents is to be regarded as a palliative at best, not to be compared with satisfactory control measures, if these are possible. Furthermore, despite the good points of modern repellents, even the best fall short of what is wanted and probably the ideal repellent does not exist. Ideally, the application of a few drops of repellent should keep off noxious insects for an indefinite period. In practice, most modern repellents act by contact, with little action at a distance, so that they only protect the area actually treated. Furthermore, the period of protection is limited to a few hours (for treated skin) and a week or two (for treated clothing). Since, however, the repellents now in use devolved from very extensive trials, it seems unlikely that very much better examples of the type will be found. It is conceivable that something quite different might be achieved by systemic repellents (i.e. substances taken by mouth) but this line of approach has yielded little success so far.

(a) Typical uses

(i) Against flying pests

Flying bloodsucking insects, such as mosquitoes, stable flies, horseflies, biting midges, etc, usually land directly on exposed skin. Therefore repellents for protection against them are normally applied directly to the skin. Many repellents can be applied undiluted, but sometimes they are made up into creams or lotions, which may be considered cosmetically more desirable.

People frequently require protection from the bites of mosquitoes and similar insects when going on holiday, especially abroad. Unfortunately, they usually expose a large area of skin, which is difficult to protect. Some of the repellent gets absorbed into the skin and more of it is rubbed off by clothing, or removed by perspiration, during exercise. For these reasons, complete protection cannot be expected for more than a few hours, even with the best repellents.

Annoyance from horseflies, midges and certain other biting Diptera is usually experienced by agricultural and forestry workers and by fishermen. The main target is the hands and face. Protection of the latter can be secured by the use of impregnated veils. The use of wide-mesh netting (1 cm^2) impregnated with repellent and draped over the hat brim to enclose the face secures protection from biting insects without greatly reduced vision or ventilation.

(ii) Against crawling pests

In contrast to flying insects, the crawling bloodsucking arthropods frequently crawl over or among clothing before they reach the skin; so that repellents to deter them can be applied to the clothing. Two quite different types of pest are concerned. In the less hygienic quarters of towns and cities, doctors, health inspectors and social workers may find that they pick up fleas or even bed bugs in the course of their visits. The other pests affect people sitting or lying on the ground in rural areas; sometimes this invites the attention of harvest mites or, occasionally, ticks. The sufferers may be agricultural workers, picnickers or soldiers on manoeuvre.

Repellents applied to outer clothing are smeared over the garments, mainly in the region of the openings, such as the neck-hole, cuffs, trouser legs, etc. Applications to clothing are generally more persistent than skin treatments; they may remain effective for a week or two.

(b) Typical substances

A 'repellent', as understood by the public, is an empirical concept. So long as the pest insects are prevented from annoying them, people do not bother to consider whether this is due to avoidance at a distance, to a strongly irritant action on contact, or to a rapidly paralysing effect. However, these various forms of protection will require rather different types of 'repellent'.

(i) Volatile repellents causing avoidance

Volatile repellents are exemplified by essential oils, one of the best being citronella, extracted from the grass *Andropogon nardus*. It contains geraniol as its chief ingredient, together with citronellol, borneol and various terpenes. Though its smell is not unpleasant, it becomes wearying after a time. Furthermore, it does not give the prolonged protection of the better synthetic repellents and accordingly is not widely used now.

(ii) Synthetic repellents acting on contact

The two synthetic contact repellents most widely used against mosquitoes and biting flies are 'DMP' and 'deet', the latter being generally more effective. Other compounds which have been used are 'Carboxide' and a mixture of diethyl benzamide and *o*-chlorodiethyl benzamide ('Kik'). For non-biting flies, derivatives of cinchonic acid have been found effective [352] (e.g. di-*n*-propyl isocinchonomerate [599]); also 2-ethyl-1, 3-hexanediol and dimethyl carbate [440].

Against ticks, *n*-alkyl cyclohexylamines are effective [35], while 2-(diethyl phenoxy) ethanol was promising against fleas [34].

For repelling cockroaches, various heterocyclic amines were outstanding, and also di-*n*-butyl succinate, which was also effective against stored food beetles [383].

(iii) 'Repellents' actually paralysing or killing

Rapidly acting insecticides, such as pyrethrum or thiocyanates, have a protecting action equivalent to a repellent. Pyrethrum, however, is not very persistent and the thiocyanates are not suitable for application to human skin. For protection against harvest mites and ticks, a rapidly acting acaricide such as benzyl benzoate is often mixed with a repellent such as dimethyl (or dibutyl) phthalate.

'Moth balls', made of naphthalene or *p*-dichlorbenzene, are sometimes spoken of as 'repellents', though they have no such action. They give protection only by fumigant action, in closed containers.

4.2.9 ATTRACTANTS

Properly speaking, an attractant should act at a distance, by causing insects to make orientated movements to its source. Substances which cause insects to aggregate after coming into contact with them are sometimes described as attractants, but more properly should be called arrestants.

Some of the substances which affect insect behaviour are endogenous chemicals, secreted by the insects themselves, usually to affect other insects of the same species. These are known as pheromones. The most important type is the sex pheromone, which is secreted by the female to attract the male. Other pheromones are known, however, which are not restricted to the sexual stages; for example, those which act as aggregating agents, or (in certain ants) markers for following trails to food.

Pheromones, though often relatively simple chemicals, can be quite specific; and they are generally very powerful. Thus, sex attractants of some lepidopterans can attract males from as much as two miles down wind; and the trail pheromones of some ants are hardly sufficient to yield a continuous chain of single molecules.

Exogenous attractants are usually chemicals characteristic of the food source or breeding material. Though not as potent or as specific as sex attractants, they can be quite impressive; for example, the aggregation of fruit flies by decaying fruit or the blowflies by carrion.

Finally, some insects display an inexplicable attraction to abnormal odours. Thus, seaweed flies aggregate around a source of trichlorethylene [456]; and, in experiments with mosquitoes, the odour of gasoline (petrol) was found to be attractive [70].

(a) Typical uses

Attractants can be used to collect insects in traps, either to destroy them or to dose them with chemosterilants. Alternatively, artificial oviposition sites can be employed to induce large numbers of females to lay eggs in them, which can then be destroyed. Even if the numbers collected in these ways are not sufficient to ensure control, the numbers caught in adult or egg traps can provide a useful assessment of natural populations. An additional possible use for sex attractants has been proposed. It has been suggested that if an excess of sex attractant were to be perfused throughout a large area (which might be possible, in view of the very small quantities which are effective) this would confuse insects in the sexual phase and prevent their finding mates. This method of control does not seem to have been shown to be feasible.

The use of attractants in control of insects of public health importance does not seem to have progressed very far. Lice are known to respond to faeces of other lice and to cloth soiled by contact with the human body. Bed bugs produce pheromone which appears to assist their aggregation. In neither case, however, are there obvious practical applications for control. Houseflies have been shown to use a sex pheromone, but it is not very powerful. They are less well provided with olfactory organs than other 'wild' flies and appear to rely a great deal on visual stimuli. Nevertheless, they do respond to some odours, which can be regarded as fly attractants; and they readily congregate on sugar, which acts as an arrestant.

The possibility of using attractants for flying bloodsucking insects like mosquitoes are being explored. It appears, however, that their response to odours is only one of a complex of chemical and physical stimuli which guides them to a blood meal [644].

(b) Typical substances

(i) Pheromones

A number of sex attractants have been isolated, identified and synthesized, and a few have been produced commercially. These include Disparlure (for gypsy moths), Codlemone (for codling moths), Medlure (for the Mediterranean fruit fly) and Grandlure (for the boll weevil). Muscalure (for houseflies) is a recent addition; it consists of (Z)-9-tricosene. Unfortunately, it does not seem to be very effective in practice [503].

The trail pheromones of pharaoh's ant are being considered as possible aids for control; they are alkaloidal in character and are known as monomorines I, II and III [506].

(ii) Food and oviposition lures

It is not entirely easy to distinguish oviposition from food-lures, since some female insects feed on protineaceous material in the breeding medium. Several workers have found the products of fermentation (alcohols, ketones, aldehydes) and those of putrefaction (ammonia, amines) to be attractive to various flies. The aggregation of *Drosophila* flies near alcoholic drinks and decaying fruit is well known. It has generally been found that a natural mixture of odours is more attractive than the individual components. Thus, alcohol and *iso*-valeraldehyde are both attractive to flies, but the best attractant of this type is molasses or malt in the early stages of fermentation [71]. Decomposing egg solids have been used as the basis of an effective bait for the American eye fly [441], *Hippelates collusor*. The active components were found to be trimethylamine, indole, ammonia and linoleic acid; but these were most effective in combination (and they were also very attractive to houseflies [442]).

A totally different type of attractant, presumably not depending on odour, is the oviposition trap for aedine mosquitoes, which consists of a strip of filter paper attached to a strip of wood standing in water. This 'ovitrap' has been found useful in population surveys.

4.3 APPLICATION OF INSECTICIDES

4.3.1 POISON BAITS

(a) Uses

Formerly, poison baits were employed against various indoor insect pests, such as cockroaches and Pharoah's ant. Their use has substantially declined with the introduction of synthetic residual contact poisons. At their best, baits are somewhat of a palliative, seldom obtaining complete eradication. Also, they are troublesome, unsightly and potentially dangerous to children and domestic animals. On the other hand, their usefulness may somewhat revive if the new synthetic insecticides become ineffective due to the emergence of resistant strains.

Poison baits may still find employment against certain garden pests (ants, earwigs, woodlice) which enter houses, but which are difficult to attack with residual poisons. They may be used, too, against houseflies in barns and stables, where some of their drawbacks do not apply. It is also possible that where low level resistance has developed in flies, the substance concerned may be more effective as a stomach poison than by contact, since a larger dose could be swallowed than would be picked up by the feet from a residual film.

(b) Equipment

For liquid poison baits (e.g. for houseflies) it is desirable to offer the liquid in an absorbent material, since insects cannot easily drink from open water surfaces, without getting trapped. A good bait dispenser can be prepared from chicken-watering devices [324]. These consist of a jar of water (1 or 2 litre size) inverted in a 15 cm circular metal trough. The trough is filled with a ring of plastic sponge, which keeps soaked with bait.

Ant baits in houses require protection from domestic animals and children. This can be accomplished by putting them in small tins pierced with holes or slits to allow the ants access to the bait.

(c) Formulation of insecticide

(i) Moist baits

A wide variety of edible materials may be used as the basis; mixtures of carbohydrates (e.g. cereals) and proteins (e.g. fish meal) are readily consumed [255]. Bran is one suitable base, mixed with about the same amount of water and with 10 per cent molasses added to increase its attractiveness. Alternatively, fish oil or edible vegetable oils may be added instead of the water and molasses.

Poison is mixed first with the dry ingredient, at an appropriate rate: 2 or 3 per cent of inorganic poisons, such as sodium fluoride, or 0·1 per cent of organic insecticides.

(ii) Dry baits

A simple dry bait for houseflies can be prepared by applying insecticides (e.g. various organo-phosphorus compounds) in acetone solution to granular sugar, so as to leave about 1–2 per cent toxicant when the acetone evaporates.

A solid dry bait may be prepared from 50 per cent sugar, 46 per cent sand, 2 per cent insecticide and 2 per cent gelatine (or better still bacteriological agar). The sugar, sand and poison are mixed and the gelatine or agar liquefied with boiling water and then stirred into the mixture. This is allowed to set on small squares of wire screen attached to pegs.

(iii) Liquid baits

These are made from water containing 10 per cent sugar, molasses or syrup. For houseflies, 0·1–0·2 per cent of an organo-phosphorus insecticide is added. A similar bait, containing 0·3 per cent sodium arsenite or 2 per cent sodium fluoride, has been used against Pharaoh's ant.

(d) Operation

Moist bait is crumbled and scattered in small patches near houses where invading garden pests are likely to frequent. It is advisable to cover the bait with boards or tiles, to prevent it being taken by birds or domestic animals; this will also attract light-avoiding arthropods.

Dry sugar bait can be scattered on dry floors of barns and stables to kill flies. In muddy or moist places (e.g. chicken houses, stables or cowsheds) the solid bait on pegs may be used.

Liquid bait may be sprinkled on clean, dry, concrete floors against houseflies. For Pharaoh's ant, the bait is mixed with cake or minced meat to form a paste and portions placed in bait tins at suitable points.

4.3.2 INSECTICIDAL POWDERS

Insecticidal dusts, like other powders (e.g. face powder!), adhere spontaneously to solid surfaces. This adhesion may be assisted by electrical charges (either positive or negative) given to the particles by frictional forces during passage through a dusting machine; but more important are the van der Waal forces at the actual points of contact. As particle size is reduced, the areas of contact decrease much less rapidly than the mass, so that smaller particles adhere more strongly. Calculations show that it takes an acceleration of about 4000 times gravity to shake off a 100 μm particle and as much as 500 000 times gravity to dislodge a 10 μm one.

Not only are finer powders more efficient at contaminating insects, but there is evidence that finely-divided contact poisons are more potent, probably because of more rapid absorption through the cuticle. On the other hand, the particles of very fine powders tend to stick together, forming aggregations which then become the equivalents of large particles. Probably an optimum size is just below 10 μm. Commonly, however, insecticide diluents are merely specified by the sieves through which they will pass. Thus, the World Health Organization specifications for dusting powders containing DDT, *gamma* BHC or malathion (WHO/SIF/16.R1; 17.R1; 22) require 98 per cent to pass 100 mesh (150 μm) [635]. Some common dust diluents in U.S.A. are stated to pass 350 mesh sieves (44 μm).

(a) Uses

At one time, stomach poisons in powder form were used to some extent against pests of public health importance. For examples, Paris green was dusted over mosquito breeding places and sodium fluoride or borax powders were used against cockroaches.

Contact poisons in powder form are used for certain domestic and medical insect pests. Probably the most important use is against ectoparasites. Thus dusts containing DDT, pyrethrum or malathion have been used for reducing widespread lousiness among prisoners or refugees in wartime or after disasters. (Other preparations are more suitable for reduction of head lice in normal circumstances.) Dust preparations are convenient for use against ectoparasites of domestic animals.

Contact insecticides in powder form are convenient for various domestic uses; against cockroaches and silverfish or for destroying wasp nests. On a larger scale, dust insecticides have been employed to destroy insects such as crickets or flies in refuse tips.

(b) Equipment (Fig. 4.4)

The earliest dust applicators for insecticides were merely perforated containers, to be used like sugar casters. Simple modern types are various plastic containers which emit puffs of dust when squeezed briskly.

Slightly more elaborate is the hand-operated dusting gun, of which various models are available and which is covered by a World Health Organization specification (WHO/EQP/4.R2) [637]. Air, driven by a hand-operated piston, is delivered to the base of a powder chamber and drives some of the dust out (usually through a coarse filter) through the discharge vent. This is usually fitted with an extension tube to direct the dust jet. The W.H.O. specification states that 'the volume of discharge at each stroke of the plunger shall be the same at all angles at which the duster may be used'. Generally, however, there is inevitably some difference in the quantity of dust discharged at different angles.

An ingenious method of dispersing a cloud of insecticidal dust through a considerable volume of space employs a blast of carbon dioxide. In its latest modification, the apparatus consists of a lead from a carbon dioxide cylinder to a T-junction, leading to two small dust containers and thence to two nozzles pointing in opposite directions [433]. Into each dust container is placed a foil-lined cardboard cartridge, 2·5 cm diameter by 14 cm long, containing micronized dust (about 5 μm particles). A short blast of the gas drives out the charges, which disperse adequately throughout a large volume e.g. 400 m^3. This method has been used successfully to control flies in dairy barns and various pests in railway wagons and aircraft.

For applying dusts outdoors on a larger scale (e.g. for treatment of refuse dumps, mosquito breeding places etc), a rotary duster is desirable. These are widely manufactured for agricultural purposes. Air is generated by a centrifugal fan and this drives the dust out of a wide nozzle. The fan is either operated by a hand-driven crank (in portable models) or by gearing to the wheels in wheeled types. (Fig. 4.4 shows a portable duster). Power-driven knapsack dusters are also used (Fig. 4.4); these can be converted to mist-blowers if desired. A fan driven by a two-stroke petrol engine propels a high velocity air-blast down a lance and dust from a container is fed into this air stream. These weigh about 8–10 kg and blow out about 0·75 up to 10 kg dust a minute. They are relatively expensive (about £300 at 1979 prices) and operate with considerable noise and vibration, so that they are only used where extensive outdoor applications are essential.

(c) Formulation of insecticide

Some early insecticidal powders were of vegetable origin and the powdered raw materials contained suitable concentrations of the active principle (e.g. pyrethrum, derris). When synthetic substances were introduced, however, it became necessary to dilute them with inert ingredients. Solid insecticides are usually mixed with some of the diluent before grinding, otherwise they tend to melt and fuse. Liquid insecticides are usually applied to the diluent in volatile solvents.

Various substances have been used as diluents, the main criteria being: (1)

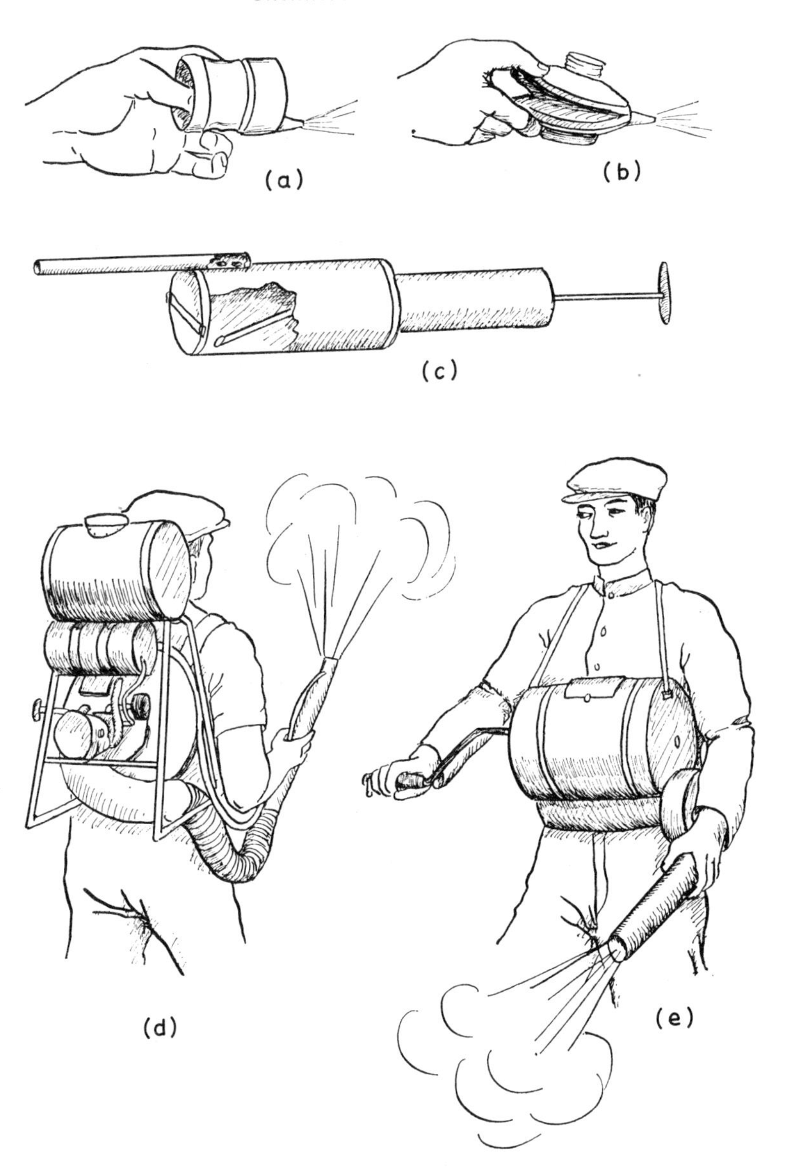

Figure 4.4 Equipment for applying insecticides in power form. (*a*) (*b*) simple hand-operated bellows types; (*c*) hand-operated piston duster; (*d*) power-operated knapsack duster (with certain modifications, the same basic equipment can be used as a sprayer); (*e*) hand-operated rotary blowing knapsack duster.

harmless to man; (2) compatability with the toxicant; (3) satisfactory particle size range (see above); (4) ease of flowing and dusting; (5) adsorptivity (this will depend on porosity – very porous particles may absorb liquid insecticides and

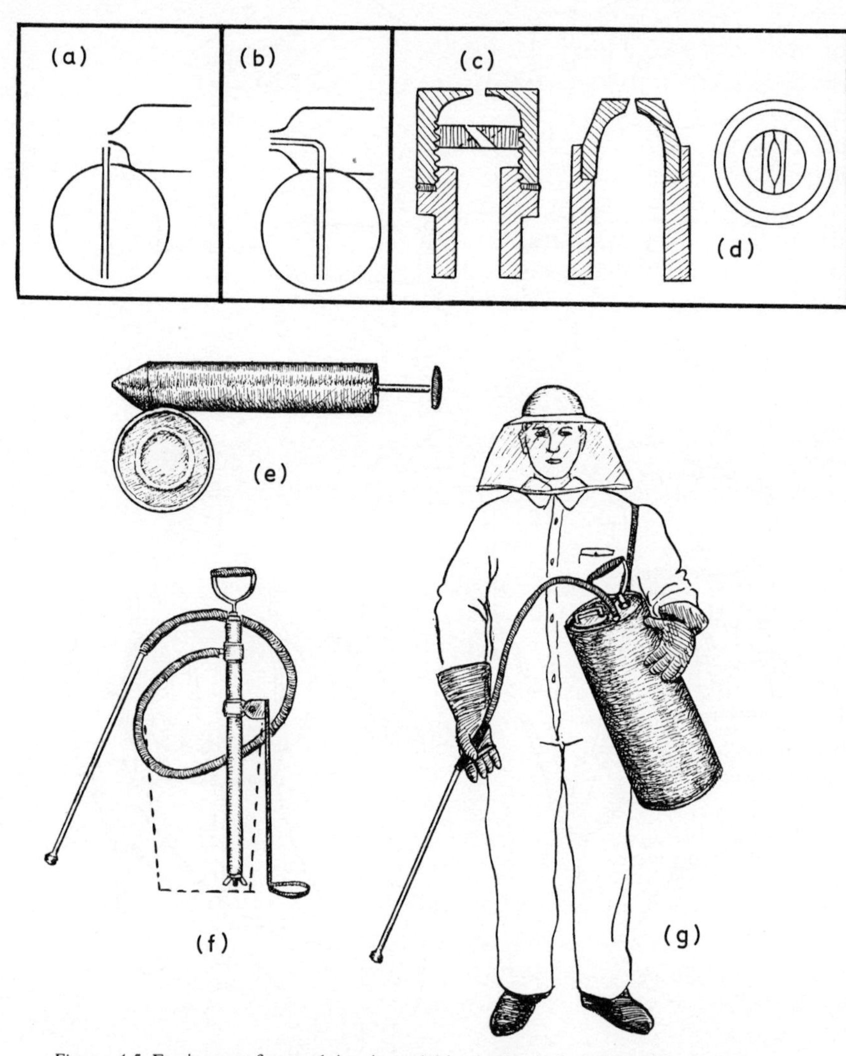

Figure 4.5 Equipment for applying insecticides in liquid form. (*a*) (*b*) two types of simple atomizing nozzle; (*c*) swirling spray nozzle, producing hollow-cone spray (section); (*d*) jet spray nozzle, producing fan-shaped spray (section and plan); (*e*) piston-operated hand atomizer; (*f*) stirrup-pump type hydraulic sprayer; (*g*) compression type (pneumatic) knapsack sprayer as used by man wearing protective clothing.

prevent their adequate contamination of the insects); and (6) price and availability.

Some of the commonest materials used for diluting contact insecticides are talc ($3MgO.4SiO_2H_2O$), pyrophyllite ($Al_2O_3.4SiO_2H_2O$) and kaolin ($Al_2O_32SiO_2.2H_2O$).

(d) Operation

For domestic uses against concealed insects, dusting guns are used to blow insecticidal powder into crevices where the pests may be hiding. For use against personal infestations of body lice, the delivery tube of the dust gun is inserted under the clothing. Mosquito larvicides in powder form can be conveniently applied by hand-operated or power-driven dust blowers. Application is made from the windward side of large breeding sites, to take advantage of the wind to carry the insecticide across the site. Since a change of wind or a blow-back can envelop the operator in the insecticidal dust, it is desirable for him to wear a face mask during treatment.

4.3.3 LARVICIDAL PELLETS, GRANULES AND BRICKETTES

Mosquito larvae breeding in water thickly set with reeds or other aquatic vegetation present difficulties of penetration to liquid or dust larvicides. A convenient and effective treatment for such sites is to scatter insecticidal pellets (0·6–2 mm) or granules (0·25–0·6 mm). These may be scattered by hand, by dusting machines, or from aircraft. Hand application can be made more accurately by the use of special applicators. One, the 'horn seeder' consists of a neoprene-treated cloth bag with a tapered metal tube attached to its lowest point. The tube contains a diaphragm for controlling flow rate. In use, the tube is swung from side to side in a figure-of-eight pattern, producing a remarkably even distributing of granules. A swath width of up to 7 m can be obtained with little effort. Another hand applicator comprises a metal cylinder of 1·3 l capacity with a trigger device which releases a measured amount of granules (usually about 0·2 g per spot-treatment).

Two types of granules may be used. One form is compounded with a relatively heavy mineral (for example, attapulgite). This falls to the bottom of water and slowly disintegrates. Provided that the water is shallow, the insecticide disperses sufficiently to kill the larvae. Another type is bound with a sticky oil emulsion, to pellets of vermiculite (an exploded, hydrated variety of biotite). These pellets are light and float on the water, gradually dispersing insecticide. Other materials that have been used include PVC, polyethylene and charcoal [615]. An additional advantage of pellets and granules is that they release the insecticide slowly, over an extended period. This is generally useful in other fields of pest control [131].

Larvicidal brickettes are also convenient for long-lasting treatment of static water breeding sites [31, 166]. Small ones (5–10 g) can be used in rot holes in trees etc; larger ones (e.g. 300 g) can be used in water butts or septic tanks, which often breed culicine pest mosquitoes. These brickettes are compounded of wettable powders containing about 50 per cent active ingredient, together with cement or plaster. Two typical formuale are as follows:

(1) 3 parts sawdust: 20 parts plaster of Paris: 7 parts wettable powder.
(2) 10 parts sand: 2 parts cement: 1 part wettable powder.

For septic tanks, the brickette must be kept at the surface, otherwise it will be buried in sludge. This can be done by fixing it with wires, or by attaching it to a wooden float.

4.3.4 AIR-BORNE DROPLETS FOR DISPERSING INSECTICIDES

(a) Nature and behaviour of air-borne droplets

Insecticides may be applied in the form of air-borne droplets, the sizes of which affect their behaviour and the type of application for which they are most suitable. The important parameter of droplet size is the volume median diameter (vmd) which divides a cloud of droplets into two equal volumes, one group smaller and the other larger than the vmd. The W.H.O. has proposed the following classification on this basis: aerosols, <50 μm; mists, 50–100 μm; fine sprays, 100–250 μm; medium sprays, 250–400 μm; coarse sprays, >400 μm [637]. In any spray cloud, droplets of different sizes occur, which can be conveniently plotted on logarithmic/probability graph paper. The following table illustrates the distribution of droplet sizes, by volume, in a typical aerosol, mist and fine spray (Table 4.2).

Table 4.2 Distribution of droplet sizes in typical aerosols, mists and sprays.

	Volume in droplets smaller than (μm)								
	10	20	30	50	100	200	300	400	450
Aerosol	0·5	16	65	99	>99·9	—	—	—	—
Mist	—	0·5	5	35	97	>99·9	—	—	—
Fine spray	—	—	0·1	0·5	3	22	56	88	>99·9

The behaviour of droplets in these different categories is related to their size, since they fall through the air at a constant velocity, which is proportional to the square of their radii (Stoke's Law). The following terminal velocities (cm s^{-1}) have been calculated for water droplets falling in still air: 5 μm diameter, 0.08; 10 μm, 0.3; 20 μm, 1.2; 50 μm, 7.5; 100 μm, 28; 200 μm, 72; 500 μm, 214. However, other factors are involved. Even moderately volatile liquids like water or kerosene do not form stable aerosols, because the droplets are constantly shrinking due to evaporation. Thus, a 200 μm droplet of water in air at 15° C and 40 per cent R.H., will evaporate completely in 63 s; a 100 μm droplet in 16 s; a 50 μm one in 4 s; and a 10 μm one in 0.1 s. Even droplets of the rather less volatile kerosene disappear quickly: a 100 μm one in 30 s and a 10 μm droplet in 0.3 s [139]. In a cloud of droplets, evaporation is somewhat reduced by local saturation of the air with vapour of the liquid; but it is evident that fine droplets do not persist long in air unless they are very non-volatile. To ensure persistence, some non-volatile oil may be added to spray formulations. Alternatively, ultra-low volume (U.L.V.) sprays of nearly pure liquid insecticides of low volatility may be produced.

(b) Sampling and measurement of droplets [380]

Because the sizes of droplets in sprays, mists and aerosols are important in their practical use, methods of sampling and measuring them have been developed.

(i) Collection

Droplets may be collected merely by allowing them to fall on to a prepared surface. This will collect only the relatively larger ones. In order to collect smaller droplets, the collecting surface must move more or less rapidly in relation to the droplet cloud. This can be done most simply by waving it through the cloud or, more scientifically, by using a mechanically rotating frame. Alternatively, the mist or aerosol can be sucked through a 'Cascade Impactor', which consists of a series of tubes ending in slits of decreasing size, opposite a series of collecting surfaces. As the air is drawn through these, its velocity increases, so that smaller and smaller particles are made to impact on the collecting surfaces.

(ii) Collecting surfaces

For relatively coarse sprays, it is feasible to add a dye and allow the droplets to fall onto a suitable paper. They will form stains, the diameters of which will be approximately related to the original drop size. This will show the numbers of drops falling on a given area and also, from preliminary calibration, a rough indication of the droplet sizes in the spray cloud.

With suitably prepared glass surfaces, droplets tend to form lens-shaped deposits. These too, can be used for approximate estimates of their original sizes.

A more direct indicator of the diameters of oil droplets in the range 10–200 μm can be obtained by collecting them in a deposit of magnesium oxide. This can be quite easily prepared by holding a glass slide over a burning magnesium ribbon. Droplets falling into this deposit form tiny craters slightly larger ($\times$ 1·15) than the original diameters.

Aqueous spray droplets are more satisfactorily collected in a small shallow tray filled with a matrix consisting of two parts light mineral oil and one part petroleum jelly. The droplets remain as nearly perfect spheres.

(iii) Measurement

This is done with the aid of a low power microscope. Either a scale in the eyepiece is used, or (more conveniently) a special graticule with circles of different diameters incised on it.

(c) Uses of sprays, mists and aerosols for dispersing insecticides

Spray droplets are mainly used to wet surfaces evenly; for example, to apply a film of residual insecticide to a wall. This does not demand very exacting size limits. However, very large droplets may fall to the ground before reaching the target; and very small droplets may not have enough momentum to impact on it and will drift away instead.

Aerosols are used to disperse insecticides through a volume of air and kill insects flying in it, therefore, they must persist in this form for an appreciable period. The droplets must be sufficiently small to remain suspended but, if they are too small ($>5\ \mu m$) they will not readily impact on the insect. The optimum size range is about 10–20 μm. Mists can combine action against flying insects with some ground deposit (e.g. attack on both adult and larval mosquitoes) or they can disperse a ground deposit more widely, by slower settling time.

(d) Generation of sprays and aerosols

The atomization of liquids has been seriously studied, since it is important for internal combustion engines and chemical engineering as well as for pesticide application. Different methods are appropriate according to the droplet size required.

Sprays are normally generated by hydraulic energy; that is, by pressure applied to a column of liquid, either by compressed air, or by pumping action. The pressure is used to force the liquid out of a nozzle, which projects an expanding sheet of liquid into the air. This breaks up into threads and then droplets under the influence of surface tension.

Aerosols are produced in various ways. The energy for breaking up the liquid may be supplied mechanically (as by the centrifugal force of a spinning disc or cylinder); or more commonly, by the shearing force of a blast of air or hot gases (as in twin-fluid atomizing nozzles). Another method depends on the vaporization of a liquified gas (as in aerosol dispensers). Finally, heat may be employed (as in smoke generators or electrical vaporizers).

4.3.5 SPRAYS AND SPRAYERS

(a) Uses

Sprayers have been used for many years to apply mosquito larvicides, especially mineral oils. More recently, the most important use in medical entomology, has been to apply residual insecticides to the walls of human dwellings.

(b) Equipment [637]

(i) Nozzles

There are three common types of nozzle used in pest control.

Fan nozzles produce a spray pattern in the form of a flat fan-shaped sheet. This is achieved by forcing the liquid through a rectangular or lenticular slit. The spray output depends on the width of the slit (at any given pressure) and this may become enlarged by erosion after long usage, especially if wettable powder insecticides with abrasive particles are used. Brass nozzles wear most rapidly and stainless steel ones are to be preferred. Ceramic ones are hard wearing, but easily broken. Whatever type of nozzle is used, it should be periodically checked for wear.

Cone or swirl nozzles emit liquid in the form of a hollow revolving cone (or, occasionally, a solid cone of droplets is obtained). This is produced by imparting tangential movement, by forcing the liquid through a swirl plate with helical slots before it emerges. The droplet size depends on the dimensions of the swirl plate and the orifice; and also on the pressure, which can range from 150 to 3500 kPa (20–500 lb in.$^{-2}$ or 1·5–35 kg cm^{-2}).*

Impact or deflector nozzles emit a stream of liquid which is either deflected by a surface close to the nozzle and at an angle to it, or by a small 'anvil' plate immediately in front of it. The former produces a fan type spray at an angle to the spray lance; the latter a complete circular spray cloud. These nozzles can be operated at low pressures (35 kPa; 5 lb in.$^{-2}$) and produce very large droplets.

(ii) Sprayer design

The power necessary to force the liquid through a spray nozzle can be applied in different ways; either directly by hydraulic pressure, or via compressed air.

Hydraulic knapsack sprayers are carried on the back. The pressure for spraying can be developed by operating a plunger pump on the spray lance (rather like a trombone slide). This produces a nozzle pressure up to 180 kPa (26 lb in.$^{-2}$). Alternatively, the lance is held in one hand, while the other works a lever which operates a pump on the sprayer body; this gives pressures up to 300 kPa (43 lb in.$^{-2}$). In either case, the spray pressure is rather irregular and the droplets produced tend to be rather large, so that the sprayer is most suitable for applying larvicidal oils at heavy doses.

Hydraulic stirrup-pump sprayers (Fig. 4.5) require a two-man team for operation. The spray liquid is mixed in a bucket in which the barrel of the stirrup-pump is immersed. One man operates the pump, which drives the liquid down a long hose (6 m) to the spray lance held by the other man. A spray pressure of

* In accordance with international recommendations, pressures are expressed in pascals, the SI unit. 1 pascal (Pa) = 1 newton per square metre (N m^{-2}) = $0{\cdot}145 \times 10^{-3}$ lbf in.$^{-2}$.

700 kPa (100 lb in.$^{-2}$) can be developed. Apart from the fact that two men are required, this involves more risk of spilling spray fluid from the open bucket; and the spray pressure may be rather uneven. However, the apparatus is relatively cheap and efficient. The W.H.O. provides a specification for this type of sprayer (WHO/EQP/3R3).

In the *compression, or pneumatic sprayer* (Fig. 4.5), air is pumped into the container above the liquid and this is used to drive it along a hose and lance to the spray nozzle. The operator controls the delivery of spray with a tap. When the pressure falls too far, the air pressure must be restored. The fall in pressure will tend to affect the degree of atomization. Some models have a constant-pressure valve in the nozzle to ensure delivery at a constant rate. Pressures from 200 to 900 kPa (30–130 lb in.$^{-2}$) may be used, according to size and construction of the sprayer. Various types are available, from small hand models containing 1–2 l, to larger sprayers (10–20 l) carried on the back. They may be made of plastic, brass or stainless steel. The metal models are obviously heavier and more expensive but they are more durable and reliable, and generally most suitable for regular use in a control programme. The W.H.O. gives a detailed specification of a knapsack compression sprayer (WHO/EQP/1R4) and describes tests of the various criteria.

A useful modification of the hand-operated compression sprayer has been described, which used compressed carbon dioxide from a cylinder to avoid the labour of pumping [578]. Stainless steel tanks, manufactured for use with soft drink vending machines, are modified by fitting Schrader valves (as on car tyres) to their lids, and also a hose, lance and a shoulder-strap. The sprayer thus produced is considerably cheaper than commercially manufactured stainless steel sprayers.

(c) Formulation of insecticide

As already mentioned, sprays are commonly used to apply insecticide to the surfaces of walls, vegetation etc, and also to apply larvicides to water surfaces. Insecticide is incorporated in a liquid carrier either in solution, emulsion or suspension.

The use of water as a carrier liquid has obvious advantages. It is cheap, safe and, being always locally available, does not have to be carried about with the equipment. Unfortunately, few contact insecticides are water-soluble, so that water cannot be used as a solvent carrier, though it is employed for preparing emulsions and suspensions.

(i) Solutions

Some of the earliest wall-spray treatments with DDT and other organochlorine insecticides were applied in kerosene (paraffin oil). Because crude kerosene is rather smelly, a refined grade was used, from which odorous aromatic and unsaturated compounds had been removed. This left saturated hydrocarbons, with reduced solvent powers, to improve which, auxiliary solvents such as xylene or cyclohexanone were sometimes added.

(ii) Emulsions

Aqueous emulsions permit the use of the cheap and widely available water to be used as a major part of the carrier liquid. Their use in applying residues to walls of dwellings has declined slightly in recent years in favour of suspensions (q.v.). They are still useful for larvicides against mosquitoes in special circumstances; e.g. in polluted water or against forms like *Mansonia* which do not live at the surface of the water. To prepare an insecticide emulsion, the insecticide is dissolved in some suitable solvent (e.g. xylene, toluene, aromatic petroleum fractions such as polymethyl naphthalenes) and this is dispersed in water with the aid of an emulsifying agent. Emulsifying agents have molecules partly hydrophilic and partly hydrophobic which tend to orientate themselves in the interface between water and an oily or fat-solvent liquid. In an emulsion, a monomolecular film of emulsifier round the dispersed droplets repels other droplets and prevents aggregation. The oldest and most readily available emulsifiers are natural colloids with molecules carrying polar and non-polar groups. These include many proteins and hydrocarbons such as casein, starch and various gums and resins. Many synthetic emulsifers are now available, built on the general plan of a polar group with a long hydrocarbon chain; the polar group being attracted to the water and the hydrocarbon chain to the oil or solvent. In some of these compounds, the polar group is non-ionic; but in others it may be a cation (alkali salt) or an anion (halide). The type of emulsifier, however, is designated according to the other ion, i.e. the one bearing the hydrocarbon chain. Thus, sodium naphthyl sulphonate is an anionic emulsifier, whereas cetyl pyridinium bromide is a cationic agent.

Emulsion concentrates are convenient for preparing insecticidal emulsions in the field. These are of two kinds: the 'mayonnaise' type and the 'soluble oil' type. The former is actually a concentrated emulsion containing very little water. It needs merely further dilution, but this involves considerable stirring. The 'soluble oil' type contains no water; it is the solvent (dispersible) phase, containing the insecticide and a suitable quantity of emulsifier. On adding it to water, a good emulsion forms spontaneously, with very little stirring. This type of preparation is probably more widely used than the mayonnaise type today, on account of its greater convenience. Strictly speaking, it should be called an 'emulsifiable concentrate' rather than an 'emulsion concentrate'; but the latter term is widely used. The World Health Organization gives specifications for (soluble oil type) emulsion concentrates of DDT, HCH, dieldrin, methoxychlor, diazinon, malathion, trichlorphon, fenthion and temephos (WHO/SIF/4, 5, 6, 11, 13, etc) [635].

The stability of a diluted emulsion depends on its liability to 'break' or to 'cream' and it is important to distinguish these two phenomena.

'Breaking' of emulsions is the separation of the two phases by coalescene of the droplets. After separation, the two phases cannot be re-emulsified by mere shaking. Badly made or unstable emulsions may break too soon, but it is sometimes an advantage for an emulsion to break fairly soon after application (e.g. on foliage, to contaminate it with oil, which might otherwise run off). The required stability can be attained by the use of a suitable emulsifier.

'Creaming' of emulsions is merely due to the oil droplets moving to the top of the liquid (e.g. cream on milk) and forming a layer there without coalescing. Creamed emulsions can be redistributed by shaking. Creaming can be prevented by forming emulsions with very small droplets which are prevented from rising rapidly by frictional forces.

(iii) Suspensions

Aqueous suspensions of insecticides are now used extensively in pest control, to a large extent displacing solutions or emulsions, for applying residual deposits. This is partly because suspensions are cheap and easy to prepare; but also the residual deposits from them on porous materials are generally more efficient. When a suspension is sprayed on to a porous wall surface, only the water sinks in, leaving all the insecticide on the surface, where it is available to contaminate insects. In contrast, when the liquid phases of solutions or emulsions sink into wall pores, they carry in some insecticide, which is therefore lost. On non-porous surfaces, however, emulsion residues may be more efficient.

Suspensions are prepared from water-dispersable powders available commercially. These 'wettable powders' comprise the insecticide, an inert carrier or diluent and a wetting agent. The proportions of insecticide vary, but preferably range between 25 and 75 per cent. The higher the percentage, the less will the final insecticide deposit be masked (or covered up) by inert ingredients. The inert diluent is necessary with solid insecticides to assist in fine grinding and also to counteract the high density of most chlorinated hydrocarbon insecticides. In a good preparation, the insecticide particles and diluent should be firmly associated; otherwise they may get separated by vibration during transport of the preparation in bulk. With liquid insecticides, the diluent also acts as a carrier. Suitable materials for diluents are natural or synthetic silicon oxides or calcium silicates. The surface-active materials (wetting agents) play the same part as emulsifying agents; that is, to prevent aggregation of the suspended particles. One of the most important qualities of a wettable powder is the stability of the suspension prepared from it. A good product must remain in suspension during the period of spraying, otherwise the concentration of insecticide applied will progressively fall as the suspension deposits. The World Health Organization gives specifications for water-dispersable powders of DDT, HCH, dieldrin, diazinon, malathion, fenitrothion, and propoxur (WHO/SIF/1, 2, 3 etc) [635].

Among other criteria a suspensibility test is described. Wettable powders must give satisfactory performance not only when fresh but after storage in bulk, possibly under tropical conditions. Therefore provision is made for accelerated storage tests, employing pressure and heat.

(d) Operation for residual spraying

Probably the most important use of spraying in public health entomology is for application of residual films of insecticide in dwellings. To control house-haunting pests (flies, mosquitoes, bed bugs, etc), the following deposits have been recommended ($g\ m^{-2}$): DDT, 2, *gamma* HCH, 0·5, dieldrin 0·5, malathion, 2. To apply these deposits, it is necessary to know the application rate of a sprayer as normally used. Given a standard machine operated at a particular pressure, the spray rate depends on the skill and experience of the operator. Usually the aim is to wet a surface thoroughly to a point just short of run off; this is normally between 4 and 8 l per 100 m^2. Spray operators should be trained to apply standard amounts and the insecticides can then be made up accordingly.

The efficiency of the operator in spraying correct dosages can be checked by pinning up filter papers at random on the surfaces to be sprayed and subsequently removing them for analysis by a chemist equipped for this work. The results will show the actual deposit and how it varies from place to place. (The only flaw in these estimates is due to the fact that spray men tend to give special attention to the test papers, giving them an extra heavy dose!) Experiments of this type have shown that deposits of even the best operators vary greatly from place to place. Furthermore, owing to fall out of spray and droplets bouncing off the wall, only about 80 per cent of the expected deposit arrives and a lot of insecticide is found on the floor below.

4.3.6 MIST GENERATORS

(a) Uses

Mist generators are often referred to as fogging machines, though a true fog consists of smaller droplets (in the range 5–10 μm). As has been stated, a wide range of equipment is available; small machines for indoor use or limited treatments and large ones for extensive outdoor use, including aerial applications. The production of large quantities of aerosol or mist may be valuable for rapid treatment of buildings or even towns, in case of epidemics of insect-borne disease. Alternatively, they may be used to secure relief from the nuisance of mosquitoes or houseflies, usually by repeated applications. In many cases, the droplet size produced can be altered by adjustments over certain limits changing an aerosol to a mist of fine spray. Such coarser applications, though fine enough to drift adequately, will eventually deposit enough insecticide to treat breeding sites or resting insects.

(b) Equipment [637]

Three types of generator can be distinguished, according to the following energy sources: (i) air blast; (ii) hot gases, with vaporization; and (iii) centrifugal force.

(i) Air blast generators

Paint-spray guns can be used to atomize insecticide, using compressed air. The nozzle uses high pressure, low volume air, flowing over an internal insecticide feed. Though quite efficient, they rather easily become blocked, especially if the spray liquid contains particulate matter.

Motorized knapsack mist generators use 1·5–2·2 kW motors to drive a fan which produces an air blast at about 100 m s^{-1}, into which insecticide solution is injected. The air-shearing force produces a mist or fine spray by regulation of the feed. The output can be varied in different models, from about 0·5–4 l min^{-1}. The air blast carries the mist considerable distances (10 m upwards, if necessary). These machines weigh about 8–12 kg; they are somewhat noisy and rather expensive.

Large vortical mist generators use slightly larger engines to drive rotary compressors, which produce an air blast at about 35 kPa (5 lb $in.^{-2}$), which is spun round the liquid feed by helical vanes. Insecticide liquid is supplied under slight pressure and the output is about 0·1–0·5 l min^{-1}. The droplet size produced can be varied to some extent, down to 10 μm. Since these are heavy machines (about 200 kg) they are either mounted on a small vehicle or towed on a trailer behind one.

(ii) Hot gas atomizers

Pulse jet atomizers operate on the principle of the Wartime flying bombs. A fuel–air charge in the combustion chamber is fired and a further charge is sucked in by the subsequent fall in pressure. Insecticide is fed into the exhaust gases from 80 such pulses per second. The solution is vaporized and condenses in the air to form an aerosol, of about 30 μm vmd. Pulse jet machines weigh about 10 kg and their outputs range from 0·2–0·5 l min^{-1}.

Exhaust gas atomizers utilize the exhaust gases from ordinary petrol engines, either on a vehical or a propeller-driven aeroplane. The combination of the shear force of the moving gases and the vaporization at high temperature, followed by condensation in the atmosphere, produces a mist in the range 40–100 μm vmd.

TIFA machines comprise a four-stroke engine which drives a fan to produce high velocity air. Petrol and insecticide is injected and a continuously operating spark ignites the mixture and vaporizes it. The output is in the region of 1·5–3 l min^{-1}, giving an aerosol of 10–50 μm droplets. These machines are rather heavy (95–200 kg) and are mounted on a vehicle or towed behind one.

(iii) Centrifugal force atomizers

Spinning-disc atomizers have been available for some years, but recent improvements have extended their uses. Insecticide solution is fed to the centre of discs spinning at speeds up to 15 000 rev min^{-1}. The powerful centrifugal force flings off droplets, of fairly constant size, which depends largely on the speed of rotation. Thus, a range from 300 μm down to 10 μm can be covered. Discs are made of plastic or metal, and may be dish-shaped to aid in directing droplet emission; also, uniformity of droplet size is improved by radial grooves ending in teeth on the periphery. There are portable models powered by dry batteries, with outputs, of 0·06–0·18 l min^{-1}. Larger models can be powered by electric supplies, with outputs of about 1·25 l min^{-1}. These weigh up to 50 kg and usually incorporate a fan to project the aerosol outward.

By multiplying the number of discs (each supplied with insecticide fluid) a large output of mist or aerosol can be produced, with much lighter equipment than the air-blast or hot gas atomizers described above.

Spinning-brush atomizers may be made of fibre, plastic or metal. Fluid is fed to the brush in much the same way as to the spinning disc.

Spinning-cylinder atomizers of wire or plastic gauze are used to provide a surface from which liquid can be discharged as small droplets. The liquid feed resembles that in other spinning devices; but, in some cases the liquid passes through two or more screens to reach the outer spinning surface. These devices are often used on aircraft and the spinning energy is derived from a propeller driven by the aircraft's slip-stream. They produce droplets in the range 25–50 μm.

(c) Formulation of insecticide

Mist blowers will disperse solutions, emulsions or suspensions of insecticides, according to requirements. With rather volatile carriers, such as water or kerosene, there is liable to be considerable evaporation, even with 'cold' mist blowers. Evaporation retardants are sometimes included to reduce the shrinkage of the droplets.

For ULV treatments, very concentrated solutions may be used and, to obtain these, various commercial solvents are available. Most of them contain about 80 per cent alkylated benzenes or naphthalenes. With liquid insecticides, adequate droplet size and dosage can often be obtained by using the undiluted technical product.

(d) Operation for ground treatments

Portable mist generators can be used inside buildings and are sometimes used to supplement gross applications by power-operated generators. Large machines are carried in, or towed behind vehicles, at speeds between 20 and 2 m.p.h. (32 and 3·2 km h^{-1}, but commonly about 5 m.p.h. (8 km h^{-1}). Doses of organophosphorus insecticides which have been used against mosquitoes range from 10 to 500 ml ha^{-1} (0·01–0·45 lb $acre^{-1}$). Treatments can only be applied successfully in very calm weather; that is, with little wind and early in the day or in the evening when convection currents are minimal.

4.3.7 AEROSOL GENERATORS

While insecticidal aerosols are normally intended to contaminate flying insects, mists are sometimes intended for wide dispersion of insecticide (e.g. to extensive breeding sites) sometimes utilizing wind-assisted drift. A wide variety of appliances is available for different types of use, ranging from small hand-held atomizers to large and expensive generators. In surveying this large range, it will be convenient to begin with smaller equipment and proceed to larger ones.

(a) Hand atomizers (Fig. 4.5)

(i) Uses

For many years a simple type of hand-operated atomizer has been used to spray insecticides indoors, mainly against houseflies. The same type of atomizer may be employed to apply concentrated high-spreading mosquito larvicides.

(ii) Equipment

The sprayer consists of a hand-operated piston, supplying the air, a liquid reservoir and a simple nozzle. In some types, the spraying is *intermittent*, corresponding to the strokes of the piston. In others, a more or less *continuous* action is possible, owing to a build-up of air pressure in the reservoir above the liquid.

There are two main types of nozzle: one in which the air and liquid jets are opposed at right angles and the other in which the liquid jet is introduced concentrically into the air jet (Fig. 4.5a, b). In both cases, the movement of air over the mouth of the liquid feed sucks up the insecticide.

The droplet size range produced is rather variable, but a large proportion should be below 50 μm. With badly designed or constructed atomizers, a proportion of large drops are produced, which fall out rapidly. A good atomizer should show no gravity deposition within a yard from the nozzle.

The World Health Organization provides specifications for these sprayers (WHO/EQP/1R3) [637].

(iii) Formulation of insecticide

For domestic use ('fly sprays'), the usual carrier is odourless kerosene and the principal insecticide is pyrethrum, to give a rapid 'knock-down'. A concentration of 0·1 per cent pyrethrins is adequate, or its equivalent of synergized pyrethrins. To this may be added a small quantity of synthetic insecticide to improve kill.

(iv) Operation

Theoretically the atomizer is intended to produce an aerosol; but to produce an aerosol effective throughout a large room would entail much hard work. There is no doubt that the use of this type of atomizer is partly as a direct, aimed, spray and only partly for its aerosol effect.

(b) Liquified gas aerosol dispensers

The prime importance of aerosol dispensers in the field of public health is for the disinsection of aircraft. Even before World War II, when the rise in air traffic began to become significant, there was concern about the possibility of noxious insects being transported to new areas by intercontinental air traffic. Many countries began to demand the disinsection of aircraft coming from possible danger areas. Theoretically this does not concern Britain, since the possibility of disease vectors becoming established here is negligible. However, some aircraft in transit (e.g. from Africa to Asia) must be treated at London Airport (see also page 163).

The aerosol dispensers used resemble those used for domestic purposes, against flies and mosquitoes and also some crawling pests such as cockroaches. (Other types of formulation are intended to give surface deposits, to protect clothing against moths or furniture against beetles.)

(i) Equipment

The design of an aerosol dispenser ('aerosol bomb') is simple. It consists of a metal can with a tube running up the centre to an orifice (0·25–0·50 mm) at the top, controlled by a simple press valve. The can is filled with 'Freon' (trichlorofluoromethane and dichlorofluoromethane) liquefied under about 3·6 kPa pressure, together with non-volatile oil, containing insecticide. When the valve is opened, the Freon acts as a propellent, rushing out and vaporizing, leaving the insecticide solution dispersed in the air as an aerosol.

The droplets produced should be in the range 1–50 μm, with not more than 20 per cent having a diameter above 30 μm.

Dispensers can be of various sizes ranging from very small ones, intended to empty completely in one operation, up to large dispensers of about 2·5 kg, which can treat up to 30 000 m^3.

The W.H.O. provides specifications both for the disinfestation of aircraft (WHO/EQP/7) and for domestic use (WHO/EQP/8).

In recent years, there has been some adverse criticism of this type of aerosol dispenser, especially in the U.S.A. It has been suggested that, because of their very wide use (not only for insecticides, but for many other purposes) the halocarbons used as propellents could conceivably have a deleterious effect on the stratospheric ozone layer, which might eventually allow excessive ultra-violet radiation to reach the earth. Some authorities regard this as extremely improbable; but the use of such aerosol dispensers in the U.S.A. has been restricted. A possible alternative is to use propane and isobutane as propellents [644a].

(ii) Formulation of insecticide

In the immediate post-war years, there was extensive testing of aerosols for aircraft disinsection, especially in Britain and the U.S.A. Results of the former were published between 1949 and 1957 [97, 208] and provided the basis for the formula recommended by the W.H.O. [627]. At the same time, unpublished research by the U.S. Department of Entomology and Plant Quarantine formed the basis of slightly different formulae. Both the British and U.S. formulae contained pyrethrins (0·4–1·2 per cent) and DDT (2–3 per cent). The pyrethrins were for rapid effect combined with safety; but at high concentrations, the natural pyrethrum was somewhat irritating to passengers, so the DDT was added to boost potency.

In the early 1960s, the W.H.O. sponsored trials of both formulae which were both found effective, applied at 'blocks-away' (i.e. after embarkation of passengers but before take-off) [351, 561].

About this time, the possibility of disinsection by fumigant action was considered. Already, some modest results with lindane vapour had been obtained; and the introduction of dichlorvos was considered very promising [524]. However, though no toxic hazard could be demonstrated, it was postulated that low concentrations might affect pilot judgement; and there were suggestions of metal corrosion by the vapour [629].

In the 1970s, the introduction of highly-potent synthetic pyrethroids obviated the need for DDT (which, in any case, had become of dubious value because of resistance). Doses of aerosols containing 2 per cent bioresmethrin [107] or phenothrin [351], at 35 g (aerosol) per 100 m^3 were found effective. Furthermore, tests of water-based aerosols, propelled by propane and isobutane, were found effective at the same dosages [562].

(iii) Operation

In aircraft disinsection, the dispensers are operated by a person walking slowly down the aisle, directing the jet upwards, but at least 0·5 m from the ceiling. In the 'blocks-away' procedure, treatment is made with the passengers aboard, between the time of door closure and take-off. Dosage is by time of operation, most dispensers emitting about 1 g s^{-1}. For aircraft disinsection dispensers are operated for 10 s per 30 m^3; for domestic purposes 3–5 s per 30 m^3 are recommended.

4.3.8 APPLICATIONS FROM AIRCRAFT

The use of aircraft is obviously restricted to large-scale operations; for example, to extensive and expensive crops. In the public health field, there are two main uses: (i) to treat very extensive breeding sites of vector or nuisance species, and (ii) to control vectors of epidemic diseases rapidly, especially in tropical towns.

(a) Types of aircraft

Both fixed-wing aircraft and helicopters may be used, according to circumstances. The former are preferred when high payload and speed are needed to perform the task efficiently. They are less expensive to buy and to maintain than helicopters, except occasionally when the site to be treated is remote from an airfield. The majority of fixed-wing aircraft used for insecticide application are small single-engined models; but twin-engined and even four-engined aircraft have been used for controlling serious epidemics. In low-level operations over towns, multi-engined aeroplanes are essential for safety.

Helicopters are used when good manoeuvreability, low speed, or low-level surveillance is required, or when landings must be made in unprepared terrain. They have the additional advantage that the down-draught from the rotor opens up dense foliage and assists deposition of the spray.

Details of 26 fixed-wing aircraft and 18 helicopters are given in the 18th Report of the W.H.O. Insecticide Committee [632]. The following summary shows the range of certain characteristics of the two types.

(i) Fixed-wing aircraft

Working speed: 130–219 km h^{-1} (80–136 m.p.h.). Stalling speed: 60–108 km h^{-1} (37–67 m.p.h.). Power: 110–450 kW (150–600 h.p.). Maximum rate of climb (loaded): 144–398 m min^{-1} (472–1300 ft min^{-1}). Load: 108–290 kg (240–2000 lb).

(ii) Helicopters

Working speed: 80–200 km h^{-1} (50–125 m.p.h.). Initial rate of climb: 137–660 m min^{-1} (450–2170 ft min^{-1}). Power: 130–600 kW (180–800 h.p.). Service ceiling: 2150–6000 m (7000–20 000 ft). Load: 226–910 kg (500–2000 lb).

(b) Equipment: tanks and pumps

(i) Tanks

Insecticide solution is carried in one or two tanks of aluminium alloy, stainless steel or resin-bonded glass fibre. In fixed-wing aircraft, these are generally mounted inside the fusilage; in helicopters, often outside. In all cases, there must be provision for rapid jettisoning of the insecticide liquid in an emergency, commonly by a valve which empties the tank in 5 s. Filters must be fitted to the filler holes of the tanks, on the suction side of any pumps and in the spray boom and nozzles (if used).

(ii) Pumps

Pumps used on aeroplanes can be driven by windmills in the slip-stream, or by electric motors. Windmills are simplest to use, but the air speed may vary and the aerodynamic drag is high. Some form of brake is necessary to stop them when spraying is finished. Electric drives are particularly suitable for ULV, when the low volume requirement does not demand more than 1 kW, which can be supplied by a 100 A, 12 V alternator.

Three kinds of pump may be used for moving the spray liquid, depending on the pressure required, the rate of delivery and the nature of the spray liquid.

(1) *Reciprocating piston pumps* operate at rather high pressure (300–1500 kPa; 43–220 lb $in.^{-1}$) displacing relatively low volumes. They are therefore suitable for ULV applications via nozzles. Their disadvantage is a liability to suffer from wear, if insecticide liquid containing abrasive particles is used (e.g. suspensions).

(2) *Rotary pumps* displace liquid by the movement of closely meshed gears or vanes, usually working at 700 kPa (100 lb $in.^{-2}$). These too, are liable to suffer from wear, because of the finely-machined tolerance of the parts.

(3) In *centrifugal pumps* liquid is fed to the centre of rotating vanes and centrifugal force expells it from an exit on the periphery of the pump casing. There are no close metal contacts, so that wear by abrasive particles is less severe. These pumps work at low pressure and high volume; so that they are particularly suitable for supplying boom and nozzle systems on aircraft.

(c) Liquid dispersing systems

(i) Boom and nozzle arrangement

Liquid from the reserve tanks is pumped to long booms attached beneath the wings. In fixed-wing aircraft, they are attached to the trailing edge of the wing and generally can be lowered for spraying or raised for take-off and landing. Hollow-cone or fan-type nozzles (page 95) are used to produce a rather coarse spray (100–500 μm vmd).

(ii) Rotary atomizers

These are used for producing fine sprays of aerosols or mists. They may be driven by windmills or electric motors. Two or more may be mounted, according to the size of the aircraft and the application rate desired.

(iii) Exhaust atomizers

Thermally-produced aerosols may be produced by injecting insecticide solution into the exhaust tubes from the engine. However, this system is becoming rather obsolete.

(d) Dispersal of solids

Dusts or granules (page 93) can be applied from the air in some circumstances.

(i) Tanks or hoppers

These must have large openings for easy filling and tightly-fitting lids. The outlet is controlled by a sliding gate operated from the cockpit.

Distributors from fixed-wing aircraft may be by 'ram air' (venturi) spreaders, or by spinning spreaders. The latter are almost exclusively used on helicopters, with the discs being driven by electric motors.

(e) Operations in aerial spraying

(i) Weather conditions

The two main meteorological conditions affecting aerial spraying are wind and temperature.

Applications are done in alternating strips (swaths), back and forth, at right-angles to the wind. To avoid too great a lateral dispersion, the wind speed should be less than 5–8 km h^{-1} (3–5 m.p.h.).

Temperature is important because, especially in hot weather the soil heats up during the day and the air just above it tends to warm, expand and rise in huge invisible bubbles. This turbulence prevents spray particles from reaching the treatment site. At night, the soil cools and the air is less turbulent; but aircraft cannot be flown at low altitudes in the dark. Therefore, in warm weather, spraying is confined to early morning and late evening.

(ii) Droplet size and performance

The behaviour of droplets of liquid in air discussed earlier (page 94) is very relevant to aerial spraying. Aerosol droplets below 20 μm vmd are quite unreliable because they drift extensively in air currents. Those in the range 20–60 μm vmd may be used against flying insects, but they are still liable to considerable dispersion and very little is deposited on the ground anywhere near the site. To be effective, such treatments must be applied in cool, calm weather.

At the other extreme, coarse and medium sprays (120–400 μm vmd) are mainly useful for ground deposits (e.g. for larviciding). The intermediate (mist) zone can combine killing of adult insects with some ground deposition. There will be a degree of drift, varying with droplets of different sizes. Thus, spraying from 10 m with a cross-wind of 2·4 km h^{-1} (1·5 m.p.h.), the lateral drift of droplets will be approximately as follows [280]:

Droplet size (μm):	250	200	150	100	40
Drift (m):	9	12	23	40	100

Using a (ULV) mist application, one might expect a ground deposit swath of 180 m, from the zone of the 100 μm droplets to that of the 70 μm ones. To ensure good coverage, swaths are made to overlap (e.g. by 120 m) so that the effective swath width would be 60 m.

Generally, ULV treatments of organo-phosphorus insecticides aim at applying a dose in the range 75–500 g active ingredient per hectare (ha), with swath widths from 40–200 m, using aircraft flying at 150–250 km h^{-1}, 10–25 m from the ground.

(iii) Flying requirements [637]

Where there is an airfield within reasonable distance from the treatment area, it should be used as the operational base. In some cases, however, it may be necessary to construct an improvised airstrip, which should be on level ground, on an elevated area rather than in a valley. The runway length should be double the normal take-off distance and parallel with the prevailing wind. For light aircraft, it should be about 20 m wide, with a further cleared area about 50 m on each side.

Aircraft should be provided with VHF radio, for communication with the ground; but sophisticated radio beacon equipment is seldom feasible. Guidance in flying the swaths is generally done by two men at the ends of the application run, who move forward one swath width on each pass, and who are provided with flags, boards or lights.

Flying accurately at very low levels is a demanding task and the pilot should be protected from making errors due to fatigue. He should, of course, be fully briefed on the nature of the operation, but should not be overburdened with administrative duties and on no account take part in loading the pesticide.

(iv) Costs

Only very approximate costs can be given, owing to the vagaries of inflation. On the basis of 1970 prices (adjusted to 1978), a small fixed-wing aeroplane equipped for spraying might cost from U.S. $30 000 to 100 000. A helicopter capable of carrying the same payload, would be about three times as expensive.

Operating costs would be in the region of U.S. $50 000 p.a. Salaries and allowances would comprise about 40 per cent of this; maintenance and repairs, 15 per cent; insurance and depreciation, about 11 per cent each. Operating a helicopter might cost 40–50 per cent more, but in some cases, the extra efficiency would cover the difference.

4.3.9 INSECTICIDAL SMOKES AND SUBLIMATES

(a) Insecticide smoke generators

(i) Uses

Smoke generators dispensing DDT or *gamma* HCH are employed to some extent in horticulture (e.g. in greenhouses), but they are not very suitable for use indoors. The main use of smokes against domestic insect pests is the burning of mixtures of pyrethrum with a slow combustion mixture to destroy mosquitoes in bedrooms in the evening. A common form is the 'mosquito coil', mainly produced in Japan. Ordinary combustion, however, destroys much of the pyrethrins present (80–95 per cent). Improved dispersion with less combustion may be achieved by adding substances which generate gas on heating (e.g. $NaHCO_3$ or $(NH_4)_2C_2O_4$ which emit CO_2). An improved pyrethrin smoke generator employs as 'reaction propellent' 3,7-dinitroso-1,3,5,7-tetra-azabicyclo-[3,3,1]-nonane, which gives off nitrogen on ignition.

(ii) Equipment

No equipment is needed.

(iii) Formulation of insecticide

(1) *A typical mosquito coil* may be made from a paste containing 20–40 per cent pyrethrum powder; 34–40 per cent combustible filler (e.g. wood flour); 25–30 per cent water-soluble gum; a trace of fungistat and a dye [362].

(2) *Pyrethrum reaction aerosol powder* comprises 50 per cent pyrethrum powder; 6 per cent piperonyl butoxide; 44 per cent reaction propellent. Suitable quantities are sealed in small plastic bags [61].

(iv) Operation

A mosquito coil may be simply ignited by a match. The reaction aerosol powder must be ignited by flameless heat, e.g. a nitrate fuse placed on the plastic bag (which may be held on a shovel).

(b) Electrical insecticide vaporizers [154]

(i) Uses

Electrically operated thermal insecticide vaporizers are available commercially, being intended for domestic insect pests, especially houseflies. They have been quite widely installed in restaurants, kitchens and food stores.

(ii) Equipment

The apparatus consists of a metal cup containing insecticide, which is maintained in a liquid condition by an electric heater controlled by thermostat. One or more units are fixed to the wall, according to the size of the room.

(iii) Formulation of insecticide

Undiluted insecticide is used, commonly DDT, *gamma* HCH or a mixture of the two. However, the use of these chlorinated insecticides is likely to be abandoned.

(iv) Operation

The vaporizing units are intended to operate continuously. The insecticide is either sublimed or emitted as a very fine aerosol. It is not clear whether the aerosol affects the insects directly or whether the action is mainly due to the deposits which build up, especially on the walls and ceiling near the vaporizers.

The quantity of insecticide given off varies according to the temperature and prevalence of local air currents, being in the region of 0·25–2·0 g per unit in 24 h. In addition, the insecticidal efficiency (and the possible toxic hazard) of the apparatus depends on ventilation. This varies considerably, since the air in a room may change completely from five to several hundred times a day, according to structure and conditions.

4.3.10 INSECTICIDAL LACQUERS [29, 153]

(a) History

In the years immediately following the Second World War, the newly introduced synthetic insecticides were tried out in diverse formulations. One idea was to mix DDT with whitewash, distemper or paint to obtain insecticidal wall decorations. Unfortunately, it was found that much of the contact action was lost by the DDT being masked by whitewash particles or embedded in the paint film. Then it was discovered that, if a sufficiently high concentration of DDT was included in certain oil-bound paints, the insecticide would migrate to the surface and become extruded as a bloom of crystals. Furthermore, this bloom would be renewed, if wiped away. Subsequently, it was found that even better results could be obtained with insecticides incorporated in certain synthetic resins, which produced insecticidal lacquers.

(b) Uses

Insecticidal lacquers are rather expensive but very satisfactory in other respects, being persistent and not unsightly. They have been most successfully used against kitchen pests, for example, German cockroaches and Pharaoh's ant in hospitals and similar institutions. In some early trials, a single application was able to control cockroaches in the galleys of merchant ships for over a year.

(c) Formulation of insecticide

Various types of synthetic resin combinations have been used, including urea–formaldehyde, coumarone–styrene, coumarone–indene and chlorinated synthetic rubber. The first mentioned is probably most widely used.

Synthetic resin alone is not satisfactory and a plasticizer is added to encourage the blooming of insecticide. The rate of release of insecticide is governed by the resin to plasticizer ratio, being faster with the softer finish containing more plasticizer. A typical formula is as follows:

Urea–formaldehyde resin	50 parts
Castor oil alkyd resin (plasticizer)	50 parts
Butanol and xylol (solvents)	50 parts
Insecticide	10 parts

Before use, an acid accelerator is added, to promote polymerization and hardening. A common one is 10 per cent sulphuric acid in butanol, one part being added to 20 parts of lacquer.

Various insecticides have been incorporated in lacquers, including DDT, *gamma* HCH, aldrin, dieldrin, pyrethrins and malathion. The liquid ones are extruded as a greasy film, not as crystals; and some compounds give off insecticidal vapour through the resin (*gamma* HCH, aldrin, malathion).

The most persistent action, however, is obtained with dieldrin or DDT. A resin containing 5 per cent dieldrin may remain highly active for over a year.

(d) Application

Insecticidal lacquer may be conveniently applied by a large paintbrush. Care is needed, because the lacquer is somewhat corrosive (owing to the acid accelerator) and also, once it has set hard, it cannot be removed by solvents or paint strippers. Splashes on skin or clothing should be avoided; but when they do occur, they should be removed at once with solvent applied by the manufacturers. This solvent should also be used to clean brushes and other equipment after use.

It is not generally necessary to apply lacquer to whole wall surfaces. The aim should be to put barriers between the insects' hiding places and their food. For this purpose, bands about 10 cm wide are painted around skirtings, door and window frames, legs of tables, entry points of water pipes, electric conduits, etc.

The lacquer may be applied to any clean, non-porous surface (hard paint, glazed tiles, glass). Paintwork less than three months old should not be treated, however, and metals should be pre-treated with a primer to avoid corrosion. It is also inadvisable to treat surfaces constantly wet or frequently washed.

4.3.11 INSECTICIDE IMPREGNATION

Residues of persistent contact insecticides may be applied to various materials, by dipping the latter in solutions (or, in some cases, emulsions) of insecticide, so that a particular amount remains after drying.

(a) Impregnation of fabrics

Towards the end of the Second World War, the underwear of British troops on active service was impregnated with 1 per cent by weight of DDT to destroy any lice they might acquire. DDT impregnation of woollen garments against clothes moths was formerly undertaken by dry cleaners, but this is now rare. More efficient is the impregnation of wool in the dyebath by hot emulsions of dieldrin, which is still widely used for carpets, etc.

(b) Impregnated suspended objects, against flies [25, 485]

Ribbons, cords and cardboard objects have been impregnated with insecticides and hung up in places infested with houseflies to kill those which rest on them. Cords 2·5–5 mm in diameter, crepe bandage or narrow strips of wire mesh (7×6 per cm) have been used for this purpose. Formerly they were usually impregnated with DDT or dieldrin; more recently, organo-phosphorus insecticides have been used, owing to the widespread resistance to chlorinated insecticides.

These cords, ribbons, etc, are usually hung from the ceiling, allowing about 1 m for each m^2 of floor (30 ft cord per 100 ft^2). The most effective treatments are parathion or diazinon, which are available commercially. Because of the toxicity of the chemicals, it is recommended that cords be prepared by experienced persons only and the directions on commercially available cords and strips should be carefully followed. While this measure against flies has been used in canteens, it may be thought rather unsightly for human dwellings; it is, however, extensively used in dairy barns (e.g. in Denmark, where farming standards are high).

Impregnated cardboard shapes, hung up to attract and destroy flies, were popular in the early days of DDT; and later, cardboard circles, impregnated with a carbamate insecticide were used. These were coloured red and printed with fly shapes, which were believed to attract flies to them [610].

4.3.12 LARVICIDES

Various insect pests with aquatic larvae may be attacked with larvicides, including blackflies and non-biting midges. Generally, the most important are the mosquitoes, which may breed in a wide variety of habitats, ranging from lakes, swamps, rivers or streams to water butts, drains or seepages.

(a) Mineral oil larvicides [361]

(i) Characteristics of larvicidal oils [432]

In 1892 the American, L. O. Howard, discovered that a film of kerosene applied to water will kill mosquito larvae breeding in it; and this method has been followed for many decades. Various unrefined mineral oils have been used as larvicides and at first the choice was quite empirical. Later research centred attention on three essential criteria: toxicity, spreading pressure and stability.

Toxicity

It is a common misconception that oil films kill mosquito larvae by suffocating them; but experiments have shown that this is seldom true. Asphyxiation of insects is generally slow and, in any case, mosquito larvae can survive for many hours by utilizing air dissolved in water. It is true that oil larvicides penetrate the tracheal breathing tubes of the larvae, but their lethal action is generally due to toxic components of the oils. These are, apparently, the aromatic hydrocarbons (the paraffinic compounds being relatively non-toxic). A high aromatic content, however, is not necessary and may confer disadvantages (e.g. too high a specific gravity).

Various unrefined oils have been used for many years and were quite effective when used at adequate doses; generally of the order of 280 l ha^{-1} (25 gal $acre^{-1}$). More recently, more potent petroleum derivatives have been introduced, which are apparently effective at a tenth this dosage (e.g. 'Flit MLO'), the composition of which has not been disclosed [39, 206].

The potency of any larvicide can be estimated from laboratory tests, in which well-grown larvae are exposed to carefully prepared films. The doses used should be lower than are necessary for field conditions. With the older conventional oils, a film 10 μm thick [corresponding to 100 l ha^{-1} (9 gall $acre^{-1}$)] should be used; while 'Flit MLO' can be tested at one tenth this rate [409].

Spreading pressure

This property, expressed in newtons per metre, is a measure of the force exerted by a film on water to overcome resistance to its spread. It is zero for non-spreading oils like medicinal paraffin, which remains in globules even on a clean water surface. Kerosene has a value about 0·01 Nm^{-1}, but a good larvicidal oil should have a value about 0·023 Nm^{-1}; and if specially good spreading (and penetration among aquatic vegetation) is required, a spreading pressure of 0·046 Nm^{-1} is recommended.

The spreading pressure of mineral oils is generally due to impurities exercising surface action and it can be reinforced by the addition of 'spread-aiders' (such as sodium alkyl sulphate or an alkylated aryl polyether alcohol). Commercially available anti-mosquito oils (e.g. 'Malariol H.S.') are now generally fortified in this way.

A simple method of measuring the spreading pressure of oils (and also the resistance of natural barriers to spreading) has been described by Adam [2].

Stability

The factors responsible for the stability of oil films after spreading are complex and not fully understood. It is known that the most stable oils are those consisting of wide and overlapping cuts with either very high or very low aromatic content and they should not contain fats or fatty acids as spread-aiders.

The quality can best be judged by an empirical field test, applying a standard quantity to a pool on a windless day and rejecting any specimen which does not form a stable film lasting at least two hours.

(ii) Equipment

The hydraulic or pneumatic knapsack sprayers, described on page 97, are suitable for applying larvicidal oils. The oils should be sprayed over the entire surface where larvae occur. For applying heavy doses of conventional oils, equipment giving a coarse spray should be chosen; while a finer spray is suitable for the lighter application of 'Flit MLO'.

The application of heavy doses of oil obviously entails more labour but it has certain advantages. The oil tends to kill vegetation at the sides of ditches, keeping them open and proving that the larviciders have done their job. Also, though the oil has no lasting toxic action, there is some evidence that it repels ovipositing female mosquitoes for a day or so.

(b) Insecticides in oil solution

(i) Characteristics of insecticide solution larvicides

The lethal does of DDT for mosquito larvae is of the order of 10 mg m^{-2}; this would be applied in 2 l of a 5 per cent solution per ha (1·43 pints acre^{-1}), corresponding to a film 0·2 μm thick. Such a small quantity is difficult to apply and the normally recommended dosage is 2·8–5·6 l ha^{-1} (2–4 pints acre^{-1}).

These thin films, though killing larvae reliably, do not have any residual action.

(ii) Equipment for applying insecticide-solution larvicide

The dosage is too light for application by knapsack sprayers but hand atomizing guns (page 104) may be found suitable. Alternatively, providing sufficient spread-aider has been incorporated to give high spreading pressure (about 0·046 Nm^{-1}), the larvicide may be applied in small doses (up to 56 g) allowing the oil to spread itself over the surface. A maximum individual dose of 56 g should be sufficient to destroy larvae over an area of 180 m^2 (the area of a circle of radius 7·5 m); this corresponds to 2·8 l ha^{-1} (2 pints acre^{-1}). Alternatively, various larvicide applicator guns are available. For example, a small squirt gun, holding a few ounces of oil larvicide, is carried in one hand and, by pressure of the trigger, ejects a jet of about 1 ml to a distance of 4·5–6 m (15–20 ft). This is sufficient to treat an area of 1·65 m^2 (2 yd^2) of water surface.

(c) Aqueous larvicides

Emulsions and suspensions of insecticides in water have been used; but since their effect is dissipated through the depth of the water, they tend to be less efficient than oil solutions, which are concentrated at the surface, where nearly all mosquito larvae spend most of their time. An exception is the *Mansonia* group, which draw their oxygen from plants, and cannot be controlled by surface treatments.

Aqueous solutions of organo-phosphorus insecticides (e.g. 0·5 ppm trichlorphon; 0·25 ppm mevinphos; 0·01 ppm parathion) have been used to control mosquitoes in irrigation water, in California [197]. Though apparently rather dangerous, these compounds become inactivated in a few days and may, in fact, be less harmful to wild life than the persistent chlorinated hydrocarbon insecticides.

4.3.13 FUMIGATION [460]

(a) House fumigation

The technique of fumigation depends, to a large extent, on the means of containing the gas or vapour. In the simplest logical situation, pests are fumigated in the infested structure, e.g. a building or ship.

In public health entomology, houses have been fumigated to destroy bed bugs, fleas, cockroaches, clothes moths, etc. Most infested houses are poor at retaining gas and only the most toxic fumigants are effective, usually hydrogen cyanide. Owing to the potential danger and the elaborate precautions required, this type of fumigation has virtually ceased.

(b) Fumigation under fabrics

In another type of fumigation, the gas is retained under plastic sheeting and this method is still widely practised for treating heaps of grain or other food products. Small (8·1 m^3:300 ft^3) tents for this purpose are available.

In the public health field, the method was represented by a device for delousing uniforms, used by American forces in the Second World War [341]. The clothing was put into a gasproof bag and a glass vial of methyl bromide broken inside it. This method has been revived for disinfesting clothing of lousy individuals at cleansing stations, prisons, etc, using ethyl formate [96].

(c) Van or chamber fumigation

A third method of fumigation involves the use of specially constructed fumigation chambers. Mobile fumigation vans are convenient for disinfestation of bedding and other household effects. In Britain, fumigation vans have been largely employed for destruction of bed bugs, in connection with slum clearance schemes; in addition, these vans can be used to eradicate fabric pests and furniture beetles.

The fumigant is usually introduced into the chamber in liquid form, usually on to an electrically heated pan to volatilize it. Provision is generally made for an air circulation system, to distribute the gas rapidly and evenly. By suitable design, the fan and duct system used to circulate the air can be diverted to extract the gas after fumigation and blow it up an exhaust chimney.

This type of van was used for many years with hydrogen cyanide and, more recently, with methyl bromide [73].

(d) Small-scale box or bin fumigations

Where articles require speedy disinfestation (e.g. clothing infested with clothes moths or by lice or fleas), they may be fumigated in a reasonably air-tight trunk or in a metal bin, such as a clean empty dustbin. Liquid fumigants with relatively high volatility are necessary and, for unskilled users, one that is safe and pleasant to handle should be chosen. For examples, ethyl formate, methyl formate or methyl allyl chloride can be used, at the rate of 300 cm^3 per cubic metre, with an exposure of 5 hours. On such a small scale, it is easy to avoid danger from fire risk when using the formates. The temperature should preferably be 20° C or above [96, 138].

(e) Use of solid fumigants

Crystals of solid fumigants such as naphthalene or paradichlorobenzene may be scattered among woollen goods before storage, to destroy any eggs or larvae of clothes moths or similar pests which might be present. The fabrics should be stored in reasonably air-tight receptacles such as trunks or suitcases. In addition, the use of a suitable container (e.g. plastic bag) will prevent entry of moths or carpet beetles, so that, after the initial treatment, the goods remain free from attack.

Large articles such as carpets may be rolled or folded up tightly with crystals between the layers. In this case, naphthalene should be used as it is less volatile than paradichlorobenzene.

(f) 'Residual' fumigation

A new interest in fumigation for pests of medical importance was provided by the compound dichlorvos, which is so highly insecticidal at low concentrations that susceptible insects such as mosquitoes or flies are killed in a normal unsealed room (provided that ventilation is not excessive) by vapour emitted from one or more dispensers.

(i) Equipment

Ordinary dispensers
Two types have been used, both allowing the dichlorvos to diffuse from them. In one form the dichlorvos (40 per cent) is fused with montan wax and moulded into cylinders (3·75 × 12·5 cm). In another form of dispenser, the dichlorvos (14 g : $\frac{1}{2}$ oz) is put into a plastic bottle (2·5 × 5·6 cm) closed with a plug. This is wedged into a plastic sheath permeable to dichlorvos. On squeezing the unit, the plug is ejected from the inner container and the dichlorvos begins slowly to diffuse through the sheath.

Mechanical dispenser
A miniature air compressor drives air slowly through a porous membrane impregnated with dichlorvos. The air stream containing the vapour is then distributed as required. Operation of this unit for $\frac{1}{2}$ h should secure disinsection.

4.4 INSECTICIDE RESISTANCE

The very wide use of modern synthetic insecticides, beneficial in many ways, has had some undesirable consequences. One of these is the emergence of resistant strains, due to the widespread and persistent destruction of normally susceptible insects, while the abnormally tolerant ones manage to survive. Emergence of resistant strains, therefore, is a selection of pre-existing genetic types, analogous to the selection of new types and varieties by natural selection. The idea that resistance can be developed in individual insects by exposure to sub-lethal doses, is a misconception.

4.4.1 GROWTH AND IMPORTANCE OF RESISTANCE [639]

Prior to the Second World War, resistant strains were rare curiosities because, until the introduction of modern synthetic insecticides, the attack on pests by toxic chemicals was very rarely on a scale sufficient to exercise appreciable selective effect on the natural population. Beginning with DDT-resistant flies in 1947, however, the phenomenon has appeared in pest after pest all over the world until well over 300 species are now involved. About a third of these are pests of public health importance and the growth in their numbers is shown by the following figures:

Year:	1947	1950	1956	1958	1960	1969	1975
No. of species:	1	15	20	50	80	100	120

These figures are approximate and it would be misleading to claim precision, since the number of cases depends on the numbers tested; however, certain conclusions may be drawn. Resistance is most likely to develop in species which are regularly and extensively exposed to the selective effects of insecticides; hence, it occurs in most of the important vector groups. For example, the 1975 figures include about 40 each of anopheline and of culicine mosquitoes, eight other biting Diptera and eight non-biting species, 11 parasites (bugs, lice, fleas), two cockroaches and nine ticks. Indeed, resistance now extends to all the arthropod genera of medical importance, except (so far) tsetse flies and *Phlebotomus* sandflies.

The numbers of cases of resistance which have been recorded, though significant, do not give an adequate picture of the seriousness of the problem. This depends on: (i) the importance of the species concerned (e.g. if it is an important disease vector); (ii) whether the degree and extent of the resistance actually prevents effective control; (iii) whether there are other, non-chemical means of control; and (iv) whether there are alternative insecticides suitable for maintaining control. The first three criteria demand appropriate information and sound judgement from experts in the field. The fourth point involves an understanding of the toxicology and genetics of resistance, which will be outlined later.

4.4.2 DETECTION AND MEASUREMENT OF RESISTANCE

In the early post-War years, it became evident that resistance posed a serious threat to the control of vector-borne disease, and it was extensively discussed at meetings of the W.H.O. Expert Committee on Insecticides (in 1956, 1957, 1959,

1962, 1968 and 1975). One of the first problems was to distinguish genuine cases of resistance from control failures due to other causes. It was soon realised that there was a necessity for standardized tests, which could be used by workers all over the world. Accordingly, a series of test methods has been devised and published in successive reports of the Insecticide Committee. (The F.A.O. became seriously interested in the same problem from 1965 onwards and has published similar tests for pests of agricultural and veterinary importance).

The intention of these resistance tests, was to combine reliability with simplicity, so that they could be used in or near the field operations, with a minimum of sophisticated equipment. The W.H.O. actually assembles complete kits for conducting the tests in Geneva and supplies them to field workers, at cost. From the results of these standardized tests, W.H.O. has a continuously updated picture of the global resistance situation. Following the 1975 Insecticide Committee meeting, the test methods were revised and the following are available: WHO/VBC/75 . . . (code number). Adult mosquitoes, organochlorine insecticides (581); adult mosquitoes, organo-phosphorus and carbamate insecticides (582); larval mosquitoes (583); biting midge larvae (584); body lice and head lice (585); adult bed bugs (586); reduviid bugs (587); fleas (588); adult blackflies, sandflies, and biting midges (589); houseflies, tsetse flies, stable flies, blowflies (590); blackfly larvae (591); adult ticks (592); cockroaches (593).

(a) General principles of resistance tests

The following principles underlie all the tests. First, the susceptibility levels of a normal population of a given species is determined with a strain of known susceptibility. This can either be collected from the field where insecticide has not been used, or obtained from a colony initiated before insecticide usage.

The actual test consists of exposing batches of insects to a range of doses and noting their subsequent mortality. In some tests, the doses are provided by exposure to different concentrations of insecticide for a standard time; in other tests, dosage is related to the time of exposure to a standard concentration. Whichever method is used, it is important to maintain conditions as standard as possible and to use insects of the same age and stage.

When the dosage–mortality data are available, they are plotted on special 'logarithmic-probability' paper, which allows for (i) the fact that effects of poisons are proportional to the logarithm of the dose; and (ii) the normal distribution of susceptibility levels within a population, with most individuals clustered round the mean. As a result of the corrections provided by the logarithmic-probability paper, dosage plotted against mortality gives a straight regression line. From such a line, representing a susceptible population, the kill to be expected from any given dosage can be estimated. On this basis, a concentration or exposure is chosen which is expected to give a very high kill (say 99·9 per cent). This 'diagnostic dosage' is used for regular monitoring of field populations. When survivors regularly appear in such tests, there is good evidence of the presence of resistant individuals in the field population. The percentage survival in such tests gives an approximate estimate of the proportion of such individuals in the wild. It does not, however, indicate the strength or degree of resistance. This can only be measured by the ratio of the dose necessary to kill a given proportion of the resistant strain as compared to the equivalent dose for the normal. It is more difficult to assess and requires an experimental investigation by an insect toxicologist.

(b) Experimental details

It is not feasible here to give a full description of all W.H.O. resistance tests. Anyone intending to conduct such tests should obtain the instructions from W.H.O. Some brief descriptions, however, will indicate the kind of operations involved.

The most scientific way of dosing insects with contact insecticides is to apply a measured quantity to each one. This is only feasible with relatively large insects, and it is the standard way of treating houseflies, blowflies, tsetse flies etc. The insecticide is dissolved in a volatile solvent (acetone, butanone) and applied to the insects with a tiny, self-filling micropipette.

Many insects are too small or too fragile for this 'topical application' method to be generally acceptable. In such cases, they are made to dose themselves by being confined with a treated surface, for a standard period. For the following types of insect, the insecticide residues are in the form of oil solutions impregnated on filter paper: adult mosquitoes, blackflies, biting midges, bed bugs, triatomid bugs, lice and fleas. Packets of the impregnated papers can be obtained from the W.H.O. in Geneva; alternatively, the impregnation is not difficult to perform, given comparatively simple equipment. Methods of retaining the insects on the impregnated papers vary in different species. With flying insects (mosquitoes, sandflies, blackflies), the special equipment shown in Fig. 4.6 is supplied with the W.H.O. test kit. Bed bugs, triatomid bugs and fleas are confined in tubes on pieces of impregnated paper, also shown in Fig. 4.6.

Aquatic larvae of mosquitoes, blackflies and biting midges are exposed to aqueous suspensions of insecticides, prepared by adding alcoholic solutions of them to large volumes of water. (In the test for blackfly larvae, special provision for careful handling and well aerated water are necessary).

The test for resistance in cockroaches is slightly anomalous. It is an early established test, which may, perhaps, be modified in the future. Deposits of insecticide are applied to the inside of glass jars by adding a small quantity of solution in a volatile solvent and rotating the jars until the solvent has evaporated. The insects are confined in the jars and observed for knock-down (i.e. paralysis). Susceptibility or resistance are judged from the times for 50 or 99·9 per cent knock-down.

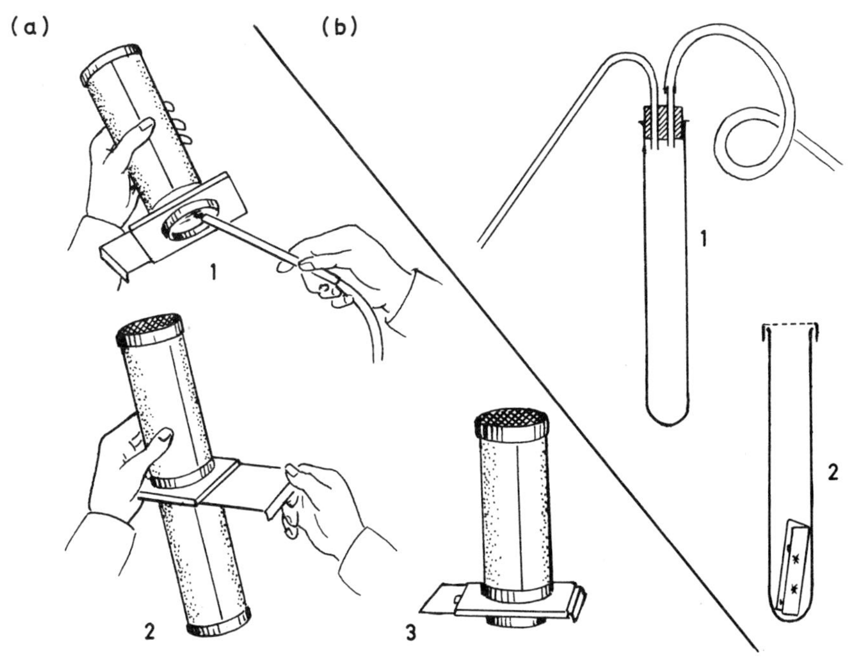

Figure 4.6 World Health Organization standard tests for insecticide resistance.
Left (a) for mosquitoes:
1. Mosquitoes collected by sucking tube are put into holding tube.
2. Holding tube is screwed on to exposure tube, slide removed, insects blown into exposure tube and slide removed.
3. Exposure (with holding tube removed). Afterwards, holding tube re-attached and procedure reversed to transfer mosquitoes to it, until examination for kill.

Right (b) for fleas and bed bugs:
1. Test tube, with sucking attachment to collect fleas. (Bugs collected by forceps.)
2. Exposure, with insects on folded, impregnated paper.

4.4.3 NATURE OF RESISTANCE [93]

(a) Toxicological aspects

Extensive biochemical research has revealed something of the physiological nature of resistance. It has been found that resistance depends on one or more mechanisms which protect the insect from the effects of the insecticide. Perhaps the most common is an enhanced detoxication system, which degrades the insecticide into harmless metabolites, which can be excreted. Various mechanisms of this kind have been elucidated. Some are specific to a particular type of insecticidal compound (e.g. the enzyme which transforms DDT to DDE; or the one which attacks the carboxyl radicles of malathion). Others are more general in action (e.g. the microsomal oxidase systems, which can metabolize a wide variety of toxicants).

Another type of mechanism which is not very specific is a barrier to penetration (possibly due to a modified cuticle). While this does not, in itself, provide a high degree of resistance, it magnifies the effects of other mechanisms. A third type of defence mechanism consists in a change in the physiological target at the poison, which renders it insensitive. In the case of the organochlorine insecticides, whose mode of action is obscure, little or nothing is known about this defence mechanism. Something is known, however, about the organophosphorus resistance which depends on a change in the acetylcholinesterase target.

In the early years of resistance to the new synthetic insecticides, it happened that specific resistance mechanisms were developed, which vitiated only groups of related compounds, to which the strain involved was said to show *cross-resistance*. It was then a simple matter to change to another type of insecticide, which would be effective until a new type of resistance developed. Thus, it was usual to class insecticides according to resistance groups, as follows: (1) DDT and related compounds; (2) HCH and the chlorinated cyclodiene compounds; (3) organophosphorous compounds (and carbamates?); and (4) pyrethroids. Unfortunately, as widespread insecticide usage continued, strains with two or more defence mechanisms coping with distinct types appeared. These were said to have *double-*, *treble-* or *multiple-resistance*. Multiple resistance is a serious matter, because the numbers of different types of insecticide are not large, so that chemical control soon becomes very difficult.

It is also possible for a strain to possess two or more defence mechanisms towards the same insecticide; this can be called *duplicate-*, *triplicate-* or *multiplicate-resistance*. The practical significance of multiplicate resistance is that when, occasionally, scientific methods of overcoming one type of mechanism has been discovered, it does not bring permanent relief.

(b) Genetical aspects

Each of the various mechanisms mentioned is genetically determined, nearly always by normal somatic genes which can be dominant, recessive or intermediate in expression. The first genetical investigations of resistance were made on strains which only possessed one defence mechanism, and usually rather a specific one. As a result, there was clear-cut segregation of the genotypes in the F_2 hybrids and back-crosses. When duplicate and triplicate resistant strains appeared, the effects were less easy to distinguish, especially as the strains used were not always genetically homogeneous. However, so far there have been no indications of modes of inheritance other than according to Mendelian principles. This fact somewhat simplifies the predictions concerning the spread of resistance genes in wild populations. Thus, in general, the development of resistance depends on: (1) the pre-existence of some individuals in the population with genes for resistance; (2) the intensity and extent of selection by a given insecticide; and (3) the nature of the resistance genes. Recessive resistance genes (e.g. DDT-resistance in anopheline mosquitoes) are slow to spread, since the heterozygote forms have no survival advantage. In contrast, dominant or semi-dominant resistance genes (e.g. HCH/dieldrin resistance in most insects) rapidly respond to selective pressure and become widespread.

It has generally been found that when resistance genes first become prevalent, they have deleterious side-effects, so that in the absence of continuing selection, they become less abundant. If, however, selection continues unabated, the resistance factor seems to become better integrated into the genetic background of the strain and then persists in the population, even if selection is stopped.

4.4.4 COUNTERMEASURES AGAINST RESISTANCE

Unfortunately, it must be admitted that, despite an immense amount of research on resistance, which has revealed a considerable insight into its nature, no one has discovered a way of restoring susceptibility to a field population which has firmly developed resistance. All that can be done is to use the standardized tests to detect the resistance as early as possible and, by further investigation, determine its characteristics (e.g. by cross-resistance studies). This will enable early planning for substitution of an alternative insecticide.

Since resistance develops in response to powerful selection, excessive use of insecticides should be avoided. Alternative methods of control (e.g. improved hygiene) should be used instead of insecticides (or in combination with reduced insecticidal use). Unfortunately, it has often happened that resistance in medically important insects has developed as a result of agricultural uses. Thus, residues from crop-spraying may contaminate mosquito breeding sites and select them for resistance. It is often difficult for public health authorities to restrict this trouble.

From time to time, people have suggested the possibility of using mixtures of two different kinds of insecticide; or, alternatively, using them one after another. It is not really clear how this could prevent the development of resistance. Laboratory tests of the plan have not been encouraging and it would be difficult to arrange for an adequate field trial. The idea might well succeed if two poisons could be found, of which one was most toxic to the individuals most tolerant of the other. Despite some early hopes, no such ideal combinations have been found.

4.5 TOXIC HAZARDS OF INSECTICIDES [384]

At the begining of this chapter, it was stated that, with the progress of pest control science, the older crude inorganic poisons were being replaced by synthetic organic compounds specifically toxic to arthropods rather than to man or higher animals. During the 1960s, however, immense quantities of organochlorine insecticides were used, under the impression that they were harmless. Indeed, they caused little harm to man or his domestic animals, and a great deal of good in saving crops and preventing disease. But, all the time, they were building up persistent residues in the environment and also in the bodies of vertebrates. Furthermore, excessive use (mainly in agriculture) was responsible for considerable, intermittent deaths of wildlife. These incidents, dramatised in Rachel Carson's *Silent Spring*, created a strong reaction in the press, the public and the government.

4.5.1 OFFICIAL SURVEILLANCE

The public concern about possible harm from chemical pesticides has stimulated official bodies to appoint numerous commissions, committees and working parties to assess and control the hazards. In the international field, the primary emphasis has been on the safety of pesticide residues in food, which are monitored by joint committees of the W.H.O. and the F.A.O. These try to recommend 'acceptable daily intakes' which are estimated concentrations believed to be harmless if consumed daily for life. The second focus of attention (reflecting current anxiety about pollution in general) is the possible harm to wildlife. This, again, is considered by working parties in W.H.O. and F.A.O. concerned with safe usage of pesticides; and also by UNESCO, as one aspect of its programme on Man and the Biosphere.

On the national scale, there are numerous bodies concerned with pesticides in all advanced countries. In Britain, the main advisor to the government is the Advisory Committee on Pesticides, under the aegis of the Department of Education and Science, but serviced by the Ministry of Agriculture, Fisheries and Food. This monitors the use of existing and new pesticides, under a non-statutory *Pesticides Safety Precautions Scheme*. New candidate pesticides are submitted to a scientific sub-committee by the manufacturers, together with appropriate physical, chemical and toxicological data. These are considered and an evaluation passed to the main Advisory Committee; the compound is then approved, or rejected; or further toxicological tests or trials may be demanded. Though voluntary, this arrangement has worked very well for a dozen years, by the smooth collaboration between industry and the committee, which comprises official and independent members, with an independent chairman.

In addition to this Committee, pesticides hazards in Britain (especially those affecting workers in the industry) are monitored by the Health and Safety Executive and the Department of the Environment in regard to wildlife hazards; while the Government Chemical Laboratory regularly analyses for pesticide residues in food etc. Currently (1979) negotiations with the European Economic Community are attempting to harmonize regulations with those of the Community. Also the Organization for Economic Co-operation and Development is drafting advice to member states on methods of evaluating hazards.

In the U.S.A., the organizations concerned with pesticide hazards include the Department of Health, Education and Welfare; the Consumer Protection and

Environmental Health Service; the Food and Drugs Administration; the National Institute of Environmental Health and others. In reviews and reports, these tend to get referred to by initials (e.g. FDA, USAA, FTC, PHS, HEW, PIB, MID etc) which one writer described as the U.S. government's 'alphabet soup'.

Despite all official precautions, it is essential for those engaged in chemical pest control to understand the nature of the hazards involved. Thus, it is important to distinguish between the dangers of acute and chronic poisoning; and to recognize the very different hazards to high risk people, whose occupation involves contact with pesticides, and the general public.

4.5.2 ACUTE TOXICITY OF INSECTICIDES

(a) Types of hazard

The dangers of acute poisoning, by ingestion or contamination by a single dose of insecticide, are relatively easy to assess. In general, people occupationally exposed to pesticides are more liable to accidents by contamination, whereas the general public are more liable to suffer from ingestion.

(i) Persons particularly at risk

The comparatively few people engaged in the manufacture of pesticides fall into the same category as others involved in making potentially dangerous chemicals. They are, or should be, under careful medical supervision and protected according to regulations with which the Health and Safety Executive are concerned.

As regards application of pesticides, two types of people are involved. There are professional pest control operators (whether employed by a commercial firm or a health authority) and there are farmers or horticulturalists who need to use large quantities of pesticide occasionally. The latter are not normally trained (and sometimes not experienced) in regard to pesticides; so they are more likely to be involved in accidents than the professionals. Fortunately, in the public health sphere, large scale applications are almost always done by professional operators.

(ii) The general public

It is generally recognized that many substances in daily use involve some risk of misuse (e.g. coal gas, disinfectants). In the same way, insecticide concentrates (including their solvents) may be dangerous to careless people, to children and to potential suicides. The numbers involved are small. Thus, deaths from poisons of all kinds in the U.S.A. are only about one-sixtieth of those from all accidents; and, at the time of maximum DDT usage (1966), accidents to children swallowing insecticide were only about 3 per cent of the total, compared to 25 per cent involving aspirin.

Another kind of accident involving the public can occur if transportation regulations are inadequate or are not observed. If dangerous insecticides (like parathion or endrin) are despatched by rail or sea inadequately packed, so that broken or leaking containers can contaminate other goods, there can be poisoning if foodstuffs are involved. An American report lists 25 such accidents from different countries between 1958 and 1969. These affected some 6000 people and there were about 500 deaths [Mrak (Chairman) Report to DHEW, 1969].

(b) Assessment of acute toxicity

The danger of acute intoxication from swallowing or gross contamination with insecticide concentrates can be judged from experiments with laboratory animals. Since tolerance varies to some extent from one species to another, the most reliable estimates of possible danger to man should be based on tests with different animals, preferably including monkeys. This is not always feasible, however, partly because of the expense and also because of the public objection to excessive experimentation on live animals. In any case, preliminary information is normally obtained with rats, batches of which are exposed to a range of doses of the poison under examination. From the resulting mortalities, the average lethal dose (LD50) is calculated statistically. Some data of this kind are given in Table 4.3, together with results of LD50 tests with some insects, to show relative toxicity. Data are obtained for oral dosing and dermal applications; and, if appropriate, inhalation. On the basis of the oral and dermal toxicity to rats, the World Health Organization published in 1975 a schedule of suggested levels of hazard, as shown in Table 4.4 [638].

(c) Prevention of poisoning accidents

People occupationally exposed to pesticides, whether making them in a factory or applying them in the field, are more likely to be poisoned by skin contamination than by ingestion. Accordingly, when they are dealing with a potentially hazardous substance, they should wear protective clothing. Various forms have been designed for spraying operators and an example can be seen in Fig. 4.5.

Accidental swallowing of insecticide concentrates is more common among children than adults, so it is very important not to transfer them to unlabelled bottles (even soft drink bottles have been used!). Empty containers should always be disposed of safely.

To guard against accidents due to breakage of insecticide containers during transit, the European Economic Community is currently (1979) engaged in standardizing methods of labelling packages, to indicate various types of hazard. Dangerous packages will include those which contain substances which are very

Table 4.3 Toxicity and potency of some insecticides (from various sources).

		Rat* (mg kg^{-1})		Adult insects (mg kg^{-1})		Aqueous (ppm)	
Common name	Other name	Oral	Contact	Fly	Mosquito	Mosquito larvae	Fish
DDT	Dicophane	180	2500	18	15	0·017	0·1
Dieldrin	HEOD	45	60	2	2	0·006	0·01–0·04
Gamma-HCH	Lindane	90	500	1·5		0·011	0·04
Bromophos	Nexion	5 700	2200[a]	5		0·0018	0·5–1
Chlorpyrifos	Dursban	160	2000[a]			0·0007	0·03–0·1
Diazinon	Basudin	250	700	5		0·0033	
Dichlorvos	Vapona	65	90	1·5			1
Dimethoate	Cygon	550	—	1·0			50
Fenchlorphos	Korlan	1 750	2000	8·0			
Fenitrothion	Sumithion	300	3000[b]	6·4	1·7	0·008	
Fenthion	Baytex	260	400	3·0	1·5	0·0026	
Iodophenphos	Nuvanol–N	2 100	2000	4·0			0·06–1
Malathion	Cythion	2 800	4100[a]	50		0·029	
Parathion	E–605	8	14	3			
Pirimephos-me	Actellic	2 000	2000[a]				1·5
Temephos	Abate	10 000	1900[a]			0·0006	
Bendiocarb	NC6897	50	> 400				
Carbaryl	Sevin	850	4000	250			
Dioxacarb	Famid	100	1950[a]				
Propoxur	Baygon	100	900	35	1·2	0·64	
Allethrin	—	920					
Biopermethrin	NRDC147	2 000	—	1·0	0·5		
Bioresmethrin	NRDC107	8 000	—	1·0			0·025
Permethrin	NRDC143	1 500		1·7	0·6		
Pyrethrin I	—	350	—	20			

* Unless otherwise indicated, thus: a, rabbit; b, mouse

Table 4.4 Levels of toxic hazard suggested by W.H.O.

		LD50 for rat (mg kg^{-1})			
		Oral		Dermal	
Class	Hazard	Solids	Liquids	Solids	Liquids
Ia	Extreme	< 5	< 20	< 10	< 40
Ib	High	5–50	20–200	10–100	40–400
II	Moderate	50–500	200–2000	100–1000	400–4000
III	Slight	> 500	> 2000	> 1000	> 4000

toxic, toxic or harmful, as well as those which are irritant, corrosive, highly flammable or explosive. The classification of 'very toxic', 'toxic' or 'harmful' will be based on the W.H.O. schedule given above. The regulations are expected to come into force in 1980 or 1981.

(d) Symptoms of poisoning and treatment

Metallic salts of arsenic, mercury and thallium, as well as the inorganic fluorine compounds, are cell poisons with deleterious effects on many enzymatic processes, in different tissues. The more advanced organic insecticides, however, are practically all specifically toxic to nerve cells and are usually without effects on other tissues and moderate doses. These nerve poisons all cause certain symptoms (headache, nausea, vomiting, dizziness) which are not specific and, indeed, can be due to causes other than poisoning. Accordingly, in the following sections, the characteristic signs and symptoms of intoxication by different insecticides will be stressed.

(i) The chlorinated insecticides

DDT is known to act preferentially on sensory nerves and appears to cause unstabilization of the polarization of the nerve axons, so that nerve impulses tend to be reduplicated. As a result, both insects and mammals show the effects of overstimulation causing continual tremors and muscular inco-ordination. One of the symptoms characteristic of DDT poisoning is paraesthesia (numbness, tingling) of the mouth, part of the face and, in severe cases, the extremities.

Less is known about the mode of action of *gamma* HCH and the cyclodiene series, the site of action of which appears to be in the central nervous system. Mammals poisoned by these compounds tend to show hyper-excitability and later suffer severe clonic and tonic convulsions, even in non-fatal cases.

(ii) Organo-phosphorus and carbamate insecticides

Considerably more is known of the mode of action of this group of insecticides, which act as anti-cholinesterase poisons, both in mammals and in insects (see pages 71–72). Accordingly, the more characteristic signs and symptoms of intoxication are those due to excessive acetylcholine. In mammals these can be recognized by excessive stimulation of the parasympathetic system. There is contraction of the pupils, profuse secretion of saliva and tears, diarrhoea, discomfort in the chest (due to constriction of the branchioles) and retardation of the heart. Additional symptoms, which are not so specific, include headache, nausea and blurred vision.

(iii) Treatment

Only the simplest remedial measures can be attempted by a non-medical person, the main one being to remove the poison as soon as possible. For a poison taken internally, an emetic may be given or, as a first-aid measure, vomiting induced by a finger down the throat. Evacuation of the gut is desirable, avoiding oily laxatives especially where chlorinated insecticides may have been taken.

After external contamination, thorough washing of the eyes or body is advisable. Artificial respiration, preferably by mechanical means, may be required for patients who have collapsed after organo-phosphorus poisoning. Medical treatment in many cases must be symptomatic, in view of our ignorance of the mode of action of many insecticides, e.g. the use of sodium pentobarbital as a rapid sedative to combat hyper-excitability or convulsions. Intoxication by the anti-cholinsterase insecticides may be more scientifically treated by atropine sulphate (1 to 2 mg) which antagonizes acetylcholine.

4.5.3 CHRONIC TOXICITY OF INSECTICIDES

The chronic toxicity of repeated exposure to pesticides depends on slow accumulation of the chemicals in the body. This is more liable to happen with chemically stable compounds than those that are more easily attacked by enzymes and are thus 'biodegradable'. The organochlorine compounds are generally more difficult for tissues to metabolize and tend to accumulate more than the organo-phosphorus or carbamate insecticides.

(a) Persons particularly at risk

People occupationally exposed to pesticides, whether making them in a factory or applying them in the field, are more likely to become poisoned by skin contamination than by ingestion. Accordingly, the relative levels of contact toxicity to rats is more relevant to this hazard than oral poisoning.

The ability to penetrate mammalian skin partly determines the danger and this to some extent accounts for the rather low toxicity of DDT by this route. Nevertheless, residues do accumulate, as shown by amounts up to 700 ppm of DDT in those workers with a long (18 years) experience of manufacturing it; this is 50–100 times the level in the general population. Such workers experienced no adverse symptoms nor did the thousands of spraymen engaged in anti-malarial house spraying during the past 20 years. This suggests that man is relatively tolerant of the toxic effects of such internal residues.

The danger of fatal poisoning by chronic exposure is relatively easy to assess (though more difficult than that of acute poisoning). However, possible harmful, though sub-lethal effects must not be forgotten. Therefore pest control operators ought to undergo periodic medical examination to detect possible lesions, especially if there is any suspicion that they are handling a dangerous class of compound.

Apart from systemic intoxication resulting from repeated exposure to insecticides, there is also the possibility of adverse skin reactions. Thus, a specific allergy is liable to occur with extracts of natural pyrethrum [504]; and a more generalized skin reaction can occur in atopic subjects by frequent exposure to many chemicals, including carrier solvents, such as kerosene.

(b) Toxic hazards from environmental residues

Whatever the problems of protecting pest control operators from chronic intoxication by pesticides, at least the dangers are fairly well defined. The possible hazards from residues in the environment, however, are more difficult to assess. There has been much anxiety about this matter, partly due to the astonishing sensitivity of modern methods of detecting minute traces of organochlorine insecticides, which have made people aware of tiny traces of them in all sorts of places. The amounts of DDT detected are of the following orders of magnitude. In the air, $0–10 \times 10^{-12}$ g; in rain water, $10–100 \times 10^{-9}$ g; in rivers, $10–1000 \times 10^{-9}$ g; in (U.K.) arable soil, $0{\cdot}01–1 \times 10^{-6}$ g; in (U.K.) orchard soil, $3–30 \times 10^{-6}$ g.

From the environment, the insecticides invade plants and thus, herbivorous animals. The latter store residues of pesticides in their fatty tissues, so that predators which feed on them tend to acquire even heavier burdens and are generally more at risk. Humans too, acquire traces of pesticide in this way; though because human food is generally washed and cooked, the residues are somewhat dissipated.

It should be realized that nearly all the anxiety about environmental residues relates to organochlorine insecticides, which tend to persist in the soil for 2–5 years before over 75 per cent disappears. In contrast, the elimination of some organo-phosphorus insecticides was found to take only 1–12 weeks. Herbicides were intermediate, requiring 1–18 months for equivalent disappearance. But herbicide residues are very low in human foods. A six year average intake of people in the U.S.A. amounted to only 0·008 μg kg^{-1} day^{-1}, compared to 0·19 organo-phosphorus compounds and 1·1 organochlorines. Among these, DDT and its metabolites nearly always predominated, until recently; but since the restrictions on its use, these quantities have been declining. However, as a result of 2 or 3 decades of DDT usage, residues in human fat have become almost universal. Numerous and extensive surveys have been made from post mortems and biopsies; and Wassermann [596] has summarized nearly 90 papers, from which the following have been calculated.

Area	*No. of papers*	*Average DDT (ppm)*
Asian countries (excluding Japan)	6	17·2 ± 1·8
East European countries	15	14·4 ± 1·2
South European countries	4	11·5 ± 2·4
North America	20	8·2 ± 0·6
West European countries	20	3·9 ± 0·3

Although these figures may be considered disturbing, they should not cause alarm, in view of the healthy DDT workers with up to 700 ppm DDT in their bodies, mentioned earlier. Furthermore, they are likely to decline, in the future, with reduced DDT usage.

Official action in restricting DDT has been almost universal, but varies from a complete ban to limited use (as, at present, in U.K.). Perhaps it is unfortunate that DDT has been banned from public health uses, which only comprise about 3 per cent of the total (so that environmental pollution is almost entirely due to gross agricultural applications). No definite evidence of harm to man has resulted from the enormous quantities used in the past; and its cheapness is a boon to poor underdeveloped countries, which cannot afford the new sophisticated pesticides, and which have serious vector-borne disease problems.

(i) Assessment of chronic toxicity

The alarm engendered by the discovery of insecticide residues in the environment and in animal bodies, as well as the reports of deaths of wildlife attributed to them, has resulted in the erection of a series of testing procedures by which all new pesticides must be scrutinized before approval. These include the following.

Chronic toxicity
Experimental animals (rats, mice, guinea-pigs, rabbits or dogs) are fed traces of the chemical at various levels, with regular measurements of weight, feeding, biochemical, haematological and general clinical observations.

Carcinogenicity
The soundest evaluation of this hazard is by life-long exposure of test animals; but good indications of the risk are provided by looking for mutant changes in bacteria and cell cultures.

Mutagenicity
Treated animals are observed for three generations.

Teratogenicity
The public concern following the thalidomide case drew attention to possible effects on pregnant females. These tests are difficult and costly.

Hazards to wildlife
Toxicity tests to fish and other aquatic organisms are involved.

CHAPTER FIVE

Bloodsucking flies

5.1 INTRODUCTION

5.1.1 THE DIPTERA OR TWO-WINGED FLIES

The order Diptera, of two-winged flies, is of prime importance in medical entomology, containing as it does bloodsucking forms, disease vectors and various nuisances. It is a large and successful insect group divided into three sub-orders at different evolutionary levels: Nematocera, Brachycera and Cyclorrhapha. The first appears as fossils in the Permian while the other more advanced groups do not appear till the early Tertiary.

Nematocera are generally slender fragile flies, with long antennae composed of many similar segments and palpi of several joints. The larvae, usually active, with well-developed heads, live in water or damp soil. They include the biting gnats and midges discussed in this Chapter.

Brachycera are usually large flies with short antennae, projecting forward, and two-jointed palpi. Only one family, the horseflies, is of medical importance and it is considered in the next chapter, on bloodsucking flies. The larvae have reduced heads and live in damp soil or water.

The *Cyclorrhapha* is the most advanced group, mostly comprising flies of compact build, often bristly. The antennae consist of three joints lying down in front of the face, with a large bristle (the 'arista') on the last segment. The palps are single-jointed.

The larvae are headless maggots, devoid of eyes or other complex organs of special sense, living in and feeding on decaying organic matter. Pupation in this group always occurs within the last larval skin, which remains to protect the pupa as a small barrel-shaped cover. When the fly emerges, it pushes off the top of the 'puparium' which splits along a circular line (giving the Greek derivation of the name Cyclorrhapha).

The Cyclorrhapha contains a number of families mainly of small flies rather difficult to define and two distinct important families: the Muscidae or houseflies and Calliphoridae or blowflies. These flies may be annoying or directly harmful to man and sometimes act as vectors of disease.

5.1.2 FEEDING HABITS OF BITING FLIES

Bloodsucking is widespread in the many families of Diptera and it is possible that this was the original habit of ancestors of the group. Some forms seize and suck the blood of other insects; but many prey on the blood of vertebrates (especially mammals and birds) and we are here concerned with these. In the more primitive sub-orders, Nematocera and Brachycera, only the females take blood, while both sexes take sugar, usually from the nectar of flowers. The sugar is necessary for activity of daily life, while the protein of blood is utilized for egg production.

The mouthparts used for piercing and sucking in these groups comprise a bundle of piercing stylets carried on a flexible labium (Fig. 5.1, page 136). The stylets are formed from highly modified mandibles and maxillae, together with a blade-like labrum and a narrow hypopharynx carrying the salivary duct. These piercing elements are more or less vestigial in the males.

In the highest sub-order of Cyclorrhapha, the bloodsucking habit of the females was abandoned, and the piercing mouthparts disappeared. Feeding was done by licking with the sponge-like labella at the tip of the labium. There was still a tendency for females to need protein for egg development, but this may be obtained from various sources (liquid excreta, milk, pollen).

Finally, a secondary development of bloodsucking has occurred in isolated groups of Cyclorrhapha. This may have developed from the habit of some flies with licking mouthparts which take blood from small wounds. Later, teeth in the labium assisted a rasping action and finally a horny piercing labellum was evolved in some genera. It will be seen that the redeveloped biting mechanism is quite different from the primitive one (necessarily so, since no piercing mandibles and maxillae are present). Also, both sexes take blood though they also take separate meals of nectar.

5.2 MOSQUITOES (CULICIDAE)

The true mosquitoes (Culicinae) are of great hygienic importance and will be considered in some detail (page 137ff.). A few related groups of no medical importance should be mentioned first, however, since they may be mistaken for mosquitoes on casual inspection. These are:

(1) The Chironomidae, or midges, often seen in summer swarming in large numbers near ponds, lakes or streams in which they breed. Occasionally, they may be so numerous as to be a nuisance. Also, their larvae sometimes get through filters into tap water and cause dismay (Fig. 5.5c).
(2) The Dixidae, or dixa midges, which may swarm in gardens, often in surprisingly cool weather (Fig. 5.5d).
(3) The Chaoborinae, or phantom midges, a sub-family of the Culicidae, less common than the others.

The adults of these insects may be distinguished from mosquitoes by (a) the absence of scales on the wings (except for some on the hind margin of those of Chaoborinae) and (b) the absence of a long biting proboscis.

The larvae, which are all aquatic, are also readily distinguished. Chironomids and dixa larvae are cylindrical, without the enlarged thorax of mosquito larvae.

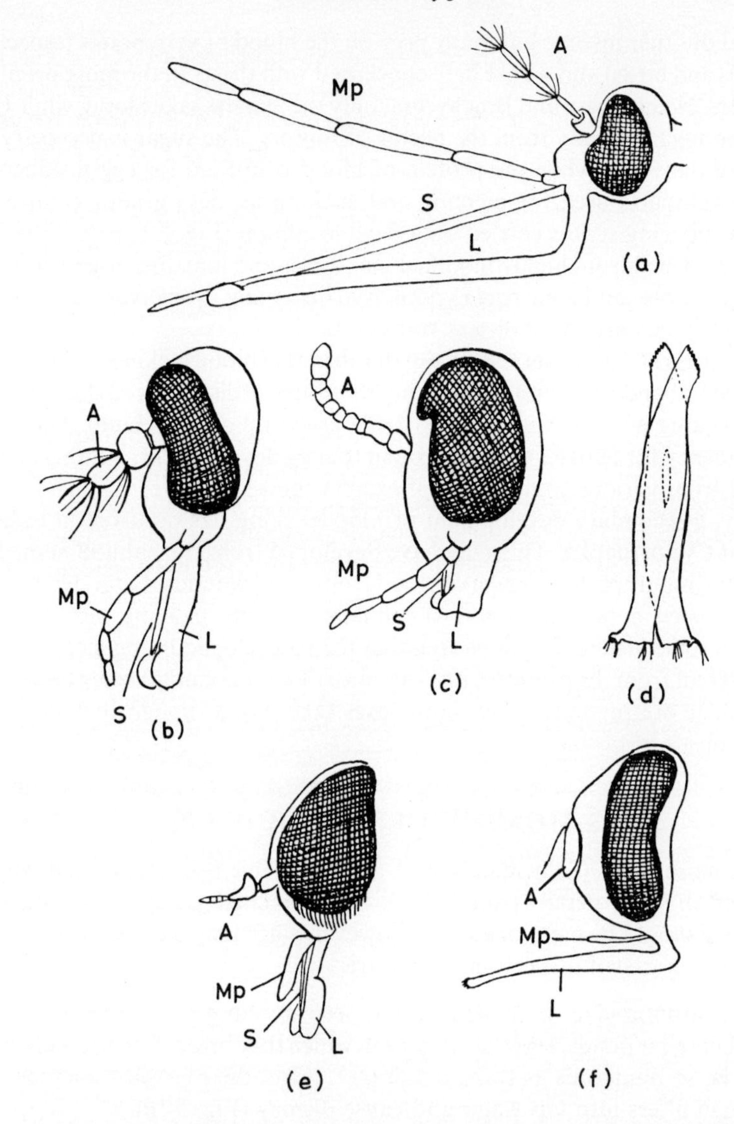

Figure 5.1 Heads of biting flies, all viewed from the left side. (*a*) Anopheline mosquito (with labium partly retracted, as in biting); (*b*) *Culicoides*; (*c*) *Simulium*; (*d*) mandibles of *Simulium* further enlarged and viewed from above to show scissor-like action; (*e*) *Tabanus*; (*f*) *Stomoxys*. In all cases: A, antenna; Mp, maxillary palp; S, stylets (mandibles and maxillae); L, Labium. (After various authors.)

The former bear two pairs of pseudopods (false legs), one on the thorax and one on the hind segment. Dixa larvae commonly lie in a U-shaped position. Phantom midge larvae are very transparent (hence their name). They have a wide unsegmented thorax like mosquito larvae but have prehensile antennae with pronounced apical spines, with which they capture their prey.

5.2.1 THE CULICINAE

The general appearance of mosquitoes or gnats is familiar to most people. They are small insects (about 3–6 mm long) with long legs, a globular head, laterally compressed thorax and a long cylindrical abdomen. The wings are rather long and narrow and are carried flat over the back in repose. The antennae are somewhat hairy in the female and bushy in the male.

There are, however, certain non-biting midges which answer to this general description and which, like the mosquitoes, have aquatic larvae. Nevertheless, it is possible to recognize the true mosquitoes in both stages, without much difficulty, if they are carefully scrutinized. The characteristic features of the adult mosquitoes are as follows:

(1) The mouthparts form a long, thin, projecting proboscis.
(2) The wings bear tiny scales along the veins as well as a fringe of them along the hind margin. The wing venation, also, is characteristic. There are six longitudinal veins of which the second, fourth and fifth are forked.

The diagnostic features of mosquito larvae are:

(1) There is a well-developed head followed by a swollen, unsegmented thorax.
(2) They breathe atmospheric air through a pair of spiracles at the hind end of the body and almost always spend most of their time at the water surface with the spiracles through the water film.
(3) There are tufts of bristles arising from many of the body segments.

(a) Main types of mosquito

Within the Culicinae there is a group of non-biting mosquitoes, the Megarhinini (or Toxorhynchitini). They are large and brightly coloured, with peculiar recurved proboscides and their larvae are predaceous, and therefore beneficial. Apart from one or two North American species, they are tropical in distribution.

The medically important mosquitoes belong to the tribes Culicini and Anophelini. The medical importance of the latter, as vectors of malaria, is well known (see page 6). Culicines are also vectors of some tropical diseases, such as yellow fever, dengue and filariasis; and they can transmit certain viral infections in the temperate zone (see page 9). Apart from this, they include some of the most seriously annoying biting forms.

Culicines and anophelines can be distinguished in all stages by the following characters, illustrated in Fig. 5.2.

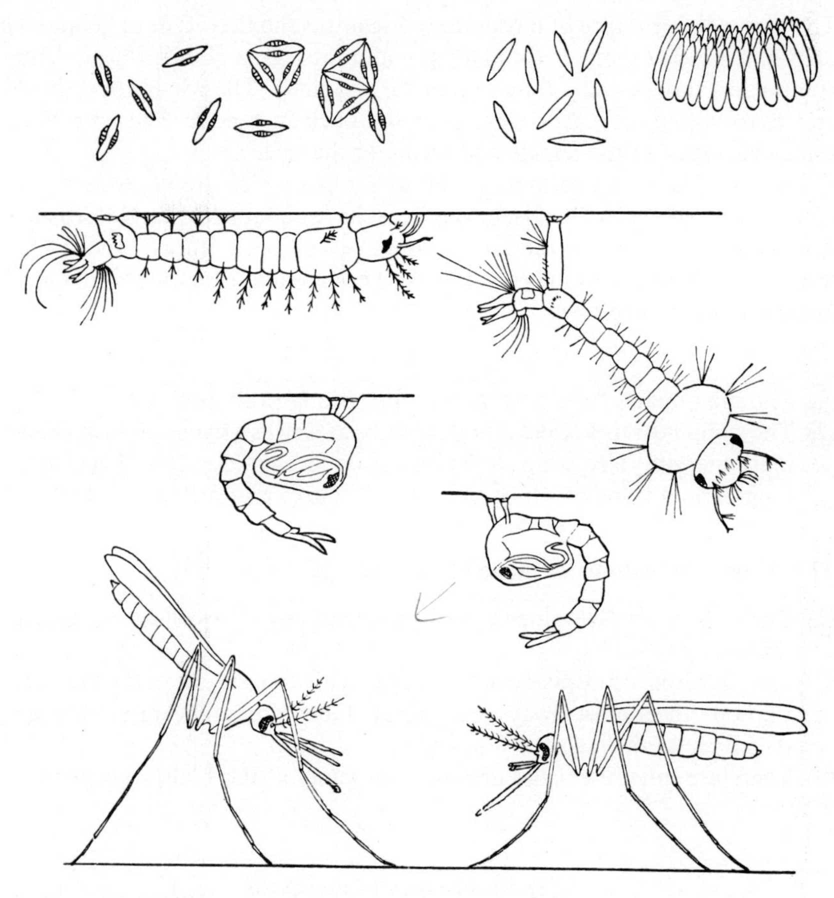

Figure 5.2 Differences between anopheline and culicine mosquitoes in different stages of the life cycle. (After Marshall, [377].)

Stage	*Anophelines*	*Culicines*
Eggs	(1) Provided with floats (2) Laid separately	(1) No air floats (2) Sometimes stacked together in egg rafts
Larvae	(1) Hind spiracles on the body surface (8th abdominal segment)	(1) Hind spiracles at the end of a tail-like tube or siphon (projecting from the 8th abdominal segment)
	(2) Larva held up horizontally beneath water film float hairs, etc.	(2) Larva hangs down at an angle with only the tip of the siphon in the surface film

Pupa	Respiratory trumpets more conical	Respiratory trumpets more cylindrical
Adult	(1) Females with palps as long as proboscis	(1) Females with short palps
	(2) No scales on abdomen	(2) Abdomen covered with scales
	(3) Rest with proboscis and abdomen in a line, at an angle to surface	(3) Rest with proboscis and abdomen forming obtuse angle, the abdomen more or less parallel with surface

It should be noted that, in the anomalous culicine genus *Mansonia*, the respiratory siphon of both the larva and the pupa ends in a toothed projection, which is used to pierce the air-containing roots of aquatic plants (see Fig. 14.1 page 521).

The anopheline group is smaller than the culicine and more uniform, with only three genera and most of the species in the genus *Anopheles*, of which 370 species are included in the book by Horsfall [285]. In contrast, 28 genera of culicines are recognized, though some are small and of restricted distribution. The main genera are *Culex* (463 spp.), *Aedes* (661 spp.), *Mansonia* (54 spp.), *Psorophora* (30 spp.) and *Culiseta* (27 spp.). The genus *Aedes* contains most of the troublesome biters, especially in the sub-genus *Ochlorotatus*.

Keys to some of the more important temperate region species of anophelines and culicines are given in the Appendix (page 522 ff.).

(b) Species complexes among mosquitoes

As might be expected on the basis of continuing evolution, the careful examination of many old 'species' reveals the existence within them of separate sub-species or even cryptic species. The usual reason why such entities had not been previously noted is that their more or less complete genetic isolation had not resulted in obvious morphological differences between them. Therefore, recognition of their specific or sub-specific status has had to wait for ecological observations and genetical experimentation.

Species complexes of this type are not uncommon among mosquitoes and some important ones occur in the temperate region. Among anophelines, there is the *maculipennis* complex, with forms now generally recognized as separate (but closely related) species: *Anopheles maculipennis*, *An. messeae*, *An. atroparvus*, *An. labranchiae*, *An. sacharovi*, *An. melanoon* and *An. sub-alpinus* in Europe (see page 146). Also, *An. quadrimaculatus*, *An. occidentalis*, *An. freeborni*, *An. earlei* and *An. axtecus* in North America (see page 149). Another complex in S.E. U.S.A. is represented by *An. crucians*, *An. bradlei* and *An. georgianus*. Finally, the *Anopheles hyrcanus* complex of the Oriental region just extends to the Eastern end of the Mediterranean with the sub-species (?) *An. hyrcanus pseudopictus*.

Among the culicines, the most important complex in Europe and N. America is that of *Culex pipiens*, *Culex fatigans* (*quinque-fasciatus*) and *Culex molestus* (*autogenicus*). Another group, which is mainly tropical but just extends to the southern temperate region, is *Aedes aegypti*, with variants (? sub-species) *formosus* and *queenslandensis* (see page 152).

Species complexes are by no means merely academic curiosities, since the different habits of the cryptic species may profoundly affect their biting propensities and their vector potential.

5.2.2 LIFE HISTORY OF MOSQUITOES

(a) Oviposition

It will be convenient to begin a description of the life history with the egg-laying. Whilst all mosquitoes breed in water, the different species and varieties have a great number of individual preferences, ranging from small accumulations of water in rot holes in trees to large lakes and from salt marshes to mountain streams. The females seeking sites to lay their eggs display remarkable adaptive instincts in selecting appropriate places. Usually the eggs are laid on the surface of water, either from the edge, from the water surface or hovering above it. But certain species (of *Aedes*) lay eggs in shallow depressions on dry land which subsequently becomes filled with water at the appropriate season.

Eggs of anopheline mosquitoes are always laid separately on the water surface though they may occur there in large numbers. Some kinds tend to group themselves in triangles and other patterns owing to surface forces exerted by different parts of the egg. Certain culicines (genus *Culex*) have the habit of stacking the eggs together as they are laid (each egg being disposed vertically) and they adhere to one another, forming raft-like masses of perhaps several hundred eggs.

(b) Egg

The eggs are spindle-shaped and about $\frac{2}{3}$ mm long. Those of anophelines are canoe-shaped with lateral fan-shaped floats which distinguish them at once from culicine eggs. The egg surface sometimes shows characteristic markings which have proved of great value in identifying certain species of the *Anopheles maculipennis* group which are otherwise difficult to distinguish.

The eggs only hatch in water and those of many species cannot survive drying for more than a few days. Certain desert living species of *Aedes*, however, lay eggs which are said to remain viable for several years without water and to hatch within a day or two of being immersed in it; and British species of *Aedes* can likewise survive 6 months in the egg stage.

The duration of the incubation period in water depends on the species and the temperature, being generally of the order of a few days in the summer.

(c) Larva

The larva has a head, a swollen thorax and a cylindrical abdomen composed of nine distinct segments. At various points there are tufts of bristles which are normally pinnate (or 'feathered') in anophelines and always simple in culicines.

The head of a culicine larva is decidedly wider than that of an anopheline, being nearly as wide as the thorax. Both types have pigmented simple eyes, short antennae and mouthparts with strongly toothed mandibles. On the labium are two pronounced groups of bristles overhanging the mouth like a heavy moustache. These are known as the 'mouth brushes'.

The larvae have two methods of feeding. For most of the time the mouth brushes are actively vibrating and cause a current of water to flow past the mouth. Small organic particles present in the water (protozoa, bacteria, algae, fungal spores, pollen, etc) become entangled in the maxillary bristles. From time to time a bolus of these small particles is passed into the mouth and swallowed. In addition to this method of feeding by filtration, the larvae can use their mandibles to nibble off bits of algae, plants and, occasionally, other larvae. For reasons to be described in a moment, the anopheline larvae feed with the head much closer to the surface film than the culicines.

Both types of larvae breathe atmospheric air from a pair of spiracles on the eighth abdominal segment. These spiracles penetrate the water film while the larvae are at the surface and when they dive they are protected by valve-like cover flaps. There is a great difference in the position of the spiracles in the two types of mosquito. In culicines they are borne at the end of a long tube-like projection called the siphon and the larva hangs obliquely head downwards from this tube with only the tip of it anchored in the surface film. (Except for *Mansonia* spp. which attach by their toothed siphons to the roots of aquatic plants and remain permanently below the surface). In anopheline larvae the spiracles are nearly flush with the surface of the eighth segment and the whole dorsal length of the larva is held close to the surface film by star-shaped float hairs and other organs. The neck is very flexible so that the larva can turn its head right round and feed on organisms in the film directly above it. Thus, the characteristic positions of culicine and anopheline larvae hanging from the surface film are quite different and can be distinguished at once by the naked eye. The anophelines move about at the water surface in a series of jerks, tail first, whereas the movements of culicines at the surface are confined to wriggling. In addition, both forms, when alarmed, will break contact with the surface film and either sink or swim downwards with vigorous wriggling movements. Eventually, however, they swim up to the surface and renew contact with the air.

The last abdominal segment bears two bunches of bristles and four finger-like blood 'gills'. The function of these organs, however, seems to be regulation of osmotic pressure and not respiration.

There are four larval stages separated by four moults. After the last moult the pupal stage begins.

(d) Pupa

The pupa is generally described as comma-shaped. The rather large dot of the comma is made up of the head and thorax while the tail is formed by the flexible abdomen. The rapid movements of this tail, which ends in a pair of paddles, enables the larva to dive with a series of jerky somersaulting movements, when alarmed. Normally the pupa, like the larva, remains at the surface film with its respiratory system in communication with the air through a pair of respiratory trumpets on the back of the thorax. (These 'trumpets' are more or less cylindrical in culicines but distinctly conical, widening apically, in anophelines.)

The fused head–thorax mass displays traces of the wings and long legs of the adult insect which are developing inside. In the centre, between the rudimentary wings, is a large air cavity which gives buoyancy to the pupa and helps it to maintain its position at the surface.

At the end of the pupal period, the newly formed adult swallows some of the air from the central bubble and this enables it to swell and burst the pupal skin. The adult emerges from a dorsal split and rests for a short time (either on the discarded skin or on adjacent vegetation) until the adult cuticle has hardened with the wings expanded.

(e) Adult

The adult has a globular head, a large part of the surface of which is taken up by the compound eyes. The antennae, which are about three times as long as the head, are somewhat hairy in the female and quite bushy in the male; this provides a ready means of distinguishing the sexes with the naked eye. In both sexes the mouthparts are elongated into a proboscis, but those of the male do not include elements capable of piercing skin to suck blood. A pair of palps is present, one on each side of the proboscis. In the female, these organs are slender and smoothly scaled; those of anophelines being about as long as the proboscis while those of culicines are from one-fifth to one-half the length. The palps of males are usually ornamented with tufts of bristles; in all British species except one they are about as long as the proboscis. The palps of male anophelines are usually clubbed at the end while those of culicines are tapering and curl upwards.

Both sexes feed on various juices of fruits and flowers but the females also take meals of blood from vertebrates. The piercing mouthparts of the female are constructed as follows: the labrum forms a long narrow projection with the edges rolled under to form the sucking tube. Underneath runs the long narrow hypopharynx. The mandibles and maxillae are long blade-like stylets curved round this tube and adhere to it partly owing to an oily fluid which binds the parts together. All these elements are carried in the trough of the elongated labium, which is U-shaped in section. The labium does not enter the wound, however, but as the piercing mouthparts penetrate the skin of the host, the labium gradually bends away from them but remains guiding them at the point of insertion 'like a billiard player thrusting with a cue'. Liquids are drawn up the food channel by the pumping action of enlargements in the pharynx. The alimentary canal runs back into the thorax where there are diversions into two dorsal and one large ventral blind sacs (the latter running back into the abdomen). There follows a sphincter valve guarding the entry to the flask-shaped midgut. When blood is taken (by female mosquitoes) it is admitted directly into the midgut; but when fruit juices are swallowed, they are stored temporarily in the blind sacs.

At the end of the midgut are the usual discharge ends of the Malpighian tubes, followed by a hindgut leading to the anus.

The thorax is narrow and deep with a characteristic hump-backed appearance. It is composed of three segments, as usual, but the delimitations between them are hard to make out. The greater part of the dorsal surface is formed by the shield-like scutum of the second segment which bears the wings.

The long thin legs are clothed in scales, sometimes forming light and dark bands of diagnostic value. The foot consists of a pair of minute claws which are simple in some genera and toothed in others.

The stances of the two varieties of mosquitoes are characteristic. Anophelines carry the abdomen in a line with the thorax and proboscis, at an angle with the surface on which they are resting, giving the impression that they are resting on their heads. Culicines, on the other hand, hold their abdomens more or less parallel to the surface; often the thorax is bent in a somewhat hump-backed attitude to achieve this. In both types the hind legs are often held in the air curving gently upwards.

The wings, as already mentioned, are ornamented with scales, not only as a thick posterior fringe, but also along the course of the veins. In action, the wings beat two or three hundred times per second and propel the insect at a rate of about 0·8–4·8 km h^{-1}. (0·5–3 m.p.h.) [321].

The flight range of mosquitoes, like that of other insects, cannot be simply specified. Some species tend to remain close to their breeding sites, while others migrate to considerable distances. Apart from this, there is a tendency (which might be expected) for larger numbers to be found near the source and fewer and fewer at increasing distances. There are published records of female (*Aedes*) mosquitoes being caught as much as 100 miles out at sea; but these are freaks due to unusual conditions. The normal limits of dispersal are most probably indicated by the results of a three year experiment during which 250 000 marked specimens of *Culex tarsalis* were released in California [24]. Most of those recaptured were taken within a radius of two miles. Beyond this, the great majority were caught down wind, within a ten mile radius. It was noted that, at low wind velocities (3·2 km h^{-1}: 2 m.p.h.) dispersal occurred in all directions. There was a slight tendency to fly up wind, presumably following host odours, with a maximum of 4·4 km (2·75 miles) in that direction. At wind speeds of 6·4–9·7 km h^{-1} (4–6 m.p.h.) captures were mainly down wind. Higher wind speeds did not increase the distances of dispersal, because high winds tend to inhibit flying.

From the practical standpoint, the relevant flight range is that of the majority of mosquitoes, rather than the long distance records of a small minority. Generally this is greater in temperate regions than in the tropics, where anti-larval measures in a circle of 0·8 km ($\frac{1}{2}$ mile) radius will usually be sufficient to protect the centre. In the north temperate regions, the corresponding distance is about 3·2–3·6 (2–3 miles) [235].

In addition to migration of mosquitoes by flight, they may be passively transported in vehicles, etc. One important result of this is that the development of regular air transport brings the risk of carrying mosquitoes or associated parasites into areas of the world where they are at present absent, although conditions may be entirely suitable for them to breed.

The abdomen, which is more or less cylindrical, is thickly covered with overlapping scales in culicine mosquitoes but only bears fine hairs in anophelines. There are ten segments, the last two being 'telescoped' inside the eighth. Protruding at the end of the female abdomen are a pair of cerci, which, however, are very small except in the genus *Aedes*. The males bear clasping organs for use in copulation. It is an interesting fact that the tip of the male abdomen rotates through 180° shortly after its emergence and remains upside down throughout its adult life.

Like certain other small flies, most species of mosquitoes perform a nuptial dance before mating. The 'dance' is performed by a number of males (from a few to several thousand, according to species) rising and falling in the air in a stationary swarm. This commonly occurs out of doors in the evening but certain species will dance in quite small cages in a laboratory and even mate without swarming. The females fly into the swarm and copulation takes place in the air; it lasts from a few seconds up to a minute. Females will pair before taking a meal and store the spermatozoa they receive in a special sac ('spermatheca') for subsequent egg-laying. A single impregnation may suffice for many batches of eggs laid over a period of many weeks.

5.2.3 QUANTITATIVE BIONOMICS OF MOSQUITOES [285]

(a) Speed of development

Mosquitoes which can be reared in the laboratory show the usual response of increased speed of development with rising temperature. The following figures are incubation periods of the eggs in days:

Anopheles quadrimaculatus: 1·6 at 28° C; 2·3 at 23° C; 4·5 at 18° C; 15 at 12° C.
Culex pipiens: 1 at 20–25°C; 3 at 18·5°C; 6 at 13°C; 9 at 10·5°C.
Anopheles claviger: 2 at 28° C; 6 at 19·5° C; 4 at 18° C; 10 at 12° C.

Total development times in other investigations were:
Anopheles atroparvus: 12–15 at 34° C; 18·5 at 24–30° C; 24 at 15–18° C.
Culex pipiens: 10 at 25° C; 45 at 15° C; 60 at 10° C.
Anopheles claviger: 4 at 19·5° C; 59 at 12° C; 76 at 10·5° C.

In nature, the relation of life history to temperature is complicated by different ways of passing the winter and by egg laying habits. In temperate climates, many mosquitoes pass the winter in the egg stage; for example, *Aedes cantans*, *A. punctor*, *A. annulipes*, *A. excrucians*, *A. communis*, *A. dianteus*. A few overwinter as larvae, such as *Aedes rusticus*, and still others hibernate as adults, including *Anopheles messeae*, *Culex pipiens* and *Culiseta annulata*.

(b) Population density

The changes in population density at different seasons of the year must vary a great deal in mosquitoes of different breeding habits. Obviously species that select specialized breeding sites such as rot holes in trees cannot have the same potentialities for increasing in numbers as forms which breed in lakes or marshes. In general, there is an enormous increase in adults during the summer months. By releasing marked mosquitoes and estimating the total population from the proportions recaptured, American workers have assessed the populations of adult *An. quadrimaculatus* to be of the order of 10 000 to the acre, in summer months in an area in Tennessee [177].

It is rather difficult to decide what are the effective checks to growth of mosquito populations. The actual area of suitable breeding water of the right pH, saline content, and so forth, may be important; in most ponds, food supply seems unlikely to be a serious limitation since the larvae can feed on so many things. Throughout life, mosquitoes are subject to attacks by different predators. The larvae are eaten by predaceous aquatic insects and by fish. (Small fish of the genus *Gambusia* have been successfully used to control them in suitable places.) The adults are caught and eaten by dragon-flies, wasps and various predaceous Diptera besides birds and bats. In the laboratory, adults may live for several months with regular opportunities for feeding, and egg-laying, but their length of life under natural summer conditions is problematical. In recent years, however, much has been learnt about the age of wild female mosquitoes, from dissections which reveal how many times they have laid egg batches [144]. The cycle of feeding and egg-laying in *Anopheles maculipennis* takes about 4 days in temperate regions around 20° C (68° F). In the vicinity of Moscow, most of the specimens caught in the summer months had completed only one or two cycles and very few had more than five or six egg batches. Apparently, the average age was about a week with very few more than 3-weeks-old.

5.2.4 NOTES ON VARIOUS TYPES OF MOSQUITO

In certain cases, the occurrence of different stages of the mosquito in various months of the year will be indicated by the formula suggested by Marshall [381], thus: E = eggs; L = larva; A = adults; H = hibernating adults. The figures in brackets give the months during which these stages may be found.

(a) Anophelines

(i) European species [60] (See Fig. 14.2, page 524)

Probably the most important group of mosquitoes in Europe are the members of the *maculipennis* complex. Because they are very similar in appearance (and can only be separated with certainty by the pattern on their eggs), they were at one time all regarded as a single species. Subsequently, it was found that distinct forms with more or less complete genetic isolation were present; and they are now generally accorded specific or sub-specific status.

Distribution
An. maculipennis (type form): Europe, except Britain, N. Scandinavia, Greece and Italy. *An. messeae*: as above, but including England and excluding Spain. *An. labranchiae atroparvus*: Europe, from Spain north of the coastal plain, eastward to the Caspian and north to 55° latitude. *An. labranchiae labranchiae*: shores of the Western Mediterranean. *An. sacharovi*: the eastern Mediterranean and Middle East to the Caspian. *An. sub-alpinus*: around the Pyrenees, Alps, Balkan mountains and the Elburz mountains of Iran. *An. melanoon*: Corsica, Albania and southern half of Italy.

Breeding sites
An. atroparvus, An. labranchiae and *An. sacharovi* tend to be coastal in preference, breeding in slightly brackish water, though they can develop satisfactorily in fresh water. The remaining species are confined to fresh water and occur inland. *An. maculipennis* and *An. messeae* larvae are found in sunlit ponds and streams, especially with aquatic vegetation. *An. sub-alpinus* is found in pools in plains below mountains. *An. melanoon* breeds in marshes and rice fields.

Feeding and resting habits
All the species prefer to feed on large mammals, including domesticated animals, and to some extent on man. The frequency with which man is attacked depends partly on opportunity; but there also seem to be some differences in preference. Thus, *An. maculipennis* and *An. melanoon* very rarely take human blood, while *An. messeae* and *An. sub-alpinus* will do so occasionally. The other species too, prefer larger animals, especially in the summer months. However, in winter, *An. atroparvus*, *An. labranchiae* and *An. sacharovi* commonly rest inside inhabited shelters, but continue to feed at intervals. If they rest in barns, stables or pig-sties, animal blood will be taken, but in human dwellings, the inhabitants will be attacked. The likelihood of feeding on man, therefore, depends on availability, so that numerous animal shelters may divert mosquitoes away from entering houses. But the type of shelter is also important, since these mosquitoes prefer rather dingy undisturbed corners and are often found resting on cobwebs. Accordingly, well-lit, clean modern houses are not very attractive to them. The

life cycle of this group, in Europe, can be roughly expressed as: E(4–9), L(4–9), A(1–12).

The other members of the maculipennis complex (*An. maculipennis*, *An. messeae*, *An. melanoon*, *An. sub-alpinus*) hibernate in the adult stage. In autumn,

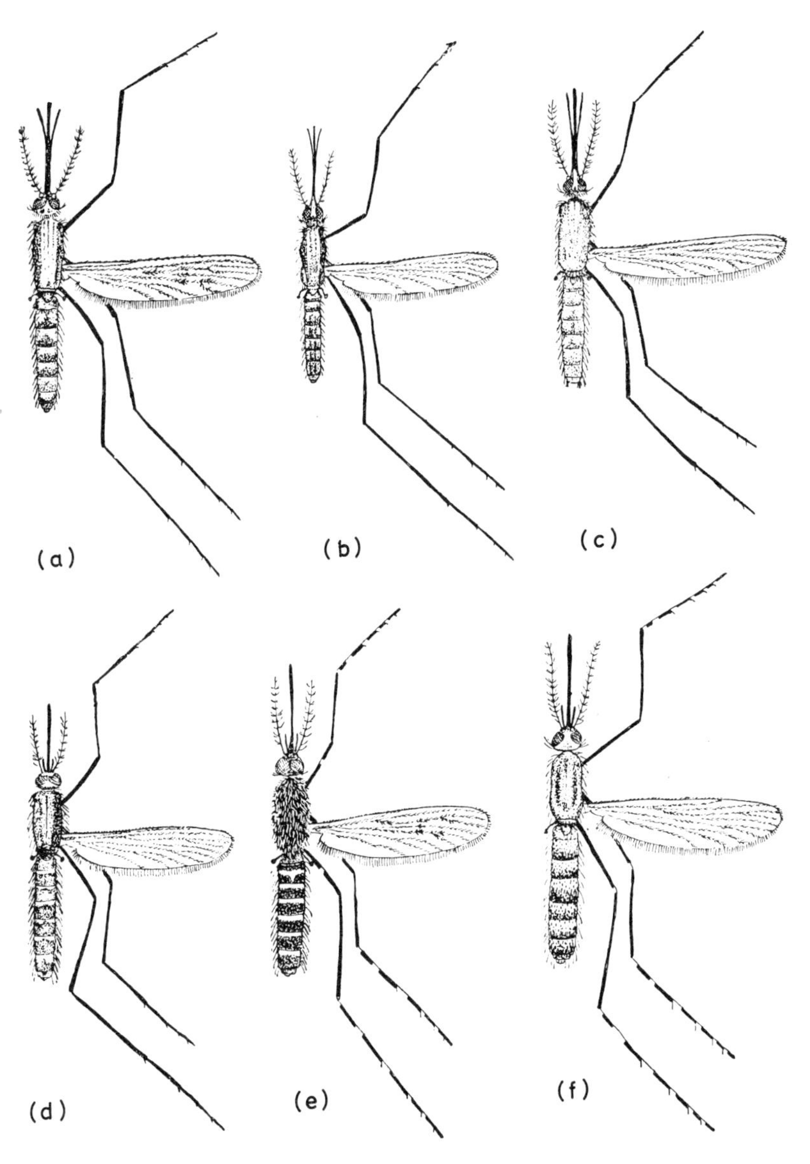

Figure 5.3 Some common British mosquitoes. (*a*) *Anopheles maculipennis*; (*b*) *Anopheles plumbeus*; (*c*) *Anopheles claviger*; (*d*) *Culex pipiens*; (*e*) *Culiseta annulata*; (*f*) *Mansonia richiardii*. (After Edwards *et al.* [162].) Approx. × 4½.

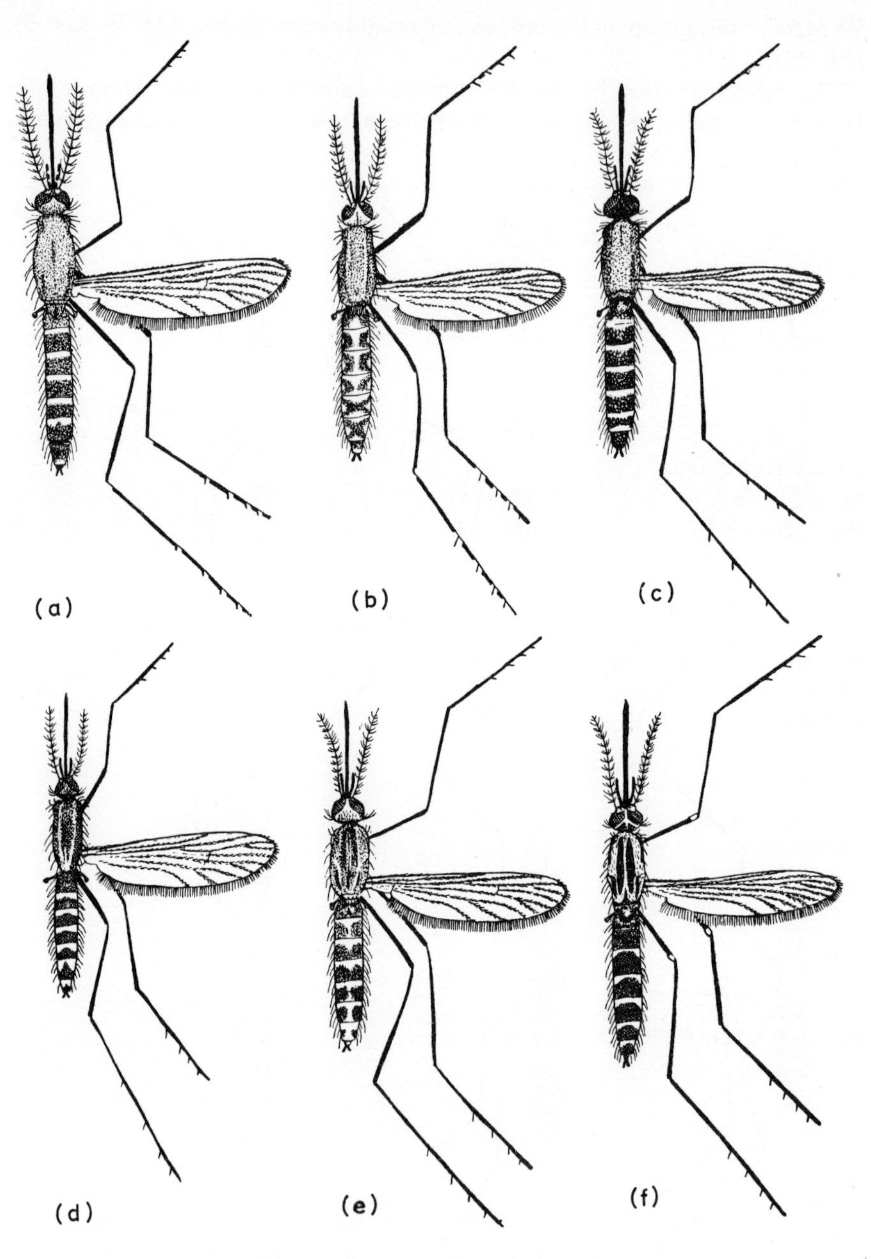

Figure 5.4 Further common British mosquitoes. (*a*) *Aedes cantans*; (*b*) *Aedes caspius*; (*c*) *Aedes detritus*; (*d*) *Aedes punctor*; (*e*) *Aedes rusticus*; (*f*) *Aedes geniculatas*. (After Edwards *et al.* 162.) Approx. × 4½.

they begin to lay down reserves in their fat bodies and cease to feed or lay eggs. During the winter, they rest immobile in uninhabited shelters, such as caves and empty buildings. Their life cycle can be expressed as: E(4–9), L(4–9), A(6–9), H(9–4).

It will be appreciated that the combination of readiness to feed on man with continued winter feeding accounted for the fact that the main malaria vectors in this group in the past were *An. atroparvus*, *An. labranchiae* and *An. sacharovi*.

An. plumbeus (Fig. 5.3) is widely distributed in Europe, south to Sicily and east to the Caucasus. It commonly breeds in the water in rot holes in trees; but sometimes in vases, flower pots and similar containers. The larvae are resistant to cold and can overwinter in this stage. The adult feeds both day and night on man or other animals near its breeding sites. People are generally attacked in parks and woodland country.

An. claviger (Fig. 5.3) also has a very wide range in Europe, down to north Africa and extending east through southern Russia. In the cooler part of this range, it will breed in stream margins, ditches and pools; but in the warmer regions (e.g. Near East) it is largely confined to underground cisterns and wells. It commonly feeds out of doors, but in England it will sometimes enter houses to bite in the spring. The life cycle in Europe can be: E(5–9), L(6–4), A(5–10).

An. algeriensis occurs over much of the range of *An. claviger*. In England, it is a marsh breeder, while in Palestine, it breeds in rivers and pools overgrown with reeds and oleander. It commonly feeds on wild animals, but sometimes on man and occasionally enters houses.

(ii) North American Anophelines [60] (See Fig. 14.3, page 525)

The *maculipennis* complex in America is represented mainly by *An. quadrimaculatus* (which is widely distributed east of a line from southern Ontario to eastern Texas) and *An. freeborni* (which occurs mainly west of the Rocky Mountains). These two were formerly the principle malaria vectors in the U.S.A. and were known as the common malaria mosquito and the Western malaria mosquito. In addition, there are: *An. earlei*, which occurs in southern Canada and northern U.S.A., *An. occidentalis*, restricted to a narrow coastal strip of California and British Columbia, and *An. aztecus* which occurs in the interior of Mexico.

Since *An. quadrimaculatus* occupies such a large area, its biology and habits vary considerably in different places. At the level of Illinois, there are 2–3 generations a year, whereas there may be 8 or more in Georgia. In the south, adults are common for 8 months of the year, but only for 4 months in the region between 10 and 14° C isotherms.

This mosquito breeds in a fairly large variety of fresh water habitats, but its preference is for large bodies of impounded water in which there is floating vegetable matter (green algae or vegetable debris). In the hot summers of the south, intermittent plant growth providing partial shade is favourable.

Adults will feed on man and most of his domesticated animals. As with the European *maculipennis* group, the choice seems to depend partly on opportunity, but with a slight preference for the larger animals. They will readily enter buildings, both houses and animal shelters and also find transient resting sites in privies, empty barrels, hollow trees and culverts. Some hibernation occurs in winter but there are records of continuous feeding at temperatures above 18° C.

An. freeborni, which very closely resembles *An. maculipennis* in appearance, prefers to breed in shallow sunlit seepage water with emergent vegetation. Common sites are irrigated meadows and rice fields. Adults feed on any available mammal, including man, but domesticated animals are preferred. They often enter buildings, particularly occupied animal quarters or human dwellings. In winter, they rest in natural or artificial shelters and cease to lay eggs. They do not hibernate completely, however, but move about to some extent and take small blood meals.

In the summer, like *An. quadrimaculatus*, they do not disperse more than a few kilometres from their breeding sites. In the autumn, however, and again in the spring, they tend to migrate long distances, perhaps 20 km. Possibly this is an adaption to the nature of much of their terrain, which is arid, so that breeding sites dry up seasonally. In the height of summer, adult life is curtailed by low humidity and there are peaks of abundance in spring and autumn.

An. punctipennis has the greatest distribution of any north American anopheline, ranging from southern Canada to Mexico and from coast to coast, except for the western great plains and the Rocky Mountains. It is a polytypic species and would probably be found to comprise a complex of species or subspecies, if more extensively studied. Its breeding preferences are the shallow borders of slow flowing sandy streams but deeper pools and faster streams are colonized if they have weedy borders. Adults feed on man and his domesticated animals. They seldom enter bedrooms to feed, but often bite people on porches in the evening and even in shade in the afternoon. In the north, these mosquitoes usually hibernate in unheated shelters (hollow trees, culverts, outhouses). In the south, they may continue to feed throughout the winter but the larvae also persist and are important for survival.

An. crucians is a south-eastern species, ranging from Massachusetts round to Texas and New Mexico and down the east coast of Central America. The preferred breeding sites are shallow pools and swamps, with abundant decaying vegetation, providing acid conditions. All stages are present during the winter, though breeding is retarded. The adult is common in the vicinity of buildings and sometimes enters them. The females will feed on man as well as domesticated animals. Nevertheless, this mosquito was never an important malaria vector, apparently because it is refractory to the parasite.

Adults of *An. bradlei* and *An. georgianus* are indistinguishable from *An. crucians*, so that their specific habits are not clearly known. The larvae of the former breed in brackish water and the latter in small shallow seepages on boggy hillsides.

An. barberi occurs over the forested part of the U.S.A. from Texas and Kansas to the Atlantic and northwards to New York. It breeds in rot holes in trees and sometimes in artificial containers. Perhaps because of its restricted breeding sites, it is never very abundant. Houses near the breeding sites may be entered and the females will feed on man.

(b) Culicines [285, 103]

The foregoing account of anophelines of the north temperate region has emphasized those with the habit of repeatedly biting man indoors. This is because that habit is important for malaria transmission, which is the most serious aspect of anophelines, since their contribution to biting nuisance and to virus transmission is considerably less than that of culicines in this zone.

With culicines, however, rural outdoor biting can be excessively unpleasant and since, many encephalitides are transmitted from wild birds, *repeated* feeding on man is unimportant. Accordingly, the habits and biology of rural species will be considered briefly as well as those of domestic and peri-domestic forms.

(i) Domestic and peri-domestic culicines

Probably the most important mosquitoes of this group in the northern temperate region are the members of the *Culex pipiens* complex. These are very widely distributed round the world, between latitude 60° N and 40° S. The group includes three fairly distinct types, generally regarded as sub-species: *Culex pipiens pipiens*, *C. p. molestus* and *C. p. fatigans* (*C. p. quinquefasciatus* of American authors). These show slight morphological differences as well as fairly distinct biological ones. In addition, there are other indeterminate races with various degrees of genetical isolation.

C. p. pipiens (Fig. 5.3) is holarctic in distribution, above 30° N. In North America, it extends up to 55° N and somewhat higher in the Old World (even up the coast of Norway to the Arctic Circle). *C. p. fatigans* encircles the world in the tropics and the warmer temperate regions. In parts of the world where the distributions of *pipiens* and *fatigans* meet, intermediate forms exist (*C. p. pallens* in the Far East and North America). These may have originated as hybrids, but there is some reason to believe that *C. p. pallens* at least can form stable populations. *C. p. molestus* has an erratic distribution, commonly associated with urban conditions, in Europe, North America and northern Asia. It is the universal form around the Mediterranean, especially on its southern and eastern shores.

All members of the complex are prone to breed in small collections of water, often in artificial containers around human dwellings; furthermore, the adults commonly enter houses. Biological differences between them are, however, of considerable importance. *C. p. pipiens* feeds almost exclusively on birds. It breeds in water open to sunlight and the females require a blood meal before laying viable eggs. *C. p. molestus* seems to be more closely connected with man and is often present in an urban environment. It breeds in drains and other water sources near houses, including underground sumps. It will readily feed on man, even in winter, as it does not hibernate. The females will lay fertile eggs without taking a blood meal. *C. p. fatigans* resembles *molestus* in its close association with man, breeding near dwellings and readily feeding on him; also it does not hibernate. It cannot, however, reproduce without a blood meal.

The foregoing habits and distribution account for the different importance of the three sub-species. *C. p. molestus* is troublesome for its biting, especially because of breeding near houses. *C. p. fatigans* is a tropical vector of urban filariasis, a problem of considerable and growing seriousness. *C. p. pipiens* is of no public health importance but its habit of hibernating indoors causes it to be accused of the bites of other species. Life histories (Britain): *C. p. pipiens*, E(4–10), L(4–11), A(5–9), H(9–4). *C. p. molestus*, E(1–12), L(1–12), A(1–12).

Aedes aegypti, the yellow fever mosquito, is an even more important insect, with a complex of the following forms: *Aedes aegypti* (type form), intermediate in colouration, common in Asia, the Pacific region and parts of America; *A. a.* var. *queenslandensis*, formerly common round the Mediterranean and southern states of the U.S.A.; and *A. a. formosus*, a very dark sub-species, restricted to the interior of Africa south of the Sahara. It happens that the degree of closeness of association with man parallels the degree of coloration, *formosus* being rather 'wild' and *queenslandensis* being most 'domestic'. In the tropics, *Aedes aegypti* transmits dengue, including the haemorrhagic form, and is a potential vector of urban yellow fever. In recent years, however, it has been largely eradicated from the northern temperate region.

(ii) Semi-domestic culicines

Certain species of mosquitoes are liable to enter houses, either to feed or to hibernate (or both) but are not so closely associated with man or his domesticated animals as the foregoing types. Several of them are prone to breed in polluted water and hence in drains, sumps, and effluents of cess pools or stables. *Culiseta* spp. larvae develop in cold water and breed in summer in northern countries, and in spring and autumn in mid-temperate regions, and in winter in the warm temperate zone. *Culiseta annulata* (Fig. 5.3) occurs over most of northern Europe and as far south as North Africa. In Britain, it is a sporadic pest in different parts of the country. In summer, it bites fiercely out of doors, in gardens and parks. In winter, it will attack man in mild weather and not infrequently enters houses to do so. Though it will breed in a considerable variety of places, it prefers stagnant and often polluted water. Occurrence (Britain): E(4–10), L(5–10), A(4–11), H(11–3).

C. inornata occurs from Canada, all over the U.S.A. to Mexico but most commonly in the West. As with other species of the genus, it frequently breeds in foul water. The females are found in winter in basements, cellars and various natural resting sites; and on mild days in early spring, vacate these places to bite, even when there is snow on the ground. *C. incidens*, with somewhat similar habits, occurs over the Western U.S.A. Neither of these species bite man readily, but they are very annoying to livestock.

Mansonia richardii (Fig. 5.3) and *M. perturbans* are either closely related or identical species. The former occurs in Europe (from Britain and Sweden to the Balkans) and the latter in North America (from Canada to the Gulf of Mexico).

Mosquitoes of this genus have an unusual life cycle. The eggs are laid underneath the leaves of aquatic plants. The larvae resemble normal mosquito larvae, except that the breathing siphon ends in a kind of tooth, which is inserted into the roots of aquatic plants and draws air from them. The plants commonly used in Britain are *Ranunculus, Acorus, Glyceria* and *Typha*. In the U.S.A., *Typha, Limnobium, Pistia, Nymphea, Pontedaria* and *Piaropus* are used. The pupae have the same habit, so that neither stage needs to rise to the surface to breathe. The mature pupae detach themselves from the plants, rise to the surface, and normal adults emerge. The adults are troublesome biters, both out of doors and in houses. Their greatest activity is at dusk and in the early part of the night.

Apart from the aforementioned mosquitoes, certain other species occasionally enter houses, either to bite or to hibernate. In Europe, these include the following: *Aedes annulipes, Ae. punctor, Ae. geniculatus, Ae. crucians, Ae. dorsalis*, and *Ae. sticticus*. In North America, they include the following: *Culex tarsalis, C. nigripalpus, C. salinarius*, and (rarely) *C. restuans*; also *Psorophora ciliata*. Most of these species will be noted later as more commonly associated with particular breeding sites.

(iii) Rural culicines [430] (Fig. 5.4)

Nuisance mosquitoes can be grouped according to their breeding preferences and, although these are not always sharply differentiated, the grouping may be convenient for seeking the probable source of a nuisance. The same general categories occur in Europe and North America.

(iv) Salt marsh culicines

Two common species tend to breed in brackish water round the coasts of Europe: *Aedes detritus* and *Ae. caspius*. Of these, the former is the more closely adapted to salty water; *Ae. caspius* will breed in fresh water in unshaded pools. *Ae. detritus* has only one generation per year in the north, but several in the south; but *Ae. caspius* has several summer broods, even in Britain. Both species can cause biting nuisances at 5–10 km from their breeding sites.

The main salt marsh mosquitoes in North America are *Ae. taeniorhynchus*, which breeds along the Atlantic and Pacific coasts, *Ae. sollicitans*, which occurs all along the eastern seaboard, and *Ae. squamiger*, which is found along the coasts of California and Mexico. *Ae. taeniorhynchus* is particularly prevalent in the south-eastern states, especially after a high tide and heavy rains. *Ae. sollicitans*, though mainly a salt marsh breeder, occurs in some brackish swamps of the eastern states. Both species, but especially *Ae. sollicitans*, migrate extensively and can cause biting nuisances many miles (30–50 km) from their breeding sites. *Ae. squamiger* also migrates to some extent in spring and invades sub-urban areas of San Francisco.

Aedes dorsalis and *Ae. melanimon* form a link between the characteristic salt marsh species and the freshwater forms. The former has a very wide distribution throughout the northern hemisphere. It breeds in a variety of unshaded sites, ranging from coastal marshes to mountain snow pools. It is an important pest of man and animals in the Western states of U.S.A. and Canada. *Ae. melanimon*, which resembles *Ae. dorsalis* and was formerly confused with it, breeds in brackish water and irrigated fields in California.

(v) Flooded meadow culicines

There is a group of species which prefers to breed in shallow fresh, unshaded water, especially flooded meadows or rice fields. Several of them are widely distributed in Europe as well as North America. They include the following: *Ae. flavescens, Ae. sticticus*, and *Ae. vexans*. In addition, there are some important species restricted to North America: *Ae. nigromaculis, Psorophora ciliata, Ps. confinnis* and *Ps. discolor*. Of these, *Ae. vexans* and *Ae. sticticus* occur over a wide band of the northern countries and are troublesome biters. *Ae. sticticus* can migrate as much as 16 km from its breeding sites. *Ae. flavescens* occurs in a more northern belt and is less troublesome.

Ae. nigromaculis is distributed over the western half of North America, from Manitoba and Washington to Texas and Mexico. It is one of the most vicious biting forms on the plains of the Middle West and in California. The species of *Psorophora* are commonly associated with rice fields, which provide excellent breeding places. *Ps. confinnis* has a scattered distribution over the U.S.A. from coast to coast. The other two species have a more eastern range. The adults oviposit on drained rice or alfalfa fields and the eggs hatch on flooding. The larvae tend to be predaceous. The adults, which are troublesome biters, will migrate 16 km from the breeding grounds.

In the genus *Culex*, there is *C. tarsalis* which breeds profusely in irrigation water, seeps and marshes. However, it is fairly catholic in taste and will also breed in ditches, rain barrels and foul seepages from slaughter houses or stables. It feeds extensively on birds and also large farm animals. An analysis of several surveys of blood meals showed 55 per cent on birds, 31 per cent on cattle, 8 per cent on horses and only 1 per cent on man. It is, however, an important vector of western equine encephalitis.

(vi) Woodland culicines

Among the species of *Aedes* there is a range from those which will breed both in open and partially shaded pools and on the margins of woods. For example, *Ae. cinereus, Ae. annulipes, Ae. excrucians, Ae. intrudens*, and *Ae. fitchii*. Others normally chose shaded pools and ditches, such as *Ae. punctor, Ae. stimulans* and *Ae. rusticus*. Finally, there are some, as *Ae. cantans*, which prefer shaded woodland pools almost exclusively. (Of these, *Ae. punctor* and *Ae. intrudens* are holarctic; *Ae. annulipes, Ae. rusticus* and *Ae. cantans* are European; *Ae. stimulans* and *Ae. fitchii* are North American).

Most of the northern species have one generation a year and pass the winter in the egg stage (or, rarely, as larvae: e.g. *Ae. rusticus*). Generally, the adults do not travel far from their breeding sites to bite, although a few (like *Ae. excrucians*) disperse out into open country.

(vii) Arctic culicines

It is surprising to learn that some of the most intense biting plagues of mosquitoes occur in the arctic region. Accounts of the suffering of men and animals (both travellers and the inhabitants) are well documented by Leif Natvig in his book on the Culicini of Denmark and Scandanavia [445]. The species concerned overwinter as eggs and hatch out when the long northern summer days melt the snow. Most of the species are holarctic, being equally troublesome in Finland, Siberia and Alaska. They include: *Ae. dianteus, Ae. impiger, Ae. nigripes, Ae. communis* and *Culiseta alaskaensis*. It is presumed that these mosquitoes normally rely on reindeer for blood meals, with possible supplements from small mammals like lemmings, and on birds. It is also possible that some lay autogenous eggs, fertile without a blood meal.

(viii) Rot hole breeding culicines

A rather specialized type of woodland mosquito is that which breeds in rot holes in trees. *Anopheles plumbeus* has been mentioned among the anophelines; and this habitat may have been the original one of *Aedes aegypti*. Other culicines with this habit are found in North America, including: *Aedes triseriatus* (Eastern states, west to Montana and south to Texas), *Ae. varipalpus* (Pacific coast from British Columbia down to southern California). In Europe, there is *Ae. geniculatus* (Southern Norway down to North Africa). Such mosquitoes are generally only a nuisance in the vicinity of their breeding sites and bite out of doors; *Ae. geniculatus* and *Ae. triseriatus*, however, will occasionally enter houses to bite.

5.2.5 IMPORTANCE OF MOSQUITOES

As pointed out in Chapter 1, mosquitoes are outstandingly important vectors of diseases, including malaria, various virus infections and filariasis. Most of these diseases occur in tropical countries. Filariasis does not extend to the northern region. Malaria, though once prevalent in parts of this zone, has been eradicated from it. The vectors remain, however, and are potentially able to transmit it from infected people coming from the tropics. (Nearly 5000 such cases entered Europe during 1971 [75]). Of the virus diseases, certain encephalitides occur in the northern temperate zone. The natural reservoirs are wild and domestic birds and some small mammals; while the vectors are various mosquitoes.

Eastern equine encephalitis (E.E.E.) occurs in the eastern states of U.S.A. and down to South America, mainly in freshwater swamp areas. It is transmitted by *Culiseta* sp. and *Aedes sollicitans*. Though many sub-clinical cases occur, it can be serious and, occasionally, fatal. Western equine encephalitis (W.E.E.) occurs in the western side of U.S.A. where it is transmitted by *Culex tarsalis* and *C. p. fatigans*. It is less severe than E.E.E. and overt cases are mainly among children. St Louis equine encephalitis has been reported from 14 of the U.S.A. and many outbreaks are in urban areas. Common vectors are, again, *C. tarsalis* and *C. fatigans*, though in Florida, *Ae. nigripalpus* and *Ae. excrucians* are involved. There appears to be a high proportion of inapparent infections, as shown by the presence of 50 per cent antibodies in half the negroes in Florida. Children and invalides are most seriously affected. More recently, Venezuelan equine encephalitis has invaded the U.S.A. In Texas, in 1971, there were nearly 2000 sick horses with this disease, but fewer than 100 human cases and no deaths. In a vigorous suppression campaign, 1·3 million horses were vaccinated and 8 million acres sprayed with insecticide from the air, at a cost of $30 million. This compares with only $35 million spent yearly on all organized mosquito control campaigns in the U.S.A. [158]. Japanese B encephalitis, which occurs throughout the Western Pacific region, exists in an unknown reservoir, but is transmitted by *Culex tritaeniorhynchus*, which breeds in rice fields. Though the disease extends northwards into Japan and Korea, incidence in those countries has declined with extensive vaccination and decline of the vector in insecticide-treated rice fields (though it is now showing signs of resistance).

It will be appreciated that the threat of disease transmission by mosquitoes in the northern temperate region is not very great. Most of the impetus behind control measures comes from the nuisance of their bites. In some places this can be very severe and accounts of one or two problem areas will be given later.

5.2.6 CONTROL OF MOSQUITOES

The various methods and materials available for mosquito control have been described in earlier chapters. The choice of suitable measures will depend on the stage of the life cycle to be attacked which in turn will depend on other circumstances.

(a) Impact of insecticide resistance

The growth of insecticide resistance has impeded the control of mosquitoes to a serious extent. The first chemicals to be affected were the organochlorine groups (DDT and the HCH/dieldrin group) both of which were inexpensive as well as being effective. While either of these groups remained effective, control was still cheap and simple but, as double resistance began to effect one species after another, more expensive and less persistent alternatives had to be introduced (or, alternatively, large scale source-reduction operations undertaken). It is obvious that the expense and difficulty of implementing such changes was most serious for impoverished tropical countries, where also mosquito-borne disease was rife; but control of nuisance mosquitoes in northern temperate countries has also been hampered.

Because resistance is dynamic and shows changes from year to year, not a great deal of weight can be placed on a statement of the present position (as listed in a 1976 W.H.O. report [639]); but it can unfortunately be said that this will indicate the minimum extent of future trouble. Thus, currently, resistance in European anophelines has been confirmed in most of the *maculipennis* group, as follows. **DDT**: *An. labranchiae* (N. Africa), *An. messeae* (Balkans), *An. maculipennis* (Turkey). **HCH/dieldrin**: *An. atroparvus* and *An. messeae* (Balkans), *An. labranchiae* (N. Africa), *An. sacharovi* (Greece, Turkey) **OP/carbamate**: *An. sacharovi* (Turkey). In the U.S.A., *An. quadrimaculatus* is resistant to both DDT and the HCH/dieldrin group, while *An. crucians* is resistant to HCH/dieldrin only.

Among culicine mosquitoes, the members of the widely dispersed *Culex pipiens* group have developed resistance to various groups. Thus, *C. p. pipiens* (including probably *molestus*) is resistant to all organochlorines throughout the warmer north temperate region, and also to organo-phosphorus compounds in France, Egypt and Israel. *C. p. fatigans* is resistant to organochlorines everywhere and in many places to organophosphates too. In the U.S.A., the following species have become resistant to all groups: *Culex tarsalis, C. peus, Aedes nigromaculis, Ae. melanimon.* Those resistant to both organochlorines: *C. salinarius, C. restuans, Psorophora confinnis, Culiseta inornata* (and *Ae. cantator* in Canada). To DDT only, *Ae. sierrensis* and *Ae. triseriatus.*

(b) Control of disease vectors [633]

So far as malaria is concerned, a large proportion of serious vectors bite humans indoors. Therefore, the application of DDT and other residual insecticides inside dwellings has been a most successful control method. The remarkable achievements of this method in tropical countries in the 25 years following the Second

World War are well known, although they have been somewhat eroded by the emergence on insecticide resistance in many vectors. This is of less importance in the north temperate region, however, since malaria transmission has been declining in the area and the use of the residual insecticides completed the eradication. While the health authorities maintain vigilance to prevent its reintroduction, it is unlikely that extensive insecticidal control of anophelines will ever be necessary.

The occasional outbreaks of encephalitis in North America and the Far East represent a possible danger. Generally speaking, however, the incidence of human cases is of a relatively low order (a few hundred benign cases a year in the U.S.A.). But, occasionally, alarming epidemics may occur (as that of V.E.E. mentioned in the previous section). In such cases, the use of insecticidal fogs (U.L.V.) applied from the air or the ground, might be appropriate.

(c) Control of nuisance mosquitoes

In a substantial part of Europe and North America, the agricultural development of land does not encourage proliferation of biting mosquitoes in serious numbers. In the extreme north, the uncultivated forests and tundra do produce vast hordes of mosquitoes (and other biting Diptera) in the summer months, as already mentioned (page 155). However, the area is sparsely populated and the only sufferers are the inhabitants, a few travellers and the staffs of meteorological or radar stations. Occasionally, aerial applications of larvicides are made in early summer; otherwise, these people have to rely on repellents and protective clothing. Further south, there are other notorious mosquito breeding sites; for example, the coastal salt marshes and, in some places, irrigated lands and rice growing areas. As examples of the type of problems and the responses, one or two such regions will be described.

(i) Mosquito nuisances in Britain [427]

Although over 30 species of mosquito are indigenous to Britain, only a few are significantly annoying to man; and these are not troublesome every year. They may be grouped as follows.

(1) Species which breed in stagnant brackish water in low lying coastal regions, especially along the south east coast from Suffolk to Dorset. Only two species are involved: *Aedes detritus* and *Ae. caspius*. Both have multiple broods from March to September and their eggs will hatch even years after being laid.

(2) Species which breed in untreated and partly treated sewage, in tanks and other containers on allotments and in gardens; notably *Culiseta annulata*.

(3) Species which breed in water in open country, lakes, ponds and flooded meadow land: *Aedes vexans, Anopheles atroparvus* and *Mansonia richardii*.

(4) Sylvan or woodland species: *Aedes cantans, Ae. punctor, Ae. rusticus* and *Ae. annulipes*.

(5) Species which breed in collections of water below ground level e.g. flooded cellars, inverts of underground railways, storage tanks etc: *Culex p. molestus.*

Remedial treatments are practised sporadically by the local Health Departments, using larvicides (e.g. malathion). But even these limited operations sometimes encounter opposition from people worried about environmental pollution.

(ii) Mosquito nuisance on French Mediterranean coast

The French Riviera, well known to holiday makers, occupies the eastern half of the French Mediterranean coast. The western half has, until recently, remained undeveloped, mainly because of the intense nuisance of biting mosquitoes. Whereas the eastern coast abuts on the *Alpes Maritime*, the western part is flat and beset with brackish swamps and lakes, which provide ideal breeding grounds for nuisance mosquitoes.

The flora and fauna of the region have been fairly carefully studied in recent decades. Three anopheline species are found there: *An. l. atroparvus, An. melanoon sub-alpinus* and *An. hyrcanus*. However, the main biting trouble is due to *Culex molestus* in urban areas and a number of rural species, especially *Ae. caspius* and *Ae. detritus*. (Also, *Culex modestus, Mansonia richardii* and *Culiseta* spp. [495]).

While the deserted stretches of the Camargue, with its half-wild horses and bulls, has some romantic appeal, the French local authorities are much more interested in making the region fit for tourists and, in the neighbourhood of Marseilles, for industrial development. Pressure for amelioration began in the 1950s and, in 1958, there was formed the *Entente Interdépartmentale pour la Demoustication du Languedoc-Roussillon*. A few years later, this was extended to cover additional districts to the Pyrenees. The activities of this organization embraced both attacks on the urban and rural pests [172].

Methods used in the rural areas

It was evident from the outset that the main attack should be on the breeding sites, since these occupied only about 100 km^2, whereas the area invaded by the adults was about 3000 km^2 (39 and 1160 $miles^2$, respectively). In the early years, DDT and HCH were used as larvicides; but later, mainly organophosphates, especially fenitrothion (at 500 g ha^{-1}) and temephos (at 100 g ha^{-1}). These were applied to large sites from vehicles; either small lorries ('Unimog') or little mobile trucks with caterpillar tracks ('Cushman'). For difficult sites, applications from knapsack sprayers were made on foot. In breeding sites with slow flowing water, larvicides were applied from stationary drums, which dispensed the liquid drop by drop. Occasionally, where the larvicides were ineffective (e.g. because of difficulty of access) fogging treatments against the adults were made by TIFA machines, using naled (at 70 g ha^{-1}), bioresmethrin (at 3–4 g ha^{-1}) or malathion (at 200 g ha^{-1}).

To supplement the insecticides, physical measures of source reduction were undertaken, including drainage and damming, according to site conditions. In addition, vegetation providing breeding opportunities for *Mansonia* was extirpated.

Methods in urban areas
The breeding sites were mostly treated manually, using chlorpyriphos in granular form, at doses of 0·02–0·1 p.p.m. Once again, when larvicidal treatments had not been effective, fogging treatments against adults was done, either with TIFA, or more often with pulse-jet machines carried on foot.

Physical measures consisted in reducing, as far as possible, suitable breeding places.

Research
Because of the problem of resistance in *Culex p. molestus* and the possibility that it could occur in other species, investigations of other measures of control have been studied. These include use of *Gambusia* predators and trials of Insect Development Inhibitors. So far, however, novel methods have not been used extensively in the field.

Cost
The 1978 budget amounted to some 12 million francs, partly paid by the central government and partly raised by local taxes. Although this seems a large sum, it amounts to only about 0·04 per cent of the money spent by the two million tourists visiting the area each year, which provides 35–40 per cent of the regional revenue.

Results
After some preliminary difficulties, the campaign soon (by 1966) achieved good control of the salt marsh breeders. However, the sites remain and require constant attention. The urban problem of *Culex p. molestus* is especially worrying, since breeding sites multiply with extensive urbanization and control is handicapped by insecticide resistance.

(iii) Nuisance mosquitoes in California [158, 202]

California is a large state 157 000 miles^2 (408 000 km^2); with extremes of climate, ranging from hot deserts to snowy mountains. Since 1960, immigration has been so great that it has been suggested that the continent may be tilting to the west! The population is now about 20 million. Land pressure has encouraged extensive irrigation of the arid regions, so that the state now contains 25 per cent of all irrigated land in the U.S.A. This has encouraged the breeding of certain mosquitoes, especially *Aedes nigromaculis* and *Psorophora confinnis*. Even more extensive nuisances are caused by *Culex tarsalis* which breeds in a wide variety of habitats, covering some 20 000 miles² (51 200 km²). These range from cool clean water in the hills to heavily polluted seepages from slaughter houses. At the same time, extensive urbanization has encouraged the breeding of *Culex p. fatigans* (*C. p. quinquefasciatus* of American authors) in drains and cess pits; and, with *C. tarsalis*, in the run-off from vegetable and fruit washing and dairy farms. In addition, brackish swamps in coastal areas favour the breeding of *Aedes squamiger, Ae. dorsalis, Ae. taenirrhynchus* and *Culiseta inornata*. Some of the species (*Ae. squamiger, Cu. inornata*) tend to be prevalent in the mild winters; whereas the irrigation breeders have two peaks of prevalence, in spring and autumn.

The state of California has a long history of active control of mosquitoes. As long ago as 1909, the eminent American entomologist W. B. Herms, together with H. F. Gray, an engineer, campaigned for official action against malarial mosquitoes. But it was not until 1914 that pressure from land development agencies, handicapped by the biting nuisance, induced the state to enact legislation for mosquito control. Mosquito Abatement Districts were designated, begining with two in 1915 and increasing to over 60 now. They are responsible for control of mosquitoes and certain other insects over an area of 40 000 miles² (103 000 km²) and spend over nine million dollars annually, raised from local taxes. Overall responsibility is taken by a board of trustees for each district, representing both urban and rural areas. Their salaried staffs include a manager, a source reduction expert and an entomologist. These superintend subordinate staff, such as inspectors, foremen and operators. In recent years, these organizations have been supplied mechanical equipment for drainage and other source reduction operations, as well as insecticide application machines.

According to a survey reported by Georghiou *et al.* [202], the usage of insecticides over the past decades has shown some significant changes, as follows (percentages):

	DDT etc	Malathion	Parathion	Parathion methyl	Fenthion	Naled
1960	26	41	28	5·3	0·1	0
1968	0·3	7·0	17	21	43	3·8
1973	0	10	16	7·2	46	2·9
	EPN	Dichlorvos	Temephos	Chlorpyriphos	Propoxur	Lethane
1960	0	0	0	0	0	0
1968	3·2	0·5	0·1	0·1	0	3·8
1973	0	0·6	0·6	4·5	9·2	3·1

Noteworthy is the complete disappearance of DDT and the decline in usage of malathion, both probably due to resistance. Fenthion had become popular in 1973 and there were rises in use of propoxur and chlorpyrifos. It is, however, doubtful whether any of these insecticides will continue to be very extensively used in view of the growing resistance of several of the more important species (*Ae. nigromaculis, Culex tarsalis* and *C. p. fatigans*). As a result, the total quantities of insecticides used seem to be declining, with a corresponding rise in the use of oiling.

	1968	1969	1970	1971	1972	1973
Insecticides	83·1	108	90·6	53·5	41·1	49·5
Oil (litres × 10^3)	9550	9500	9200	11600	14000	19300

These difficulties emphasize the need for source reduction, aided by appropriate farming practices. Thus, breeding can be reduced by correctly-timed irrigation and drainage; and wet pastures should not be grazed, since breeding can occur in water in hoof prints.

(iv) Anti-mosquito operations in Sardinia [46]

Sardinia is a Mediterranean island of about 24 000 km^2 (9000 miles^2, or slightly larger than Wales). The island has a long history of endemic malaria and, in 1946, there were 75 000 cases in a population of about a million, 1000 being new infections. At this point, an attempt was made to exterminate the anopheline vectors (mainly *An. labranchiae*) by the then newly-introduced DDT. An organization known as ERLAAS (*Ente Regionale per la Lotta Anti-Anofelica in Sardegna*) was sponsored by the Rockfeller International Health Division and the Italian government. This conducted a vigorous campaign from 1946 to 1950, the island being split up into divisions, sectors and smaller units which were treated at weekly intervals in the summer with DDT larvicide, while the winter resting sites were sprayed with the insecticide. A small international headquarters staff administered up to 6000 field workers (temporarily expanded to 35 000 for some extensive site reduction works). The total cost was 6742 million lire, or rather more than $10 million. The campaign ultimately failed to exterminate the vector mosquito; but malaria was eradicated and there were other benefits. An interesting account of the project, which cannot be adequately described here, was later published by ERLAAS director J. A. Logan [358]. (A contemporaneous campaign in Cyprus [86], then a British colony, has not been similarly documented.)

At the conclusion of the extermination attempt, some of the ERLAAS staff were absorbed into a new permanent organization CRAI (*Centro Regionale Antimalarico e Anti-insetti*) administered by the Italian High Commission for Health and Hygiene. The main work of its staff is to continue the control of anophelines and other mosquitoes initiated by ERLAAS. A substantial amount of the mosquito control is achieved by drainage and canalization, to obviate prolific breeding sites in the meandering rivers. Larvicidal treatments are also made with DDT, Paris Green, HCH and fenthion.

CRAI has its headquarters at Cagliari, with about half a dozen senior staff. Its annual budget (over 28 years) averages 875 million lire (about $1 million). The organization and its achievements are described in an illustrated booklet, published in 1970.

5.2.7 MOSQUITOES TRANSPORTED BY AIRCRAFT AND THEIR CONTROL

The various aspects of mosquito control discussed earlier in this chapter relate to work in the areas where the mosquitoes breed. A special problem, however, concerns the accidental transport of mosquitoes in aircraft, the possibilities of which have, to some extent, grown with the enormous expansion of this form of transport. Inspections of aeroplanes arriving in different parts of the world have been made at various times since the late 1930s. Houseflies and mosquitoes were the most common insects of public health importance discovered and, among the mosquitoes, *Culex p. fatigans* and *Aedes aegypti* were most common [175].

The situation has somewhat changed with the introduction of large jet-propelled aircraft, with pressurized cabins. Owing to the high altitudes reached and the consequent low temperatures, the baggage compartments are now much less favourable for insect transportation; but the very large cabins would still seem to offer opportunities for insects, though there seem to have been few recent studies.

One possible danger, which has been evident for many years, is the chance of introducing an *Aedes aegypti* infected with yellow fever virus into Asia, from Africa or America. Another danger is the possibility of assisting the spread of an efficient vector, or of an insecticide-resistant strain of one into new territory.

Methods of disinsecting aircraft to meet this problem are discussed in pages 105–106.

5.3 BITING MIDGES (CERATOPOGONIDAE)

5.3.1 DISTINCTIVE FEATURES (Fig. 5.5)

The very small midges concerned belong to a family intermediate between the Chironomidae and the Simuliidae. They are more or less intermediate in build, having shorter legs and broader wings than the Chironomidae, but a more slender body than the Simuliidae. On the other hand, the non-biting males resemble these of the Chironomidae and the Culicidae in having bushy antennae, whereas those of the Simuliidae are bare in both sexes.

The females all possess sharp biting mouthparts, though only a proportion of them (belonging to three genera) are known to attack birds and mammals. (Species of one genus attach themselves to the wings of butterflies and suck blood from the wing veins.)

Most of the species which attack warm-blooded animals belong to the genera *Culicoides* or *Leptoconops*, the latter being adapted to warmer climates and absent from Britain. The two genera can be distinguished as follows.

Median vein on the wing forked and connected to the anterior veins by a distinct cross-vein. Fine setae on the wing interspersed with larger setae. Female antenna with 14 segments. *Culicoides*

Median vein unforked and with no cross-vein. Only small setae (microtricia) on the wing. Female antenna with 11–13 segments. *Leptoconops*

A number of species of *Culicoides* occur in Britain, including *C. impunctatus*, *C. nubeculosus*, *C. obsoletus*, and *C. punctatus*, which are widely distributed, though localized. Also, *C. vexans* and *C. heliophilus* are found in the south, while *C. halophilus* and *C. maritimus* are coastal in distribution [162]. Several British species have been recorded in Denmark, especially *C. impunctatus*, *C. vexans* and *C. punctatus* [448]. *C. obsoletus* and *C. pulicaris* are also found in coastal Poland [536], together with *C. fascipennis*, *C. punctatus* and *C. reconditus* are pests in the Ukraine [647].

In coastal districts of Italy, species of *Leptoconops* constitute a severe nuisance in the summer, especially *L. irritans*, *L. kerteszi* and *L. bezzii* [47]. *Leptoconops irritans* and *L. kerteszi* are also troublesome in the Camargue region of the South of France, together with *Culicoides circumscriptus*, *C. maritimus* and *C. nubeculosus* [495].

In North America, *C. furans* is an important pest of the Eastern States and Gulf coast of U.S.A.; *C. sanguisuga* occurs in northern woodlands and *C. haematopotus* in Georgia; *C. biguttatus* and *C. pilliferus* in Michigan; *C. obsoletus* in the mid-West and West; and *C. tristriatulus* and *C. alaskaensis* in Alaska [618]. *Leptoconops kerteszi* and *L. torrens* are annoying biters in California, being known as the Bodega black gnat and the valley black gnat, respectively [549].

5.3.2 LIFE HISTORY [271, 323, 549]

There is considerable diversity in the breeding sites chosen by *Culicoides* species, there being only one feature in common. The sites are all very wet for at least part of the year and most are continuously so. Various types of larval habitat can be distinguished (some six different types have been noted in Britain); and it is important to identify the breeding grounds of the most annoying species. Thus, acid bog, beset with moss (*Sphagnum*) and jointed rush (*Juncus articulatus*) favour breeding of *C. impunctatus* in Scotland and Denmark. Freshwater swamps with water level with the surface encourage *C. cubitalis* and *C. pallidicornis*. Submerged swamps favour *C. pulicaris* and *C. punctator*. Bare mud at lake margins, especially if fouled with animal excreta, is suitable for *C. nubeculosus*. Acid grasslands are suitable for *C. cubitalis* and *C. obsoletus*.

An important site is provided by salt marshes, which favour *C. maritimus* and *C. halophilus* in Britain [323], *C. furans* in the eastern U.S.A., *C. tristriatulus* in Alaska [618], *Leptoconops irritans, L. kerteszi* and *L. bezzii* in Italy [47].

As regards soil type, *L. irritans* and *L. kerteszi* are found in silt-clays, while *L. bezzii* occurs in almost pure sand [369]. On the other hand, in California, *L. kerteszi* is generally found in the surface layers of sandy or porous soils with adequate moisture; whereas *L. torrens* prefers deep loci (46–76 cm) in soils with over 40 per cent clay, which tend to crack in the dry season [609].

In addition to these extensive sites, a few species of biting midges select micro-habitats, such as animal dung in fields, debris in tree rot holes and decaying fungi.

(a) Egg stage

The eggs are laid singly or in small groups (not more than 100) sometimes in echalon. Females are said to lay more prolifically in groups than singly, which tends to produce local aggregations of eggs.

The eggs are cigar shaped, about 0·3–0·5 mm long and covered with longitudinal rows of mushroom-shaped processes, giving a ridged appearence. Pale when laid, they darken within half an hour.

In most species, the eggs hatch in a few days; but in a few, the eggs enter a diapause and hatching is delayed for several months. Thus, in *C. vexans*, this lasts from June to October, while the breeding sites are likely to be dry.

(b) Larval stage

The larva has an oval brownish head and a long, cylindrical, segmented body which is dull white in colour and translucent. It is devoid of appendages, lacking even the false legs or pseudopods typical of chironomid larvae, but moves with a wriggling eel-like motion.

Many larvae subsist on micro-organisms (algae, fungi, bacteria) or on detritus; but some species are carnivorous and can be reared on free-living nematodes.

It has been noted that various species are found in different kinds of habitat. These are chosen by the ovipositing females, as the larvae cannot travel far. The actual conditions are not necessarily essential (for example, salt marsh species can be reared in fresh water). Presumably the separation prevents too intensive competition. This is assisted by different species living at different depths in the soil. *Culicoides impunctatus*, *C. pallidicornis* and *C. obsoletus* are concentrated in the top 2 cm, *C. cubitalis* in the next 5 cm and *C. heliophilus* occurs below this. *Leptoconops torrens* occurs between 30 and 60 cm below the surface, but it is not known where they reside in wet weather.

Presumably, lack of oxygen must present a problem for larvae in muddy soil, unless they are near the surface. All members of the family are devoid of spiracles and respiration takes place through the cuticle.

(c) Pupae

Mature larvae wait until they are uncovered by water, or crawl out of it to pupate; pupation does not occur if they are kept covered with water. When the pupae are flooded, however, they can free themselves and float to uncovered soil.

The pupae are provided with a pair of respiratory trumpets on the back of the thorax, like mosquito pupae, and they breathe atmospheric air. As usual with nematocerous flies, the adults finally emerge from a longitudinal split on the back of the thorax.

(d) Adults

The adults are all very small, dark-coloured flies, with a wing span of 2–3 mm (Fig. 5.5). The head bears well-developed compound eyes and moderately-long antennae, hairy in the female and bushy in the male. The mouthparts are considerably shorter than those of mosquitoes but they include similar elements, which, in the female are likewise adapted for piercing and sucking. They comprise: labrum, mandibles, maxillae and hypopharynx; a pair of jointed maxilliary palps and an elastic labium. (Fig. 5.1). The labium does not enter the wound, but behaves like its counterpart in the mosquito, guiding and supporting the piercing elements. Feeding is rather rapid and only occupies about three minutes. The males have similar mouthparts, but the mandibles are shorter and weaker.

The feeding preferences of the females are obviously important, but detailed information about them is difficult to obtain. It seems that each species has a preference for one or more warm-blooded hosts, but will occasionally feed on others. Many species will attack man and some of the more troublesome ones have been mentioned earlier. Apart from choice of host, species may vary in the part of the body selected (e.g. *C. tristriatulus* attacks the head; *C. furens*, the legs).

Periods of activity also vary. Most species of *Leptoconops* and a few *Culicoides* bite in the day; but most of the latter are most active at dawn and dusk. Winds over 3·5 m.p.h. (5·6 km h^{-1}) and temperatures below 10° C inhibit flying. The flight range of a species is important in determining how far control measures have to be extended. Some remain very close to their breeding grounds (e.g. *C. nubeculosus*, *C. impunctatus*); others fly further (*C. furens*, 800 m, *C. tristriatulus*, 6·4–8 km). These midges rarely bite indoors.

5.3.3 QUANTITATIVE BIONOMICS [323]

A few species of *Culicoides* have been reared in the laboratory. The durations of the various stages (in days) were as follows: *C. impunctatus* at 16–19° C: E, 7–20 (av. 14); L, approx. 150; P, 5. *C. obsoletus* at 16–19° C: E, approx. 17; L, approx. 100; P, 5. *C. nebeculosus* at 15–20° C: E, 3–6; L, 20–60 (summer), 40–125 (autumn); P, 4–7. In the field, development was considerably slower, there being usually only one generation a year. Near Liverpool, adults (and presumably egg-laying) of *C. impunctatus* are at a peak from mid-May to July. Early larval stages are found in the soil in summer and overwinter as fourth stage larvae, which pupate in the following April. In the same area, *C. obsoletus* adults have two peaks of abundance, in May–June and September–October. This is probably a second summer generation [162].

In Italy, *Leptoconops irritans* is mainly troublesome in June–July, while *L.*

kerteszi occurs in greatest numbers in late September and early October [47, 271].

In California, *L. torrens* is abundant for a 6 week period beginning in early or mid-May [549].

5.3.4 IMPORTANCE

In West Africa, a biting midge (*Culicoides grahami*) transmits a filarial worm, *Acanthocheilonema perstans*, to man; but it is not a very harmful parasite. In the West Indies, *C. furans* transmits another filarial worm, *Mansonella ozzardi*. In the North Temperate regions, however, biting midges are not vectors of human disease. On the other hand, they have been incriminated as vectors of various animal parasites. In Britain, these midges transmit *Onchocerca cervicalis*, which causes the disease known as fistulous withers in horses [271]. In the U.S.A., blue tongue disease of sheep appears to be spread by *C. variipenis*.

The main importance of biting midges to man is the great annoyance of their bites during their seasons of prevalence. Records of landing rates of 1000 per hour are not uncommon in some places; while, for most people, five bites per hour are considered the acceptable limit. The irritation caused by the bites can last for days, or even weeks. Scratching aggravates the pruritus and may lead to bacterial infection and slow-healing sores. A common result is a moist open lesion that 'weeps' exudate for a week or two and finally heals with a definite red scar.

Probably the worst areas in Britain are in moorlands in Scotland, where the vast numbers in early summer may actually prevent outdoor work. It has been suggested that the backward condition of croft farming in western Scotland could be ascribed to these midges; and analogous suggestions concern the eastern seaboard of the U.S.A. (More frivolously, the Scottish midges combined with the kilt have been mooted as the origin of the Highland fling!). Attacks of midges certainly tend to discourage tourists from certain fishing, walking or climbing resorts.

5.3.5 CONTROL

Control of biting midges can be attempted in several ways: by protection from or temporary destruction of adults; by physical measures against breeding grounds; or by chemical control of the larvae.

(a) Measures against the adults

Personal protection can be obtained by wearing relatively wide mesh veils, impregnated with repellent (see page 83). In tropical countries, where midges may enter buildings, they have been excluded by screens treated with DDT or repellents; but this is not really relevant to temperate conditions.

For temporary control of an outdoor area, adults can be destroyed by insecticidal fogging (see page 101). This could be worth undertaking before a large outdoor gathering.

(b) Physical measures against larvae

(i) Salt-marsh species

In Florida, some success has been obtained by impoundment of water over the mud flats where the midges breed, so that they are constantly covered with a layer of a few inches (5–8 cm) of water. If fresh water is used, this may have the effect of controlling brackish water mosquitoes as well. Alternatively, construction of a dike to exclude the sea, followed by filling with earth, may be considered. The level must be raised a considerable amount, however (75 cms) [355]. In Jamaica, inadequate consolidation resulted in the replacement of one species (*C. furans*) by a worse pest (*C. bequerti*).

(ii) Floodwater species

In California, the valley black gnat, *L. torrens*, has been steadily attacked by two methods. Regular annual discing of the soil renders it unattractive for oviposition by the females. Alternatively, where water is available, properly-timed irrigation prevents the emergence of the adults, since it prevents the soil cracking, which facilitates their escape.

(c) Insecticidal treatments against larvae

The breeding grounds of several annoying species are fairly extensive, so it is important to identify the exact loci in which the larvae proliferate. Estimation of numbers in the soil may be made from samples (e.g. 7·5 cm diameter, 5–7·5 cm deep) which are sieved and then the larvae floated up in concentrated magnesium sulphate (20–30 per cent). This may produce anything up to 30 larvae; but the method is rather laborious. Alternatively, emergence traps, about 0·4 m^2 may be placed over likely terrain when the adults are expected to appear. A simple trap consists of an inverted flower pot, with a plastic funnel over the hole in the base protruding into a glass jar.

Applications of insecticide in the 1960s usually involved DDT, HCH or dieldrin at rates of 0·56–2·24 kg ha^{-1} (0·5–20 lb $acre^{-1}$) in different parts of the world. More recently, with the disquiet about the persistent residues of the organochlorine insecticides, various organo-phosphorus compounds have been used; for example, malathion at 2·2 kg ha^{-1} (2 lb $acre^{-1}$) or diazinon or fenthion at a quarter that rate [501]. Some laboratory tests with *Leptoconops kerteszi*, showed that the most potent of these compounds were: chlorpyriphos-methyl, parathion-methyl, chlorpyriphos, dichlorvos, fenthion, phoxim and naled [201].

5.4 BLACKFLIES (SIMULIIDAE)

5.4.1 DISTINCTIVE FEATURES (Fig. 5.5)

Blackflies can be fairly easily recognized from their general appearance. They are small, hump-backed blackflies, with broad wings, of which the anterior veins are considerably thicker than the remainder. They vary from about 2–6 mm in length and are thus smaller than mosquitoes but larger than midges. The body is stouter and the legs relatively shorter than those of midges or mosquitoes. The antennae, though many-jointed, are quite short and bare in both sexes.

The family is relatively homogeneous in appearance and breeding habits. While a number of generic names have been proposed, several of them could be submerged in the two main types, *Simulium* and *Prosimulium*. About 1300 species are known, 116 in America [10] north of Mexico and about 20 species occur in Britain [162]. Of these, only a small proportion are important as vectors of disease, or from the nuisance of their bites.

5.4.2 LIFE HISTORY [541, 645]

(a) Oviposition

Blackflies breed in unpolluted, rapidly-flowing water, which can range from small streams to huge rivers. The females generally lay their eggs in large masses, sometimes in a gelatinous matrix, on leaves of aquatic plants, or twigs, stems or stones which are submerged. A few species, however, such as *S. arcticum* or *P. mixtum*, drop their eggs into the water, dipping in the abdomen [623].

(b) Eggs

The eggs have a triangular, cushion shape. White when newly laid, they become brown in a few hours and darken further before hatching. Eggs laid in spring and early summer, hatch in 3–7 days and there may be several generations in a year. Those laid in late summer hatch in the following spring. Single generation species, however, lay eggs in the spring which hatch in the following autumn or winter.

(c) Larvae

The larvae, about 10–15 mm long, are characteristic. There is a well-developed head, with pigmented eye spots and small antennae. There are two appendages above the mouth, each bearing a fan-like spray of bristles. These are apparently specialized parts of the labium, and their purpose, like the mouth brushes of mosquito larvae, is to gather small particles of food. When not being used, the bristles close up like a fan. The larvae, like those of mosquitoes, are non-selective

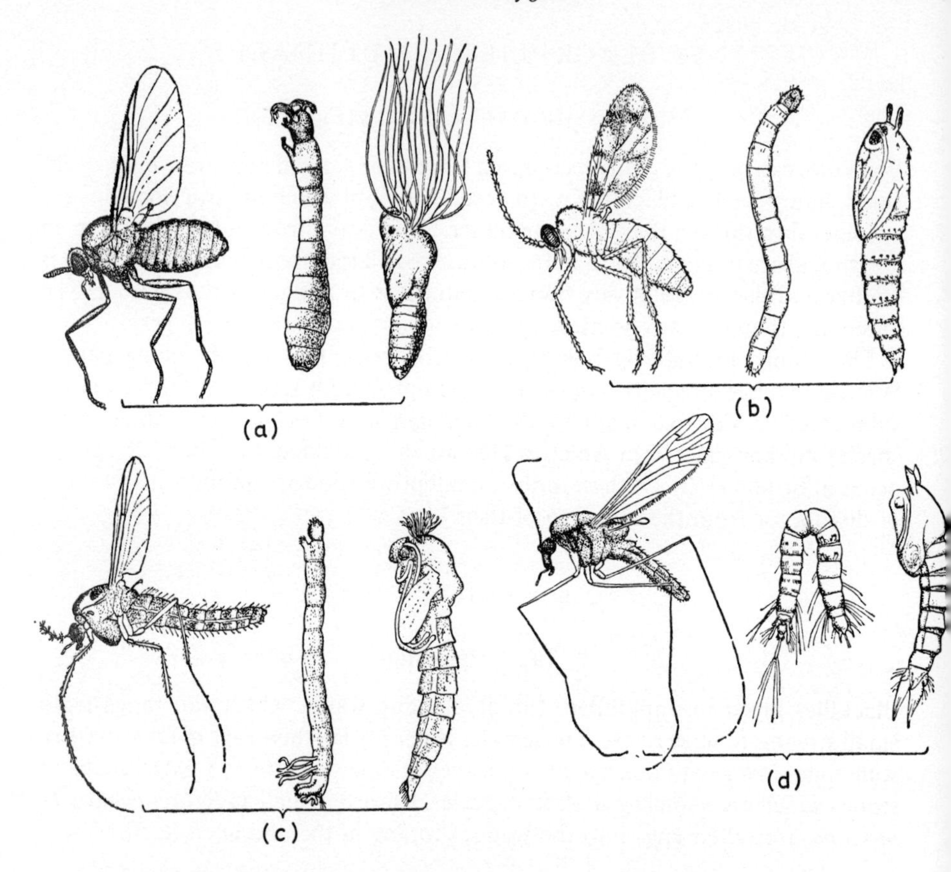

Figure 5.5 Midges with aquatic larvae. (*a*) and (*b*) biting; (*c*) and (*d*) non-biting. (*a*) *Simulium ornatum* (after Smart [542]); (*b*) *Culicoides impunctatus* (after Hill, [271]); (*c*) *Chirononomus* sp.; (*d*) *Dixa* sp. (after Seguy). (*a*) A × 7, L and P × 4; (*b*) A × 10, L × 7, P × 16; (*c*) A and P × 4, L × 3; (*d*) A × 10, L and P × 7.

feeders and swallow particles of silt or mud, which pass through the gut unaltered. Small micro-organisms, however, are digested.

The rest of the body is shaped somewhat like a truncheon, the thorax being the handle. There is a proleg on the thorax and a misnamed 'sucker' at the end of the abdomen; both are furnished with tiny hooks and are used in locomotion. The larvae also spin silken threads over submerged leaves and stones and progress with a looping motion. Sometimes they let go and hang downstream on a silken thread, in which case they either climb up it or drift downstream to a new site. This habit can be used to assess wild populations of blackfly larvae, since they readily colonize artificial objects put into infested streams. Hollow metal cones (20 cm high, with a 10 cm base) painted white, have been found most successful for this purpose [622].

Respiration is carried out by finger-like gills extruded through the anus. The larvae are sensitive to oxygen lack and can only survive in well-oxygenated water. There are six larval stages.

(d) Pupae

Before pupation, the larva spins a slipper-shaped cocoon, with the 'toe' pointing upstream. The pupa is roughly conical and immobile. From the thorax on each side, there is a short stalk bearing a group of respiratory filaments, about as long as the pupa itself. When the adult fly is formed, air is collected within the pupal skin, until finally the skin splits and the fly escapes and floats up to the surface in a bubble of air.

(e) Adults

The general appearance of the adults has already been briefly described. The mouthparts are even shorter and broader than those of *Culicoides*, but similar in appearance and function (see Fig. 5.1). The males emerge first and remain close to the breeding site. Females of the more important vector species and pests require a blood meal before their eggs can mature, but a few species are autogenous. Both sexes take nectar from flowers for energy.

Unlike midges, blackflies have a considerable range of flight. Ten miles (16 km) is quite common and some species can cover the countryside for longer distances. Thus, *S. arcticum* has reached pest proportions at a distance of 140 miles (225 km) from its breeding site. It is probable, however, that some of the flight is wind assisted.

The biting activities of the adults seem to be most intense at dawn and dusk, when the air humidity is high and there is little wind. Humans are attacked on any exposed skin (face, arms or legs) and they can be bitten through clothing where it fits snugly against the body. Those species which attack cattle, commonly bite the thighs, belly, nose and ears.

The adults, in general, are not very long lived. Three to four weeks seems a reasonable estimate for most species, although some individuals may live for a few months.

5.4.3 QUANTITATIVE BIONOMICS

The females lay up to several hundred eggs. The eggs of *S. ornatum* hatch in 5–6 days at 16°C and the larval period occupies 7–10 weeks, with a pupal period of about a week. The complete life cycle may extend to nearly 3 months in summer, but considerably longer over the winter months, owing to a greatly extended larval period (up to 6 or 7 months). The larval period of the European *S. colombaschense* extended for 20 days at 20–25°C while the pupal stage lasted five days. The Canadian *S. venustum* reared at 18–20°C passed the larval stage in 3–4 weeks and the pupal stage lasted 5–7 days [190].

In Britain, most species have two and sometimes three generations a year, the spring brood being sometimes different in size and morphology from the summer one [162]. In southern Europe, *S. colombaschense* has only one generation a year, with maximum adult emergence in April–May; but other species (e.g. *S. erythrocephalum*) have two or three generations annually [648].

Most north American species (*S. venustum*, *S. arcticum*, and *S. vittatum*) have two or three annual generations and overwinter in the egg stage (*S. vittatum*, occasionally, as larvae). *P. mixtum* has only one generation a year and overwinters as larvae.

5.4.4 IMPORTANCE [588]

Over large parts of the tropics, blackflies are vectors of the parasitic worm *Onchocerca volvula*, which causes nodules on the body, skin degeneration and, after invading the eye, can lead to blindness. Although human onchocerciasis does not occur in the north temperate zone, related parasites can be transmitted by blackflies to domestic animals. Thus, *S. erythrocephalum* and other species have been shown to transmit *Onchocerca cervicalis* to cattle and horses in the U.S.S.R.; *S. ornatum* transmits *Onchocerca gutterosa* among cattle in S. E. England [164]; and *S. venustum* is a vector of *O. faliensis*, a parasite of wild and domestic ducks in Canada. Blackflies also transmit various species of haemosporidia to birds, including ducks and turkeys. Among the virus diseases, Eastern Encephalitis has been recovered from blackflies in the U.S.A.; and although they may not transmit it to man, they maintain the reservoir of infection among wild birds.

(a) Bite nuisance

The greatest importance of blackflies in the north temperate region is from their bites. Though always annoying, the bites can become intolerable when vast swarms occur and the after-effects can be severe. Even in recent times, some extensive outbreaks have been recorded. Thus, in 1961, the attacks of *S. erythrocephalum* in Hungary caused extensive suffering [568]; and again, in 1970, a total of 2600 people required medical attention for the bites, and 240 went on sick leave for about 5 days [569]. In Yugoslavia, a severe outbreak of the notorious Goloubatz fly, *S. colombaschense*, was recorded in 1950, and of *S. erythrocephalum* in 1965 and 1970. During the latter outbreak mentioned, some 2000 people sought medical attention [650].

Other troublesome species in northern Europe (Britain to N. W. Russia) are *S. equinum*, *S. ornatum* and *S. reptans*. Also, *S. truncatum* is a nuisance in Norway [494].

In North America, the most troublesome species are *S. venustum* which is holarctic and occurs from Alaska to Greenland and south to Texas and South Carolina; also *Prosimulium mixtum*, which occurs in the N. E. of the U.S.A. and eastern Canada [10].

Attacks on animals can be as serious as the human cases, since the animals are unable to protect themselves. As a result of the severe shock occasioned by the attacks of thousands of these pests, many animals die. (It was formerly thought that suffocation by occlusion of the respiratory passages was responsible; but *post mortem* evidence suggests that death is due to leakage of fluid from the capillaries into tissue space and body cavity).

In the past, great losses of livestock were due to swarms of *S. colombaschense* breeding in the Iron Gates region of the Danube. In 1934, nearly 14 000 cattle died in this way [32]. Since 1964, however, a hydroelectric dam has flooded much of the breeding site and the species has become much less common in that area [649].

In America, *S. arcticum* is a serious pest of livestock in the western prairie states (Alaska to Manitoba and south to Colorado and Arizona). Other, less serious pests, include *S. pecuarum*, which killed many livestock in Oklahoma, Illinois and S. Carolina in the 1930s; *S. meridionale* (Alabama) and *S. vittatum*, which also occurs in Europe.

5.4.5 CONTROL

(a) Mechanical larval control

On a small scale, some relief from blackfly breeding can be obtained by removing submerged vegetation and debris from streams, to eliminate good sites for larvae. An unusual outbreak of *S. austeni* in Dorset in 1970 was dealt with in this way.

On a larger scale, the larval populations of streams and small rivers can be reduced by artificial increase in turbidity. It has been shown that where the turbidity of the water is high (80–475 mg dry substance per litre) half the larval population leave their substrate in 10 minutes, and after 20–30 minutes, few remain. Trials in the U.S.S.R. have shown that increasing turbidity every 4 days in heavily infested streams will result in good control [588].

In large rivers, the construction of a dam will eliminate breeding places in the flooded area; though new sites may be formed in the spillways.

(b) Insecticidal control

From about 1950–1970, there was extensive use of DDT for controlling blackflies. Aerial application of insecticidal fogs have been used to give brief reductions during periods of intense biting; and somewhat longer control by spraying riverine vegetation to kill resting adults after emergence. However, the main value of insecticides has been to destroy the larvae; and there was one example of a species (*S. neavei* in Kenya) being eradicated from a large area by this means.

The insecticide is applied near to the source of each river or tributary. Either a liquid preparation is added gradually, or (more rarely) an impregnated brickette is hung in the water. The turbulence of the water disperses the insecticide, which is carried downstream and can destroy all the larvae for many miles. It was found that maintaining a concentration of 0·1 p.p.m. DDT for 15 minutes was highly effective [300].

Unfortunately, the use of DDT has been greatly reduced or completely abandoned because of anxiety about environmental pollution and its harmful effect on wildlife, especially fishes. In the U.S.A., the safer methoxychlor has been used in the same way, at double the dose; but it seems likely that future control will mainly depend on biodegradable organo-phosphorus and carbamate compounds. For controlling the onchocerciasis vectors in Africa, the following method has been found effective. A large quantity (10–20 l) of concentrated temephos solution is rapidly dumped from aircraft into the headwaters of a river, to give a concentration of 0·05 p.p.m. This gives effective control for up to 50 km downstream [343].

(c) Biological control

Ecological studies of blackflies have shown that the larvae are attacked by various parasites, especially mermithid nematodes and microsporidia. There are also predators such as leeches and larvae of caddis flies; also, various flying predators attack the adults. Although there has been some interest in such organisms for control purposes, no practical demonstration has shown much promise.

(d) Repellents for adults

Adults may be repelled by certain chemicals, and a Russian investigation indicated the following skin treatments as being effective: carboxide (dihexamethylene carbamide) 0·04–0·08 mg cm^{-2}; benzamine, 0·08–0·2 mg cm^{-2}; deet, 0·14–1·4 mg cm^{-2} [376].

Presumably, impregnated wide mesh veils might be effective (page 83).

5.5 SANDFLIES (PHLEBOTOMINAE, PSYCHODIDAE)

5.5.1 DISTINCTIVE FEATURES

Flies belonging to the family psychodidae are small, very hairy insects, not more than 5 mm long. The term sandflies is used for the phlebotomine psychodids; but it is neither very exact (since ceratopogonids are sometimes miscalled sandflies), nor very descriptive, since they have no special association with sand. They occur mainly in warm or tropical regions, but extend up to 45° or 50° N, which includes southern Europe (especially the eastern portion) and part of the U.S.A.

There are two sub-families (though some regard them as full families): the phlebotominae, with blood-sucking females which can transmit diseases, and the psychodinae, or moth flies, which are medically unimportant (Fig. 6.2 b) but can become a nuisance in large numbers (see pages 212 and 387). The two types can be distinguished with certainty from the second longitudinal vein on the wing, which has a single fork in the psychodinae and a double fork in the phlebotominae. Generally, however, the wings are so densely covered with hairs that the veins are obscured, unless the wings are mounted specially on a microscope slide. But the two forms can usually be distinguished in nature from the position of the wings at rest. In the phlebotominae, they are held vertically above the body; whereas, in the psychodinae, they are folded down over the body like the sides of a roof. The wings of both forms are almond-shaped, but those of the phlebotominae are narrower.

5.5.2 LIFE HISTORY [479]

(a) Oviposition

Sandflies deposit their eggs in dark, damp recesses; in neglected cellars, in animal houses, in wild animal burrows, in crevices in walls and even in the tunnels of termite nests. The females lay their eggs over an extended period, extruding them one at a time, with occasional long pauses.

(b) Eggs

The eggs are oval, about 0·35 mm in length, pale when laid but subsequently darkening. The shells are decorated with a reticulated pattern.

(c) Larvae

The newly-hatched larva is only about 0·5 mm long, but doubles its length before moulting. There are four larval stages and the last one reaches a length of 4–6 mm.

The larva has a well-developed head, with biting mouthparts which are used to triturate the food. This consists of decaying organic matter (vegetable debris, faeces, and small dead animals). The yellowish body consists of three thoracic and nine abdominal segments, of which the first seven abdominal ones bear prolegs. The body carries numerous characteristic bristles, feathered and ending in a thickened head ('matchstick hairs'). At the end of the abdomen are four long caudal bristles (two in the first instar). The first stage larva has one posterior pair of spiracles and later stages also have another pair on the mesothorax.

(d) Pupae

The pupae of sandflies are about 3 mm long and club-shaped, convex dorsally. The posterior segments are covered with the cast skin of the last larval stage and attached to the substrate in a vertical position. The presence of the old larval skin, with the prominent caudal hairs, assists in identification.

(e) Adults

The general form of the adult is shown in Fig. 6.2 b. The sexes can be distinguished with the naked eye from the shape of the abdomen. In the female, this is blunt; but in the male, the presence of prominent clasping organs makes the tip appear to be tilted up.

The female has biting mouthparts, somewhat similar to those of ceratopogonids and simuliids; and blood meals are taken after the skin is punctured with the same snipping movements. The males do not take blood meals, but feed on nectar etc, as will the females too, on occasion. Although many sandflies will feed on man, none is confined to human hosts. Apart from other mammals (dogs, rodents) birds and even reptiles may be attacked. Feeding is mainly nocturnal, but they will feed in dark locations by day. Like mosquitoes, sandflies pass through several (3 or 4 ?) gonotrophic cycles; i.e. the digestion of the blood meal and maturation of eggs before laying. This occupies about 3 days at 28–30° C and nearly 10 days at 18° C.

Sandflies do not fly far from their breeding sites; generally not more than 50 m. They usually progress by a series of short 'hopping' flights. They normally rest in cool sheltered places, since they do not well withstand hot arid conditions.

5.5.3 QUANTITATIVE BIONOMICS

Females lay batches of from about 15–40 eggs. Various workers have studied the development of sandflies, especially *Phlebotomus papatasi*, (summarized by Perfil'ev) [479]. In Malta, during the summer the duration of stages of *P. papatasi* were: E, 9; L, 24+; P, 10; average total, 47 days. In Uzbekistan, at 22–28° C, total development was 44–53 days. At Tiflis, at 22–27° C, E, 10–17; L, 27–71; P, 8–41. Under similar conditions, for *P. major*, E, 9–13; L, 21–31; P, 5–10. Another worker found a shorter development for *P. sergenti* under similar conditions, of 37–44 days.

5.5.4 IMPORTANCE

Sandflies are vectors of several diseases, including visceral leishmaniasis (kala azar), cutaneous leishmaniasis (oriental sore) and sandfly fever (a benign viral complaint). All these can occur in southern Europe, though their incidence has declined in recent years [505, 651]. The main vectors in the Mediterranean region are *Phlebotomus perniciosus* and the closely-related *P. ariari* in the west, *P. major* and *P. papatasi* in the east, and *Sergentomya minuta* around the coastal region. Dogs and wild rodents form a reservoir of leishmaniasis and possibly of the sandfly fever. However, the latter may be exclusively human and persist through transovarial passage in the sandflies.

In the U.S.A., some six species of sandflies are known, but they are not of medical importance [493].

Apart from disease transmission, sandflies can cause annoyance from their bites. Severe reactions can occur in sensitive people, producing an effect known in the Middle East as 'harara'.

5.5.5 CONTROL

(a) Measures against larvae

Since sandflies do not fly far from their breeding sites, it would seem possible theoretically, to control them by treating or eliminating such places. However, the foci are scattered and not easily located, so that only a modest degree of control can be achieved. Moist litter and rubbish in the neighbourhood of houses should be removed. Cracks and crevices containing damp organic matter should be sealed up or at least treated with insecticides.

(b) Measures against adults

House-spraying with residual insecticides, especially DDT, is particularly effective against sandfly adults, which are very susceptible and do not seem to have developed resistance. The various anti-malarial spraying campaigns in the Mediterranean and Middle East were found to have an additional bonus in the drastic reduction of sandflies and, in certain areas, the substantial elimination of leishmaniasis. Unfortunately, where DDT spraying was discontinued, either because of malaria eradication, or because of vector resistance, sandfly prevalence has returned.

Temporary reduction of sandflies can be achieved by using insecticidal aerosols or burning pyrethrum candles in the bedroom at night. Sandfly bites outdoors can be prevented to a large extent by repellents, such as dimethyl phthalate or diethylamide. Also, wide mesh nets, treated with repellents and worn over the head, have been found effective.

5.6 HORSEFLIES ETC (TABANIDAE) (Fig. 5.6)

The Tabanidae are practically the only members of the sub-order Brachycera which are harmful to man and his domestic animals. Because of their large size and troublesome biting, they have been given a variety of common English names, including horseflies, deer flies, clegs, breeze flies, greenheads and mango

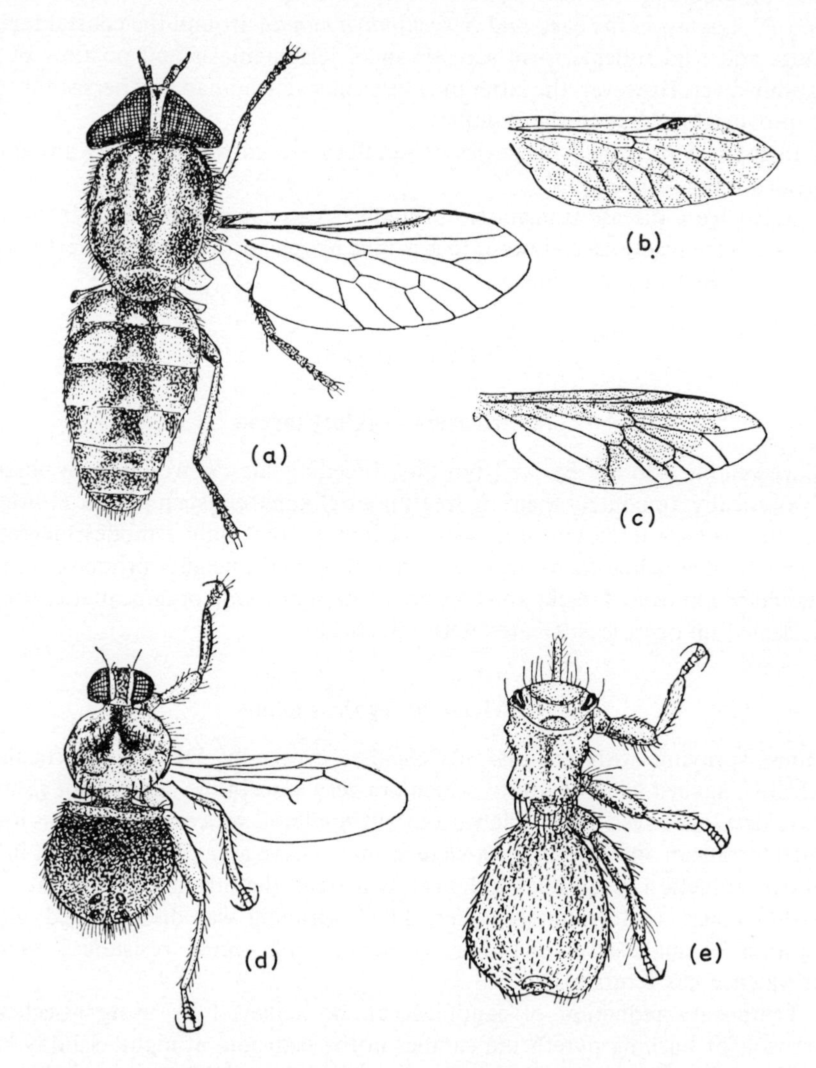

Figure 5.6 Biting flies. (*a*) *Tabanus bromius* (female); (*b*) wing of *Haematopota pluvialis*; (*c*) wing of *Chrysops sepulcralis*; (*d*) *Hippobosca equina* (female); (*e*) *Melophagus ovinus* (female). (After Edwards *et al.* [162]) (*a*) × 3½; (*d*) × 4; (*e*) × 7.

flies. The family contains about 3000 species and is widely dispersed, from the tropics to the arctic and from sea level to some 3000 m above. Members of three sub-families take blood: the Chrysopinae, the Tabaninae and the Pangoninae. The last group have exceptionally long proboscides adapted for collecting floral nectar, as well as biting mouthparts but they are rarely troublesome and will not be mentioned further.

5.6.1 DISTINCTIVE FEATURES

The Tabanidae are quite unlike any of the other biting Diptera considered so far, being rather large, robustly-built insects, more like blowflies than the tiny blackflies or the fragile gnats or midges. On the other hand, they resemble these members of the sub-order Nematocera in that only the females take blood meals, using piercing labra, mandibles and maxillae, which are entirely absent in the more specialized biting Cyclorrhapha. The antennae are intermediate between the two groups, being reduced to only three joints (though with traces of segmentation on the last one) and without the characteristic large bristle (arista) of the Cyclorrhapha.

The larval stage is also intermediate between the active larvae of Nematocera and the headless maggots of higher flies. Tabanid larvae have heads, but they are much reduced and can be retracted into the thorax. The pupae, like those of Nematocera, are not formed within the last larval skin (puparium) as in the houseflies and blowflies.

There are three important genera, which may be distinguished as follows.

1. Spurs on the tips of the tibiae of the hind legs. Antennae with three long segments. Wings folded over the abdomen at rest, like sides of a roof. Wings often with a central dark band. (6–10 mm long) *Chrysops*
No apical spurs on the tibiae. Second antennal segment shorter than the others. Wings held horizontally at rest at an angle of 30°. (2)
2. Wings mottled with grey. (6–10 mm long) *Haematopota*
3. Wings clear, but sometimes with a brownish tinge near the front border. (10–25 mm long) *Tabanus*

5.6.2 LIFE HISTORY [410, 455, 457]

(a) Oviposition

The females choose moist or semi-aquatic breeding sites, ranging from the muddy margins of ponds or sluggish streams to damp earth or even rotting logs. They lay rather large batches of eggs (100–1000) on projecting vegetation or other objects above the breeding sites. The eggs are stacked closely together in a layer, sometimes with another tier on top.

(b) Eggs

The eggs are banana-shaped, 1·0–2·5 mm long, white when laid and darkening to a greyish brown.

(c) Larvae

The larvae are legless grubs, with small dark heads and an 11 segment body. Each of the first seven abdominal segments bears a ring of pseudopods, or fleshy processes, possibly connected with locomotion. The skin is creased with longitudinal striations and is white, cream or brown in colour. The last segment bears a single protrusible breathing tube or siphon, ending in a pair of spiracles. These communicate internally with two longitudinal air trunks running along the body.

Upon hatching, the larvae usually drop into the water and bury themselves in mud. They spend their larval life in wet mud or shallow water. There are three to eight larval stages, the first of which is gregarious and takes no food but subsists on yolk from the eggs. The older stages (of *Tabanus* and *Haematopota*) are carnivorous and prey on small Crustacea, worms, snails and insect larvae. If kept in captivity, they often become cannibalistic.

(d) Pupae

Pupation occurs in earth or drier mud around the margins of ponds. The fused head and thorax of the pupa displays the rudiments of the adult appendages; and two ear-shaped respiratory trumpets can be seen. The abdominal segments bear rings of short bristles and the tip of the abdomen carries a cluster of six tooth-like spines. The adult finally emerges through a longitudinal split in the back of the pupal skin.

(e) Adults

The adults are rather handsome flies, with large eyes which are often irridescent with beautiful gold, green and blue tints. (These fade in dead specimens.) The antennae are rather well developed for the group (Brachycera) and are carried projecting from in front of the head. The mouthparts comprise the same piercing elements as those of midges or mosquitoes. These are piercing labrum, mandibles and maxillae. There are two rather short two-jointed palps and an elastic labium which is curved back when the piercing stylets are thrust into the skin. The labium ends in a pair of licking lobes, or labellae, covered with fluid-absorbing grooves (as in the housefly).

Because they do not take blood meals, the males are not often seen, except on vegetation resting after emergence. The females fly vigorously at speeds up to 14 $m\,s^{-1}$ (31 m.p.h.) and may range several miles from the breeding sites. The males commonly fly high, but sometimes swoop down to take water from pond surfaces while in flight. Both sexes visit flowers for nectar and specimens are often found to be covered with pollen when collected.

The females apparently do not take blood meals until they are fertilized and most species require a meal to produce viable eggs. The flies approach silently and rapidly. Species of *Haematopota* often approach waist high; those of *Chrysops*, about head high and many *Tabanus* species prefer to bite the legs.

5.6.3 QUANTITATIVE BIONOMICS

Many British species hatch from the eggs in 4–12 days, according to temperature. The larval stage is somewhat prolonged and, in temperate climates, the fly passes the winter in this form; sometimes, indeed, two winters may pass before development is completed. Pupation occupies about 1–3 weeks. British species do not have more than one generation per year [162].

In Arkansas, *Tabanus lineola* was found to hatch in 4 days, with a larval period of 49 days, pupation for 8 days and a pre-oviposition period of 9 days, giving a total of about 70 days for a generation [535].

5.6.4 IMPORTANCE

(a) Disease transmission

The large blood meals taken by tabanids (one species was calculated to take 0·359 cm^3 [572]), together with the habit of resuming feeding on another host if disturbed, render them capable of transmitting pathogens mechanically. In the U.S.A., it appears that *Francisella tularensis* causing tularaemia, can be transmitted in this way by *Chrysops discalis*. It has also been suggested that tabanids can act as vectors of anthrax in the same way. In West Africa, *Chrysops* may transmit the filarial worm *Loa loa* and, in parts of the Far East, tabanids spread 'surra' a highly fatal disease of horses and camels.

(b) Biting nuisance

In temperate regions, tabanids are mainly important because of the nuisance of their bites. In man, these can be painful, both when inflicted and (in sensitive individuals), in subsequent reactions. Animals, too, can suffer severely. Estimates of blood losses during persistent attacks, range up to 300 cm^3 blood per day [484]. Furthermore, the constant annoyance and irritation of the bites can substantially reduce the milk yield of cattle.

In Britain [162], *Haematopota pluvialis* is, perhaps, the commonest species and is quite common near its breeding grounds in low-lying marshy country. *Tabanus bromius* is probably next most prevalent. These two species are also troublesome in the south of France [495] and also in Czechoslovakia [108], together with other species including *Chrysops* spp.

In America [261], the main pests of the Eastern states are *Tabanus lineola* and *Chrysops atlanticus*. Also in the east, but widely distributed west to the Rocky Mountains, are *C. discalis, T. atratus* and *T. stygius*. On the Pacific coast, *T. punctifer* is common and species of *Haematopota*, though this genus is not well represented in America. In the southern states, *T. quinquevittatus* is a pest; in the north and in Canada, *T. affinis* and *T. septemtrionalis* are troublesome.

5.6.5 CONTROL

Control of tabanids presents many difficulties. The breeding grounds are diffuse and extensive. Larval populations in Canada have been estimated at about 4840 per acre or 3 million per mile^2, so that larvicidal treatments would be expensive. Some promising results have, however, been obtained by temporary flooding of salt marshes to destroy early stages of coastal deer flies (*Chrysops* spp.) in eastern U.S.A. [6]. Other types of site may be drained.

Aerial spraying of insecticides against the adults was tried several years ago in Canada. It was found that the application of 0·5 lb acre^{-1} (560 kg ha^{-1}) of lindane temporarily eliminated the insects from open country, but not from woodland [72]. The oiling of the surface of ponds from which adults have been observed to drink has also been recommended.

Cattle may be given some relief from *Tabanus* spp. by the provision of sheds in which they can shelter, since these sun loving insects do not like to enter such places to feed. In Czechoslovakia, it has been noted that *Haematopota* and *Chrysops* commonly rest in bushes and shrubs, so that clearing these from grazing fields may benefit cattle [108]. Persistent attacks on humans may perhaps be prevented by repellents.

5.7 THE STABLE FLY AND ALLIES (STOMOXYS CALCITRANS, ETC)

Nearly all the highly developed flies such as the housefly and the blowflies have specialized mouthparts in the adult stage, which are adapted for licking up exposed liquids. The parts used by gnats and horseflies for piercing skin are completely atrophied. Nevertheless, a few members of the Muscidae have developed the bloodsucking habit (possibly via the habit of drinking blood from small wounds; see page 135). These species pierce the skin by the labium, which is modified into a horny beak instead of being a soft proboscis. This horny, beak-like proboscis is carried horizantally and projects in front of the head, providing a ready means of distinguishing these flies from the non-biting muscids. The modified proboscis and the bloodsucking habit occur in both sexes.

The most important insects of this group are the tsetse flies, vectors of sleeping sickness in man and the fatal disease of horses and cattle, known as nagana; however, they are confined to Africa. In the temperate region and, indeed world wide, the most annoying pest of this type is the stable fly, *Stomoxys calcitrans*. This readily bites both man and cattle as well as domestic animals such as dogs. Somewhat similar to this are flies of the genus *Haematobia*, which accompany cattle in the fields and bite them from time to time. *H. irritans* is known as the horn fly, from its habit of resting on cattle at the base of the horns. It is more completely adapted to parasitic life than the stable fly, since it remains close to the cattle except for short excursions to lay its eggs. It rarely bites man. *H. stimulans* is rather less specialized; it leaves the cattle more readily and does not rest on them at night. It is also more prone to bite man; for example, the legs of people milking cows [239]. *H. irritans* has been introduced into North America, where it has spread widely and become a serious pest; whereas *H. stimulans* is restricted to Europe. The adults of these species may be distinguished as follows.

1. Proboscis long, projecting a head's length in front of the head. Palps short and thread-like. *Stomoxys calcitrans*

2. Proboscis short; palps club-like and shorter than proboscis. *Haematobia stimulans*

3. Proboscis short; palps not club-shaped and as long as the proboscis. *Haematobia (Lyperosia) irritans*

5.7.1 STOMOXYS CALCITRANS

(a) Distinctive characters

This fly is very similar in general appearance to the common housefly and, since it is sometimes encountered indoors, it has been called the biting housefly. As already mentioned, the proboscis projecting in front of the head, which is even visible from above, enables it to be easily distinguished from the common housefly. The fourth wing vein also does not bend sharply, as in *M. domestica* (Fig. 6.1).

(b) Life history [471]

(i) Oviposition

The females lay eggs in horse, pig or calf manure, in decaying straw or in various types of rotting vegetation. They breed very well in the bedding of farm animals, especially if it is soiled by dung or urine. Under hot conditions (27° C) the females mate when 6 days old and lay eggs 1–2 days later. About 10 batches of some 35 eggs may be laid over a period of 12 days. In cooler conditions, the oviposition period is more extended and the batches may contain about 100 eggs.

(ii) Egg

The eggs are about 1·1 mm long and 0·2–0·3 mm thick, narrowly oval and slightly pointed at one end.

(iii) Larva and pupa

The larvae resemble those of the housefly in general appearance and habits. After the usual three stages, reaching a length of 11 mm, they pupate in the same way, usually leaving the breeding material for a drier zone. Larval and pupal development periods are greatly extended at low temperatures and probably represent the usual method of passing the winter.

(iv) Adult

After emergence, the adults expand their wings, and the proboscis, which is folded backward in the pupa, extends forward and hardens. The flies are able to feed a few hours after emergence. Both sexes take blood regularly and, while they will also take sugar solution, they thrive best and survive longest on blood meals. Although cattle are apparently the preferred source of blood, *Stomoxys* will also feed on many different animals, including horses, pigs, dogs and men. Like cattle, men are often bitten on the legs, socks or stockings being no protection. In feeding, the proboscis is directed downwards and the fly rasps a way through the skin by everting the small labella at the tip; this exposes the file-like teeth on their inner surfaces. Often several tentative incisions are made before a good blood supply is discovered. Undisturbed feeding takes about 15 min, but the flies are often brushed away by the victim and return to make another attempt. The weight of a full meal (about 25 mg) is roughly three times that of the fly.

Stable flies feed only in daylight and the need for meals increases with temperature. In hot weather (25° C) the blood is digested in 12–24 h and meals are taken every 1–3 days. In cool weather digestion may require 2–4 days and feeding may be delayed for 10 days. Below 9° C the flies make no effort to find food and remain quiescent. Although the need for food is less, the final result of prolonged low temperature is starvation. Accordingly adult flies seldom survive the winter in unheated buildings, although in warm stables and cowsheds, intermittent feeding may continue until the following spring.

The habits and behaviour of stable flies differ in several ways from those of houseflies. They are rural in distribution, being virtually absent from large towns. In warm weather they may be encountered in the open and they are common in stables and cattle sheds; less frequently, they enter houses. They spend more time resting than houseflies, possibly because they do not need to quest so much for food. After a meal many of them rest in animal houses, usually choosing the darkest places, high up on the walls or ceiling. Nearly always, they rest with the head upwards and the body inclined at an angle to the surface. (In contrast, houseflies often rest head down, in a crouching position with the body parallel to the surface.) Activity begins with daylight when stable flies tend to explore the lower parts of the animal houses and seek the cattle. Most of their flights are quite short (less than 1 m) and are mainly concerned with seeking food or mates.

Mating, like that of *M. domestica*, originates with visual stimuli and male stable flies often attempt to mate with each other or even jump on to small dark objects.

Over the temperature range, activity begins at 10° C, is maximum at 28° C and heat paralysis sets in at 42·6° C. The range is thus somewhat narrower than that of *M. domestica* (see p. 211).

(c) Quantitative bionomics [340]

At 16° C (60° F)	E 5	L 34	P 19
At 20° C (68° F)	E 2	L 16	P 14
At 25° C (77° F)	E 1·5	L 9·5	P 6·5
At 30° C (86° F)	E 1	L 6	P 5

Minimum development times therefore range from 7 weeks at 16° C to 12 days at 30° C, in all cases being slower than *M. domestica*.

Normal adult life is unknown but flies have been kept alive for 10 weeks in the laboratory.

(d) Importance

The stable fly has been accused, especially in the U.S.A., of taking part in the transmission of poliomyelitis; but the evidence is very tenuous.

When they are numerous, stable flies are very objectionable from their biting habits, not only in the open but in cool weather when they may enter houses. Dairy farmers find that their bites may irritate cows sufficiently to reduce milk yield.

(e) Control

(i) Larval control

To prevent or reduce breeding, the sides of hay and straw stacks should be kept vertical and the tops rounded and thatched so that the rain drains off. Loose straw and chaff should be burnt or scattered; the important thing is to prevent accumulation during the summer months of rotting straw, chaff or other vegetation (or especially, a urine-soaked condition) [51].

Larvicidal treatments suffer from the same drawback as in the control of houseflies, namely the bulk of material to be treated in order to reach the larvae. However, a 1:3 creosote–water emulsion applied at the rate of 1 gall/85 ft^2 (4·8 l/10 m^2) was effective in controlling breeding in decaying marine grass deposits in the U.S.A. [537]. More recently, some experimental trials of insect development inhibitors have been made, applied in this way [643] or given by mouth to cattle to produce insecticidal dung [245]. The latter method, however, is more likely to be effective with pure faecal breeders, like *Haematobia* spp. or even *Musca domestica*.

Biological control by parasites or predators has been considered, for example with breeding in poultry dung [347]; and one author suggests a combination of juvenoids and parasites [642].

(ii) Adult control

In Germany, stable flies have been excluded from cattle sheds by gauze screens, combined with window traps [230]. In Florida, adult traps have been used, the flies being attracted by carbon dioxide emission and ultra violet lamps and then killed by an electrified grid [525]. More recently, a simple trap consisting of two plexiglass panels, at right angles vertically, treated with permethrin [394] has been used. In the past, residual organochlorine compounds (DDT [241], chlordane [470, 551]) have been applied in or near breeding grounds to kill emerging adults. Alternatively, DDT has been used to spray the walls of cattle barns or milking sheds. Since these compounds cannot now be used for this purpose in many countries, other insecticides must be employed. In Norway, a commercial preparation of 6·5 per cent trichlorphon, painted on the wooden walls of cattle sheds, gave good control; but not on limewashed walls [551]. In Malta, residual sprays of tetrachlorvinphos every 9 weeks, gave good control of both *Stomoxys* and *Musca*; but it was emphasized that this treatment should be combined with good farm hygiene to reduce breeding [244]. Another alternative to DDT which has been recomended is 1 per cent fenchlorphos [123].

For rapid kill of adults without residual action, an application of insecticidal dust, at 88 g/100 m^3 by dust gun has eliminated both houseflies and stable flies from large barns [433]. Outdoors, ultra low volume sprays containing resmethrin or primiphos-methyl can achieve the same result [113].

5.7.2 HAEMATOBIA IRRITANS

The horn fly is about half the size of the stable fly, being only about 4 mm long; but in colour and general appearence, it resembles it. When the horn fly is at rest on an animal, the wings lie flat on the back, folded rather closely; but when it is biting, the wings are spread and the insect stands almost perpendicularly.

(a) Life history [391]

(i) Oviposition

The eggs are laid in small groups (3–7) on cow droppings, usually under the sides of the 'cake', or on the soil underneath.

(ii) Eggs

The eggs, 1·3–1·5 mm long, are reddish brown in colour and not easily seen on the dung.

(iii) Larvae

Horn flies invariably breed in dung, the physical and chemical nature of which has been carefully studied. Warm and wet when passed, it cools and dries, forming a dry upper crust; subsequently, the temperature in the upper layers rises again, due to bacterial action. Since it contains a rich supply of proteins, carbohydrates and minerals, dung invites the attention of numerous coprophagous insects, and no less than 64 species have been recorded as breeding in it.

Hornfly larvae occur in the upper layers of the dung cakes but pupate in the drier earth beneath. Those of *Haematobia stimulans* emerge from the dung and pupate a short distance away. Both species, of course, pupate inside the last larval skin or puparium.

(iv) Adults

Hornflies are obligate parasites of cattle that remain on their hosts night and day. They rest on any part of the body that cannot easily be reached by head or tail: often at the base of the horns (hence the name), but also behind the shoulders, on mid belly and low down on the hind legs. Mating takes place on the host and the females leave only briefly to lay their eggs. Because of their close association with cattle, hornflies only invade new territory when the latter are transported to it.

(b) Quantitative bionomics

Various studies of the life history have been summarized by McLintock and Depner [391], from whose paper the following data have been abstracted (Stage durations in days).

At 18·2° C		L 12·8	P 14·3
At 18·6° C		L 10·9	P 13·9
At 24·4° C		L 5·8	P 7·1
At 25° C	E 0·82	L 7	P 5–7
At 32° C		L 3·7	P 3·9
At 35° C	E 0·47		

(c) Importance

Haematobia flies are not, so far as is known, disease vectors; and their importance lies in the annoyance and disturbance of cattle, which impairs their digestion, with consequent loss of flesh and milk production.

(d) Control

(i) Larval control

To discourage breeding, cattle droppings can be scattered by raking over pastures, so that the dung dries quickly and kills any larvae present. The action of dung beetles which disperse and bury the dung, also reduces breeding. This form of biological control is most practical when dung beetles can be introduced into new territories where their natural enemies are absent (e.g. in Australia).

Alternatively, insecticides with low mammalian toxicity may be added to cattle feed or salt licks [411], so producing dung lethal to the larvae. Recently, methoprene has been used for this purpose [36, 412].

(ii) Adult control

Control may be obtained by spraying cattle with residual insecticides; but since the organochlorine insecticides have been restricted, this is less feasible although methoxychlor dust may be used [411]. More satisfactory is the provision of tetrachlorvinphos-treated sacks or bags, hung on chains over gateways, so that the cattle can use them as back rubbers. Pyrethrins, iodophenphos or trichlorphon have also been used [444].

The possibility of eradicating stable flies by the method of releasing sterilized males has been under consideration in one or two countries [112, 336]. One big drawback to the application of this method to stable flies is that (in contrast to the mosquitoes) the released males would temporarily add to the biting nuisance; so that public reaction might be hostile.

5.7.3 PUPIPARA

These flies are a remarkable group, apparently related to the muscids, but showing degenerative adaptions to a parasitic life, as bloodsuckers of mammals or birds. Their bodies are flattened and leathery and the feet bear strong claws for gripping the host. They have developed the habit of giving birth to fully developed larvae, which pupate at once, usually among the fleece of the host.

The Hippoboscidae includes some winged forms, such as *Hippobosca equina*, the forest fly found in the New Forest, England; also various deer flies which occur in north America. Another winged form is *Pseudolyncha canariensis*, the pigeon fly, which is mainly tropical but occurs in the southern U.S.A. One of the deer flies, *Lipoptena depressa*, casts its wings when established on a new host. Finally, *Melophagus ovinus*, the sheep ked (or louse fly) is completely wingless.

(a) The sheep ked, Melophagus ovinus

The sheep ked is a curious deformed-looking insect, which crawls among the fleece like a spider (Fig. 5.6). The whole life cycle is spent on the host. The keds are found near the skin, by parting the fleece; often several hundred may be present. The pupae are found in the shoulder, thigh and belly regions.

(i) Quantitative bionomics [567]

The pupae require 19–23 days before emergence in summer (Wyoming, U.S.A.) and 40–45 days in winter, in sheep kept out of doors. The female requires 14–30 days to reach maturity and then reproduces a single larva every 7–8 days, probably for about 4 months.

(ii) Importance

Although heavy infestations cause irritation and annoyance to sheep, normal infestations do not seem to have much effect on growth and development.

Like several other species of hippoboscids, sheep keds may bite man; but the trouble is rare except among people shearing sheep. Other biting forms are the forest fly, some deer flies and the pigeon fly. The bites are somewhat painful and the clinging tenacious insects can be rather disturbing.

(iii) Control

In Poland, sheep have been treated with fenchlorphos powder or emulsion [473]; other treatments employed include diazinon, dioxacarb and fenitrothion [150]. In the U.S.A., 1 per cent coumaphos powder, applied at the rate of 80–100 mg active principle per head was most effective [483].

(b) Other families of Pupipara

The Nycterividae and the Streblidae are parasitic on bats, the latter family being confined to the tropics and warmer regions.

CHAPTER SIX

Houseflies and blowflies

6.1 INTRODUCTION

6.1.1 THE HYGIENIC IMPORTANCE OF NON-BITING FLIES

Their habit of breeding in putrefying matter has led many of the higher flies into an interesting and sometimes important association with various animals (wild and domesticated) and with man. Although some maggots feed on living plants, it is possible that the original breeding material of the group was decaying vegetation, which is still the normal source of many fruit flies, seaweed flies and many others. Then some forms found that vegetable debris enriched with protein from excreta of higher animals, provided a better breeding material. Such is the normal site of larvae of the stable fly. From this, animal dung became the preferred breeding medium of many species, including the housefly; but the facility of sometimes using decaying vegetable matter was still retained. Other species like the lesser housefly, became more restricted to faecal matter.

The adults of these scavenging flies usually lay large batches of eggs (and a few give birth to young maggots) on the chosen site. The young larvae, which are generally gregarious, then browse on the abundant food supply. The decomposing organic matter provides a rich source of proteins and amino acids, to digest which the larvae have evolved suitable enzymes. They have also acquired the ability to rid themselves of excess nitrogen by excreting ammonia and nitrogen-rich compounds such as allantoin.

The dung-feeding forms show more or less definite preference for particular types of faeces (the preference being exercised, of course, by the mother seeking a suitable breeding site). The quantity and condition of the breeding material, especially the water content, are important. Some species tend to breed in isolated dejecta in the fields, while others prefer large masses of manure in farmyards.

In addition, to strictly coprophagous species, there are some which begin as scavengers and later feed on other, smaller maggots.

The family calliphoridae, or blowflies, discovered another protein-rich breeding material in carrion, although the less specialized members of the group also breed in manure. Then, in some species, the eggs are laid on festering sores or wounds, or on excrement-soiled parts of living animals. The maggots develop on the body and enter the flesh, sometimes making such a large wound that the animal dies. This invasion of living animals by fly larvae is known as myiasis (see pages 224–226). There are still more specialized forms which are able to penetrate uninjured flesh and develop in it; but human cases do not occur in temperate regions.

It will be noted that the habits of higher Diptera bring them into contact with man or his domestic animals to different degrees. Some exist mainly in wild country, but may invade human rural settlements, to breed in refuse or discarded offal. With others, the natural ecology is such that human activities provide ideal opportunities for breeding and adult food. There are many species associated with cattle, breeding in their dung and taking protein as adults from animal sweat, blood from small wounds, etc; and some species have evolved into bloodsuckers (see page 182). These forms are sometimes described as *symbovine*.

Other species have penetrated closer to man, attracted by his rubbish or to slaughter houses, even in towns. Some too, find the shelter of human dwellings favourable, either for hibernation (so-called 'swarming houseflies'), or as a more or less permanent refuge. Above all, of course, the common housefly is an example of such *synanthropic* flies; the lesser housefly is a somewhat less prevalent example. This chapter considers these flies in two groups; (a) those which are likely to be encountered as adults and (b) those likely to be troublesome in the larval stage. Group (a) will be considered in Sections 6.2–6.5 and group (b) in Sections 6.6–6.10.

6.2 THE HOUSEFLY (MUSCA DOMESTICA)

The housefly is world-wide in its distribution and everywhere lives in close association with human dwellings. There are, however, slightly different races living in temperate, warm and hot climates; the characteristic forms being known as *M. domestica domestica, M. domestica vicina* and *M. domestica nebulo*, respectively. They differ in depth of coloration (darkness being associated with cool conditions) and the width of the '*frons*' between the eyes of the male. These forms are not separate species; they intergrade, and even the extreme types are quite inter-fertile. Observers have noted somewhat different habits of flies in the tropics. For example, they rest out of doors at night and they are more prone to visit and breed in human faeces, if accessible. Yet one cannot say whether these differences are innate, or whether they represent a similar reaction to different environments [519].

6.2.1 DISTINCTIVE FEATURES

The common housefly is usually about 6–7 mm long, with a wing span of 13–15 mm. The thorax is grey with four longitudinal stripes, most clearly distinct in front (Fig. 6.1). The fourth vein on the wing bends sharply forward and nearly meets the third vein at the wing margin. The sides of the basal half of the abdomen are yellowish buff and sometimes transparent, especially in the male. A central longitudinal band broadens at the back to cover the final segments.

6.2.2 LIFE HISTORY

(a) Oviposition

It is convenient to commence a description of the life history with the mature female seeking a suitable site to lay her eggs.

Flies will breed in a large number of substances (ranging from snuff to spent hops!) of which the only common factor seems to be a moist, fermenting or putrefying condition. Typical examples are (a) the excrement of various animals (pig, horse, calf, man), (b) rotting vegetable matter, especially with a high protein content (seeds, grain), and (c) the heterogeneous mixture which constitutes garbage. Some types of breeding material are more favourable than others. The matrix must not be too dry or too wet; thus, on the one hand, horse droppings in the field are unsuitable since they soon become desiccated, whereas bucket latrines are too wet and seldom contain housefly larvae. Where the egg-laying female can exercise a choice (on a farm for example) she shows a predilection for certain kinds of dung. Horse dung is preferred if it is fresh but it rapidly loses its attraction. Pig dung is very attractive for a week or more and is a common breeding material. Calf dung is sometimes and cow dung very rarely infested [334, 614]. Battery-type poultry houses, in which a considerable quantity of poultry dung accumulates, frequently cause excessive fly breeding; but the dominant species is usually the lesser housefly, *Fannia canicularis*.

In cities there was formerly much breeding in manure stacks near stables, but such sites have very greatly diminished with the replacement of horse-drawn traffic by the motor-car. There are, however, occasional nuisances which develop from riding stables near built-up areas. At present, fly nuisances in cities are mainly due to excessive infestation of municipal refuse tips or other large accumulations of decaying organic matter.

The females seek out suitable breeding material largely by olfactory organs which are located on the antennae. If these are amputated the females lose their discrimination between various breeding materials [348, 577]. There is also evidence that they confirm their judgement by tasting the material they have selected. The female flies usually prefer to lay eggs on breeding matérial exposed to light; they do not often fly into dark places in search of suitable sites. But having landed on the pile of material, they often crawl into small crevices and, by the use of their ovipositor, deposit the eggs still further into the mass. The value of this behaviour is probably that the eggs will be protected from desiccation by exposure to dry summer air.

During a single day a fly, if undisturbed, may lay the whole batch of eggs which are mature in her ovaries, usually about 100–150. This may be repeated four or five times at intervals during her adult lifetime.

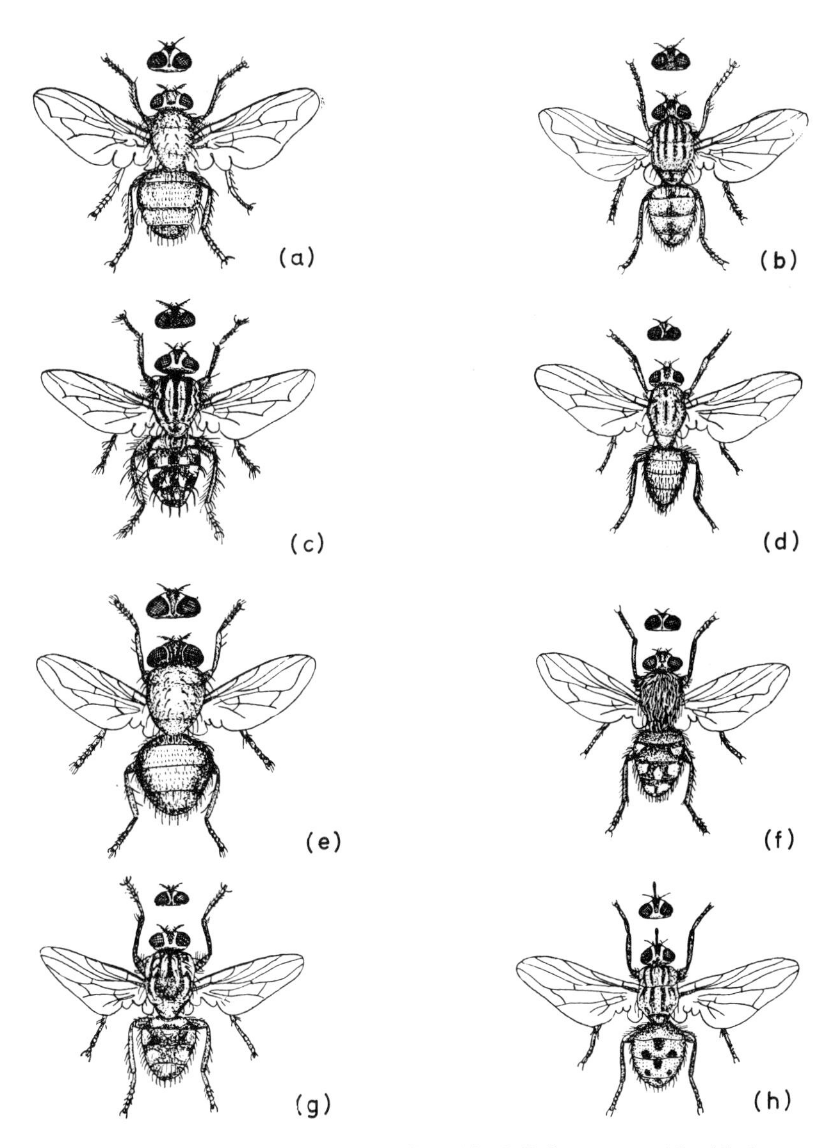

Figure 6.1 Some houseflies and blowflies. (*a*) *Calliphora vicina* (the bluebottle); (*b*) *Musca domestica* (the housefly); (*c*) *Sarcophaga carnaria* (the flesh-fly); (*d*) *Fannia canicularis* (the lesser housefly); (*e*) *Lucillia sericata* (a greenbottle); (*f*) *Pollenia rudis* (the cluster fly); (*g*) *Muscina stabulans*; (*h*) *Stomoxys calcitrans* (the stable fly). All specimens are females with the heads of the males above. (After Graham-Smith [216].) (*a*) (*c*) × 1½; (*e*) × 2; (*f*) (*g*) × 2½; (*b*) (*h*) × 3; (*d*) × 3½.

(b) Egg

In appearance the eggs of the housefly are rather like minute pine kernels; they are cylindrically oval, white bodies about 1 mm long. The incubation period varies from about 8 hours to 2 days, depending upon temperature, and there are a number of records available for different air temperatures. It should be remembered, however, that owing to fermentation, the matrix of the breeding material will be considerably warmer than the air.

When the embryological development is complete, a longitudinal split develops in the egg shell and the young larva emerges.

(c) Larva

The young larva is repelled by light and burrows into the food material, which it consumes. As already mentioned, the putrefying substances in which fly maggots breed generate heat through bacterial action. In large heaps of manure or refuse a temperature of 70° C is often reached, which is too high for insect life; but the temperature falls off towards the periphery. The maximum temperature and the steepness of the gradient depend on the nature and bulk of the fermenting material, the tightness of packing and the external air temperature. The fly maggots choose a layer according to their preference which is remarkably high [142]; they are frequently found in manure or refuse at a layer where the temperature is 45–50° C. Under such conditions the larva grows quite rapidly and in the course of its larval life moults three times and there are three larval stages. These three stages are similar in general appearance and habits. The body consists of thirteen segments arranged to form a shallow cone (Fig. 6.3). The mouth is at the point of the cone but there are no ordinary insect mouthparts (mandibles etc); instead there is a pair of vertical black hooks inside the mouth which articulate with a small black internal skeleton. At the posterior end of the body are two spiracles which communicate with the simple respiratory system. These two spiracles, which are borne on small tubercules and are readily seen, are of considerable importance in identifying the maggots of different kinds of fly. The cuticle of the maggot is thin and flexible but it is fairly tough.

Towards the end of larval life, the maggot changes somewhat in colour, becoming creamy rather than white. This is due to the deposition of fat reserves in the body wall and to other internal changes. At this point the larva ceases to feed and leaves the hot, damp fermenting part of the breeding materials. Whereas the larva can live and develop satisfactorily at quite high temperatures, the pupal development must be passed in a cooler environment. If larvae are prevented from leaving the hot fermenting material there is a considerable mortality among the pupae at temperatures above 40° C [143]. A larva developing in a manure stack usually descends to the ground round the periphery of the heap where it burrows into loose soil. If the ground happens to be very hard the maggot sometimes wanders away as much as 50 yards in search of a suitable spot. Finally the larva prepares to pupate by contracting to a smooth ovoid shape and casting the last larval skin, which dries and hardens in this form. As it hardens, the old skin gradually darkens in colour to a deep chestnut brown. If fly-infested material is buried beneath a covering of earth, the larvae will usually burrow their way up until they are not far below the surface before pupation. To prevent their emergence the covering of earth must be thick (about 2 feet) and well consolidated.

(d) Pupa

Inside the brittle ovoid skin cast by the larva and called the 'puparium' lies the pupa in which the rudiments of the anatomy of the adult fly can be discerned (Fig. 6.3). The very great reorganization of tissues necessary to change a larva into an adult are begun towards the end of larval life and they are carried to completion in the pupa.

When the adult fly is ready to emerge, it presses against the front of the puparium which splits in a perfectly regular manner characteristic of most higher types of Diptera. A circular split develops round at the level of the sixth segment and two lateral splits run forward from this, leaving a split cap at the front of the pupa case which is easily detached and leaves the way clear for the emerging insect.

During the emergence from the pupa case and from the earth in which it is usually buried, the adult fly makes use of a very peculiar organ on the head. This is an evertible bag which arises from a crevice between the eyes. By alternately inflating and deflating this tiny bag, the insect clears a course for itself through the loose soil. When it finally emerges to the air, the inflated bag (or 'ptilinum') is withdrawn into the head and is never used again.

(e) Adult

The relatively large wings of the adult fly are constricted by the pupal skin and, on emergence, are in a somewhat crumpled condition. One of the first acts of the young adult fly is to expand the wings so that it can fly. This is accomplished by the internal hydraulic pressure of the blood before the adult cuticle has hardened. Newly emerged flies can often be found on or near their breeding ground. Apart from the crumpled wings in process of expansion, they can be recognized by the pale colour of their not yet hardened cuticle. When the hardening process has finished, the fly has reached its final condition, no further growth is possible.

Some aspects of the fly's anatomy are important in relation to its habits and these will be briefly considered. The division of the body into head, thorax and abdomen is very obvious because of the marked constrictions separating them. The most prominent features of the head are the large compound eyes which contain about 4000 facets and testify to the relative visual acuity of the housefly. The antennae of the housefly, like those of other flies of this type, are reduced to short stumpy appendages consisting of two small and one large segment, the latter bearing a characteristic bristle.

The antennae, as already mentioned, carry olfactory organs which enable females to detect breeding medium and increase male reactions to female odour. In addition, both sexes are attracted by certain odours, presumably associated with food. Generally speaking, a mixture of odours is more attractive than any single component. One attractive mixture is a malt extract with addition of a little acetal, alcohol and skatole [71]. More recently, it has been shown that decaying egg solids are attractive to both houseflies and the American eye-fly, *Hippelates collusor*. The active components were trimethyl amine, indole, ammonia and linoleic acid; and these were more attractive together than separately [442].

The olfactory organs of flies, however, appear to be less well developed than in related flies and blowflies. It has been suggested that the housefly, in its limited environment, has less need of accute olfactory perception than, say, the blowflies, which display astonishing powers of detecting carrion at a distance. The housefly seems to make up for its lack of olfactory perception by restless, inquisitive behaviour, combined with acute eyesight [611]. Any new object (e.g. a black tile) introduced into a cage of flies, will attract numerous visits, especially during the first 20 min [437]. Flies are also attracted by aggregations of other flies (even dead ones, or black marks simulating them) [612]. These habits assist flies in discovering new sources of food, or locating those frequented by other flies.

It is not known whether flies can distinguish colours, although this faculty has been demonstrated in bees. There have been intermittent investigations of the 'colour preferences' of houseflies giving a rather confusing picture, since some of the conclusions appear contradictory [258]. Perhaps the explanation is that flies are more sensitive to short wavelength light, so that blue appears brighter than red. A disturbed fly, seeking escape towards a light, will then choose a blue source; whereas flies allowed to settle quietly seem to prefer dark sites and often choose red in preference to other colours.

The mouthparts of the fly are so very highly modified from the primitive type that it is impossible to recognize their affinities by mere inspection. It appears that the lower lip (or labium) has become changed into a flexible proboscis. This proboscis is hinged in the middle so that the lower part can be either folded up or extended downwards to probe the ground. The sides of the proboscis turn forwards to form a groove in which the food channel lies and the upper lip (labrum) with certain other elements, forms the roof of the groove. At the end of this apparatus are two sucking lobes, joined along the centre, each lobe being traversed by fine suction canals which collect liquid towards the central feeding channel. There are no mandibles or any other parts for piercing or biting so that it is clear that the fly can only feed on liquid food. If a fly is dissected it will be found that the food channel or pharynx, after entering the head, runs back through the neck into the thorax where it divides into two. The lower, more direct, branch continues backwards and leads to a blind sac, the 'crop', in the abdomen.

The upper branch, which begins with a muscular valve, continues as the main intestines, which lie coiled in the abdomen and finally communicate with the anus. Besides these organs there is a pair of long thread-like salivary glands lying in the body cavity and running forward to discharge into the food channel in the proboscis.

The primary food needs of houseflies are carbohydrates (mainly sugars) for energy, and protein, which the females especially require for egg production. Water is also necessary, particularly as flies salivate and pass liquid faeces. Sugar, however, is the most urgent need and flies die more quickly in its absence than when deprived of water. Convenient sources of sugar are found in human dwellings, which also provide sources of protein (e.g. milk). Many other substances may be visited and tasted by flies, such as fresh faeces and perspiration.

Curiously enough, the most important tasting organs of the fly are on its tarsi. A hungry housefly walking over sugar (especially if it is slightly moist) can immediately detect it and will extend its proboscis at once, to feed. Very often, a tiny speck of sugar, or a small droplet, will not satisfy a fly, which then begins a curious circling behaviour, running rapidly round in expanding spirals [436]. This instinctive habit has probably developed as a convenient way of finding adjacent food sources.

As already stated, flies can only swallow liquid food. They can, however, 'lick' over solid foods, using their salivary juice (and apparently also some regurgitated fluid from a previous meal) to liquefy them. The liquid is drawn up the proboscis by sucking movements of part of the pharynx, and passes straight into the crop. If they are undisturbed, flies will feed until their crops are full to bursting point. Subsequently this liquid is regurgitated and appears as a large pendant drop from the end of the proboscis. Drops of liquid are slowly extruded and withdrawn several times and finally the liquid passes through the muscular valve into the main intestine for digestion. It seems quite possible that the object of this regurgitation is to discharge into the liquid a further quantity of salivary juice.

During the process of regurgitation, it quite often happens that a drop of the fly's vomit falls from the end of the proboscis to the surface on which it is resting. These vomit drops, being sweet, are attractive to other flies which readily feed on them. Even more attractive are such traces of other flies on dry sugar or starchy foods, which become partly liquified. It can be readily shown that lumps of sugar contaminated by feeding flies are more attractive to other flies than clean sugar. The marks of these 'vomit spots' can quite often be traced on places frequented by flies, together with darker specks that mark deposition of the fly's faeces. In addition to this way of contaminating their environment, flies defaecate very frequently; probably about every 5 min. This can be demonstrated as follows. Some well-fed flies are confined on clean paper in a Petri dish. After a few minutes this is held under an ultraviolet lamp and the paper will be seen to be covered with flecks of faeces, which fluoresce a bright blue colour, though scarcely visible to the naked eye.

The thorax of the fly is almost entirely devoted to locomotion. It is compact and rather robust and a dissection of this region reveals the relatively powerful muscles which activate the wings indirectly by distortions of the thoracic frame. These muscles constitute 10 per cent of the weight of the body. In flight the wings are set beating at the rate of nearly 200 strokes per second. They beat downwards and forwards and then vertically (with edge uppermost) and move back. This motion supports the fly in the air and moves it forwards at the rate of 4 or 5 m.p.h. [367].

The landing behaviour of flies has been studied in some detail. When an object is moved briskly towards a (tethered) flying fly, the two front legs are raised on each side of the head and the hind legs are extended backwards. This reaction can be evoked by spinning a spiral pattern in front of the insect, which (as with the human eye) gives the impression of an approaching object [571]. More practical studies concern the choice of surfaces for resting. Apart from the matter of colour or brightness (already discussed) flies tend to rest on rough surfaces (straw, sacking) rather than smoother ones (wood, concrete) and least on shiny polished materials (tiles, metal) [16].

Entomologists are sometimes asked 'How far does a housefly fly?' It is as difficult to give a simple answer as it is to the question 'How far does an Englishman travel to work?' Actually, the distances might not be so very different! After some early British experiments, numerous trials were done in the U.S.A., and more recently in Egypt [570], Italy [520] and Japan [454]. Flies marked with dyes or radio-active tracers have been caught up to 5 miles away in 24h, and later, up to 20 miles. For practical purposes, however, the exceptional cases are much less important than the average; and it appears that the bulk of flies in towns are unlikely to move more than a mile or two from the breeding source. On the otherhand, flies in open country will disperse more widely in search of human dwellings. A migratory fly is much influenced by attractive objects it may encounter, so that large numbers congregate in areas of defective sanitation. They will also follow vehicles along roads, especially farm carts or refuse-collecting vehicles.

The legs of the housefly are well developed and normal in appearance. The feet consist of a pair of claws by which the insect is able to grasp roughened surfaces and a pair of pads (the 'pulvilli') which are covered on their ventral surface with innumerable closely set hairs. These hairs secrete a sticky substance and it is by virtue of this that flies are able to walk on highly polished surfaces even when they are vertical or upside down. During the daytime, flies are active and spend much time visiting horizontal surfaces where they may encounter food; e.g. tables in houses or floors of animal houses, dairies, etc. At night, they retire to the upper parts of rooms to rest [312, 586]. When seeking places to rest, flies are especially attracted by narrow strips of material, dark lines or edges. They are especially prone to rest on vertically hanging cords (e.g. lamp fittings) and this habit is exploited in the common sticky fly-paper trap and in insecticide-impregnated cords (see pages 114–115).

The abdomen contains the main digestive and reproductive organs. At the end of the abdomen of the female is a long thin ovipositor, formed of reduced segments which, when extended, equals the rest of the abdomen in length. When it is not in use, it is retracted into the posterior end of the abdomen, the segments being telescoped one within the other so that only the tip is visible from the outside.

Flies become sexually mature a day or two after emergence and thereafter mating can frequently be observed [121]. The male seeks the female primarily by sight (since he will attempt to mate with other males or even small dark objects). In addition, a sexual odour emitted by the females stimulates his lust [102]. Normally, the male springs on to the back of a resting female, grasps her with his legs and strokes her head. The female may reject him by kicking backward; as a result, many too ardent males may be found with frayed wings. If the male is accepted, however, the female introduces her ovipositor into his genital opening and union is effected. The act may last from a few moments to several minutes. The females begin to lay their eggs about four days after copulation.

Adult flies live for about a month in summer; under cooler conditions in winter, they have been kept alive for nearly 3 months.

6.2.3 QUANTITATIVE BIONOMICS

(a) Speed of development

Like other insects, the housefly is profoundly influenced by temperature in its speed of development. There are a number of records of experiments at different temperatures which illustrate this fact, but unfortunately they are mostly unreliable because the temperature measured was that of the air above the breeding medium. Owing to putrefactive processes, the latter is almost always above air temperature, the amount depending on the bulk of the breeding matter. Experiments with flies reared in small quantities of breeding materials kept at different air temperatures gave the data set out in Table 6.1.

Table 6.1 Observations on the speed of development of *Musca domestica* at different temperatures.

	Average durations of different stages (days)							
	16° C 60° F	18° C 64° F	20° C 68° F	25° C 77° F	30° C 86° F	35° C 95° F	40° C 104° F	Reference
Egg	1·7	1·4	1·1	0·66	0·42	0·33	—	[406]
Larva	11–26	10–14	8–10	6·5	4·5	3·5	5·0	[143, 261]
Pupa	18–23	12	9–10	6–7	4·5	4	4	[331, 576]
Total	32	23	19	11	8	6	9	[328, 576]

Generally, the development occupies about 2 or 3 weeks in an English summer and a month or more in spring or autumn. Added to this must be a period of 3 days to a fortnight (according to temperature) for mating and maturation of the eggs. Taking 3 weeks as an approximate average for the complete life cycle during the summer, it is likely that there are about 4–6 generations during the breeding season.

With the advent of cooler weather in the autumn, the life cycle increases in length and there is less and less outdoor activity of the adults. As a result, new breeding grounds get less and less attention from females seeking sites to lay their eggs. The numbers of flies fall off enormously in the winter and their method of passing the winter has long been a point of controversy. It seems most likely that the great majority of flies die off in the autumn and that breeding diminishes. Probably the species continues by slow breeding in manure and kitchen refuse in stables, barns and similar sheltered spots. In the early summer there is certainly a vigorous recolonization of all available breeding sites.

(b) Checks to population growth

Various calculations have been made of the numbers of progeny deriving from a pair of houseflies in a single season, assuming that everyone survived and reproduced. Estimates vary from five billion to 190 000 000 billion. While this is more of an exercise in arithmetic than biology, it does serve to emphasize the rapid powers of multiplication of the fly and equally the importance of natural checks on population growth. Some estimates of natural population changes over the period of a single generation, have been made in the Caribbean, in connection with trials of chemosterilants. The indications were that the natural population of flies was remarkably stable, with increases of about × 1·3 in favourable seasons, and decreases of about × 0·8 in unfavourable seasons. With populations artificially reduced by the chemosterilants, the growth rate rose to × 5, presumably taking advantage of unused breeding material [604]. Fairly high rates might, perhaps, be expected in cool climates, when flies reappear after virtual disappearance in the winter.

The causes of mortality in nature are little understood. One important factor must be the food supply. Heavily infested horse manure often contains several hundred larvae to the pound. But when very large numbers are reached, the larvae compete for food; many are stunted and become undersized adults while others die before reaching maturity. Certain nutritious foods can support very large numbers of maggots. A piece of liver weighing 5 g (just under $\frac{1}{6}$ oz) provides food for up to 70 maggots. Above this number they become stunted and some die; but half of them survive if 300 of them are crowded on this small piece [590]. It is, however, doubtful whether overcrowding of this type is the main check to population growth; if it were, one would expect to encounter many stunted flies in nature, but this is not the case. Perhaps the main restriction may be that few mature female flies in towns actually find suitable breeding material.

Houseflies suffer from depredations of a wide variety of natural enemies. In the egg and larval stages, they are preyed upon by various arthropods but especially by mites (*Macrocheles muscaedomesticae*) [592] and certain ants. The larvae are also sensitive to pathogenic bacteria, including *Bacillus thuringiensis*. The pupal stage may be parasitized by various chalcid Hymenoptera [345, 346, 438]. The adults are sensitive to a few staphyloccocal bacteria, microsporidia (*Octosporea muscaedomesticae*) and various fungi. The best known of the fungi is *Entomophthora* (*Empusa*) *muscae* which causes flies to die in exposed places, their bodies swollen with mycelia of the fungi; this is not uncommon in damp autumn weather.

Scarcely anything is known about the quantitative importance of these natural enemies and, in most cases, it is difficult to see how they could be used for biological control. Tentative experiments have been in mass rearing of *Macrocheles* mites and rather more practical trials with *B. thuringiensis* (see page 205). Another possibility is the fungus *Beauvaria* sp., spores of which can be produced on a large scale and used against insects.

Some vertebrates (frogs and lizards) and some arthropods (spiders) prey upon the housefly, but it is very unlikely that they exert a profound effect on the fly population. A number of different mites and the pseudo-scorpion *Chermes nodosus* are occasionally found clinging to the bodies of flies but it is difficult to decide whether they are at all harmful or not; probably they merely use the fly as a convenient vector for dissemination.

6.2.4 IMPORTANCE

(a) Transmission of disease

Flies and blowflies which breed in media teeming with bacteria, including pathogens, might be expected to carry some of them from the larval stage to the emerging adult. The matter has been investigated by various workers, especially B. Greenberg. It seems that the hostile environment of the larval gut, combined with the antagonism of the normal flora, keep the numbers of pathogens low; and during metamorphosis, there is further suppression, so that 20 per cent of houseflies and 37 per cent of greenbottles were found to be sterile on emergence [224].

The main risk of flies as vectors depends on contamination of the adults. Although they could theoretically carry pathogens on their feet, the amount of contaminant on such small objects would be very small and liable to desiccation, which would eliminate much of the risk. The main hazard arises from flies feeding on liquid containing pathogens, which can often happen. Again, theoretically, vomit drops could transmit infection; but these arise from liquid stored in the crop, which is mainly limited to sugary solutions. Therefore, deposits of excreta constitute the main source of transmission. Furthermore, experiments have shown that bacteria such as *Salmonella typhimurem* can multiply in the mid- and hind-gut and can be passed out at intervals of up to at least a week or so [227]. It was also found that the greatest amount of proliferation occurred with pure cultures of *S. typhimurem*. When mixed with other organisms (as in mouse faeces) the multiplication was less; and there was distinct antagonism with a specific organism (*Proteus mirabilis*) which eliminated infective faeces in about 2 days.

There is no doubt that, in warmer climates, houseflies do take part in disease transmission. Control of flies has been shown to reduce enteric infections (Shigellosis) in southern U.S.A. [602] and in Italy [429]. In Egypt, the housefly and its relative *Musca sorbens* swarm over children's faces in the villages and transmit the germ of trachoma. On the other hand, the medical status of flies in northern Europe is more dubious. In the early years of the century, epidemics of infantile diarrhoea were suspected, on good evidence, of being fly-borne; but this is now much less common [265]. Not only does water-borne sanitation make human faeces much less accessible, but there is evidence that flies are not readily attracted by it, in cooler climates [573]. There are, of course, other sources of harmful bacteria; e.g. slaughter houses, animal faeces, soiled napkins or dressings in hospitals; and in general no one should willingly permit fly contamination of foods (especially that of young children).

6.2.5 CONTROL MEASURES

The principles of housefly control have nearly passed through a complete cycle. Before the Second World War, it was considered that insecticides (which consisted essentially of fly sprays) were merely local palliatives and that effective, lasting control required improved sanitation. The introduction of the modern synthetic insecticides offered an easy and effective method of controlling adults; but the steady growth of resistance steadily eliminated these, so that there has been a substantial return to the original attack on the breeding sites.

However, the current status of resistance varies from place to place, according to the local usage of insecticides. In certain places, some of the synthetic insecticides may be still effective; and since their simplicity encourages people to use them, their potentialities must be considered.

(a) Control with insecticides

(i) Impact of resistance

Development of resistance has grown furthest in industrialized countries with severe fly problems (e.g. the U.S.A. and Japan); while in Europe, heavy usage of insecticides on farms has been notable in Denmark. The succession of insecticides introduced and then abandoned because of resistance has been documented by J. Keiding in Denmark [313] and G. P. Georghiou in California [200]. The organochlorines were discarded about 1952–3. Then a series of organo-phosphorus compounds were tried and resistance was also developed, some of them suffering from cross-resistance to compounds used earlier. (Diazinon, malathion, coumaphos, fenchlorphos, fenthion, naled, dimethoate and dichlorvos.) Finally, the introduction of the improved highly effective pyrethroids soon produced a resistance in areas where they were used intensively.

It has been said that the series of failures outlined above is not universal. In Britain, for, example, the mild climate and relatively few fly nuisances have prevented excessive use of insecticides. Some forms of resistance occur (certainly to DDT and HCH [90]) but recent details are not available. It is probably significant that records of insecticides used on refuse tips by local authorities (mainly for fly control) show a substantial decline in the use of malathion (7000 kg in 1968–9 to 148 kg in 1973–4).

Because the situation varies from one area to another and is constantly changing, it is impossible to give precise and concrete recommendations of insecticides and dosages; but certain guide lines may be set out.

(1) It is unlikely that any organochlorine insecticide would be effective in any industrial country where fly control has been practised; however, some of the organo-phosphorus compounds may still be effective. Wherever possible, an entomologist should be asked to assess the situation and recommend the best available material.

(2) Usage of organo-phosphorus insecticides should be kept to the minimum necessary for controlling severe fly nuisances at peak periods. Since prolonged insecticidal action is most likely to produce resistance, residual treatments to *all* surfaces should be avoided. Instead, insecticide should be applied to selected areas where flies tend to rest at night; or to wall edges, wires, cords or tapes suspended from the ceiling. Suitable compounds and concentrations are as follows [631].

Diazinon,	1–2 per cent; 0·4–0·8 g m^{-2}
Dimethoate,	1–2·5 per cent; 0·4–1·6 g m^{-2}
Fenchlorphos,	1–5 per cent; 1–2 g m^{-2}
Malathion,	5 per cent; 1–2 g m^{-2}
Naled,	1 per cent; 0·4–0·8 g m^{-2}
Tetrachlorvinphos,	1–5 per cent; 1–2 g m^{-2}

(3) Pyrethroids are still effective in most places (although pre-existing DDT-resistance predisposes the flies to acquisition of pyrethroid-resistance). The development of stable, persistent pyrethroids (e.g. permethrin) may offer a temptation to use them as residual sprays. However, this may easily accelerate the development of resistance; and it is suggested that their use is best restricted to the periodic use of aerosols during the fly season [314].

(4) There are considerable possibilities of using toxic baits of organo-phosphorus compounds to counteract resistance which might come if they are used as residual sprays. Their use needs to be combined with reasonably good sanitation and a reduction of attractive food supplies for the flies. The inclusion

of a good attractant (see page 85) might add to the efficacy of baits. Various types have been used.

(ii) Larvicides

The effective use of larvicides must surmount some formidable difficulties. The larvae are generally restricted to small pockets in a large mass of rubbish or manure so that it is very difficult to reach them without using excessive quantities of material. Modern synthetic insecticides should be avoided as, even if temporarily successful, their use as larvicides is particularly liable to induce resistance. On the other hand, the new insect development inhibitors have shown some promise as larvicides. Although they are not immune to the development of resistance, experience in Denmark has shown that many multi-resistant strains do not have automatic cross-resistance to diflubenzuron [15].

(b) Control by parasites and predators

(i) Arthropod parasites

The stage in the life cycle most vulnerable to parasites is the fully grown larva, about to pupate. While a wide range of hymenopterous parasites have been recorded from houseflies, species of two genera are most common in all countries. These are *Spalangia* and *Muscidifurax*, both belonging to the family Pteromalidae. These and other hymenopterous parasites are by no means confined to *Musca domestica*; they attack various dung-breeding fly larvae and are site-specific rather than host-specific.

The practical impact of such parasitism is not easy to assess. An investigation in Denmark suggests that at least 12·4 per cent of manure-breeding houseflies are parasitized [438]. Although this does not seem impressive, the mortality comes after substantial losses in early life stages, when fly maggots are very vulnerable to predators, adverse physical conditions and competition. It is, however, difficult to see how the parasites can be directly used for control. Practical trials of releases have not been encouraging. At least, the benefits of parasitism should be encouraged. For example, dung in chicken houses and barns should not be cleared out *completely* at one time, so that a residual population of parasites remains. Also they should not be killed by heavy doses of chemical larvicides.

(ii) Predators

Dung-breeding Diptera are most vulnerable to predators in the egg, or early larval stages. Later, they congregate in inaccessible recesses of the breeding medium; and they excrete strong concentrations of waste products (allantoin, ammonia) which tend to protect them from would-be predators.

Among the predators of eggs and first stage larvae are various histirid and staphylinid beetles. One of the former, *Carcinops pumilo*, appears to be an important regulating agent in deep-litter poultry houses in Britain [50]. Predators of another type, which may attack young larvae, are the well-grown larvae of some other Diptera, such as *Ophyra leucostoma*, which may actually reduce fly maggot populations [226].

Most authors agree, however, that the main predators of fly eggs and young larvae are mites, especially *Macrocheles muscaedomesticae*. Two or three other species of *Macrocheles* are found on cattle dung, and also *Glyptholaspis confusa*. In poultry houses in Kentucky, *M. muscaedomesticae* was prevalent together with *Fuscuropoda vegetans*, the former feeding mainly on fly eggs and the latter on the young larvae [510]. Laboratory experiments have indicated high levels of control of young larvae, if the mites were present to the extent of 20 per cent of the number of fly eggs. The life cycle of *M. muscaedomesticae* is short; about 4 to 5 days [18]. The female lays about 1–3 eggs a day over a period of about 2 weeks. Unfertilized females produce only males; fertilized ones, mainly females. They consume about 3 eggs or larvae a day.

Problems of practical utilization again arise, especially as the ecology of the mites is not well understood. Larvicidal treatments which tend to be especially harmful to the mites should be avoided. Experimental studies have shown that some compounds (ronnel, diazinon, malathion) are relatively much more toxic to fly larvae than to mites;while others are about equally toxic to both (fenthion, trichlorphon, propoxur, DDT, lindane).

(iii) Competitors

Human and animal faeces have attracted a considerable variety of scavenging insects, including dipterous larvae, which tend to compete with houseflies. The situation is further complicated by the fact that the presence of some competitors may change the physical condition of the medium, perhaps unfavourably for houseflies. Thus, larvae of *Hermetia illucens* breed in human faeces in pit latrines and tend to liquify the medium, making it unsuitable for fly larvae. In the early days of housefly resistance, insecticide treatments of privies killed the *H. illucens* but not the fly maggots. Thus the insecticide had the effect of actually *increasing* fly breeding.

(c) Miscellaneous biological control measures

A variety of other methods of controlling flies have been suggested and investigated in the laboratory; but the relatively few field trials performed have not shown much practical promise. Microbial control by fungi (e.g. *Enthomopthera*), bacteria (e.g. *Bacillus thuringiensis* [81]) or viruses have made little progress. The use of chemosterilants, which appeared promising in the 1960s, would seem too hazardous for practical use.

Sterile male release, a very expensive method, is generally only possible for serious pests, with isolated populations and sparse numbers. Although there have been a few trials with flies [508], they would not seem at all suitable candidates for this technique.

(d) Urban fly control

(i) Sanitation

Household garbage and, even more, the refuse from restaurants, contains a proportion of decomposing material, attractive to houseflies and blowflies, in which they can lay eggs and develop. Accordingly it is very desirable to keep the dustbin area clean and tidy and to use dustbins with properly fitting lids, which are kept in place. Further information on household refuse as a source of insect nuisances is given in Chapter 9 (pages 378–384).

(ii) Larvicides

In the past, commercial 'tip dressings' of undisclosed composition were used; but they were seldom very effective. Sodium cyanide solution (3 gal of 1·5 per cent per yd^2) was recommended, but is too dangerous and troublesome for extensive use. The modern synthetic insecticides have never been entirely satisfactory and they should be avoided because they are so prone to induce resistance.

As a larvicide for use by the individual householder, paradichlorbenzene crystals, used at the rate of 2 oz per dustbin, should give control for a week or two.

(iii) Adulticides

Probably the most widely used control measure for flies in domestic premises consists of aerosols based on pyrethroids sometimes synergized and sometimes with various organo-phosphorus additives. In view of the high efficiency of the newer, potent pyrethroids, however, it is doubtful whether these additives are justified.

For large-scale fly control in urban areas, periodic U.L.V. sprays containing pyrethroids have shown promise. There have been several successful trials of bioresmethrin under difficult urban conditions; it has been applied by mist generators in Cairo [368] and Monrovia [379] and from helicopters in Saudi Arabia [295].

Currently, plastic strips impregnated with dichlorvos are popular; despite some alarmist reports, there is no evidence of toxic danger to man. However, they are only completely effective in conditions of minimal ventilation. By themselves, they would be unlikely to induce resistance; but where the organo-phosphorus insecticides have been extensively used in the past, the local flies may have acquired cross-resistance to dichlorvos.

(iv) Mechanical methods

Under this heading may be mentioned fly papers, fly traps and fly 'swatters', which were formerly widely used, but are now seldom seen. They are rather unsightly and somewhat inefficient. Fly papers, however, can act as indicators of population trends and may be useful in pest control trials [498].

Since flies breed out of doors, it would seem logical to prevent their invasion of houses by screens. In practice, however, this is difficult and expensive. Weatherproof wire screens (10 mesh, 32 swg, is suitable; see page 40); it must be fitted outside all windows and this reduces light and ventilation and precludes the use of hinged windows which open outward. Two pairs of swing doors separated by a trap porch are required for external communication, to prevent entry of flies. These difficulties restrict the employment of fly screens; although they may be worth considering for special places such as restaurant kitchens, food factories (including jam, which attracts wasps) and portions of hospitals.

An ingenious method of preventing the entry of many flies is to provide a curtain of air moving across a doorway. Experiments show that an air speed of 7·62 m s^{-1} (1500 ft min^{-1}) will exclude about 75 per cent of flies. Higher velocities were more effective, but objectionable to people passing through the barrier [382].

(v) Electrocution

Several devices are marketed for killing flies (and other flying insects) by electrocution. High voltage electricity (4500 V at 10 mA) is supplied to alternate wires of a grid, and insects settling on the latter are immediately incinerated (with a sharp hiss and smell of burnt insect). Usually, a mercury vapour lamp is fixed behind the grid with the object of attracting the insects; and a tray is placed below it to collect the electrocuted insects.

(e) Rural fly control

(i) Mechanical measures: screening

On fly-infested farms, it may perhaps be worth considering the advisability of screening living quarters, or at least the kitchens. It is worth noting that relatively coarse 18·75 mm ($\frac{3}{4}$ in.) netting or hanging bead screens, will exclude quite a proportion of houseflies; this is sometimes useful for protecting temporary structures such as dining tents or field latrines.

(ii) Sanitation

On a farm, several sorts of dung may be available for fly breeding of which pig dung and horse dung are preferred by the fly. Some breeding will occur in the layer of straw and dung in stables and pigsties, but if these are cleaned out at reasonable intervals, the main breeding sites will be the dung heaps. Manure stacks beside stables and on allotments are also likely to be infested on a large scale. Whereas the flies from a badly kept manure heap on a farm will usually only annoy the occupants, the same nuisance in stables or market gardens in a partly built-up area will spread to adjacent households.

(iii) Biothermic method

Manure piled in a heap retains the heat generated in fermentation so that the great bulk of the dung is too hot for fly maggots which are accordingly restricted to a layer below the surface. If the pile is tightly compacted the heating is more uniform and persists for a longer time. A roughly cubical stack should be built with vertical sides patted tight with a shovel. New additions of horse dung (which is attractive to flies only while fresh) should, if possible, be buried in the hot centre of the pile under dung which has been exposed for some days [514]. This method is more successful with horse manure than pig dung because the latter does not ferment so quickly nor reach so high a temperature.

(iv) Maggot traps

With some preliminary care and expense, fly breeding in manure heaps can be permanently reduced. The method is based on the tendency of fly maggots to leave fermenting manure and burrow in the soil to pupate. To prevent this, maggot-trapping manure stands can be built. The simplest form consists of a concrete base surrounded by a small moat of water. If the manure is kept tightly stacked as described above the maggots will migrate from it and will be trapped in the water of the moat. A more complex type consists of wooden slats or a wire frame to hold the manure over a shallow trough filled with water. The manure is stacked tightly and occasionally watered in hot weather and the migrating larvae fall into the water as before.

The following measures have also been suggested but they are of doubtful practical value:

Manure attractive to flies (horse, pig) can be covered with cow dung which is hardly ever infested. Unfortunately cow dung is scarce in the summer fly season since the cows are out grazing in the fields.

Manure spread in thin layers to dry before stacking is scarcely attacked; but this is a laborious method and uncertain in a damp British summer.

Chickens allowed to scratch over a dung hill will eat numerous maggots; but it is doubtful if this biological control is highly effective.

(v) Covering breeding material

An effective fly control measure, which admittedly involves some trouble and expense, is to cover up a breeding site completely. This prevents the emergence of flies from maggots already present and denies other flies access for oviposition. Suitable breeding sites are manure stacks on farms, which may be covered with good sacking or, better, tarpaulin [575]. The method has also been found effective on a large mass of decomposing waste from a tomato cannery. This was covered by plastic sheeting measuring 100 ft × 144 ft. It abated a severe fly plague and was expected to last at least 2 years [387].

(vi) Larvicides

Larvicidal treatments are handicapped by the danger of poisoning farmyard animals and by the possibility of impairing the manurial value of dung. This rules out some highly poisonous persistent chemicals, including arsenicals. On the other hand, fly maggots are very tolerant and are usually protected by the great bulk of the dung, so that large doses of insecticide are usually required to kill them [538].

Among the older recommendations are the following treatments (per ton of manure): 4·5 l creosote oil; 1–1·5 kg boric acid; 2·3 kg borax; 0·7 kg paradichlorobenzene. More recently, 28 g thiourea per 25 ft^3 manure has been found effective against flies resistant to chlorinated insecticides [311].

(vii) Systemic treatments

Various insecticides (mainly organo-phosphorus compounds) have been used to prevent fly breeding in chicken dung by adding them to the food or drinking water of the birds. There have been comparative trials of different compounds, some promising results with dimethoate, and with tetrachlorvinphos (including observations of effects on the hens) [534]. As mentioned, however, treatment of breeding sites is rather liable to provoke fly resistance to organo-phosphorus compounds. The possibility of using insect development inhibitors in this way should be considered.

Biological control, by adding spores or *Bacillus thuringiensis*, has shown some promise. Thus, mash containing 500 mg kg^{-1} spores ($25–75 \times 10^9$ g^{-1}) prevented fly breeding in chicken dung.

(viii) Adulticides

Remarks and precautions regarding the use of sprays and residual treatments for fly control have already been made.

(ix) Baits

Poisoned baits may control flies in some cases, even when residual treatments fail. If applied properly they can be used in farm buildings without harming animals or contaminating milk. They may be prepared as (1) dry baits, for scattering, (2) liquid baits, (3) paint-on baits, (4) solid mixtures for 'bait-stations'. Some of these are commercially available; but, with a little trouble, they may be prepared from concentrates of malathion, diazinon, ronnel, dichlorvos, trichlorphon, naled, Dicapthon or dimetilan.

(1) Dry baits are prepared by mixing 1–2 per cent insecticide with sugar (which should be discoloured by admixture of a little lamp-black to prevent its accidental consumption). Dry sugar bait is applied from a shaker-top can or glass jar on *dry* floors, window sills and other places where flies gather. Usually about 56g per 1000 ft^2, but if flies are numerous, heavier applications may be advisable.

(2) Liquid baits are made from water containing 10 per cent sugar, molasses or syrup and 0·1–0·2 per cent insecticide. They may be sprinkled on impermeable surfaces (clean concrete) at the rate of 5 l/100 m^2. Alternatively, they can be offered to flies in bait dispensers made from chicken-watering units modified by inserting a cellulose sponge in the trough, to prevent its becoming clogged with dead flies.

A long-established liquid fly bait, still occasionally used, may be prepared by adding a teaspoonful of 40 per cent formaldehyde to a cup of milk-and-water. This is offered in a saucer to the flies.

(3) Paint-on baits are prepared in a sticky form for application to vertical surfaces, such as posts, fences or the walls of animal pens. They are useful where dirty or muddy floors preclude the use of scattered dry or liquid baits. A 1–2 per cent mixture of insecticide with molasses, corn-syrup or a thick sugar-water slurry, makes a satisfactory mixture to paint on surfaces where flies rest. The treatment should last a week or more.

(4) A convenient solid 'bait-station' may be prepared by fastening a 10 cm × 10 cm (4 in. × 4 in.) square of screen wire to a wooden handle about 15 cm (6 in.) long. The wire is coated with bait composed of 50 per cent sugar, 46 per cent sand, 2 per cent insecticide, 2 per cent gelatin (or better still bacto agar). The sugar and sand are thoroughly mixed. Boiling water is used to liquefy the gelatin or agar, with stirring, and the liquid is then poured over the sugar–sand mixture and stirred thoroughly, adding the insecticide. A putty-like consistency should be achieved, adding more water if necessary, and the mixture is then spread over the screen wire squares and allowed to dry.

In use, the wooden handles of the bait-stations are pushed into the soil, round the edges of animal pens or into the manure of poultry houses, so that the baits are held in a vertical position just above the surface. They should be used ~ 3–4·5 m apart, where flies are numerous. They must, of course, be kept away from small children or animals.

(x) Cords and ribbons

In some regions cotton cords or strips treated with organo-phosphorus insecticide have given extended control of flies, when hung from the ceilings of animal houses. They are normally used at the rate of about 1 m for each m^2 of floor area and are mainly effective in killing flies resting on them at night. Flat strips and ribbons are apparently rather more attractive than cords to flies seeking resting sites; and vertically hanging cords or strips are more attractive than those in shallow loops.

Cords and strips impregnated with diazinon and parathion are available commercially in some countries. Because of the high toxicity of these compounds, rubber gloves should be worn while hanging them up and all directions on the packages followed carefully.

6.3 THE LESSER HOUSEFLY AND RELATIVES

6.3.1 DISTINCTIVE FEATURES

After the common housefly, the lesser housefly, *Fannia canicularis*, is the fly most frequently encountered indoors. Its abundance is sporadic, being most common in rural areas especially in the neighbourhood of poultry farms. *F. canicularis* is distinctly smaller than the common housefly, *M. domestica*, and the fourth vein on the wings curves gently away from the third instead of bending sharply towards it (Fig. 6.1). Also, the wings in repose are held parallel (and partly folded over one another) instead of in a diverging position as in *M. domestica*. The flight of the lesser housefly is also characteristic. It spends much time circling beneath pendant lamps or similar fittings, making sudden erratic turns.

6.3.2 LIFE HISTORY

Fannia canicularis can breed in various forms of moist decaying organic matter. It breeds very prolifically in large accumulations of chicken faeces and the ovipositing females will choose this in preference to other kinds of dung, possibly on account of its strong smell of ammonia.

The eggs are about 1 mm long, banana-shaped, with a pair of wing-like longitudinal ridges to assist flotation in a liquid medium. The larva is very characteristic in appearance (Fig. 6.3). It is flattened and each segment bears a number of tail-like processes of uncertain function. There are three larval stages and the final one reaches a length of 6–7 mm. The puparium, as usual, is formed of the last larval cuticle.

Most of the adult flies seen in houses are males, although females are quite common near breeding places. As already mentioned, the males spend much time circling aimlessly under pendant fittings, on which they occasionally rest, usually head downwards. At night, they rest on walls, ceilings and fittings in the upper part of rooms.

Experiments have shown that the biological temperature range of the lesser housefly is lower than that of the common housefly [328, 449]. Thus:

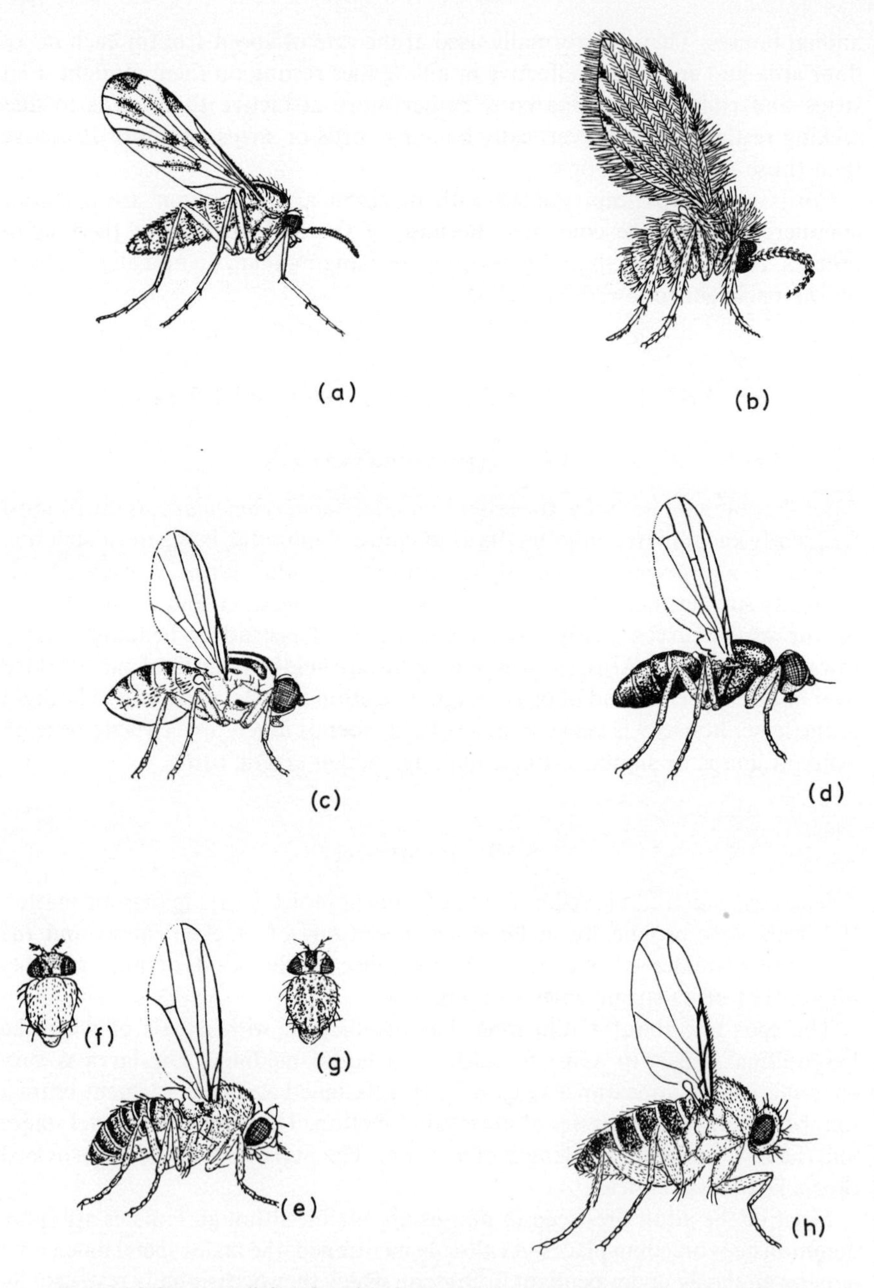

Figure 6.2 Some small flies found in houses. (*a*) *Anisopus fenestralis* (the window gnat); (*b*) *Psychoda alternata* (a moth fly); (*c*) *Thaumatomya notata*; (*d*) *Piophila casei* (the cheese skipper); (*e*) *Drosophila funebris* (a fruit-fly); (*f*) back of thorax of same; (*g*) back of thorax of *Drosphila repleta*; (*h*) *Megaselia scalaris* (a phorid). (*e*), (*h*) after Smart [542], the remainder original. (*a*) × 4; (*b*) × 10; (*c*) × 12; (*d*) × 7; (*e*) × 8; (*f*) × 14.

	Initial movements	Optimum	Heat paralysis
M. domestica	6·7° C	33–34° C	45° C
F. canicularis	4·2° C	24° C	40° C

There is also a suggestion that outbreaks of lesser houseflies occur earlier in the year. In the winter (Massachusetts, U.S.A.) *F. canicularis* has been found in all stages (except the egg) in poultry manure with minimum air and manure temperatures of 3 and 10° C respectively. In the laboratory at 27° C the speed of development is: E, 1 ½ –2; L, 8–10; P, 9–10; pre-oviposition, 4–5. Total (egg to egg), 22–27 days.

6.3.3 IMPORTANCE

Intense breeding in poultry manure at large modern poultry farms has been the source of many severe nuisances due to this fly.

The habits of the lesser housefly rarely bring it in contact with human food and it is therefore not a vector of enteric diseases. It has, however, been discovered occasionally causing intestinal or urinary myiasis.

6.3.4 CONTROL [215, 556]

(a) Operational control

Nearly all the serious outbreaks of *F. canicularis* in Britain are due to accumulations of manure in intensive poultry breeding farms. Many breeders allow the droppings to accumulate in pits, to save frequent disposal, and this provides an ideal breeding medium. The suitability of the chicken faeces can be reduced in two ways: by making it drier or wetter. Use of the deep litter system may somewhat improve the manure, but adds considerably to labour. Alternatively, in the 'lagoon' system, the dung is allowed to drop into a tank of water, the level of which is controlled.

(b) Adulticides

The most satisfactory method of adult fly control has been the use of numerous hanging cords impregnated with contact insecticides. The fly strings are prepared from white absorbent cotton 2·5 mm diameter used at the rate of 6 m per 5 m^2 floor space in the chicken houses. They should be impregnated with fenchlorphos or diazinon, dipping the strings into 25 per cent emulsion concentrate and allowing them to dry. Great care should be exercised in handling these concentrated poisons. Rubber gloves should be worn during impregnation and also in handling the strings when dry (see Gradige [215] for full details).

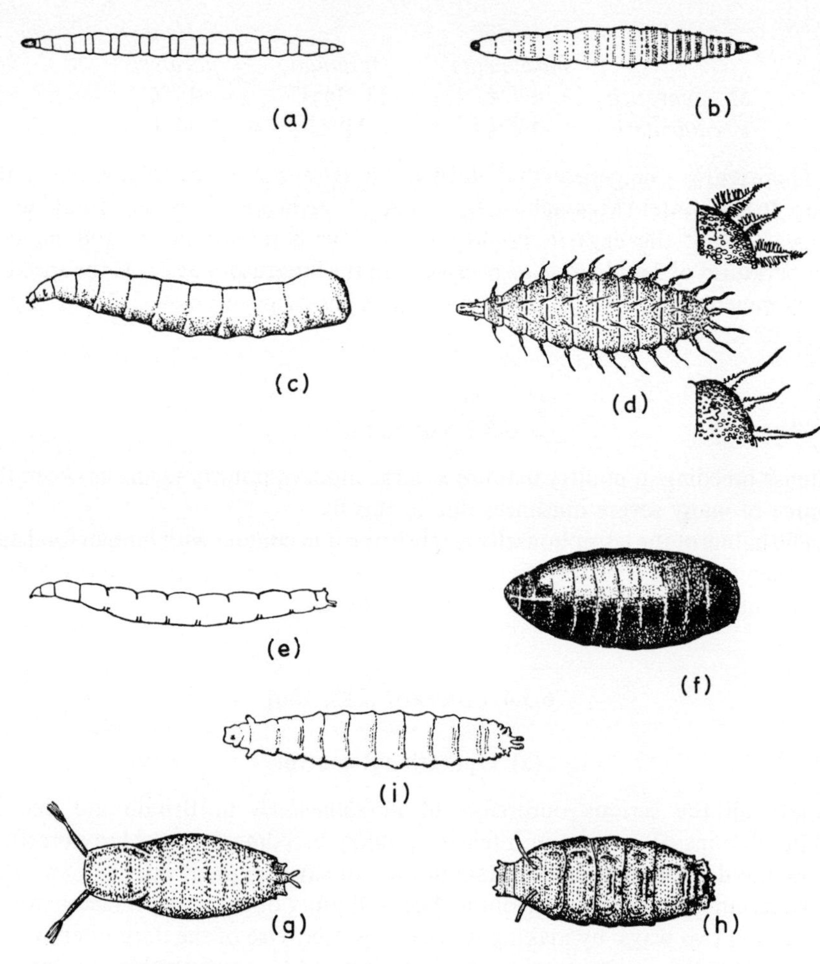

Figure 6.3 Some larvae and puparia of flies. Larvae: (*a*) *Anisopus*; (*b*) *Psychoda*; (*c*) *Musca domestica*; (*d*) *Fannia canicularis* with enlarged drawing of part of the posterior end of same (below) and of *F. scalaris* (above); (*e*) *Piophila casei*; (*i*) *Drosophila melanogaster*. Puparia: (*f*) *Musca domestica*; (*g*) *Drosophila repleta*; (*h*) *Megaselia scalaris* (a phorid); (*g*) and (*h*) original, remainder after Smart [542]. (*a*) × 3; (*b*) × 3½ (*c*) × 4; (*d*) × 7; (*e*) × 5; (*f*) × 6; (*g*) × 7; (*h*) × 6; (*i*) × 9.

6.3.5 THE LATRINE FLY

Fannia scalaris is known as the latrine fly from its habit of breeding in the site indicated. It is very rarely encountered indoors, but may sometimes be found in primitive privies, where the larvae are living in the faeces. It may be distinguished from *F. canicularis* as follows:

Larvae. The larvae resemble those of *F. canicularis* except that the tail-like processes are more distinctly branched (see Fig. 6.3). Also the body segments of

F. canicularis are crossed by median folds on the ventral side, whereas those of *F. scalaris* are smooth.

Pupae. *F. canicularis* has an anal opening hardened by a raised fold forming a raised 'V' with the point forward; while *F. scalaris* has an oval anus, with no V-shaped fold.

Adults. The tibiae of the middle pair of legs of *F. scalaris* each have a distinct tubercle, which is absent in *F. canicularis*.

6.3.6 OTHER RELATED SPECIES [226]

Fannia incisurata is closely related to *F. scalaris* and has similar bionomics. It is holarctic in distribution and has been recorded as the most abundant species breeding in cesspools near Paris. It also breeds in excrement, cadavers and debris in bird's nests.

F. femoralis occurs in the temporate regions of the Western U.S.A. In California, it is sometimes the predominant species breeding in chicken dung, especially during winter. During the summer, it tends to be suppressed by parasites.

6.4 SWARMING HOUSEFLIES [587]

6.4.1 DISTINCTIVE FEATURES

'Swarming' in the sense used here, has nothing to do with the highly specialized swarming of bees, nor of the dancing swarms of gnats and midges associated with mating. In the sense used here, swarming merely implies aggregation of large numbers of flies prior to hibernation. This may cause trouble in certain species which invade dwellings at this time. During the summer months, they live out-of-doors and are seldom noticed. Under natural conditions, most of them hibernate in dry sheltered places, as under loose bark or in hollow trees. In some rural or semi-rural areas, however, they may invade houses or other buildings and it appears that certain houses are chosen year after year. Most probably this is due to accidents of favourable location. Suitable places for hibernation are roof spaces and unheated and unoccupied rooms; also belfries and lofts above church meeting halls, etc. Sometimes they enter living-rooms, which is eventually fatal to them as a rule, because the high temperature keeps them active and depletes their food reserves. The nuisances due to these flies are especially prevalent in autumn, when they invade the houses, and in spring when they leave. In churches and meeting halls, which may be only heated at intervals, the hibernating swarms may be temporarily aroused to a semi-torpid activity and cause consternation by dropping down on to people in this state. Finally, another source of trouble is that flies may fall into roof cisterns in considerable numbers.

There are two distinct types of 'swarming' flies: (1) large (8–10 mm) muscid flies of the species *Pollenia rudis, Musca autumnalis, Dasyphora cyanella*, and

Muscina stabulans; (2) the much smaller (3 mm) chloropid fly *Thaumatomya notata*. Their swarming habits are somewhat different. The muscid species mentioned (which may occur separately, or together in various proportions) tend to aggregate on sunlit faces of buildings during warm days of autumn. As the temperature falls at night, they tend to crawl into crevices, sometimes under tiles and perhaps into spaces under the eaves or open windows. The following day they may emerge and sun themselves; but finally they remain permanently inside until the process is reversed in the following spring.

The small yellow swarming fly, *T. notata*, tends to hover in the lee of buildings, away from the prevailing wind. In cool weather they will invade rooms through open windows and crawl about on the ceiling. Usually, upper stories are most liable to invasion. They cause annoyance from dropping off the walls and flying up again; also the swarming areas may be heavily contaminated with excreta.

(a) Recognition of species

Pollenia rudis (Fig. 6.1), the cluster fly, is not unlike a large housefly in general appearance, although it holds its wings folded over the back, in repose, not at a diverging angle, like the housefly. It is brown in colour and the thorax bears numerous golden hairs (which tend to become rubbed off, however). The abdomen has a dark median line and shifting irregular reflections. *P. rudis* is the most common of the swarming houseflies and it is especially objectionable from the sickly sweetish smell emitted by large clusters.

Musca autumnalis (formerly *M. corvina*), the raven fly, resembles the housefly but can be distinguished by the males (of which some are nearly always present in swarms). Both flies have a yellow abdomen with a central longitudinal stripe. In *M. autumnalis* the stripe is well defined and the yellow area very bright; whereas the stripe in *M. domestica* is less defined and the yellow more dusty pale. In *M. autumnalis* males the eyes practically touch, whereas those of *M. domestica* are separated by a strip about one-sixth of the head width.

Muscina stabulans (Fig. 6.1) also resembles a rather large housefly, but can be distinguished by the fourth longitudinal vein on the wing, which is gently curved forward towards the third vein, not sharply angled, as in *Musca*.

Dasyphora cyanella is very similar in appearance to the common greenbottle flies of the genus *Lucilia*. It can be readily distinguished, however, by the thorax, which has two longitudinal dark marks on the upper surface.

Thaumatomyia notata (Fig. 6.2) is a small yellow fly with black markings.

6.4.2 LIFE HISTORIES

(a) Pollenia rudis

This fly has an interesting and unusual life cycle. The females lay eggs under dead leaves, etc, on the soil. The larvae hatch in about a week and seek certain species of earthworm, pierce them and live parasitically inside. In the original observations of the life cycle, in France, the worm *Allolobophora chlorotica* was involved [315]. In North America the worm *Eisenia rosea* has been attacked. There is no information as to the host worm in Britain, although both species occur here. During its growth the larva (which resembles an ordinary maggot) extrudes its hind end through the skin of the worm, usually near the mouth; thus the larva's spiracles gain access to the air. Eventually, when the host is largely consumed, the larva leaves it and apparently may attack other worms as predators, from the outside. Eventually, when fully grown, the larvae pupate in the soil.

Near Paris, apparently, there are two generations a year; but around Washington D.C. the warm summer accelerates development to the space of 4–6 weeks and there are several summer generations.

(b) Musca autumnalis

The females lay eggs on tiny stalks on patches of dung in the fields. The larvae feed on the dung and eventually pupate in the soil. The adults associate with cattle and horses in the fields in the summer months, feeding on secretions and sweat. Though unable to pierce the skin, these flies will readily drink blood from small scratches or the punctures of true bloodsucking flies. Introduced into North America about 1950, it has spread widely and become a troublesome pest of cattle, being known there as the face fly. It can act as a vector of the filarial worm *Thelazia rhodesii*, which infests the eyes of cattle, sometimes causing blindness.

As a swarming housefly, *M. autumnalis* is troublesome in spring, when it leaves houses, as well as in the autumn, when it enters.

(c) Dasyphora cyanella

This species also breeds in cow dung scattered in the fields. The females lay batches of 25–30 eggs, just below the surface of the manure. In south-west Scotland, the eggs take 1–3 days; larvae about 4 weeks; pupation 3–4 weeks; total about 8 weeks. Adults may be seen in the fields from May to November, after which they hibernate.

(d) Muscina stabulans (Fig. 6.1)

The adults of this holarctic species are always found in the vicinity of human settlements, in rural areas; they are rare in towns. They will develop in human faeces, both scattered and in privies; in other faeces and in decaying vegetation, they breed less readily. The young larvae are scavengers, but older ones attack and feed on other maggots. Development takes 31 days at 16° C, 18 days at 21° C and 14 days at 28° C.

(e) Thaumatomyia notata [481]

The eggs are laid in the soil around the roots of grasses and the young larvae attack and feed on the root aphis, *Pemphigius bursarius*. There are two generations per year, the second of which normally spends the winter as pupae in the soil, which emerge in the spring.

In a warm autumn, however, adults emerge from many of these pupae, and these constitute the swarms of tiny flies which seek shelter in houses when the cool weather arrives. Most of these adults, however, do not survive hibernation.

6.4.3 CONTROL OF SWARMING HOUSEFLIES

No substance is known which will repel these flies for several weeks and prevent them entering a building. On the other hand, the insecticides lethal to the ordinary housefly will effectively kill the hibernating species. The problem is one of application.

(a) Lofts, barns, roof spaces, belfries, etc

In these situations, where there are no furnishings or decorations to spoil, insecticidal smoke generators (see page 111) can be used satisfactorily. Those containing DDT, *gamma* HCH or a mixture of the two are generally satisfactory. Insecticidal smokes leave very little residual deposit, except on horizontal surfaces; therefore it is advisable to employ them when the flies have moved into their winter quarters. Application during strong winds should be avoided, as many roofs are surprisingly porous and the smoke may be blown away too rapidly. Smoke penetrating between tiles and through similar crevices may be beneficial in driving out flies lodged in these places. Smoke generators should be ignited with sensible precautions against fire; generally speaking this presents no difficulty. In addition, open cisterns in roof spaces should be covered during their operation.

(b) Large rooms, halls, etc

A space spray of 1·3 per cent pyrethrins (or its synergized equivalent) in light oil should be dispensed from a mechanical aerosol generator at the rate of 0·5 l per 1000 m^3.

(c) Domestic rooms

A household aerosol based on pyrethrins is convenient and effective, whether applied from a liquefied gas aerosol dispenser or a hand atomizer. It is advisable to warm up cold rooms before using the aerosol, to increase activity of the flies.

A useful method which obviates the collection of dead flies is to collect them directly with a vacuum cleaner and then to draw into the bag a little DDT or HCH dust and leave for a few hours before emptying.

6.5 FLIES FROM SEWAGE WORKS [582]

Certain kinds of small flies breed in large numbers in the filter beds at sewage works (see Chapter 9). Normally they are harmless and even beneficial to the proper operation of sewage filtration. Occasionally in the summer months, however, conditions favour enormous proliferation of these flies and large swarms are produced which become a nuisance in areas adjacent to the sewage works. The species most liable to cause trouble are *Psychoda alternata* and *P. severini* (tiny black 'moth flies') (Fig. 6.2) and *Anisopus fenestralis* (a rather large 'window gnat') (Fig. 6.2). The presence of large numbers of these flies is objectionable from their liability to get into the eyes, nose or mouth. Apart from being a nuisance out of doors, both types (but more especially *Anisopus*) may invade houses in considerable numbers. A still more unpleasant feature of *Anisopus* is the readiness with which it lays masses of eggs on any moist object. Damp foods may often have to be discarded for this reason.

It appears that insecticides and methods used to control adults of the ordinary housefly will destroy these unpleasant little flies. But their enormous numbers in the limited nuisance areas reduce the value of such measures to doubtful palliatives. More radical methods at the breeding sites are discussed in Chapter 9.

6.6 BLOWFLIES (CALLIPHORIDAE)

6.6.1 DISTINCTIVE FEATURES

The morphological character distinguishing the family Calliphoridae (e.g. from the Muscidae) is the presence of a row of bristles on the hypopleuron below the hind thoracic spiracle. While this systematic criterion is somewhat subtle, many blowflies can be easily recognized by their general appearance. The well-known bluebottles, greenbottles and the grey flesh fly, being representatives of the family in Britain.

(a) Calliphora

Calliphora spp. are the common bluebottles, which are dull metallic blue and rather bristly, about 11 mm long with a wing span of about 25 mm. *C. vicina* (formerly *erythrocephala*) is the common species in Europe and North America. It was sometimes confused with *C. vomitoria*, which prefers more mountainous areas, especially round chalets and hotels, among various refuse. *C.vicina* has a face with reddish jowls bearing black hairs; whereas *C. vomitoria* has black or grey jowls, with reddish-yellow hairs.

In the forest zones of eastern Europe, the related *C. uralensis* is common; and in North America, the group is also represented by *C. terraenovae*.

(b) Lucilia

Lucilia spp. are the common greenbottles. They are rather variable in colour, from metallic bluish green to copper colour. They are less bristly than *Calliphora*, about 10 mm long with an 18 mm wing span. *L. sericata* (Fig. 6.1) is most common and the most important species (being responsible for most attacks on sheep). *L. caesar* is also prevalent. Distinction of the species depends on numbers of 'acrostical' bristles on the back of the thorax.

(c) Sarcophaga

The flesh flies (Fig. 6.1) are greyish, with a striped thorax and a chequered abdomen. Their length is about 11–13 mm with a wing span of about 20 mm. *S. carnaria* occurs in Europe and northern Asia. *S. haemorrhoidalis* is almost world wide in distribution and is common in North America, south of Canada. In Europe, it is rare or absent in northern countries such as England or Denmark. It is less frequent in houses than some other genera, but often frequents dustbins during summer.

(d) Phormia and Protophormia

The adults of these genera resemble *Lucilia*, but are somewhat larger and darker. *Phormia regina* is dark green with yellowish-green and purple areas; *Protophormia terraenovae* is almost black, with a bluish or greenish-purple lustre. The larvae develop chiefly in carrion, but may cause myiasis in mammals or birds. Adults are frequent on certain fruits and flowers.

(e) Callitroga (Cochliomyia)

C. macellaria occurs throughout North America. The adults are attracted by carrion and garbage and are common in the vicinity of slaughter houses; they are also found feeding on wild parsnip and other plants. The larvae sometimes are involved in myiasis, but less commonly than *C. hominovorax*, the screwworm fly, of Central America which penetrates up into Texas and Florida. A vast campaign of sterilized male release was mounted against this fly and it was exterminated from the Florida peninsula, although there are reports of re-invasion.

6.6.2 LIFE HISTORY [129, 141, 149, 218]

(a) Oviposition

The adult blowfly begins oviposition about 3–7 days after emergence. At 24° C, the comparative pre-oviposition periods for various genera are: *Calliphora*, 4–5 days; *Lucilia*, 4–5 days; *Phormia*, 6–7 days.

The breeding requirements of blowflies are best filled by decaying matter of animal origin. Cadavers are the preferred breeding site of most blowflies, but some are also able to develop in suppurating wounds or sores on living animals. Others will breed adequately in animal faeces or even, occasionally, on decaying vegetable matter.

Blowflies have highly developed olfactory organs, which are very efficient in enabling them to locate breeding sites. During the summer, the exposure of meat or carcases of animals soon attracts female blowflies, seeking to lay eggs. Large egg masses are laid, sometimes comprising as many as 200; and if many blowflies are present, excessive numbers accumulate and many of the larvae cannot complete development. Nevertheless, the body of a small animal weighing only 16 oz will provide sufficient nutriment to breed up to 4000 blowflies. Generally speaking, blowflies are reluctant to enter buildings unless attracted by the presence of flesh; in this case they often enter by a sunlit window or door and leave soon after laying their eggs. Sometimes, however, *Calliphora* will oviposit at night, although *Lucilia* apparently does not.

Certain substances are particularly attractive to blowflies. Waste entrails at slaughter houses are especially attractive and can become heavily infested with eggs soon after removal from the body. Blowflies also show preference for cut surfaces of carcases and also the kidney region, which *Lucilia* in particular is prone to choose for egg-laying. Liver, on the other hand, is not attractive for the first 3 h after removal from the carcase, although it will readily become infested thereafter.

(b) Egg

The egg period varies according to the age of the egg when laid and the temperature and humidity; most commonly, eggs hatch in 12–16 h. If suitable material on which to oviposit is not available, the gravid females retain the eggs as long as possible. In such circumstances, eggs hatch in a very short time; both *Calliphora vicina* and *C. vomitoria* have been observed to deposit living, newly hatched larvae. This is the normal practice in *Sarcophaga*.

(c) Larva

For the first few hours after hatching from the egg, the maggots are more or less indifferent to light. Thereafter they tend to avoid it and burrow into crevices in the meat. The larvae feed on the liquefied tissues, dissolved partly by proteolytic bacteria and partly by enzymes secreted by themselves. They grow rapidly and moult three times. Shortly before pupation, most blowflies cease feeding and tend to migrate away from the feeding medium. Migration usually occurs at night. It has been suggested that this dispersion helps to avoid the concentrated attack of parasites and predators. The larvae travel over the ground until they find suitable loose soil in which to burrow and pupate. Usually this is within ~6 m; but if the ground is hard or waterlogged, they may travel 25–30 m.

(d) Pupa

Pupation usually occurs in the top 2 in. of soil (sometimes a little deeper). If soft earth is not available they will pupate in cracks in walls, under sacks or other objects. *Phormia* larvae, unlike those of other blowflies, often pupate on the surface of the breeding medium, unless it is very wet or exposed to bright light.

(e) Adult

Blowflies are strong fliers and range considerable distances from the breeding sites. In country districts, populations of *C. vicina* up to 2500 ha^{-1} have been estimated in peak periods. In urban districts, aggregations occur in areas attractive to the blowflies, such as refuse and garbage, especially that from butchers, fish shops and, above all, slaughter houses.

Blowflies are most active in warm sunny weather. In cool cloudy weather all of them (but especially *Lucilia*) tend to disappear to resting sites. These are usually on vegetation adjacent to the attractive sites (*Calliphora, Lucilia*) although *Phormia* tends to disperse more widely. In warm weather, *Phormia* often rests on sunlit walls.

The various blowflies vary in relative prevalence at different times of the year. Normally, all of them hibernate, in the soil, as fully grown larvae. Adults of *Calliphora* begin to emerge first, usually in April, but sometimes as early as late January. *Phormia* adults are observed next and finally *Lucilia* adults begin to emerge in May. In midsummer *Lucilia* become considerably more abundant than flies of the other genera; but, in late autumn, the *Lucilia* decline first and *Calliphora* adults are last seen.

(i) Speed of development

At 24° C single cultures on liver. The two figures for L are feeding period + post-feeding period. *C. vicina* E, 0·15–0·3; L, 4–5 + 2·5–7·5; P, 8–13. *L. sericata*: E, 0·5; L, 4–5 + 4·5–9; P, 7–9. *P. terraenovae*: E, 1; L, 6 + 1; P, 5–7. Total development in competition in slaughter-house refuse. *Calliphora*, 19–22; *Lucilia*, 13–19; *Phormia*, 9–16. *Sarcophaga* (data from another source) at 21° C, 25; at 25° C, 17; at 29° C, 13.

6.6.3 IMPORTANCE

In nature, blowflies perform a useful function in disposing of carrion; nevertheless, their habits render them annoying or troublesome to man in several ways.

(a) Transmission of disease germs

Blowflies are very commonly heavily contaminated with micro-organisms. One investigator, who recorded an average of 27 000 organisms on houseflies, found 3½ times as many on bluebottles and 14 times as many on flesh flies. In many cases pathogenic organisms are involved. Thus, blowflies having access to carrion and offal in slaughter houses have been shown to be contaminated with bacteria *Salmonella* and *Clostridium* [222, 223, 225]. Their chances of transferring these to human food about to be consumed, however, is not very high. There is also evidence that *Lucilia* sp. may carry poliomyelitis virus (especially during epidemics) and it has been suggested that this presents an opportunity for transmission of the disease, if these blowflies visit small wounds [450].

(b) Losses due to 'fly-blown' meat, etc

Meat or fish intended for human consumption must be protected from oviposition of blowflies, otherwise it will become 'fly-blown'. At slaughter houses, bacon-curing or fish-curing establishments this may be a source of economic loss due to condemning of 'blown' meat or fish. In butchers' shops or fishmongers' there is an additional risk of unfavourable reactions of potential customers. In the domestic larder there is not only the wastage of spoiled food but, where cooked meat is concerned, the possibility of swallowing live eggs or maggots; this will be considered in Section 6.7.1.

6.6.4 CONTROL [220]

The major breeding sites of blowflies, at least in urban areas, are waste products; either domestic refuse or the offal and other waste of slaughter houses. Accordingly the problems of control are considered in Chapter 9. Other aspects considered here are the protection of food and avoidance of fly-attracting odours.

(a) Fishmongers' shops, etc [220]

Protection of the fish may be ensured by plastic containers, refrigerated showcases, etc, and this will also tend to reduce fly-attracting odours.

General hygiene involves daily washing of the floors, taking care that no fish waste is left in corners or trapped in internal drain covers. This washing is facilitated by hard smooth floor surfaces. Offal buckets should be removed at least once a day.

Insecticides should not be used when food is exposed (even resident deposits on walls are inadvisable since they may result in partly paralysed flies falling on to the goods offered for sale). After removal of foods, any flies present can be destroyed by use of an aerosol (see pages 105–106). Afterwards, all surfaces on which food is placed should be scrubbed and hosed down.

Hygiene is also essential outside the premises. The floors of yards and outhouses should be of concrete sloping towards drains and should be hosed down at least once a day. Drains, of course, must be kept clear and functioning.

Fish waste and offal should be kept in bins with tightly fitting lids, preferably kept in an outhouse. The contents should be removed daily and certainly at least every two days. The outside of the bins (and both sides of the lids) should be sprayed with 5 per cent DDT suspension, weekly in the main fly season. Walls and fences should be sprayed with the same insecticide at fortnightly intervals.

Fish boxes should be scrubbed out with detergent when empty, rinsed with hypochlorite solution and allowed to dry. (Wooden boxes need 1 litre commercial hypochlorite in 13 litres water; aluminium boxes only 0·1 litre in 13 litres.)

(b) Bacon-curing establishments [426]

At nearly all stages of curing, bacon is in danger from the blowflies *Calliphora* and *Lucilia* as well as (to a lesser extent) from the fly *Piophila casei.* Some protection is afforded during immersion in brine and, during transportation, from hessian wrappings. The main period of risk is just before storing, when the sides of bacon are hung unprotected in the stores.

Since it is the practice to dust bacon sides with pea-flour at this point, the addition of a non-poisonous insecticide to the pea-flour was tried and found effective. The effective mixture was pyrethrins and piperonyl butoxide at the rate of 8 and 80 mg ft^{-2}, respectively. The required amounts, incorporated in the pea-flour, did not affect the flavour of the bacon.

(c) Domestic premises

Indoors the only measure necessary against blowflies is the use of safes (or refrigerators) to protect meat, fish or game from them. In the back yard, careful attention to dustbin hygiene will prevent the attraction of blowflies to the house.

6.7 MYIASIS [299, 652]

By this term is meant the invasion of living tissues of man or animals by fly maggots. In most cases the infestation is accidental or facultative, but there are certain groups of flies which always pass their larval stage parasitically in the body of an animal; these 'bot flies' constitute a veterinary problem and will not be considered further.

Various parts of the body may be concerned in myiasis, especially the intestinal tract, various natural cavities lined with mucous membrane and open wounds or sores. The flies liable to cause facultative myiasis are distributed among a considerable range of families with members having the common habit of breeding in decaying vegetable or animal matter and adults which shelter in houses. (The common housefly, however, is rarely implicated, it does not breed in materials likely to be eaten or drunk.) Since the flies concerned are obnoxious in the larval form it is desirable to be able to identify them in this stage. A key to these larvae is given in the Appendix (page 530).

6.7.1 INTESTINAL MYIASIS

There is a considerable number of records of fly maggots being vomited from the stomach or being evacuated from the bowels. All of the offending maggots normally feed on decaying organic matter and the usual mode of entry is by swallowing eggs or young larvae on uncooked food such as cold meat, cheese, fruit, etc. Very often, of course, the insects are destroyed and digested, but sometimes they survive and give rise to symptoms such as stomach pains, nausea, vomiting and diarrhoea with discharge of blood.

Certain accounts describe astonishingly prolonged disturbances attributed to intestinal myiasis, in which patients have been said to pass maggots with the faeces over a period of months or even years. In view of these prolonged attacks, it has been suggested that the maggots must be capable of developing and breeding in the intestines, either by a paedogenesis (non-sexual reproduction by larvae) or by completing the life cycle and by the adult flies mating inside the intestines. Such speculations have given rise to considerable doubts, especially in view of the fact that the flies concerned are very liable to oviposit in faeces so that their presence in stools might be explained by contamination of the receptacle. Sceptical writers have even doubted the possibility of larvae ever passing through the intestines alive, since in experiments with various animals, fly larvae have almost always been killed after a short period in the gut. However, it is difficult to refute the observations of certain competent observers of these cases and the position may be summarized as follows.

Eggs and larvae of certain flies are sometimes swallowed with food and may persist alive for some time, giving rise to acute gastric or enteric disorders. Cases have been reported where live larvae have been evacuated or vomited at intervals over long periods. Lack of oxygen and other conditions render it difficult to understand how larvae can exist for very long periods in the bowel: therefore such instances should be carefully investigated to determine whether the stools or vomit could have been contaminated after evacuation. Alternatively it may be possible that some habit of the patient renders repeated infestation likely. The species of fly concerned may give a clue to this.

A totally different explanation of myiasis of the lower bowel is infestation *per anum*. People using primitive privies or latrines, especially if constipated or with sores in the peri-anal region, are liable to attract flies when at stool. The flies might have opportunity to oviposit and presumably this might give rise to myiasis if the larvae entered the anus.

6.7.2 URINARY MYIASIS

There have been reports of maggot infestations of the bladder and urinary meatus, although this is less common than intestinal myiasis. The flies concerned are mainly *Fannia* spp. and *Musca domestica.* In investigating these conditions, the same caution is needed to exclude contamination of urine. (The author has once or twice been brought clothes moth larvae alleged to be passed in urine but which probably fell from the patient's bedding into the receptacle.)

6.7.3 MYIASIS OF WOUNDS

In Britain, maggot infestation of wounds and sores in man is happily very rare. A number of British soldiers have, however, experienced it abroad in transit to hospital after being wounded. Actually the presence of maggots in wounds is not necessarily serious, provided the infestation is removed at the right time. They tend to feed on necrotic tissues and may expedite healing. Sterile maggots have even been used in surgery for treating tubercular abscesses and complex fractures. However, the natural unregulated infestation is not to be recommended and is certainly very distressing for the sufferer.

In peace time the chief sufferers from wound myiasis in Britain are animals, especially sheep. The flies mainly responsible are *Lucilia*, *Calliphora* and *Sarcophaga.* Infestations may not depend on actual wounds, since *Lucilia* females are attracted by faeces in soiled fleece in the crutch of the sheep. Maggots develop first in the soiled fleece and later break the skin and make a festering wound which attracts further blowflies. The eventual result, if the infestation is not checked, is the death of the sheep.

6.8 THE FRUIT FLIES [152]

The small flies of the family Drosophilidae naturally breed in various fermenting materials, especially decaying fruit. Infestations sometimes occur in vinegar or pickle works, breweries or fruit canning factories. Some species frequent dwelling houses and show a remarkable propensity for finding and hovering about ripe fruit, alcoholic beverages and other substances giving off odours resembling their natural breeding materials.

Two important species are *Drosophila repleta* and *D. funebris.* Considerably more biological data exists for *D. melanogaster*, which is less common in Britain but has been widely used for genetical experiments in many countries. It seems likely, however, that the life cycles of the various species would not be greatly different. The following data are relevant to *D. melanogaster* [357].

The females lay from about 400–900 eggs at the rate of some 15–25 per day. The larvae hatch and burrow into the breeding material. There are three larval stages separated by moults and finally the larvae crawl out of the food to pupate, preferably in loose soil. The total development takes about 30 days at 15° C; 14

days at 20° C; 10 days at 25° C; and 7½ days at 30° C. (*D. repleta* is apparently very similar.) The average adult life ranges from 13 days at 30° C to 120 days at 10° C.

6.8.1 DROSOPHILA REPLETA (Figs. 6.2 and 6.3)

This species is believed to be tropical in origin but has been steadily extending its range in recent decades. The first record in Britain is in 1942 [116] but it is already quite common in London and other localities. It breeds in rotting vegetables and is quite often troublesome in canteens, restaurants, hospital kitchens and so forth. Owing to the predilection of the adult for feeding on faecal matter combined with a habit of alighting on white articles in human houses (plates, tablecloths!) this fly may involve a certain danger of disease transmission [560], especially in such places as hospitals. Owing to its very small size the insect is less noticeable than the common housefly.

The simplest method of control would seem to be to trace the breeding site and remove and destroy the decomposing vegetable matter concerned.

6.8.2 DROSOPHILA FUNEBRIS (Fig. 6.2)

This fly has the habit, like certain others (Phoridae), of breeding in sour milk curds. Milk bottles which are left with residues in the bottom provide suitable breeding sites and the maggots often develop to the pupal stage before the bottles are returned to the dairy. The puparia (which look somewhat like brown seeds) are tightly cemented to the wall of the bottle and cannot easily be dislodged. It quite often happens that these puparia are not cleaned out by the bottle-washing process and are overlooked when the bottles are refilled with fresh milk. The result is often a complaint to the local health department which sometimes results in legal proceedings against the milk vendors.

6.9 PHORIDAE (Figs. 6.2 and 6.3)

The adults of the small flies in this family seldom come to notice, but, as mentioned above, the puparia are quite often discovered in improperly cleaned milk bottles. Phorid larvae breed in different kinds of decaying organic matter and some have frequently been reported as breeding in dead snails. There are several accounts of intestinal myiasis due to phorid larvae.

One of the phorids, *Megaselia halterata*, is a common pest of cultivated mushrooms. While this is mainly a horticultural problem, it is worth noting that the adults are sometimes produced in such vast numbers as to constitute a nuisance to people living near to the mushroom growers.

6.10 THE 'CHEESE SKIPPER' (PIOPHILA CASEI) (Figs. 6.2 and 6.3)

The larvae of the fly *Piophila casei* were formerly not uncommon breeders in the softer and riper varieties of cheese, in which some people considered them to be proof of excellent quality rather than a revolting infestation. An occasional result of this rather medieval outlook was intestinal myiasis accompanied by acute gastric symptoms.

In addition to ripe cheese, the flies will infest other proteinaceous foods such as ham or smoked meat or fish. Sometimes they cause large infestations in meat curing factories.

6.10.1 LIFE HISTORY [329, 509]

The females lay eggs in various sized batches in crevices in or near a suitable food supply. They may be laid on wrappers or on muslin covers. The larvae on hatching find their way to the food and burrow into it. Unlike certain beetle larvae which feed on dried meats and similar substances (*Necrobia; Dermestes*) these larvae are not confined to the superficial layers but may burrow deep into the food. In infested hams, for example, they may penetrate deeply especially alongside the bone. There are three larval stages and the maggots reach a length of about 8 mm.

The larvae are known as 'cheese skippers' from their curious habit of projecting themselves a short distance through the air if they are disturbed. This is achieved by bending the body double and gripping the edge of the last abdominal segment with the mouth-hooks. The muscles are then tautened and, when the grip of the mouth-hooks is released, the body extends violently and projects the maggot as much as 15–20 cm in the air or about 25 cm horizontally.

The fully grown larvae leave the food and pupate in dark corners or crevices nearby. A puparium like that of the housefly is formed of the last larval skin. The adults live for about 5 days at 24° C and about 3 weeks at 15° C on the average, the maximum observed being 39 days. The females can begin oviposition about 10 hours after mating. They lay about 140 eggs (maximum 500) over a period of 2–4 days.

(a) Speed of development

At 27–32° C the incubation of the egg takes about a day and the larval period 5 days. Pupation occupies 8 days at 25° C and 6 days at 30° C. Thus the total development can occur within a fortnight under very hot summer conditions, but about 3 weeks is more likely in the British summer.

6.10.2 RESISTANCE TO ADVERSE CONDITIONS

The larvae are remarkably resistant to certain adverse conditions. Thus, they can survive several hours in an air temperature of 51° C, although immersion for 2 min in water at 54·5° C is lethal. Towards extreme cold, the half-grown larvae are more resistant than other stages; some have survived −15° C for 64 h.

Fully grown larvae will remain alive for 6 months at 9–10° C and the adults may live for up to a month, without food, under these conditions.

CHAPTER SEVEN

Parasites

The pests to be considered in this chapter comprise several quite different types of insects as well as mites and ticks. They are all parasitic on man and are grouped together because this habit causes them to be of medical and hygienic significance. They will be arranged, primarily, according to the type of control measures they require, which depends on the closeness of their association with the host. Pests such as the itch mite, the crab louse and head louse live in continual proximity to the skin and require personal disinfestation to destroy them. On the other hand, fleas and bed bugs, which visit the host only at intervals, demand the disinfestation of furniture and buildings. A compromise will be attempted between this empirical arrangement according to methods of control and certain other considerations such as the relative importance and systematic relationships of the pests concerned.

7.1 PARASITIC INSECTS

7.1.1 THE BED BUG (CIMEX LECTULARIUS)

(a) Distinctive features

The bed bug *Cimex lectularius* is a quite anomalous member of the order Hemiptera. There is considerable variation in form among the members of this large group, which includes the big tropical cicadas, the ubiquitous aphids, the scale insects, the leaf hoppers and aquatic forms like the 'water boatman'. All of them, however, have in common the piercing and sucking type of mouthparts, and the vast majority are vegetarian and use these mouthparts to suck out the sap of plants.

Many bugs are serious agricultural pests. A small proportion of species attack other insects and suck out their vital juices and an even smaller number have adopted the very specialized habit of living on blood sucked from birds and mammals. (This habit, as we have seen, developed at least twice in the evolution of the Diptera.)

There are several families of Hemiptera which include bloodsucking members; the most important being the Cimicidae, to which the bed bug belongs, and the Reduvidae, which includes certain tropical forms not found in Britain ('cone noses', 'kissing bugs' and 'assassin bugs'). The Cimicidae includes two species of *Cimex*: *C. lectularius*, the common bed bug, and *C. hemipterus*, the tropical bed bug. The latter is somewhat less efficient in its adaptation to environment, especially in cooler climates; so that whereas *hemipterus* is not likely to spread into temperate climates, *lectularius* is common in many regions all over the world. Certain other Cimicidae are parasites of bats and such birds as swallows and martins. This seems to suggest that *Cimex* became adapted to man during a cave-dwelling period of his prehistoric past.

The adult bed bug, *C. lectularius*, is a flat, oval insect about 6 mm long. As with other insects displaying only partial metamorphosis, the nymphal stages resemble the adult. The similarity is augmented by the fact that the adult is virtually wingless, except for a pair of small oval scales, representing vestiges of the forewings (Fig. 7.1).

(b) Life history [317, 68]

(i) Oviposition

Bed bugs live away from their host in crevices in the furniture or walls of rooms in which people sleep. It is in these harbourages that the eggs are laid, often in considerable numbers. As the female deposits them, they are covered with a thin layer of quick drying glue which cements them to the surface on which they are laid. This glue fastens the egg down permanently, so that the empty shell remains fixed in position long after hatching.

(ii) Egg

In shape, the egg is somewhat like a rubber teat, the open end being covered by an egg-cap. In size it varies from 0·8–1·3 mm long and 0·4–0·6 mm broad. At a certain point in incubation, the eyes of the young insect are visible through the shell as two pink spots. Hatching is accomplished by the insect breaking off the egg-cap which usually falls off like a manhole cover.

Unhatched eggs are pearly and opaque, whereas, after hatching, the empty egg shells are opalescent and translucent.

(iii) Nymph

The habits of nymphal bed bugs resemble those of the adults. During the day they remain hidden in cracks and crevices, which is facilitated by their flattened shape. At night they may emerge, and from time to time they seek and find a sleeping person (or animal) from which to take a blood meal.

There are five nymphal stages, all of them approximately resembling the adult in shape. The cuticle of the abdomen of nymphs is relatively thinner, however, and displays the colour of the partly digested blood inside; whereas the adult cuticle is stiffer and a mahogany brown in colour. All stages, including the adult, are white to pale straw colour immediately after moulting (or after hatching) but the harder parts of the cuticle darken to an amber or mahogany colour in a few hours.

The sizes of all stages of bugs vary considerably, not only because of individual variation but because of the great expansion of the abdomen after each meal. (The quantity of blood taken at a meal may be from 2½ to 6 times the bug's original weight.) The approximate sizes of the various nymphal stages, before feeding, are as follows [249]: Stage I 1·3 mm, II 2·0 mm, III 3·0 mm, IV 3·7 mm, V 5·0 mm. Each nymphal stage requires one full meal of blood before it proceeds to the next moult. The feeding and many other habits of the nymphs are substantially the same as in the adult stage and will be considered in the next section.

(iv) Adult (Fig. 7.1)

The head of the bug bears two fairly well-developed antennae and a pair of rather inefficient compound eyes (containing about 30 facets, mostly on the dorsal side). The mouthparts, like those of most Hemiptera, consist of stylet-like mandibles and maxillae resting on a groove on the top of a three-jointed labium; the upper lip or labrum overlaps the beginning of this 'proboscis'. Normally the proboscis is carried bent backwards under the head, but it is extended forward when the bug prepares to feed. Comparatively little is known about the stimuli which cause bugs to seek and find the host under natural conditions. There is no experimental evidence to show that a bug can perceive a source of food at a greater distance than 2 or 3 in. At close range, warmth seems to be the predominant stimulus; a hungry bug will follow and attempt to probe a test-tube of warm water. It may be that they get into close proximity by frequent random wandering [507]. Most bugs are found in the bed frame or wall close to the sleeping host, although it is true that some may be found in quite distant parts of the room.

During feeding, the stylets pierce the skin while the labium curves away from them, retaining contact at skin level, as in a biting mosquito. The external stylets (mandibles) are very slender needle-like structures which serve to pierce and lacerate the wound and each has a row of about twenty teeth near the tip. The inner pair of stylets (maxillae) are much stouter and longer; they are grooved on the inner side and, just in front of the mouth, they are locked together to form two canals, dorsal and ventral. The broader dorsal channel is the food tube up which the blood is drawn by the action of a pharangeal pump in the head. The lower canal, which is much narrower, conducts saliva into the wound. The saliva is secreted by glands in the thorax which discharge into a tiny pumping chamber which in turn injects the juice into the base of the saliva canal.

As a rule, the bug takes about 5–10 min to suck a full meal of blood. Occasionally this may be prolonged if it has chosen a site from which it is difficult to draw blood. Old adult bugs take longer than younger ones to feed and very senile ones are sometimes incapable of it, though they can pierce the skin with their mouthparts.

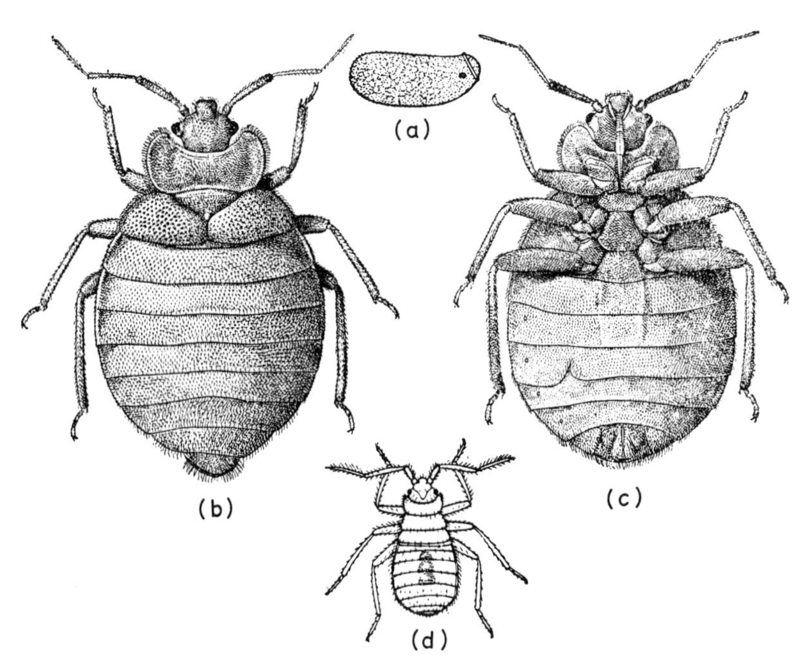

Figure 7.1 Cimex lectularis (the bed bug). (*a*) Egg; (*b*) adult male (dorsal view); (*c*) adult female (ventral view); (*d*) first-stage nymph. From [68]. (*a*), (*d*) × 15; (*b*), (*c*) × 10.

The gut of the bug consists of a narrow tube, running back to the abdomen, where it expands into a large sack-like crop, followed by a coiled midgut and a short rectum. As usual, Malpighian tubes discharge into the front end of the hindgut.

The bug's intestines usually appear to be a dull purple colour owing to partly digested blood inside. The faeces vary from a dark brown or black, viscous liquid to pale straw-coloured matter. The faecal deposits result in the characteristic speckled appearance shown by bug-infested walls, especially in the neighbourhood of harbourages.

Some further points of external anatomy await consideration. The first segment of the thorax is prominent; it bears on each side a characteristic leaf-like expansion which curves forward nearly as far as the eye. The legs are well developed but normal. The feet bear claws which can grasp quite minute irregularities and thereby climb up a vertical surface such as paper or wood. They cannot climb *clean* glass or polished metal, but a slight film of dirt or corrosion gives them sufficient foothold to climb slowly. They can walk upside down on rough paper and similar surfaces, but are not very secure and sometimes fall off. Bugs appear to run quite fast and, at least with forceps, are difficult to catch; actually their average speeds are quite low [newly-hatched bugs travel about 20 cm min^{-1} and adults about five times as fast, i.e. 1/100 and 1/25 m.p.h. (0·016 and 0·064 km h^{-1})] [249].

The greater part of a bug's life is spent in a state of immobility, usually in corners or cracks about a room, or in furniture. They avoid light and are most likely to emerge from their harbourages at night. A light left burning all night will discourage activity but not prevent it; moreover the period of maximum activity is not soon after dark but shortly before dawn. There seems to be some unknown rhythmic factor which induces activity [397]. In cool weather bugs may remain in the same crevices without emerging for as much as a month at a time.

The abdomen of the bug has an almond-shaped outline, the tip being more pointed in the male and more rounded in female. The thickness varies enormously according to the time elapsing since the last meal; the inflated thickness immediately after feeding alters slowly to a wafer-like form after protracted starvation. The abdomen also extends in length (telescopically) after feeding, to accommodate the bulky blood meal.

When they are disturbed, bed bugs emit a characteristic odour due to the secretion of so-called 'stink glands'. These lie on the back of the abdomen in nymphs but are replaced by a large gland under the thorax in the adult. It is sometimes claimed that bug-infested houses can be recognized by the smell, but in the writer's experience, most badly infested houses emit a disgusting mixture of odours characteristic of lack of hygiene. The smell emitted by bugs in a test-tube is not particularly unpleasant.

The genital organs and mode of fertilization are unusual. The male has a sickle-shaped intromittant organ curving round the tip of the abdomen pointed to the left. This 'penis' is not introduced into the egg pore of the female but into a curious organ with an opening underneath the abdomen on the right. In copulation, which does not last long, the male mounts the female and curls his abdomen over the right side of hers and inserts the penis into the copulatory pore. This leads into a bag-like organ and the sperms inserted bore their way through the walls of the bag and migrate through the body cavity to the oviducts. They enter these and fertilize the eggs as usual. The eggs pass normally down the oviducts to the egg pore.

(c) Quantitative bionomics [303]

The development and proliferation of bed bugs are mainly dependent upon the prevailing temperature and the available food supply. The two factors are interdependent, since the need for food is regulated by the temperature. Nearly all our knowledge on this subject relates to artificial laboratory conditions in which the bugs were reared at various constant temperatures and offered a meal at different intervals by being placed close to a host.

In nature, bugs are exposed to fluctuating temperatures and are compelled to find their own way to the host before they can feed. It is probable that the constant laboratory temperature is equivalent to the mean of a variable one, provided that the variations are not very gross. It is clear, however, that the artificial feeding is much more favourable than under natural conditions; it results in quicker development, more eggs and a lower mortality in the laboratory than in natural infestations.

(i) Effects of 'normal' temperatures

Rate of development

The lowest temperature at which bed bugs will complete their life cycle is 13° C. Above this point, the speed of development increases in relation to the temperature in the usual way. Table 7.1 gives the average incubation periods for the eggs and the times for complete life cycles at various temperatures, assuming regular opportunities for feeding.

Table 7.1 Effects of temperature on speed of development and resistance to starvation of bed bugs.

Temperature (°C)	(°F)	Average in days: Speed of development: Incubation	Complete cycle	Resistance to starvation: Males	Females
28	82·4	5·5	34·2		
25	77·0	7·1	46·0		
23	73·4	9·2	61·6	85 (136)	69 (127)
18	64·4	20·2	125·2	152 (260)	143 (225)
15	59·0	34·0	236·7	—	—
13	55·4	48·7	Not completed	338 (470)	360 (565)
7	44·5	No hatch	—	220 (386)	286 (465)

After Johnson [303]. The figures in brackets are the maximum periods observed.

The length of adult life depends upon temperature. With frequent opportunities of feeding, they live from $\frac{3}{4}$–$1\frac{1}{2}$ years at normal room temperatures (18–20° C), about 15 weeks at 27° C and about 10 weeks at 34° C [317].

Feeding

Bugs do not move about spontaneously and will not seek food at temperatures below 9° C. At higher temperatures, they will feed at various periods after moulting (or hatching) ranging from about 6 days at 15° C to 24 h at 25° C.

The interval between one feed and the next depends, in the adult, primarily on the rate of digestion (which, of course, is regulated by temperature). The young stages, however, normally take only one full meal in each instar; the next meal is not usually sought until after the subsequent moult, the time of which also depends on temperature. At moderate temperatures (18–20° C) the nymphs feed at about 10 day intervals and the adults feed weekly. At 27° C the corresponding periods are approximately 4 and 3 days respectively.

Starvation

The length of time which bugs will survive starvation depends to some extent on other factors besides temperature. The figures in Table 7.1 were obtained under the favourable conditions of high humidity (90 per cent R.H.) and undisturbed rest in darkness; they refer to adults, which are more resistant to starvation than the younger stages. Therefore, these data are likely to exceed the periods of survival of bugs under natural conditions, which must be often adverse in one way or another.

It should be noted that the optimum temperature for survival without food is 13° C; below, as well as above this temperature, conditions are less favourable.

(ii) *Effects of adverse temperatures*

Cold

Bed bugs are fairly resistant to short periods of low temperature. The adults are killed by 2 hours exposure to − 17° C or 1 h at − 18° C; but 1 h at − 17° C kills only about 25 per cent [317]. Recently fed bugs are more susceptible than partially starved ones. Two hours exposure at − 15° C killed 76 per cent of eggs.

Very prolonged moderately low temperatures (0–9° C) such as might occur in unheated rooms in Britain during the winter, are unfavourable and result in considerable mortality of eggs and young stages. The eggs die in 30–60 days and the average life of first and second instars is 100–200 days under these conditions.

Heat

Eggs of the bed bug are killed by an exposure to 45° C for 1 h, or to 41° C for 24 h. The thermal death points of adults are 1° C lower in each case. Atmospheric humidity is without effect in these relatively short exposures [396].

Prolonged moderately high temperatures (say 34° C), though not immediately fatal, are injurious and cause a mounting mortality in successive generations, which may eventually cause the extermination of a population [301].

(iii) *Food supply and egg production* [303]

As already remarked, bed bugs will feed on a variety of warm-blooded animals and they develop and proliferate normally on these hosts. Quite often, bug infestations occur in animal quarters of hospitals, in zoos, etc, in spite of the fact that many small animals such as rodents catch and eat a good number of them.

The data relating speed of development to temperatures given above, refers to optimum nutrition (i.e. bugs able to feed as often as they wished). If feeding is delayed or interrupted (though not to the point of actual death by starvation) the development of the bug is correspondingly prolonged.

The food supply available to the adult has a profound influence on egg production in the female. At 23° C if the bugs are fed at 10–15 day intervals, there is a latent period of about 2 days and then approximately 8 eggs are laid over a period of 2–3 days. With more frequent feeding, the latent period is reduced; until finally, with two feeds per week, egg-laying is almost continuous, at the rate of about three per day. Towards the end of the bug's life, egg-laying becomes reduced and there is an increasing proportion of sterile eggs. The total number of eggs laid by a female in the course of the adult life, at 25° C, averages 345, of which about 5 per cent will be sterile.

(iv) Humidity

Under normal conditions, humidity has only a slight influence on the biology of the bug. Low humidity (< 10 per cent R.H.) or very high humidity (99–100 per cent R.H.) both have deleterious effects resulting, for example, in reduced hatching of eggs.

Under conditions of starvation, the lower humidities definitely curtail survival.

(d) Propagation

(i) Dissemination

Where two bedrooms adjoin, it is quite likely that bugs may travel from one to another and cause a new infestation, especially in ill-constructed houses. Bugs have been accused of spreading along a whole terrace of houses in this way, but there is very little definite evidence of the distances to which they will migrate. The fact that in an infested house the bugs are almost exclusively confined to bedrooms should cast doubt on the more extravagant rumours of their colonizing powers.

Infestation of new houses is almost exclusively due to bugs being carried to them passively. Owing to the fact that they do not remain long on the host, they are not often carried on people's bodies or clothing. Occasionally, however, bugs are carried on outer clothing (overcoat, collars, hats, etc.) on to which they have presumably crawled when the article in question was hung against an infested wall. Movable articles such as suitcases, are not infrequently infested and, during the Second World War, bundles of bedding taken to air-raid shelters were certainly responsible for dissemination. Under normal circumstances, the risk is somewhat greater in hostels and lodging houses where there is frequent occupation by transient visitors, who may bring bugs with their luggage. This is rather likely to occur with immigrants from hot countries where bugs are more prevalent.

Another method of bug migration is in timber and other structures taken as firewood, or for other purposes, from infested houses which are being demolished.

(ii) Population growth

There are obvious difficulties which prevent the study of natural bug infestations over a long period. What knowledge we have depends on calculations from laboratory experiments based on temperature records in one or two representative bedrooms. The probable trends at different times of the year have been partly checked by observations on the composition of natural infestations observed on various occasions.

A very great deal depends upon prevailing temperature. The calculations of Johnson [303], which were relevant to Britain but could apply to other temperate climates, are based on the assumption that an infestation begins in winter or early spring with, say, 40 adult bugs (20 females). They will not attempt to feed until the temperature exceeds 15° C (mid-May); thereafter eggs will be laid which will result in another generation of adults, emerging in mid-August and onwards. Throughout the summer there will be progressive multiplication, depending on temperature and feeding facilities, until one to three thousand bugs of all stages are produced. The subsequent history in the following autumn and winter depends very much on whether the room is heated or unheated. In the latter case, feeding will cease in October and all the eggs, and perhaps 80 per cent of adults and nymphs, will die of starvation during the long chill winter. By next spring relatively few bugs, mostly adults, will survive, and the cycle will be repeated.

In a room constantly heated in the cold part of the winter (e.g. a bed-sitting room) the bug population will continue to feed and proliferate during the winter months and a much larger population will be present at the beginning of the next spring.

(e) Importance

(i) Objections to bugs

Bed bugs feed exclusively on blood. They are adapted for living parasitically in human houses, feeding on man. However, they are not highly specialized parasites and will readily feed on other mammals or on birds.

Despite a considerable amount of experimental work, the bed bug has not been shown to be a regular disease carrier; it is certainly not so in Britain. Objection to the bug is partly on account of the unpleasant irritation (and consequent loss of sleep) caused in some people by the bites; but to a large extent it is an aesthetic abhorrence of what is regarded as a loathsome creature. This is augmented by the fact that bug infestation is usually associated with low hygienic standards. This is important; the spread of bugs into new housing estates drives away the more squeamish householders and depresses the standards of hygiene. To this extent, the bed bug may actually be a cause of slums as well as a characteristic feature of them.

(ii) Prevalence

Considerable progress in eradication of the bed bug from Britain was made during various slum clearance schemes in the 1930s [393]. Subsequently, the introduction of residual insecticides, especially DDT, facilitated the destruction of bugs and further progress has been made, though not to the point of complete extermination (see Table 1.1, page 19). Further data available from surveys of a large servicing company, show a roughly stationary position between 1967 and 1973, with 109–189 cases a year [125]. Over this period, bug incidents were only about a third of those concerning fleas. In Denmark, the period 1965 to 1975 showed a disturbing increase from about 25 to nearly 150 annual incidents reported to the state Pest Control Laboratory [238]. It is disappointing, therefore, to find that the progress towards eradication in Europe has been arrested. Perhaps this may be due to the emergence of bug resistance to insecticides, which occurred in tropical countries in the early 1950s [89], but was not reported in Europe until relatively recently.

(f) Control [68]

(i) Detection

The first essential is to ascertain the extent and location of an infestation. Since bugs will normally be hidden in crevices during the day, an electric torch is desirable to inspect harbourages. The spots of excrement already mentioned and clusters of tiny eggs may provide evidence of present or past infestation. To discover whether live bugs are still present, it is useful to spray likely cracks with a pyrethroid aerosol. If this penetrates adequately, it will irritate the bugs and drive them out of their harbourages.

(ii) Minor infestations

Small invasions of bed bugs can occur in any household; for example, as a result of buying second-hand furniture; however, these should not be allowed to persist. Nearly all advisory articles stress the value of household hygiene. It is doubtful whether bugs are actually killed by the soap and water recommended, but some bugs and eggs may be killed by scrubbing. The essential importance of good housekeeping is the early discovery of the bug infestations, which are much easier to eradicate in the early stages.

When such a light infestation has been discovered, the infested article should be throughly cleaned and then sprayed with a household insecticide. A pressurized aerosol containing pyrethroid or dichlorvos should be adequate. In case these measures have not completely eradicated the bugs, the room in question should be searched periodically for a few weeks, with further treatment as necessary.

(iii) Major infestations

Control of large bug infestations has come to depend almost exclusively on the use of residual insecticides. The principal areas of treatment are the bed frame, the surrounding walls, and the furniture. Special attention is given to bed springs, slats, mattresses and cracks and crevices in the walls and floor near the bed. In severe infestations, the undersides of windows and door casings, pictures and mouldings, and where present, wallpapers are treated.

Except where resistance to DDT has been encountered, sprays of 5 per cent emulsion or kerosene solution are the treatment of choice. DDT suspensions prepared from wettable powders are effective, but tend to leave white smears on dark walls or furniture. It should, however, be noted that in many countries bed bugs have become resistant to DDT. Some of these are still susceptible to HCH (used at 0·5 per cent); but often the bugs are resistant to this too. In this case, either an organo-phosphorus compound or a pyrethroid must be used. Malathion can be used at 1 or 2 per cent and is as safe to use as DDT; but in some areas, resistance has developed to this compound too. Alternatives are sprays containing ronnel (1 per cent), trichlorphon (0·1 per cent), propoxur (1 per cent) or dichlorvos (0·5 per cent). The last mentioned has no residual action and may need to be used two or three times. Synergized pyrethrum sprays (0·2 per cent) or one of the new synthetic pyrethroids may be effective; the more stable compounds such as permethrin may have good residual action.

The treatment could, theoretically, be undertaken by the householder; but most people would not wish to do what is really a job for experts. Therefore, either the local Health Department should be asked for assistance or a private disinfesting company should be employed.

Applications should be made with hand sprayers or compressed air sprayers that produce a wet spray. All potential harbourages in the room should be treated. Sheets and blankets should not be treated, but laundered or dry cleaned. Mattresses may be sprayed lightly, but never soaked; and the same should apply to upholstery. On walls, base boards and floors, spray should be applied to the point of run-off; that is about 1 litre per 25–50 m^2 (1 gallon per 135–270 yd^2).

Treatment should be done early in the day so that it will dry before the room is again used for sleeping.

7.1.2 REDUVIIDAE

If the Cimicidae constitute a late stage in the specialization of bugs for parasitism, the Reduviidae represent an earlier stage in the process. They are less degenerate in possessing two pairs of functional wings, though some species do not fly much. The sub-families Reduviinae (assassin bugs), Piratinae (corsairs) and Harpactinae are predaceous on other insects. They pierce these with their proboscides, which contain the same piercing and sucking elements as the bed bug. A proteolytic enzyme is injected, with toxic properties to paralyse the prey, and the juices are sucked out. To the extent that these bugs prey on other (possibly harmful) insects, they are beneficial; but if they are handled, they will prick the human skin, causing a painful reaction not unlike a wasp sting.

The Triatominae are a more specialized sub-family, which has become adapted to blood-sucking parasitism of vertebrates. This group is characteristic of the New World and is particularly prevalent in South America, where various species act as vectors of a serious trypanosomal infection, Chagas' disease. These bugs will feed on a wide variety of wild animals and birds, the former (only) acting as reservoirs of the infection. The extent to which they act as vectors to man, depends on the degree to which different species colonize human dwellings and thus bite man frequently. Such infestations are common in rural areas in South and Central America but not in the U.S.A. This is partly due to the habits of the native species, and partly due to the better construction of the houses in North America, which provide fewer harbourages for the cone nose bugs.

(a) Diagnostic characters

The general appearance of cone nose bugs is shown in Fig. 7.2. They are generally rather large bugs, the adults being about 25 mm long. The proboscis projects from the head, bending underneath it. In the specifically blood-sucking cone nose bugs, the reflexed portion of the proboscis is slender and straight; while in the predaceous sub-families, it is generally stout and arched underneath the body (Fig. 7.2, a and b). The two antennae are fairly similar to those of bed bugs; but the eyes are better developed. The wings are held folded flat on the back in repose, covering most of the abdomen. The basal portion of the fore-wings is somewhat leathery. The legs are well developed for rapid walking.

(b) Cone nose bugs (Triatominae)

(i) Life history [518]

These insects tend to form colonies (or rather, aggregations, since they are not social insects) in the habitat of their host; as in a bird's nest, a lair or in some species, a human dwelling. In the southwestern U.S.A., infestations are often found in the nests of wood rats (*Neotoma* spp.). To some extent, these hosts restrict population growth by feeding on the bugs.

Eggs are laid in crevices, like those of bed bugs, which they somewhat resemble in shape, though larger and sometimes pink in colour. The young hatch as nymphs, resembling the adults and with similar habits. Thus, they hide in crevices in the day time and at night may come out in search of a blood meal, which is very large and makes the abdomen swell enormously. Digestion of the meal takes about a week before they seek another feed; but if they fail, they can resist long periods of starvation. There are five nymphal stages and only one full meal is

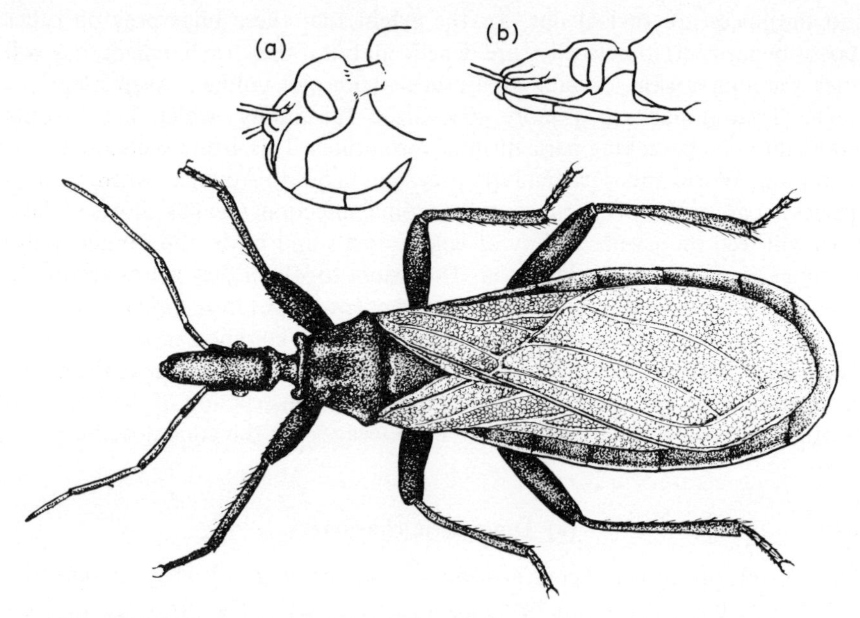

Figure 7.2 Triatoma protracta (original). Above, profiles of (*a*) the head of a predaceous bug (*Reduvius*); (*b*) a cone nose bug (*Triatoma*). (After Ryckman and Ryckman [518])

required in each. The rudimentary wing pads appear in the last nymphal stage and are expanded in the adults, which are very long lived.

The adults are the mobile stage and colonize new sites. They can fly, generally at night in warm weather. These are the forms liable to invade houses in North America. Colonies of some species may be set up indoors, especially in dwellings ill constructed, with harbourages for the nymphs.

(ii) Notes on some important species

Triatoma sanguisuga
This blood-sucking cone nose is most widely distributed in the U.S.A. It occurs in the southern states from the Atlantic seaboard to Arizona and as far north as southern Illinois. The life cycle, under simulated natural conditions, is about 3 years [232].

Triatoma protractor
The Western bloodsucking cone nose is widely distributed in southwestern U.S.A. There are three sub-species. *T. protractor protractor* of southern California and Nevada, extending west to part of New Mexico. *T. p. woodi* is found further east and south in southern New Mexico and southwest Texas. *T. p.*

navajoensis occurs further north, in Colorado, eastern Utah and northern New Mexico [517].

The life cycle of *T. p. protractor* in California can be completed in 1 year; but normally nymphs overwinter and complete their development in a second year.

Triatoma gerstaeckeri
This species is found in southern Texas and eastern New Mexico.

(iii) Importance

Disease transmission
Surveys have revealed that several triatomine species in the U.S.A. are naturally infected with *Trypanosoma cruzi*, the pathogen of Chagas' disease. Infection rates of 20–78 per cent have been found, but nevertheless, only two cases have been recorded of human infections, both in Texas [458]. There appear to be two reasons for this. Endemic Chagas' disease is characteristic of regions with high domiciliary infestations of triatomine bugs; whereas in the U.S.A., most invasions are restricted to wandering adult bugs. (A few reports of permanent infestations of *T. sanguisuga* have come from southern Illinois, apparently feeding on pets [158].) The other reason for North American immunity depends on the mode of infection, which largely relies on infected bug faeces being scratched into the skin, or invading a mucous membrane. Two of the North American species studied were seen to be reluctant to crawl over the human skin while (or after) feeding, unlike many dangerous species, thus lessening the chance of infection [486].

Bite effects
The bites of triatomines are generally not painful, a useful character in a bloodsucker, to avoid disturbing the host. However, many people who have been bitten suffer painful after-effects. These follow the usual pattern of sensitization (see page 472), so that the initial bites cause little or no reaction, but later bites cause more and more severe reactions, until subsequently a partial immunity may develop.

(iv) Control [158]

In countries where cone nose bugs are liable to set up infestations inside human dwellings, the usual defence is to spray the interior with a residual insecticide. HCH has been widely used effectively; but resistance to it is spreading and persistent pyrethroids are being substituted. Treatments should extend to domestic animal shelters, especially chicken houses, which are often adjacent to the dwelling.

In the U.S.A., domiciliary infestations are unlikely to occur in well-built houses, since they provide insufficient harbourages. The main problem in areas where the bugs are common, is the intermittent invasion of adult bugs from infested foci nearby. A search should be made for any squirrel dens, wood rat lodges or other possible sources of the bugs within about 100 yd (90 m) from the house; these should be destroyed.

If persistent trouble is experienced, the house should be carefully screened and examined for any possible entry (defective floor boards, unused chimneys, etc). Pet bedding should be checked and the animals themselves examined during the summer for bugs in their fur.

(c) Assassin bugs and others

(i) Life history

The general outlines of the life histories are similar to those of triatomines. There are nearly always five nymphal stages and the development is rather slow. *Reduvius senilis* in a laboratory colony, laid an average of 85 eggs per female and complete development showed great variability, ranging from 1 to 3 years. *R. sonorensis* likewise ranged from 157 to 564 days, with a mean of 416.

(ii) Notes on common species

Assassin bugs (Reduvinae) [518]

Reduvius personatus. This bug occurs in southern Europe, often in human dwellings, where it has been noted preying on bed bugs. It was introduced into the U.S.A. probably in the 19th century, and spread across the country from the eastern states. In the warmer regions, it breeds out of doors.

Other species. *R. senilis*, the tan assassin bug, is found in the hot deserts of the southwestern U.S.A. and Mexico. *R. sonorensis* occurs in California and Arizona; *R. vanduzeei* inhabits the cooler coastal regions and higher mountains of the same general area.

Corsairs (Piratinae) [497]
These insects are ground dwellers, living under logs and preying on miscellaneous insects. Occasionally, they fly into houses, attracted by lights at night. They tend to make a buzzing or squeaking noise when handled and can inflict very painful bites [386].

Wheel bugs (Harpacterinae)
Many harpacterinae are harmless and beneficial as predators of plant feeding insects. One species, *Arilus cristatus*, is known as the wheel bug, because of a

serrated longitudinal ridge on the thorax, which resembles part of a cog-wheel. This can cause sharp bites. It occurs in the eastern and southern U.S.A., ranging westward into New Mexico.

(iii) Control

The infrequent and unpredictable nature of the bites of these bugs makes control non-feasible. Since the bites are merely inflicted in self-defence, after rough handling, people should brush them away with care. Treatment of the bites is symptomatic (see page 474).

7.1.3 FLEAS (SIPHONAPTERA)

(a) Distinctive features

(i) General

The fleas constitute a small order, only about 1000 species being known, but they are widely distributed about the world. Systematically they are a compact and isolated group, rather similar in form and without obvious affinities to other insects. The thorax is in a primitive segmented condition and never develops wings (though transitory wing buds may be seeen in the pupae of some fleas). However, fleas display complete metamorphosis and are generally believed to have diverged from the primitive Diptera.

All fleas, in the adult form, are parasitic on warm-blooded animals. This is no doubt responsible for their degeneration, involving loss of wings and reduction or loss of eyes. The body is compressed laterally; it is 'streamlined' and covered with backward-directed bristles. These modifications assist in moving through fur or feathers on the host's body. Another well-known characteristic of fleas – their jumping powers – assists them in reaching their hosts and no doubt helps them to escape on certain occasions.

(ii) Identification of common genera and species

A key to the following common genera is given in the Appendix (pages 531–532): *Pulex, Ctenocephalides, Ceratophyllus, Nosopsyllus, Spilopsyllus, Leptopsylla, Xenopsylla.*

The genus *Ctenocephalides* includes the cat and dog fleas, *Ct. felis* and *Ct. canis*. They may be distinguished as follows.

Forepart of the head (from front to 'crown' above antennae) longer than it is high. First teeth of genal comb nearly as long as the second. *Ct. felis*

Forepart of the head as long as it is high. First teeth of genal comb only about half as long as second. *Ct. canis*

(b) Life history [482] (Fig. 7.3)

(i) Oviposition

Female fleas lay their eggs rather indiscriminately, either in the fur, feathers (or clothing) of their host or in the host's sleeping place. If they are confined to a box or tube, they will lay them quite freely on the bottom.

(ii) Egg

The eggs are small, pearly-white, oval objects, without the definite egg-cap of bug or louse eggs. They are of the order of $\frac{1}{2}$ mm × $\frac{1}{3}$ mm long, which is rather large for such small insects, and they may be distinguished with the naked eye.

When laid they are slightly sticky and may adhere to the pelt, plumage or clothing of the host. However, they are quite readily brushed off and they often fall on to the ground in or near the sleeping place of the host.

(iii) Larva

The larvae which hatch from the eggs are tiny, white, legless grubs. They are about $1\frac{1}{2}$ mm long when newly emerged and eventually grow to about 5 mm in length. They are more or less cigar-shaped with a definite head, three thoracic and ten abdominal segments. The head bears a pair of single-jointed antennae, but eyes are entirely absent. The mouthparts include a pair of toothed mandibles, short brushlike maxillae and a labium, the last two bearing pairs of short sensory palps.

The larva feeds on the miscellaneous organic debris to be found in the sleeping place of the host animal (particles of food, faeces, etc); but an especially valuable element in the diet is the partly digested blood in the excrement of adult fleas.

The thoracic segments of the larva are similar to the abdominal ones; all of them carry rings of bristles (about 6–8 short ones and a similar number of long ones on each segment). A pair of spiracles is present on the thorax and on each of the first eight abdominal segments. The last abdominal segment bears a pair of peg-like processes which are used to thrust the animals forwards. With the aid of these and the above-mentioned bristles (which are directed backwards) the larvae progress quite rapidly with a kind of wriggling movement.

Flea larvae are very sensitive to loss of moisture and die on prolonged exposure to dry air. However, they normally live in a protected microclimate which may be quite different from the general climatic conditions. Thus the air in a rat-hole, or under dirt and debris in an ill-kept sleeping place of a domestic animal, may be much more moist than the air of a room.

The larvae pass through three larval stages and when they are fully grown they spin silken cocoons in which to pupate. these cocoons are irregular tent-like covers which often incorporate particles of dirt from their surroundings which

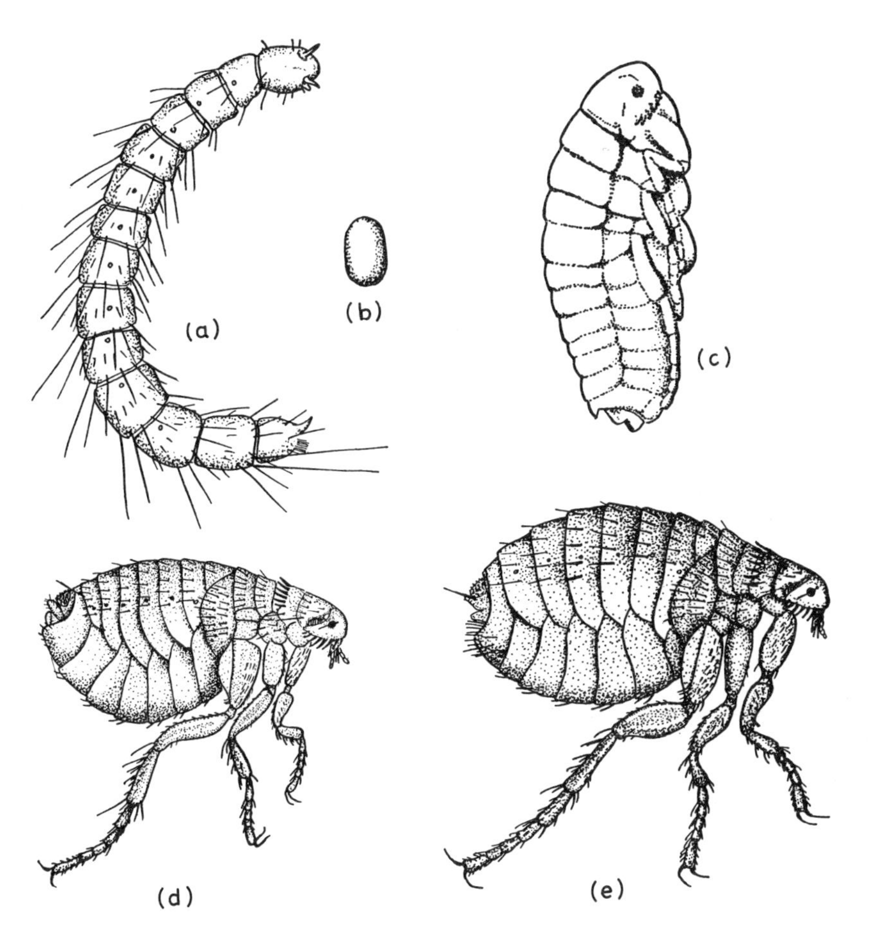

Figure 7.3 Life cycle of a flea. (*Ctenocephalides felis*, the cat flea.) (*a*) Larva; (*b*) egg; (*c*) pupa; (*d*) male adult; (*e*) female adult. (*a*), (*b*), (*c*) after Séguy [530]. (*d*), (*e*) partly after Herms [261]. All × 20.

may serve the purpose of camouflage. In the cocoon the insect changes into the pupal form.

(iv) Pupa

The pupa is of an ordinary type with the limbs free (i.e. not glued to the body). The general shape and appendages of the adult flea can be distinguished. From a creamy white, the pupa gradually darkens to a brownish colour, just before moulting to the adult stage.

(v) Adult

The adult flea does not emerge from the cocoon at once but remains quiescent for an indefinite period. It is stimulated to emerge and begin an active existence by vibrations which indicate the presence of a possible host. This is a useful adaptation for fleas which live in burrows or nests of migratory animals and it also explains why hordes of fleas sometimes attack people entering houses which have been empty for a considerable period. (The progeny of fleas of an earlier infestation, having reached the resting stage in cocoons under rubbish and in cracks in the floors, come out in response to the vibrations of people walking about near them.)

The shape of the adult flea is very characteristic and unlike that of any other insect. As already mentioned it is strongly compressed laterally and is usually easiest to examine when the flea is lying on its side.

The head is helmet-shaped and in some groups it is divided by a furrow, at the level of the antennae. The latter are short club-shaped appendages which can be tucked into a fold on the head when not being used. The antennae of certain male fleas are used to grasp the female during copulation. Eyes may or may not be present. They are rudimentary in form, which is not unusual in parasitic insects. It is uncertain whether they correspond to the ocelli of larvae or are degenerate compound eyes.

On the underside of the front of the head there is present, in many species, a row of short, broad, black spines, rather like a pronounced moustache. A similar row is sometimes present on the back of the first thoracic segment. These are simple but important aids to identification; they are referred to as the 'genal comb' and the 'thoracic comb', respectively.

In the adult stage, fleas take periodic meals of blood from mammals or birds. Nearly always, there is one warm-blooded animal on which they prefer to feed; but, whereas some species are virtually restricted to this particular host, others are more catholic in taste. Some, for example, attack a number of related animals, such as small rodents or nesting birds; others pass from one host to another according to chance contacts as from a rabbit to the fox which has killed it, or from a dog to its master. As might be expected, fleas are more ready to feed on an abnormal host if they are hungry.

The mouthparts, in both sexes, are modified for taking meals of blood exclusively. The piercing elements are the grooved labrum and a pair of sharp, sword-shaped mandibles, the concave inner sides of which form the sucking channel, together with the labrum. The maxillae are shorter, leaf-like structures with sharp points; however, they take no part in the piercing process, but carry four-jointed sensory palps. The labium is short and this too bears a pair of jointed palps which lie alongside the piercing elements, but separate from them during feeding.

Piercing is due to the probing action of the mandibles which are saw-edged at the tip. Saliva, from a short hypopharynx, runs down a groove on their lower edges into the wound. As usual, the blood is drawn up by a sucking-pump in the throat. There is no crop but a kind of gizzard with backward-directed spines, which is of prime importance in the transmission of plague. Normally the blood passes directly into the enlarged midgut and a part of it passes into the rectum and may be discharged at once, undigested.

The three thoracic segments are all freely movable and more or less similar. All the legs are highly modified for leaping and the feet also bear claws for clinging. On ordinary flat surfaces, fleas do not walk very well; but they can travel quite rapidly among fur or feathers. Leaping is useful for reaching the host or for moving from one host to another. The powers of leaping of fleas are sometimes exaggerated, however; they cannot jump upwards more than about 12·5–15 cm.

Different fleas vary in the proportion of their time spent on the host. Those species adapted to animals which regularly return to the same sleeping place tend to make relatively short visits to the host for food and spend much time in the nest, lair, kennel or bed. Other fleas are adapted to spend most or all their time on the host, although they can move to another host's body, if opportunity arises. This is obligatory for bat fleas, for example; and the habit is shared by certain 'sticktight' fleas (e.g. the rabbit flea, in Britain).

The plates of the abdomen overlap smoothly. Like the thoracic segments, each one bears a row of backward-directed bristles. The eight dorsal plates each bear spiracles. At the hind end is a plate covered with small bristles and sense organs. Also, in the male can be seen the clasping organs of the genitalia.

The sexual organs are fairly complex and will not be described here. One important fact, however, must be mentioned. The female receives the spermatozoa into a blind pouch at the end of a tube leading from the vagina. This pouch (the 'receptaculum seminis') can be seen inside the female in specimens cleared and mounted as microscopical preparations. Its shape is used as a method of identification.

In copulation, the male takes up a position underneath the female and grasps her abdomen, which lies above him, with his antennae.

(c) Host preferences and distribution of important species [482]

(i) Pulex irritans

The primary host is man, but it is sometimes found on various domestic animals (dogs, cats and various farm animals) on which it may rarely feed. Pigs are an exception, since *P. irritans* will feed on them readily and will breed in profusion in pigsties. The human flea is also found on certain British wild animals, including the fox, badger and hedgehog.

P. irritans is cosmopolitan and has been for so many years that it is impossible to be certain of its original source.

(ii) Ctenocephalides felis

Ct. felis breeds prolifically in association with the domestic cat and is also found on various wild animals of the cat family. It will also feed on dogs and fairly readily on man. Less commonly it occurs on rats, mice and other small mammals. It is cosmopolitan in distribution.

(iii) Ct. canis

The dog flea may attack other domestic animals (cats, rabbits, etc.) or man. It occurs on various wild animals related to dogs, such as the fox. It is cosmopolitan, but somewhat rare in the tropics.

(iv) Ceratophyllus gallinae

The 'hen flea' is, in fact, a common parasite of many wild birds, especially those which build nests in rather dry situations. It occurs in northern Europe, and part of Asia, but not on the wild jungle fowl from which our poultry were derived. It appears, then, to be a bird flea without strong host preference, which happens to thrive on the domestic fowl. When hungry, this flea may attack man.

(v) Ceratophyllus columbae

The pigeon flea is related to the hen flea, but in contrast it is restricted in its choice of hosts to the domestic pigeon and the rock dove. Like *Ct. gallinae* it will occasionally bite man.

(vi) Nosopsyllus fasciatus

The primary host of the European rat flea is the brown (or Norwegian) rat. It will infest other rodents (e.g. the black rat and house mouse) but is less willing to attack other mammals. Thus, it never occurs on cats (although the cat flea will occur on rats). *N. fasciatus* rarely bites man.

Originating in northern Europe and Asia, it has now spread all over the world.

(vii) Leptopsylla segnis

Adapted to the house mouse, this flea occurs on other rodents, both commensal (rats) and wild. It can be induced to feed on man, but virtually never does so in practice. It is cosmopolitan.

(viii) Xenopsylla cheopis

The tropical rat flea occurs on commensal rats and also on various wild rodents.

It will very readily bite man, which accounts for its special importance as a vector of rodent diseases (especially plague) to man.

It is distributed widely in the tropics; but in Britain it is restricted to rat-infested areas near large seaports, to which it has been introduced by shipping.

(ix) Spilopsyllus cuniculi

The rabbit flea is a parasite of wild and domestic rabbits. It is one of the most sedentary of British fleas commonly attaching itself in the region of the ears. Sometimes it transfers itself to animals which hunt rabbits (e.g. to foxes and even domestic cats). It occurs throughout Europe.

(d) Quantitative bionomics

Investigation of the quantitative biology of fleas presents considerably more difficulty than similar studies on lice and bugs. The effects of simple environmental variables such as temperature and humidity are complicated by other factors which are either obscure or difficult to define. For example, certain common species like *Pulex irritans* are very difficult to rear under apparently favourable laboratory conditions. Again, the resistance of a flea to starvation may be greatly influenced by the nature of their surroundings; fur or cloth, into which they can crawl and remain quiescent, is more favourable than a bare glass tube in which they continually hop about and exhaust themselves. For these reasons, there are widely divergent data given by different authors and sometimes wide variations are observed in one investigation (e.g. survival of certain starved fleas of wild rodents in Russia were observed to vary from 4–292 days at 13–24° C and 6–369 days at 1–15° C) [579].

Additional complications ensue from some inevitable idiosyncrasies of the various important species. Much of the existing work has been done with *X. cheopis*, because of its importance as a plague vector.

(i) Temperature

High temperature has the usual effect of speeding up the development, the frequency of feeding and other processes, and of reducing the length of period of resistance to starvation. Some data are presented in Table 7.2. The figures relating to development are taken from the most favourable averages recorded by Bacot; prolongation in other experiments is assumed to be due to unsuitable food, etc.

To the larval and cocoon periods must be added the incubation time of the egg. This seems to be about a week at moderate room temperatures and correspondingly more or less at other temperatures; precise data are lacking. As a rough guide, it may be estimated that the complete development of many domestic fleas occupies about a month in the British summer. The development of fleas indoors

Table 7.2 Effects of temperature on speed of development and resistance to starvation (at 90 per cent R.H.) of fleas.

Temperature		Average in days				
		Speed of development			Resistance to starvation of *Nosopsyllus fasciatus*	
		Larval stage		Cocoon stage		
(°C)	(°F)	*Pulex*	*Nosopsyllus*	*Nosopsyllus*	Males	Females
30	85·6	13	16	5	12·6 (26)	13·0 (25)
23	74·5	19	20	10	11·1 (26)	12·1 (23)
18	64·5	38	30	20	22·2 (39)	15·7 (27)
10	50	105	100	50	—	—

Figures for development taken from data of Bacot [23], for starvation from Leeson [344]. Figures in brackets are maxima.

in the winter must depend, as with the bed bug, on the temperature maintained.

In a warm environment, with regular opportunities to feed, the adult flea lives for 2–4 months. At low temperatures the insect is sluggish and, under these conditions, the adult life may be prolonged to over a year.

(ii) Food and egg production

The food of the larva can vary in quality and, on an unsatisfactory diet, the development period is prolonged, irrespective of the temperature. As already mentioned the partly digested blood present in the faeces of the adult is an important ingredient in the diet of many larval fleas. The absence of this can be partly made good by addition of some other form of organic iron compound.

The food of the adult, being blood, is less variable in quality, but the frequency of opportunities to feed is an important factor. At warm room temperatures, fleas will feed daily if given the opportunity, and some species which spend much of their life on the host's body (*X. cheopis* and *Ct. canis*) apparently feed even more often; but under cool moist conditions they can survive for long periods with meals at 1- or 2-monthly intervals. Below 13° C they do not attempt to feed.

Very variable results have been found for the resistance of fleas to starvation. The figures given in Table 7.2 are for favourable humidity but the temperatures are rather high and the fleas were kept in bare glass tubes. At 8–10° C the maximum periods observed by Bacot were: 125 days with *Pulex irritans*, 95 days with *Nosopsyllus fasciatus* and 38 days with *Xenopsylla cheopis* [23].

As a general rule, egg production is dependent upon feeding and, as with most mosquitoes, eggs are not laid until after a blood meal. About 4–8 eggs are laid after each feed and the total number finally can amount to several hundred.

(e) Importance

(i) Transmission of disease

Fleas are the vectors of two serious diseases: plague and murine typhus, which are essentially rodent infections which can be carried to man by rat fleas. Obviously the danger depends on the readiness with which the rodent fleas will bite man. Many fleas of wild rodents, as well as the European rat flea *Nosopsyllus fasciatus*, are very unwilling to feed on man; therefore they are never directly the cause of plague epidemics. In contrast, the tropical rat fleas of the genus *Xenopsylla*, willingly bite man and when they occur on the commensal black rat together with a source of infection, they constitute a potential danger of plague.

In Britain, both *X. cheopis* and the black rat are restricted in distribution and there is virtually no risk of a plague epidemic. In the past (14th century Black Death and 1665 plague of London) there have been epidemics in Britain. The black rat was probably more prevalent then but the identity of the flea vector is uncertain.

Fleas can also act as vectors of various tapeworms of domestic animals and rodents and these sometimes get transmitted to humans. The eggs of the tapeworm, in the faeces of the mammal, are swallowed by flea larvae (which live in dirt and debris). Inside the flea, the cestode develops to the 'cystercercoid' stage. Return to the vertebrate intestine is accomplished when a parasitized flea is swallowed. This readily happens when domestic animals or rodents attempt to kill the fleas by biting through their fur. On rare occasions, dead fleas may be swallowed on food or other objects by people (especially children) living in close contact with infected pets. A well-known example is the worm *Dipylidium caninum*, the vertebrate host being the dog (sometimes the cat) and the intermediate host being the flea *Ctenocephalides canis*. The rat and mouse tapeworms *Hymenolepis* sp., normally transmitted via rodent fleas, are also occasionally carried to humans.

(ii) Annoyance from bites

Types of reaction [44]

Persistent attacks of fleas, of whatever species, may cause irritation and loss of sleep. People differ greatly in their reaction to bites, some being hardly affected, others suffering a great deal. Infants, in particular, suffer from a papular urticaria and sometimes scratching leads to secondary infections, such as impetigo.

The primary reaction to flea bites is allergic in nature, induced by a constituent of the flea's saliva. The cause of sensitization appears to be similar to that due to mosquito bites. A person exposed to regular flea bites for the very first time may show no reaction; but if the attacks are continued, sensitization occurs. The first result is an itching papule, surrounded by reddening and swelling. This may persist for a few days. If bites are continued, an immediate skin reaction (in the form of a weal) occurs but disappears within an hour or so. Continued exposure often leads to desensitization, resulting first in the loss of the delayed (and more troublesome) reaction; and later the primary response may also fail to appear.

(iii) Prevalence

Annoyance from bites in temperate countries is now nearly always due to cat or dog fleas [266]. In the Middle Ages, when housekeeping standards were low, human fleas were no doubt common, as they are now in certain rural villages in the Near East [533]. Even in 1899, however, half the fleas collected in German infestations were dog fleas [270]. By 1976, of 2294 fleas identified in Denmark, 90 per cent were cat or dog fleas (about 50:50), 9 per cent bird fleas and only 1 per cent human fleas [136]. A recent British survey confirms these proportions [125].

Although the human flea is declining into a rarity, cat and dog fleas are becoming more prevalent, at least in Denmark. Infestations reported to the state Pest Infestation Laboratory have risen from less than 10 a year to over 2000 [136]. In Britain there is some evidence of a similar trend. In both countries, reports of infestations are lowest in winter and build up to a sharp peak in September before declining.

(f) Control

Measures against fleas may be of two distinct kinds; protection against infestation of the person by adult fleas and eradication of breeding foci in houses.

(i) Protection

One or two fleas may sometimes be acquired in travelling or in visiting crowded places in certain urban districts. This may be unpleasant for the individual, but it is not serious provided that his home is kept clean, for it is very unlikely that an infestation will develop there. This occurrence is so rare for the majority of people that it is not worth taking regular precautions against fleas. However, certain people concerned with medicine or hygiene have to make regular visits to many flea-infested houses in the course of their duty. Such people may find it advisable to use preventive measures. Perhaps the most satisfactory, simple method is to apply repellents to the clothing around such areas as the socks and trouser ends, cuffs and neckband. Fleas need to pass these zones to reach the undergarments. Drops of repellent (e.g. dimethyl phthalate) smeared over the hands can be transferred to the garments by firm rubbing. A thorough application, though invisible, should remain effective for a week or two. Health Inspectors or workmen, who may have to enter very heavily infested rooms, should, in addition, wear gum-boots. These ensure protection from fleas jumping up from the floor.

An ounce of 10 per cent DDT dust sprinkled in the underwear weekly will prevent fleas lodging there, but may not kill them sufficiently quickly to prevent them biting.

(ii) Eradication

An occasional flea bite is a minor discomfort, not to be compared with the constant attacks of an infestation in the home. Fleas cannot infest dwelling houses which are regularly swept and cleaned *throughout*; for the larvae from eggs laid by an occasional adult flea stand no chance of developing on a clean floor. The decrease in prevalence of fleas during the past 30 or 40 years is probably due to improved housekeeping and the widespread use of the vacuum cleaner.

If persistent trouble with fleas occurs in a building, there is probably an infestation (i.e. fleas are breeding) on the premises. The first essential is to find out the type of flea which is prevalent; this will give a clue to the breeding site. It is even more important to extirpate the brood than to kill the adult fleas. The use of insecticides should be combined with thorough domestic cleansing, as there may be more than one breeding site; but if the cleaning is conscientiously done, no focus will escape attention.

Human fleas

These usually infest and breed in bedrooms. If the infestation is heavy it may be advisable to dust the floor first with insecticide. The carpets and rugs are removed and cleaned or at least beaten. Bedding should be laundered. The floor should be carefully swept or cleaned with a vacuum cleaner. If the latter is not available, it is advisable to force up fluff and dirt from between the floor-boards with a knife. The floor should finally be scrubbed.

Cat or dog fleas

The majority of complaints of fleas from well-kept houses can be traced to pet animals. They normally breed in the basket or kennel where the pet sleeps; but there may be other breeding sites in armchairs, etc. In addition, a number of fleas may be present in the animal's fur. Accordingly, both the animal and the breeding sites must be treated.

Fleas on the animal's body are best dealt with by dusting a finely powdered insecticide into the roots of the fur. A simple shaker type dispenser or puff duster is convenient. Particular attention should be paid to the back, neck and top of the head. About one tablespoonful of dust is needed for a medium-sized short-haired dog; and the amount can be adjusted according to the size or length of hair.

Suitable insecticide formulae are 0·5 per cent *gamma* HCH; 1 per cent pyrethrins; 0·2 per cent pyrethrins plus 2 per cent synergist; 5 per cent malathion.

HCH or malathion should not be used on cats, which tend to lick them out of their fur.

The breeding sites should now be traced and should receive treatment. Rugs and pillows should be thoroughly shaken and then either washed or sprayed with insecticide. Straw and old rags or rubbish should be burnt. The basket or kennel should be thoroughly cleaned and, in bad infestations, sprayed with insecticide. Suitable formulae are emulsions containing 1 per cent malathion; 1 per cent *gamma* HCH.

Chicken fleas

Chicken farmers may be annoyed by chicken fleas. The hens may be treated individually with powder applications as described for pets (above). A heavy dusting of the chicken houses with 5 per cent malathion may also be necessary (allow 1 lb to 40 ft^2).

7.1.4 HUMAN LICE

(a) Distinctive features

(i) General

Human lice belong to a small degenerate order of insects, all of which are bloodsucking ectoparasites of mammals. Eyes are reduced or absent and the mouthparts are highly modifed for piercing and sucking. Their bodies are greyish, leathery and translucent. The legs end in single claws, articulated so as to grip the host's fur. No traces of wings are developed (Fig. 7.4).

(ii) Genera and species concerned

The three varieties of human lice are all sucking lice of the family Pediculidae which contains the lice of men and monkeys. The members of this family are unique in possessing eyes, which distinguishes them from other sucking lice.

Human lice belong to two genera, which are quite easily separated as follows:

1. All legs about equally strong (but anterior legs of male stouter than those of female). Abdomen about twice as long as it is broad. *Pediculus*

2. Foreleg slender with long fine claw; mid and hind legs strong with thick claws. Abdomen compressed and broader than long. *Pthirus*

Pthirus (often wrongly spelt *Phthirus*) is represented by only one species, *P. pubis*, the crab louse or pubic louse. There are two forms of *Pediculus*, which are closely related and which have generally been regarded as sub-species: *P. humanus humanus*, the body louse, and *P. h. capitis*, the head louse. Anatomically they differ principally in averages of size and proportion, but the extremes of all measurements overlap. Therefore, it is impossible to assign an individual specimen to either variety with certainty, although that should be possible with quite small samples, provided that they are measured under comparable conditions. Unfortunately, many of their dimensions vary greatly with the time elapsing since feeding, so that measurements are unreliable unless that can be standardized. Generally, body lice are about 10–20 per cent larger than head lice. In addition, head lice tend to have relatively thicker antennae, deeper indentations between abdominal segments and less well-developed abdominal musculature.

The two forms are rather similar in physiological characteristics but the body louse form is definitely more efficient (more eggs, longer life, greater resistance to starvation) at least under experimental conditions. However, the most important difference between the two forms lies in their habits, which induce one form to inhabit the scalp and the other to live amongst the underwear. It seems not unlikely that the *capitis* form was the original type adapted to man before he began to wear clothes and that *corporis* is a new form adapted to the more difficult life among movable (and removable) clothing.

Recent observations on some lice from Ethiopians with double infestations of head and body lice, have shown that populations of the two forms remain quite distinct under circumstances which should allow interbreeding [95]. This is a good argument for regarding them as separate species. Certainly, the two forms present quite different problems in public health, as will be seen.

(b) Life history [99]

(i) The head louse

Oviposition

The females lay eggs on the hairs of the head near the scalp. The lower part of the egg, when it is laid, bears a quantity of adhesive material which glues it to the hair on which the female deposits it. This cement is extremely persistent and retains the empty egg shell long after the young louse has emerged. Thus, the hair of an infested person will contain numerous 'nits', some of them viable, but many empty and harmless. 'Nits' may often be seen several inches down the hairs, but these have usually been carried out by the growth of the hair, in which case they are certainly empty or dead. No method is known of dissolving this cement without harming the hair or the scalp; certainly the idea that vinegar or acetic acid will attack it is false.

Egg

The egg of the louse is approximately oval and rather large for the size of the insect, being about 0·8 × 0·3 mm. Viable eggs are pearly yellowish-white and opaque; after hatching they become translucent and opalescent. With a little practice, hatched and unhatched eggs can be discriminated with the naked eye, or with a low-power magnifying glass.

At the top of the egg is an egg-cap which falls off like a manhole cover when the insect emerges. An area on this cap bears a variable number of cell-like air

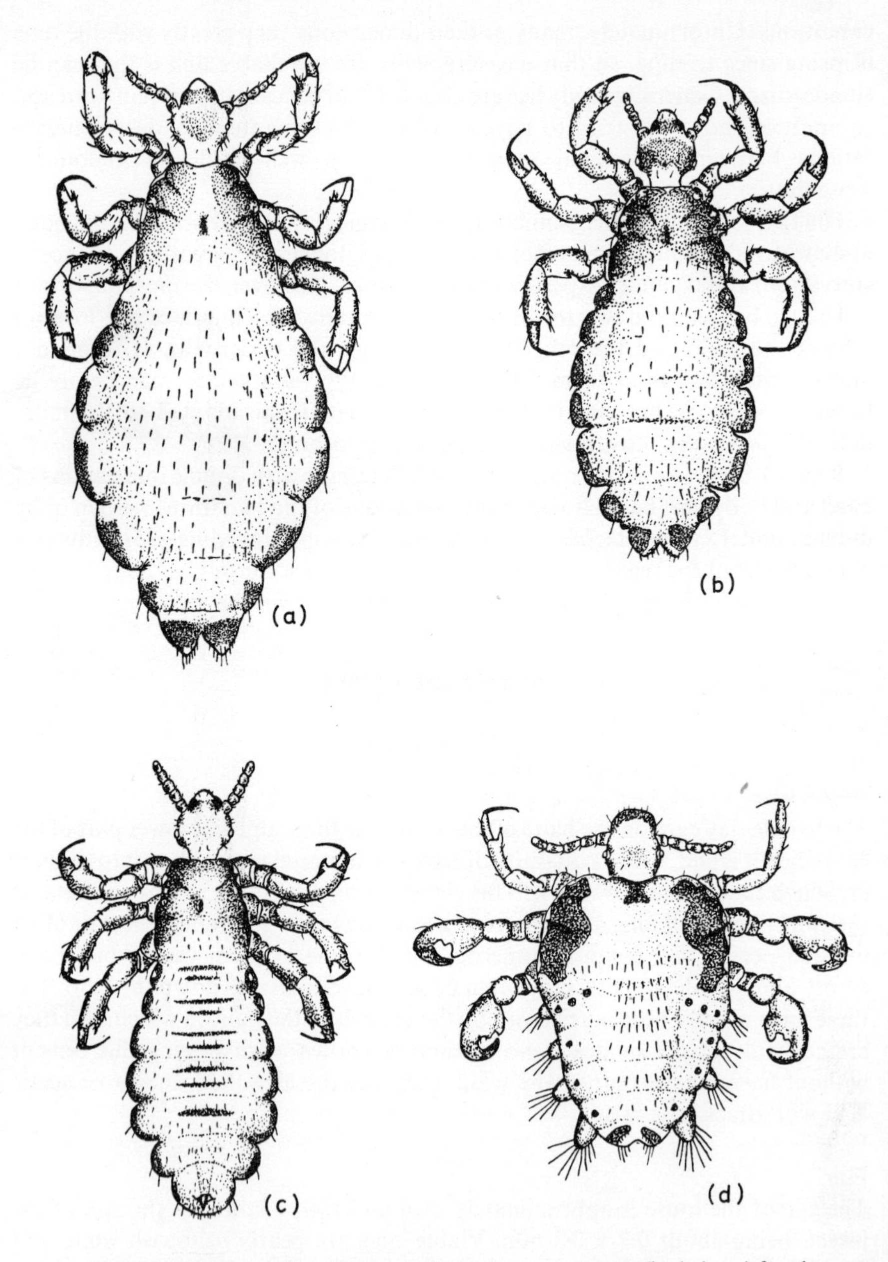

Figure 7.4 Human sucking lice. (*a*) *Pediculus humanus corporis* (body louse) female; (*b*) *P. humanus capitis* (head louse) female; (*c*) *P. h. capitis* male; (*d*) *Pthirus pubis* (crab louse) female. (After Ferris [180]. All × 22.

cavities pierced at top and bottom so that atmospheric air can reach the embryo in the egg. Towards the end of the incubation period, the dark eyes and other structures of the young insect can be seen through the shell.

Nymph

The cuticle in all stages of the louse is partly transparent so that some of the internal organs can be vaguely seen in living specimens. Over most of the abdomen the cuticle is flexible and allows the volume of the insect to increase considerably after a meal of blood.

The first stage nymph is straw-coloured. It is able to feed shortly after emergence and the first blood meal is visible through the cuticle, giving the young insects the appearance of tiny rubies. Later, digestive processes cause the blood to blacken and the gut in older lice is usually purplish-black. Feeding occurs at fairly frequent intervals, at least twice daily.

There are three nymphal stages and after the third moult the insect reaches the adult stage. Anatomically, the nymphs are very similar to the adults, the chief differences being (1) size and proportion, (2) absence of external sexual organs in nymphs.

Adult

Adult lice vary in colour from a dirty white to greyish-black. The pigmentation is most pronounced in the harder parts of the cuticle. The colour of the adult louse, which cannot be altered, is largely an adaptation to the background during nymphal life. Lice from blondes, therefore, are paler than those from people with black or brown hair [84].

The head bears a pair of antennae which are together waved from side to side with a characteristic testing motion as the insect advances. The eyes are poorly developed, which is compatible with the retiring parasitic habits of the louse. The mouthparts are anomalous in being withdrawn into a pouch in the head so that they are not normally visible. The mouth of the pouch bears recurved teeth which, in feeding, are pressed into the skin of the host to anchor the head in position so that the piercing mouthparts can enter with ease. These comprise three superposed stylets, the exact homologies of which are not known. The upper and lower stylets, which are paired in origin, form the food channel; the intermediate stylet carries the duct from the salivary glands. As already remarked, the saliva causes irritation and resultant scratching may lead to skin infections.

The thorax of the adult louse bears no trace of wings. The legs are powerful and curiously modifed for clambering among hairs. A strong terminal claw grasps the hairs by apposition against a peg-like projection higher up the leg. Lice are not very rapid walkers; a distance of 23 cm in 1 minute has been recorded for body louse adults at 20° C, and head lice, though said to be slightly more active, would not greatly exceed this. Both types of louse are repelled by light and walk away from it.

On the abdomen, the hardened plates of the cuticle are restricted to lateral lobes and some narrow strips on the back of the male; the remainder is flexible and leathery. The hind end of the abdomen of the male is pointed, with the anus and sexual orifice just dorsal to the tip. In the female, the abdomen ends in two triangular projections which can be distinguished with the naked eye and this enables the sexes to be distinguished with ease.

In copulation, the male gets beneath the female and grasps her hind legs with the (specially enlarged) claws of his front legs. He curves up his abdomen and inserts a very complex penis into her genital aperture. The union persists for a considerable time and united pairs often may be found walking about and feeding normally.

After mating, the female will lay viable eggs for a variable period up to 3 weeks.

(ii) The body louse

The life history of the body louse is very similar to that of the head louse; the principal differences (apart from quantitative ones, which will be discussed shortly) being in certain habits.

The eggs are glued to fibres of the clothing in the same way as those of the head louse are cemented to hairs. Occasionally the body louse attaches its eggs to the body hairs. There is a tendency for large numbers of eggs to be laid in restricted areas where lice congregate, especially along the seams inside of the garment next to the skin.

The hatching of the eggs, the feeding, growth and moulting of the nymphs and the mating of the adults occur exactly as in *capitis*. All stages of body lice spend most of their time on the clothing, especially on the inside of the undergarment next to the skin. They are gregarious, being attracted together by the smell of each other and of their excrement, and they tend to congregate along the seams of garments, as already remarked.

Body lice probably prefer to visit the skin to feed when their host is sitting or lying quietly. Apart from their visits to the skin, they wander about to a considerable extent among the clothing (especially the older nymphs and adults) and they are particularly active when their host becomes hot. However, smaller and smaller proportions of the louse population are found on successive garments away from the skin. Only on very heavily infested individuals can lice be seen walking about on the outer clothing.

(iii) The crab louse

Relatively little is known about the detailed biology of *Pthirus* because it is a difficult and unpleasant insect to rear in captivity. The life history is not greatly different from that of *Pediculus*.

The egg, for example, is similar but smaller with a somewhat more convex cap. It is laid on a body hair, to which it is glued like the egg of *Pediculus*.

There are three larval stages leading to the adults. All stages are very much more sedentary than head or body lice. They tend to settle down at one spot, grasping hairs with the legs of both sides of the body, inserting the mouthparts and taking blood intermittently for many hours at a time. The legs are adapted to grasping rather large hairs and, in the position adopted, the adult prefers hairs rather widely spaced (compared with the dense hairs of the head). This may partly explain the distribution of the crab louse which is most commonly found on the hair in the pubic and peri-anal regions. It is also found on other hairy parts of the thighs and abdomen and sometimes in the axillae, eyebrows, eye lashes and around the scalp.

(c) Quantitative bionomics

Lice are different from most other insects in being comparatively uninfluenced by climate. Their habitat, close to the human skin, ensures for them an environment with a warm, fairly constant temperature (about 30° C) and moderately high humidity. They have few enemies and abundant food supply. Adaption to this easy life has resulted in lowered resistance to unfavourable conditions. Although they can survive short exposures to cold and heat, they will not grow or develop at moderate or cool temperatures but require conditions like their normal habitat. They soon starve to death in the absence of a host.

The susceptibility of lice away from the host depends on their degree of independence; the sedentary crab louse is most sensitive, while the body louse, which may be removed nightly with clothing, is most resistant.

(i) Head and body lice

The results of a quantitative study of the two races under similar conditions is given in Table 7.3. In these experiments the lice were kept in breeding boxes and worn on the body for various proportions of the 24 h. Lice living permanently on the body have an incubation period of 8–9 days and a nymphal life of similar length. The complete life cycle thus occupies about 18 days, but this may be prolonged if the lice are periodically removed from the body (e.g. in clothing removed at night). It will be observed that the speed of the development is exactly the same in both races. In other respects, however, the *corporis* race is more efficient, at least, in captivity. Under these conditions, body lice show much less mortality during development, longer adult life, more eggs per female, a greater proportion of which hatch and a greater resistance to starvation.

The environment of the head louse is very equable and the insect is able to feed at any time. The body louse, on the other hand, can only feed undisturbed when the host is resting. It has been suggested that its larger size is connected with the necessity for taking relatively infrequent large meals. Certainly it needs to be rather more resistant to starvation than the highly susceptible head louse.

During the Second World War, methods of mass-rearing body lice were developed in the U.S.A. At first, the lice were kept in jars in incubators and fed daily on paid donors. Rearing took 9–11 days and the adults lived 22–23 days with the females laying 4–5 eggs day^{-1} [134]. A remarkable advance was then achieved by adapting colonies to feed on rabbits. These animals differed considerably in their individual suitability as hosts, only 7 of 97 tested being favourable. Later, the lice became better adapted to this strange host. Studies of the rabbit-fed lice were made with incubator temperatures of 29·4, 30·8 and 32·2° C. These gave nymphal development times of 10–12·8 days and adult lives of 14–18 days. The females laid 2·6–2·8 eggs day^{-1}, with 79–86 per cent hatch, so that 34–36 eggs/female were viable [183].

Table 7.3 Bionomic data obtained with head and body louse strains.

	Averages for							
	Head lice				Body lice			
Worn on body (hours per day)	24	12	2 × 3	3	24	12	2 × 3	3
Incubation: days at 30° C	8·5	—	—	—	8·5	—	—	—
Nymphal period (days)	8·5	12·2	17·6	23·0	8·3	12·8	17·5	24·3
Development mortality (%)	15	35	32	97	0	9	9	64
Male adult life (days)	10	—	—	—	20	30	30	13
Female adult life (days)	9	22	17	—	20	21	25	15
Eggs per female per day*	7·5	4·3	2·6	—	11·1	5·7	3·4	1·7
Total eggs per female*	57	56	22	—	110	98	73	15
Hatch (%)*	88	76	64	—	94	91	78	0
Fatal starvation (h) at 23° C	—	55	—	—	—	85	—	—
Fatal starvation (h) at 30° C	—	<24	—	—	—	45	—	—

Data from Busvine [85].
* Over period of maximum egg production.

(ii) The crab louse [451, 475]

The two published investigations of crab louse bionomics date from 1918. Fortunately, they generally agree and are supported by a recent unpublished study (I. Burgess, M.Sc. Dissertation, London). The egg incubation lasts 6–8 days at skin temperature and each nymphal stage about 4–6 days. The total nymphal development took 13–17 days and the adults lived rather less than a month. One female was observed to lay 26 eggs at about 3 per day. Of 200 adults removed from the body, only one survived 24 h.

(d) Parasitology and abundance

(i) Head and body lice

Growth of louse populations [99]
If any infestation begins with a single pregnant female, the subsequent population growth can be calculated from the data in Table 7.3. The populations to be expected after 3 months, under moderate and under rather unfavourable conditions, have been estimated on this basis. The numbers of active stages calculated were 4000–5000 and 400–500, respectively. Now these figures are very rarely encountered in natural infestations even amid a large number of chronically lousy people. Since there are no parasitic or predatory enemies of lice, it must be assumed that the principal checks to population growth are the actions of the infested person in removing and destroying the parasites. The efficiency with which these measures are done depends upon the hygienic standards of the infested individuals, the facilities available and to some extent on their sensitivity to louse bites.

Numbers of lice on infested people [364, 476]
Examinations of people infested with either head or body lice have always revealed that the majority of lousy individuals carry quite small numbers of lice, of the order of a dozen or so. Smaller and smaller numbers of people are found with populations up to several hundreds, while infestations of one or two thousands are very rare.

Liability to infestation
The difference in habits of the *capitis* and *corporis* races which induces one to live on the scalp and the other to live amongst the underwear, results in two quite separate problems of public health. These problems are most distinct in the more modern countries where improved standards of hygiene (especially the regular laundering of underwear) have relegated the body louse to a minor problem in peace time. The people infested by it are mainly vagrants; transitory inhabitants of the casual ward or common lodging house. Others are solitary, indigent folk; the worse cases are usually elderly and often infirm.

The head louse, on the other hand, is disturbingly common in some industrial areas. It attacks quite a different section of the population, mainly girls and young women, often of quite cleanly habits. Sometimes, indeed, the parasite has benefited from misguided vanity when the regular combing of the hair has been avoided in order to preserve the 'set' of a permanent wave.

We owe a great deal of our knowledge on the bionomics of head lice to the investigations of the late P. A. Buxton and K. Mellanby. Buxton studied the numbers of lice found in crops of hair taken from the occupants of hospitals and jails in different parts of the world. Irrespective of locality or race, the general trends were consistent; and these have subsequently been shown to hold good in Britain. Very briefly the conclusions were these: there is a strong positive correlation between weight of hair (which is the best measure we have of length of hair) and infestation. On account of their generally longer hair, girls and women are more liable to infestation than men and boys. There is a negative correlation with age, children being more infested than adolescents, who are more infested than adults. The heaviest rates of infestation are found in groups showing the highest proportion of infested people.

Comparing the proportions of children infested in families of different size, it was found that the percentage increased with the number in family. An examination of the figures for recruits to one of the women's services revealed that infestation was most common among the less intelligent groups as revealed by intelligence tests. These two accounts illustrate the association of head louse infestation with poverty and ignorance.

Transmission

The ways in which lice travel from one person to another are still largely a matter of conjecture. Transmission by fomites must occur to some extent, but this is limited by the comparatively short survival of lice away from the body. It seems that infestations are usually due to regular association of a moderately intimate nature with an infested person.

Children occasionally acquire head lice at school, probably during play. Lice readily spread through the family group and it is a common experience of school medical officers that children, disinfested at school, are repeatedly reinfested at home. Young children or adolescents outside the school age group, and not subject to frequent inspection, are often responsible for this.

Body lice may be acquired from bedding or furniture recently used by an infested person (e.g. the beds in common lodging houses). However, lice are not found in large numbers or for very long afterwards in such bedding or furniture. The most favourable circumstances for rapid and universal spread of lice is when people sleep in their clothing and huddle together for warmth (e.g. trench warfare in winter).

(e) Prevalence [217]

(i) Head lice

The available evidence shows a very high level of incidence in Britain early in the Second World War and then a sharp decline following vigorous counter measures [366, 400]. The next two decades showed a steady improvement in head louse infestation of children, as seen in Table 1.1 (page 19). About 1970, however, there began a serious rise in incidence, to the point when one estimate of the numbers was about 400 000 in children and perhaps an equal number in older and younger people [385]. Since then, the estimate has been revised to about

250 000 children in England [147]. There have been similar disturbing reports from other countries. Thus, the Danish Pest Infestation Laboratory recorded negligible enquiries before 1969, but these later rose to about 150 per year in 1975 [238], and there has been concern about the rising incidence in the Netherlands, France [339], in Israel [350] and in Canada [283]. In four cities in the U.S.A., rates of 2–20 per cent were recorded, with an average of 7 per cent for boys and 10 per cent for girls [307]; while, in 1976, the sales of pediculicides suggested that there may be about 6 million cases in that country [14].

(ii) Body lice

Surveys in some parts of the world in relatively recent times have revealed body louse infestations in 50 per cent or more of some rural populations. In more northern countries, infestations are considerably rarer than in hot countries with low standards of hygiene. However, in some Yugoslav villages, infestations were noted in 16 out of 24 families in 1966 [198]; but control measures since have been improved. In southern France, rates of 5–15 per cent were observed in vagabonds and North African immigrants in 1955 [447]. No data are available for northern Europe and U.S.A., but it is generally believed that infestations are largely confined to vagrants.

(iii) Crab lice

Statistics on crab louse incidence are especially difficult to obtain. Observations at a hospital clinic in England suggest that incidence has been rising. Among some 10 000 patients examined, the percentage infested rose from 0·8 per cent in 1954 to 3·2 per cent in 1966. The incidence was highest in girls aged 15 to 19 and men of 20 and over [181].

In Denmark, requests for advice on crab lice grew from 1 or 2 in 1968–9, to 10 in 1973 and 15 in 1974 [238].

In view of the sedentary habits of *Pthirus* and its normal occurrence in the pubic region, it is likely that crab lice are often transmitted from one person to another during sexual intercourse. However, it is possible that it may spread in other ways (towels, water closets, bedding, etc) by loose hairs dropped by infested persons. These must account for the occasional infestations of infants' scalps and the eyelashes of individuals who have not pubic infestation.

(f) Importance

In this section it will be shown that the relative importance of the various sorts of lice depends on circumstances. The body louse is most dangerous as a disease vector; but only under conditions of general lousiness which, in advanced countries, are only likely to occur as a result of warfare. Hence, the scientific attention to the louse problem was accentuated at the times of the First and

Second World Wars, but interest declined in peace time. Head and crab lice, indeed, persist as unpleasant hygienic problems in most civilized communities; but because they are both difficult and unpleasant to work with, they have not been widely studied. The recent upsurge of incidence in both insects, as well as the threat which insecticide resistance poses to typhus control, have re-awakened interest in all types of lice; and the various problems raised were extensively studied at two symposia; in Washington in 1972 [462] and at Minneapolis in 1976 [459].

Human lice can transmit certain dangerous diseases. They are the vectors of two rickettsial diseases (exanthematic typhus and trench fever) and they can carry a spirochaete infection causing a relapsing fever. These diseases are absent from Britain and north-western European countries. It appears that, under experimental conditions, both head and body lice can be used to transmit these diseases in the laboratory, but the major epidemics of the past have been all associated with widespread *corporis* infestation. Furthermore, the areas where they remain as endemic diseases are places where *corporis* infestation is not uncommon; whereas *capitis* infestation is still comparatively frequent in disease-free zones.

Very little is known about the importance of *Pthirus pubis* as a transmitter of human disease. However, it is rarely very abundant and lives a very sedentary life so that it is unlikely to spread disease rapidly through a community.

All types of human louse cause the usual irritation resulting from the bites of bloodsucking insects. The pruritus is due to the saliva of the insect and there are many degrees of reaction to the bites from slight irritation to severe urticaria. Very long-continued infestation results in a thickening and pigmentation of the skin known as 'vagabonds' disease'. The bites of *Pthirus pubis* also frequently cause characteristic small blue spots, about 0·2–3·0 cm diameter, with an irregular outline. These appear some hours after the crab louse has bitten and persist for a number of days.

The irritation caused by louse bites may be severe enough to cause loss of sleep. Scratching of the bites frequently cause abrasions which become infested with *Staphylococci* and other organisms which normally occur on the skin, leading to impetigo, furunculosis or eczema. Impetigo in school children is not infrequently due originally to *capitis* infestation and cannot be satisfactorily eliminated until the parasites are removed.

Apart from these unpleasant results of louse infestation, the insect itself occasions considerable disgust and feelings of shame which, unfortunately, sometimes lead to infestations being hidden or denied. These retrograde actions may be due to the belief that certain people are 'breeders' of lice and are therefore incurable. Another source of shame and concealment is that *Pthirus pubis* is commonly considered to be a venereal infestation, though it may be acquired in other ways.

(g) Control

The primary difficulty in finally eradicating louse infestations is the problem of reinfestation. Lousiness seldom occurs in isolation. With head lice the family is the usual infested group; with body lice the infested group is a number of people living under unhygienic conditions, varying from a few tramps or a number of prisoners in a primitive jail, to a host of refugees. Individuals deloused by evanescent methods soon become reinfested from their associates. This trouble is augmented by the tedious nature of the older delousing methods which made it impossible to disinfest large numbers in a short time.

Modern anti-louse insecticides are an improvement in two respects: (i) they are persistent and give some protection from reinfestation, (ii) they are quick and easy to apply.

As already remarked, the different habits of head and body lice and the different types of people infested give rise to two quite distinct public health problems; and this affects the choice of control measures.

(i) The head louse

Insecticides

To be successful, an insecticide preparation for head lice must be acceptable to the patient; otherwise there will be efforts to avoid treatment and substances applied under protest will be washed off as soon as possible. In contrast, an effective application which is invisible and virtually odourless will be acceptable, and the patient may facilitate treatment of other members of the family suffering from the same complaint. Where the majority of infested people are girls or young women, an aqueous preparation has been found least objectionable. An emulsifiable concentrate containing 1 per cent *gamma* HCH in alcohol is diluted with 5 parts of water before applying to the roots of the hair. In America, an emulsion concentrate known as NBIN has also been used. It contains 68 per cent benzyl benzoate, 6 per cent DDT, 12 per cent benzocain and 14 per cent 'Tween 80' (an emulsifier). Unfortunately, recent years have seen the emergence of resistance to DDT and HCH among head lice in Britain, Denmark, France, Hungary, Canada and the U.S.A. [639]. In case of resistance to the organo-chlorine insecticides, the use of 0·5 per cent malathion lotion has proved highly effective [385].

It may be considered worthwhile giving small quantities of the insecticide for home use by other members of the family who may not be available for treatment.

Combing, etc

In early days head lice infestations were treated by shaving the head or at least cutting the hair short. This drastic procedure is no longer essential, but in very bad cases some shortening of the hair may be desirable.

None of the insecticidal substances described (in fact, no known substance) will dissolve the cement that binds the nits to the hair without damaging the latter. Thus they may cure infestations but leave dead or empty nits in the hair. This may be regarded as unsightly or may confuse health inspectors who may not be able to distinguish live and dead nits. Therefore it is desirable to remove them.

Nits can be removed by thorough combing with a fine-toothed *metal* comb, of the type made by Messrs Saker. Combing alone may be undertaken to cure an infestation; but it must be done very thoroughly. Since this takes at least 20 min, and gives no protection against reinfestation, it is not recommended.

(ii) The body louse

Disinfestation of individuals entering a clean environment

Under civilized conditions, the usual problem of body louse control is to deal with relatively small numbers of people who come to public assistance hostels, charitable institutions or prison. In such situations, dusting with insecticide is unsuitable, since the obvious signs of treatment will be strongly resented; furthermore, the toxic action may be slow, so that poisoned lice may be seen crawling on treated people for some time. Since there is no danger of reinfestation, a rapid treatment giving no protection is adequate; for example fumigation or heat treatment. While individuals are having their clothes disinfested, the stray lice on the body should be removed by shaving off body hair and bathing with plenty of soap.

Fumigation. The easiest treatment is to fumigate the clothing (or bedding, where necessary) with ethyl formate in a metal bin or a plastic bag. A dose of 2 ml l^{-1} (2 oz ft^{-3}) should be used for a short (1 h) exposure; but for 5 h or longer, 0·5 ml l^{-1} ($\frac{1}{2}$ oz ft^{-3}) is adequate [92]. The short treatment could be given while the infested person takes a bath, while the longer fumigation could be used overnight. The cost is low and the time to load and unload short. On removal, there is only a slight smell of ethyl formate, which soon disperses. The inflammability and toxicity of ethyl formate are no greater than those of benzene, so that small quantities can be handled with ease.

Heat treatment. Disinfestation by hot air is described on page 46. Some local authorities in Britain find it convenient to use steam disinfestors, although the heat generated is excessive and certain articles of clothing could be spoilt.

Control of widespread infestation by body lice [628]

In circumstances of general lousiness, it is essential to reduce the infestations as quickly as possible. The measure must be quick, safe and effective and provide some protection from rapid reinfestation, which would otherwise occur. Owing to the danger of disease, people will accept treatments which would otherwise be resented.

The ideal way to deal with this emergency is to apply insecticidal dust to the underwear of every member of the community as quickly as possible. If the lice are not resistant to DDT, a dust containing 10 per cent DDT is the insecticide of choice. Where there is DDT resistance, dusts with 1 per cent malathion, 1 per cent *gamma* HCH, or 1 per cent permethrin would be effective. The dust can be applied to the underwear by the individuals, using a sifter-top can, with a dose of about 30 g (1 oz) per person. It is necessary to spread the dust evenly over the inner surface of the garments next to the skin, with special attention to the seams. The seams and folds of other garments and also the socks should be similarly treated.

For the mass treatment of large groups of people, the clothing need not be removed, and hand-operated dusters or motor-driven air-compressors with as many as 10 duster heads, may be used. About 50 g of powder is blown into the clothing, through the neck openings, up the sleeves and from all sides of the loosened waist of trousers. In delousing women, an extra quantity may be introduced down the neck of the dress and the application at the waistline omitted. The socks, head covering and the inner surfaces of other garments, as well as the bedding should also be treated.

(iii) Crab lice

As with head lice, the earlier treatments involved shaving or cutting of the infested hair; but this should not be necessary. The simplest and most satisfactory method is the application of one of the modern insecticides and a repeat treatment one week later to kill any young nymphs hatched from unaffected eggs. DDT 10 per cent dust, *gamma* HCH 0·6 per cent dust or a 2 per cent DDT or 0·2 per cent *gamma* HCH emulsion should be suitable. In treating pthiriasis of the eyelashes, it is best to use a Vaseline ointment containing pyrethrins and remove the insects with fine forceps, having applied cocaine to the conjunctiva if necessary.

The centuries-old use of mercuric ointment for crab lice cannot be recommended. Its insecticidal value is low and there is some risk of mercurial poisoning.

7.1.5 LICE OF DOMESTICATED ANIMALS

Two types of lice are parasitic on animals: biting lice (order Mallophaga) and sucking lice (order Siphunculata). The Mallophaga have chewing mouthparts and feed on pieces of feather, fur, sebum and fragments of scurf from the host's body; some of them also nibble the skin and drink the blood that exudes. Most, but not all of them, are parasitic on birds. The Siphunculata, which are all parasites of mammals, have highly specialized mouthparts for piercing the skin and sucking blood.

The host preferences of both types of lice are rather rigid. The entire life cycle is spent on the host; and lice rarely, except accidentally, transfer to a different species. If they are removed from the host, lice die within a few days. This is understandable with sucking lice, for they die of starvation. But it is also true of bird lice, which cannot survive away from the host, even if given feathers to feed on.

(a) Sucking lice (Siphunculata) [180]

There are about 200 species known and these are grouped into four families. The following list includes some of the more common species.

(i) On dogs

Linognathus setosus (Fig. 7.5). This is a parasite of the domestic dog in many parts of the world. It is usually a parasite of long-haired game dogs and occurs on the back, flanks and at the base of the tail. Somewhat like *Pediculus* in form, it lacks eyes and has a wider abdomen. Approximate lengths: male 1·5 mm; female 2 mm.

(ii) On pigs

Haematopinus suis (Fig. 7.5) occurs on wild and domestic swine. It is one of the largest of the Anoplura, the female averaging about 5 mm and the male 4 mm in length.

Infestations of *H. suis* sometimes have a surprising result. The eggs of this louse are cemented to hog bristles in the same way as those of human lice are glued to hairs. Occasionally such bristles are used in the manufacture of hair brushes. They may escape notice for some time until sudden discovery gives rise to alarm. They appear to be 'nits' of some super head louse; but of course they are dead and harmless.

(iii) On horses

H. asini is rather like *H. suis* but smaller (female 3 mm; male 2 mm) and with a proportionately larger head.

(iv) On cattle

Linognathus vituli, Solenopotes capillatus and *Haematopinus eurysternus.*

(b) Biting lice (Mallophaga)

Biting lice are small or very small (0·5–6 mm long) flat-bodied insects of rather characteristic appearance. There are about 1700 species known, divided among four families.

The breeding habits and life history of a biting louse are not very different from those of sucking lice. The eggs are cemented to the bases of feathers or hairs and incubate there. The whole life is spent on the host's body and transmission seems to occur normally during direct bodily contact (e.g. mother bird and fledglings in a nest).

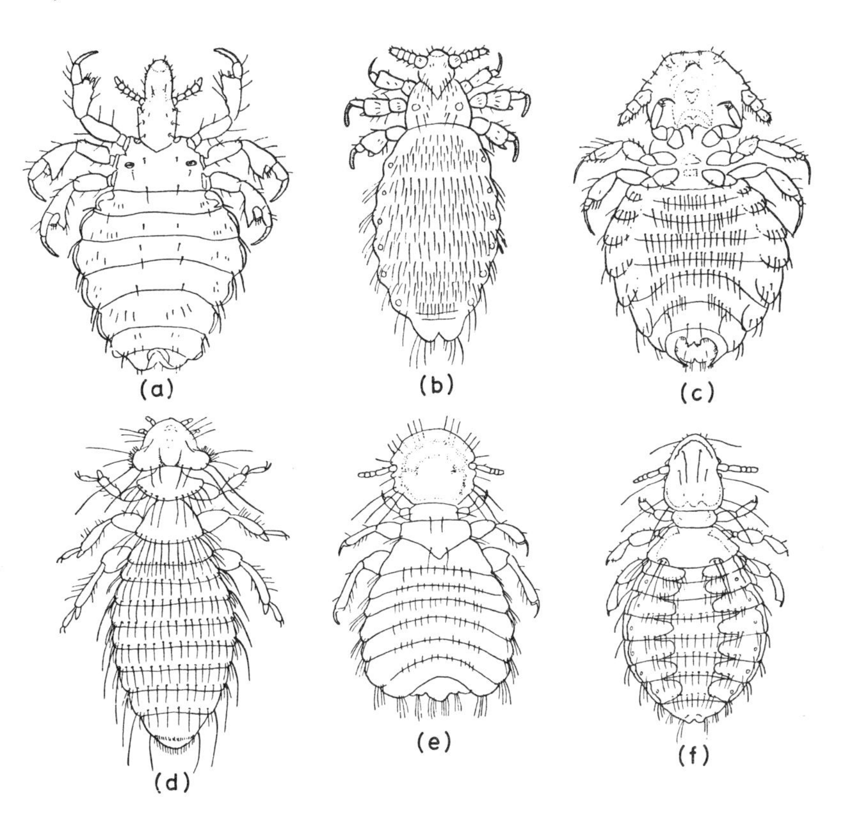

Figure 7.5 Sucking and biting lice of animals. (*a*) *Haematopinus suis*; (*b*) *Linognathus setosus*; (*c*) *Trichodectes canis*; (*d*) *Menopon gallinae*; (*e*) *Goniodes gigas*; (*f*) *Lipeurus heterographus*. (After Séguy [530].) All ♀ except (*e*). (*a*) × 8; (*b*) (*c*) (*d*) and (*f*) × 23; (*e*) × 14.

The following are the most common parasites of some domestic animals:

(*i*) *On poultry*

Menopon gallinae and *M. pallidulum, Goniodes dissimilis, G. gallinae* and *G. gigas. Lipeurus heterographus* and *L. caponis*. Domestic poultry are quite frequently infested with biting lice. The 'dust-baths' taken by fowls and other birds are apparently taken largely to destroy these parasites. When birds become badly infested, bare areas of skin may be seen where the feathers have been eaten through and have fallen out. The presence of numerous parasites causes considerable irritation which disturbs feeding and rest, and results in a very poor condition of health.

(ii) On dogs

Trichodectes canis (Fig. 7.5) occurs on dogs. Unlike *Linognathus* it commonly infests the head and neck.

(iii) On horses

Trichodectes equi and *T. pilosus*.

(iv) On cattle

Trichodectes bovis often heavily infests cattle on the withers, root of the tail, neck and shoulders.

(c) Control of lice of domestic animals

The simplest effective method of destroying these parasites is to apply a finely powdered insecticide to the roots of their fur or feathers. Any of the preparations recommended for use against animal fleas could be employed in the same way. For treating a large flock of turkeys or hens, the following rapid method could be tried. An agricultural dusting machine which emits jets of dust at intervals along an arm is laid on the ground, so that the jets point upwards and towards the flock. The birds are then driven through the dust cloud [207].

Treatments for lice of domestic animals do not require the same degree of attention to kennels, pens, etc, since the immature stages do not live away from the host.

7.2 PARASITIC AND HARMFUL ACARI

The insects owe their enormously successful proliferation to the development of wings. The other terrestrial arthropods almost all belong to the miscellaneous assembly of the Arachnida; and this contains only two large groups, the Araneae or spiders and the Acari or mites and ticks. Apart from a few poisonous forms, the spiders are generally beneficial as predators of insects. But the Acari contain a number of parasites of vertebrates, some of whom can act as disease vectors. Vast numbers of them, however, occur as harmless scavengers or predators of tiny creatures. The group is very heterogeneous, but it is possible to make a few generalizations. Most of them are small (less than 5 mm long) but a few mites and many ticks are larger. Except for a few primitive forms, they have no trace of segmentation and the body is not divided up by constrictions (as in insects and spiders). The larva has three pairs of legs, but the (one or more) nymphal stages and the adults have four pairs. The mouthparts consist of a pair of chelate *chelicerae* and two short palps.

7.2.1 HUMAN PARASITIC MITES

(a) The itch mite (Sarcoptes scabiei)

(i) Distinctive features

All mites tend to be small, generally in the range 0·2–2 mm long, with limited powers of independent movement in the parasitic forms. These characters apply to the itch mite, which belongs to the family Sarcoptidae, members of which are parasitic in or on the skin of birds or mammals. The adult females, which are the largest, are ovoid with a maximum length of 0·4 mm (weighing about 0·01 mg). The legs are stumpy, the front pairs ending in suckers and one or more of the hind pairs ending in long bristles (Fig. 7.6).

Sarcoptes scabiei infests man; but nearly identical mites can be found in the skins of various domestic and wild animals. People who (perhaps by reason of their occupation) have intimate contact with animals which are infected, may sometimes acquire transitory infestations from them; but these mites from animals are unable to lodge permanently on man. It seems that, in spite of their very similar appearance and life history, they have physiological idiosyncrasies adapting them to life on particular hosts. Analogous cases are met with among insects, the various strains being known as 'biological races'. On the grounds of small anatomical differences between them, the races of *Sarcoptes scabiei* are classed as varieties, as var. *hominis*, var. *equi*, etc.

(ii) Life history [402]

Oviposition

The mature female is the largest form of the scabies mite and decidedly the easiest to demonstrate. She forms meandering burrows in the horny layer of the skin which can be recognized with a little practice. These burrows are much more common on certain parts of the body than others. The examination of a considerable number of patients gave the figures of distribution shown on p. 274 [33, 304].

It will be observed that a high proportion of cases of scabies could be diagnosed by examination of the hands and arms. The sites where the mites burrow are not necessarily areas where the cuticle is thin; in fact the reverse is nearer the truth.

As she proceeds along the burrow the female lays her eggs.

Site	Percentage of mites found in Men (886 *cases*)	Women (119 *cases*)
Hands and wrists (excluding palms)	63·1	74·3
Palms of hands	—	7·5
Elbows	10·9	5·9
Feet and ankles	9·2	9·8
Penis and scrotum	8·4	—
Buttocks	4·0	1·1
Axillae	2·4	0·9
Knee	0·7	0·3
Navel	0·4	0·1
Chest	0·1	0·1
Breasts	—	0·5
Other sites	0·8	0·5

Egg

The egg of the itch mite is smooth, whitish and glossy, its dimensions, when laid, are about 0·17 mm × 0·09 mm and it increases slightly during development; this size is relatively large compared to the fully grown female which is only about 0·36 mm × 0·26 mm. When burrows are probed with a needle to extricate the female mites (see page 282), eggs are sometimes removed adhering to the needle.

Larva

The first stage, which emerges from the egg, is known as the 'larva'. This resembles the later stages in general appearance, but is exceptional in having only three pairs of legs, whereas the older mites have four pairs. The two anterior pairs of legs bear suckers and the last pair end with a bristle.

Sometimes newly hatched larvae may be found inside a burrow and they may make some attempt to dig into the floor. Normally, however, they soon emerge and burrow into the skin nearby; or else they may descend into hair follicles. Eventually they moult to form the first nymphal stage.

Nymph

The 'nymphs' are stages with four pairs of legs like the adults; both the hind pairs terminate in bristles. The nymphs may be found in short burrows or in hair follicles, like the larvae. Apparently the immature stages spend most of their time in such retreats, but they sometimes come out and travel to new sites.

After a certain time, the nymphs moult again to produce *either an adult male or a second-stage nymph*. The latter undergoes yet another moult before the adult female emerges.

Adults

The adults are shown in Fig. 7.6. The females are considerably larger, especially after fertilization, when they reach a size about double that of the males. Like the

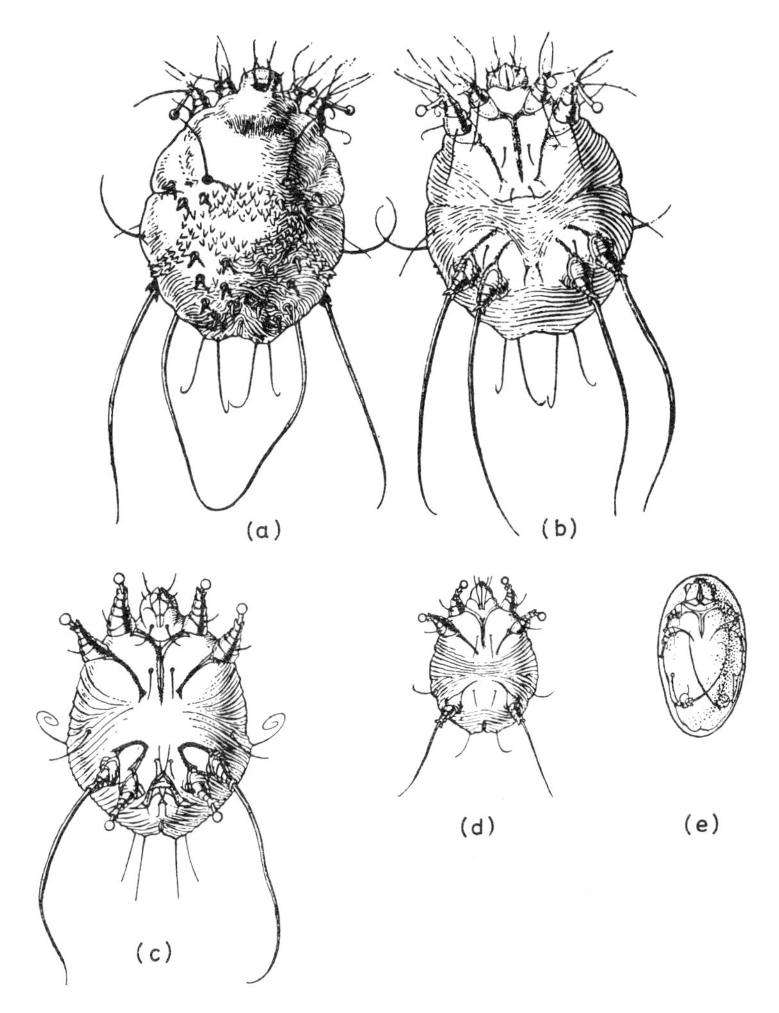

Figure 7.6 Sarcoptes scabiei. (*a*) adult female (dorsal view); (*b*) the same (ventral view); (*c*) adult male (ventral); (*d*) larva (ventral); (*e*) mature egg. From Mellanby [402]. (*a*), (*b*) × 100; (*c*), (*d*), (*e*) × 125.

nymphs, their two hind pairs of legs bear bristles; whereas the last pair of legs of the male bears suckers.

The 'head' which is broadly attached to the body, bears a pair of toothed 'chelicerae', like insect mandibles, and a pair of short palps; there are no eyes or other organs of the higher senses. The food apparently consists of the horny part of the human skin and, perhaps, certain secretions in the hair follicles (for the younger stage).

The body of the mite has a shape somewhat like that of a tortoise, rounded above and flattened below. The skin is whitish and covered with striations. On the ventral surface can be seen thickened bars in the cuticle to give support for the articulation of the legs. The short stumpy legs appear to be only capable of limited movements; nevertheless, the mites can walk fairly well over flat surfaces with the aid of the sucker-bearing legs. An adult female travels over the skin at the rate of about 2·5 cm min^{-1} (1 in. min^{-1}).

All stages of the mite show a great inclination to burrow into the epidermis shortly after being placed on the skin. They usually seek our furrows or cracks in the skin and begin to dig in them immediately. The creatures use their jaws and the front two pairs of legs for digging. The latter grip with their suckers and tear at the skin with a cutting edge on the last joint so that the animal gives the appearance of digging with its 'elbows'. The bristle-bearing legs may be used to afford purchase or leverage during the digging operations. On the back of the mite are numerous spines and small projections, directed backward; these may serve to anchor the body in position in the burrow when necessary. Different stages of the mite take from 15 min to an hour to bury themselves in the skin. The immature stages and the adult males make only short burrows, whereas the females, after fertilization, form the long meandering burrows already mentioned as characteristic of a scabies infection. The speed at which these burrows advance is very variable, sometimes the animal progresses as much as 5 mm in a day and at other times she only moves about $\frac{1}{2}$ mm. The adult males are the most active forms of the mite and often leave their small temporary burrows seeking the females.

The sexual organ of the male is ventral while the copulatory orifice of the female is dorsal. When the male enters a burrow containing a virgin female, coition is effected by the male assuming a position facing in the opposite direction to the female; then he tilts his body until his posterior under surface rests on her back end. Copulation may occupy only a few minutes, after which the male leaves the female, and apparently does not mate again.

The female has a separate oviduct leading to an egg pore on the ventral surface of the body. Eggs are laid at intervals along the burrow and remain there until they are hatched.

(iii) Quantitative bionomics

Life history on the host

The itch mite, which spends most of its time buried in the skin, is even less subject to vagaries of climate than the louse. Consequently it is possible to give single estimates for periods in the life history, under the average conditions of the human skin. These figures are the most reliable modern observations of a life history difficult to trace with accuracy [259].

Stage	Duration	
Egg stage	3–4 days	
Larval stage	3 days	
First nymphal stage	3–4 days	
Second nymphal stage	3–4 days	*Female mite:*
Unfertilized adult	?	Total, 14–17 days
Maturation after copulation, before oviposition	2 days	

The male mite lacks the second nymphal stage, so that its development will last only 9–11 days.

The life of the adult female has not been followed through, but it seems probable that she lays rather more than two eggs daily. This apparently continues for one or two months.

A high proportion of the eggs of *Sarcoptes* hatch; probably about 90 per cent. During development, however, there is a high mortality so that, even under favourable conditions for the mite, only about 10 per cent or less reach maturity.

Survival and behaviour away from the host [401]

Adult female mites removed from the host's skin will walk about fairly rapidly at temperatures above 20° C. They can climb up polished surfaces such as glass without difficulty. They show no reactions to light, either positive or negative, but they show some tendency to walk from cooler zones to warmer ones.

Little movement occurs below 20° C, and at 15° C they fall into a chill coma.

Sarcoptes, like *Cimex*, survives best at about 13° C. Given moist conditions (90 per cent R.H.) about 50 per cent may survive for 1 week and 5 per cent for 2 weeks at this temperature. Warm dry air is soon fatal; no mites have survived an exposure of 2 days at 28° C and most of them die within 24 h.

The mites are also killed quite easily by short exposures to higher temperatures. A temperature of 49° C appears to be fatal in 10 min and 47·5° C in 30 min.

(iv) Parasitology of scabies

Transmission [403]

The high mortality of the younger stages of the mite make it fairly certain that only the ovigerous females are normally successful in colonizing a new host. Artificial infestation has been done on many occasions by transplanting an adult female, but attempts with the younger stages have failed. Once the adult female has formed her burrow on the new host, she goes on producing progeny regularly and usually sufficient survive to set up a new infestation.

The helplessness of *Sarcoptes* away from the host at temperate room temperatures and its poor powers of survival under warmer conditions, greatly impair its chances of indirect infestation by formites, etc. This has been substantiated by experimental evidence and by the questioning of infected patients. It is clear that the great majority of sufferers acquire scabies by intimate association with an infested person, usually by sleeping in the same bed with them. Scabies has acquired some opprobrium as a 'venereal complaint'; but its best chances of spreading are by the regular contacts of marriage and family life (e.g. where children sleep together or with their mother). Therefore it is more reasonable to regard it as a familial infection. There is an unknown amount of infection by more transitory contacts, as between children at play or hand-holding or caresses of adults at dances, cinemas, etc. Finally there is a small percentage of cases due to infection by fomites. (An indication of the frequency is given by 300 experiments in which bedding used less than 24 h previously by a scabetic patient was slept in by an uninfected volunteer. Transmission occurred in four cases only. With very high parasitic rates the chances were considerably increased; but such infestations are rare.)

Population growth [403]

Estimates of populations of *Sarcoptes* and their growth and decline must be made by counts of the ovigerous females. Other stages are difficult to find, and it is not possible to determine how many of them are present on an infested individual. However, the number of females can be determined with some accuracy by a careful examination of the patient's body; and this serves as an indication of the magnitude of the total mite population.

If the numbers of progeny of a single fertilized female are estimated according to the life history data, assuming no mortality during development, it can be calculated that about 100 adult mites would be present after 1 month and several thousands after 2 months. Large numbers are very rarely encountered in actual infestations. A series of 886 scabetic recruits examined during World War II were found to carry 9978 mites, an average of only 11·3 per man. Over half the patients had under 6 mites and only 3·6 per cent had over 50 [304]. It is clear that there is some drastic check to population growth.

The growth of mite populations in the human skin has been studied by artificially infesting volunteers. These experiments were abnormal in one respect, for these volunteers did not resort to self-medication, which common practice has a drastic but rarely exterminating effect upon many natural infestations. It was observed that the progress of the mite community in a person infested for the first time was very different from that in reinfested individuals. The reason for this is that the human skin becomes sensitized to the presence of the parasites and sets up an acute reaction inimical to them.

In an original infestation (by a single ovigerous female) the second-generation adult females begin to appear after 3 or 4 weeks. During this period, the presence of the mite and her progeny causes no symptoms. Thereafter, the population grows, slowly at first and later at a faster rate, for about 2 months. At the beginning of this period, sensitization occurs and there follows a gradually increasing skin reaction which is accompanied by more and more severe irritation. The reaction of the skin, which is characterized by erythema and the formation of small vesicles, suppresses the activity of the mites and, in addition, the irritation results in a number of ovigerous females being scratched out of their burrows and destroyed.

About 3 months after the original infestation, when there may be a few hundred female mites present, the reactions of the host become so intense that the mite population begins to decline sharply. It was not possible to study the further course of infestations in the volunteers because, at this point, all of them were suffering severely from secondary infections. It seems likely that a small fluctuating residual mite population would remain for some time, and sometimes, perhaps, die out spontaneously.

The course of an infestation in a person who has suffered from scabies before is very different. Such a person will have been sensitized to the presence of the mites, provided that he has been infested for several months. (Although sensitization begins after a month, it is only fully developed after about 6 months.)

In the first place, it is very difficult to reinfest sensitized people. The skin reacts strongly within a day or so of implantation of an ovigerous female, and the surrounding area becomes intensely itchy. Even if a mite colony is established,

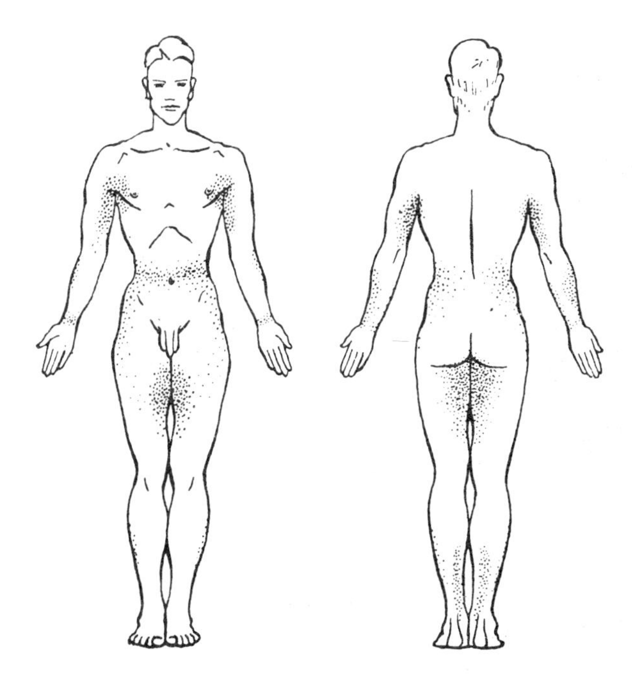

Figure 7.7 Scabies 'rash'. (N. B. the rash does *not* correspond with the sites of election of the mites.) From Mellanby [402].

the number of parasites never reach very high figures (usually less than 25). Quite often these infestations die out spontaneously. To some extent, therefore, infestation by *Sarcoptes* results in a partial immunity, apparently due to an antibody function.

The course of an infestation in an ordinary citizen is complicated in most cases by the use of medicaments as soon as the irritation begins (about a month after infestation). These substances, sometimes ineffective and often improperly applied, usually reduce the population of mites to a moderate, fluctuating figure. Sensitization continues to develop and the mites either remain at a low density for months (or years) or else die out. It will be appreciated that where the population of ovigerous females is as low as a dozen or so, the number destroyed by scratching may be decisive.

(v) Prevalence

It is difficult to compare the prevalence of the disease today with its occurrence in earlier times owing to former uncertainties in diagnosis and confusion with other skin complaints. Figures for the present century show a rise during the war of 1914–18 followed by a decline and another rise beginning a few years before the Second World War. By the 1941 the incidence had risen until certain hospitals in urban areas recorded 2·5–4·0 per cent of patients as suffering from scabies when admitted for other diseases [399]. On these data it was suggested that the general average among the British population might not be very much below 2 per cent. According to some figures based on examination of over 20 000 school children annually, the rate in a northern city fell from about 0·82 per cent in 1944 to 0·18 per cent in 1947. In two other British cities, the decline continued till 1953 and 1957, respectively. However, in more recent years, there has been a disturbing rise in incidence of scabies in Britain, where the reported cases among school children rose from about 4000 p.a. in 1963–5 to over 14 000 in 1970 [13]. A similar trend has been observed in Czechoslovakia [461] and in various parts of the U.S.A. [173, 526].

The reasons for this renewed wave of infection are not quite clear. It has been suggested that regular fluctuations of infection occur, because of the semi-immunity caused, so that the disease subsides until a new generation of susceptible people arises. Rather more probable, however, is the recent spread of sexual laxity in young people, which is also probably the cause of the steep increase in venereal diseases which has been observed in several countries [11]. The association of scabies incidence and war could be similarly attributed to the effects of conscription on behaviour and morality; but in this case, the increase in medical inspections must bring to light many cases normally undetected.

Factors influencing incidence

In the mid-19th century in Vienna, young men were observed to be seven times more often infected than women [256]. This was attributed to the way in which apprentice boys were herded together in overcrowded dormitories. In recent investigations, however, young women have been found to be more often infected than men, possibly because they more frequently share beds with girl friends or members of their families.

Bed sharing, indeed, seems to be the only social factor responsible for increasing infection. Neither affluence, intelligence, obvious cleanliness, nor large families are clearly implicated [233, 308].

(vi) Importance

The association of scabies with major wars in this century has had consequences which parallel those regarding lice. In both cases, there was heightened interest among physicians and scientists, which subsequently waned in peace time, and as with lice, the present resurgence of infestation has renewed attention to the problem, which was exhaustively discussed at a symposium in Minneapolis in 1976 [459]. At these meetings, further information was presented on the epidemiology, the clinical findings, the diagnosis and the treatment; but nothing new was mentioned on the biology of the parasite.

The reactions of various hosts to infestation by *Sarcoptes* are quite different. In animals, large numbers of the parasites give rise to sarcoptic mange; in man they cause the quite different disease known as scabies, in which unpleasant symptoms may be due to relatively small numbers of the mites.

At a certain stage, the infestation of the human skin by the itch mite gives rise to more or less severe irritation. This causes the sufferer to scratch himself vigorously and, very commonly, to resort to widely advertised 'skin trouble' ointments. The constant scratching is liable to break the skin and allow the entry of micro-organisms. Pustules, boils, ecthymata, impetigo or eczema may result, which, if neglected, will require prolonged medical treatment. Attempts at self-medication are, for several reasons, seldom successful; and they may give rise to severe dermatitis, if such things as sulphur ointment are used too frequently.

Scabies then is a distressing affliction often attended by unpleasant sequelae. Although it is commonly associated with a lack of cleanliness, there is certainly no direct causal relationship, for a scabetic infestation will persist in a scrupulously clean person. Fúrthermore, no evidence has been found among samples of recruits to the army that the sufferers from scabies were less intelligent than normal.

(vii) Diagnosis and cure of scabies

Diagnosis

It is difficult or impossible to discover a primary case of scabies in the first 3 or 4 weeks after infestation, before sensitization has developed. Thereafter, a typical case is fairly easy to recognize, in the absence of complications (i.e. secondary skin infestations). The patient complains of severe irritation which becomes intolerable at night, often preventing sleep. An examination of the body reveals a rash of follicular papules, characteristically developed in certain parts of the body as shown in Fig. 7.7. This rash is apparently due to sensitization to the mite, aggravated by the patient's scratching. It is important to remember that the rash does *not* correspond to the sites selected by the mature females to form their burrows. In the latter regions, the burrows can often be distinguished as tiny meandering greyish-white tunnels, often broken open by scratching. Around them may be found vesicles about the size of wheat grains. Small crops of these vesicles may sometimes be found in other sites, unconnected with the burrows; they often occur on the hands or feet of small children, for example.

A person who has suffered from scabies before, for a sufficient time to develop sensitization, responds quite differently to a second infestation. Local reactions (erythema and oedema) occur within a day or so in the region where an invading female has burrowed. Often the unfavourable skin reaction and the scratching of the sensitized host are sufficient to exterminate a secondary infestation. However, where a sensitized person is being repeatedly reinfested (e.g. by regularly sleeping with an infested individual) a fresh infestation may become established. When this happens, general signs and symptoms of scabies usually develop all over the body in about a week or so.

Owing to the constant scratching induced in the later stages of scabies, secondary skin infections are common. Conditions such as ecthymata, impetigo or eczema may obscure the typical signs of scabies. The mites will not develop in close proximity to septic lesions and, in any case, they are usually scarce in advanced scabies. Nevertheless, it is important to diagnose scabies, since the irritation, and hence the liability to skin diseases, cannot clear up until the parasites have been extirpated.

The best proof of scabies is to extract and demonstrate the adult female mite. (If the patient is shown the creature under a microscope, it will ensure his co-operation.) Extraction of the mites is not difficult. First, the burrow of a mature female is sought in the most likely spots (e.g. between the knuckles, in folds of the wrist and elbow, etc). Then the burrow is gently pricked open, working towards the end where the mite can usually be distinguished as a dull white spot. In this way, it is usually possible to extract the mite undamaged and adhering to the needle. A watchmaker's eyeglass is useful for the operation; it leaves the hands free.

Treatment

Bathing. Scabies can be cured without bathing, but since many patients are somewhat dirty, a bath before treatment is desirable for hygienic reasons, and necessary where secondary skin infections are present. Scrubbing, which was formerly considered essential, is unnecessary.

Medication. A mite-killing substance is applied to the whole surface of the body, excluding the head. Thorough treatment is essential and is best done by a reliable nurse or orderly, although sometimes it may be possible for the patient to be properly treated at home. Ointments may be rubbed on by hand and liquid preparations are best painted on with a soft flat paintbrush about 5 cm (2 in.) wide. After treatment with an ointment, the patient can dress at once. The more greasy ointments tend to soil the underwear somewhat, and could, perhaps, be improved by using a 'vanishing cream' type of base. After treatment with an emulsion (e.g. benzyl benzoate) the body is allowed to dry for 5 or 10 min in a warm room. Whatever the treatment, the patient is instructed not to take a bath for at least 24 h.

Of a considerable number of substances used for treatment, the following appear to be most reliable.

(1) HCH (lindane) is probably the most widely used medicament in Britain and the U.S.A. at present; but, because it is a chlorinated hydrocarbon, it cannot be used in some European countries. As regards its toxicity, the following calculations provide some assurance. Assuming that an adult applies 30 g of a 1 per cent solution, systemic absorption of say 10 per cent would introduce 30 mg, corresponding to about 1 mg kg^{-1}. This dose should not involve risk; but in case of slow excretion and prolonged penetration, repeated treatments might cause a build up internally. This should be avoided; and, in any case, should not be necessary.

(2) Benzyl benzoate has been very widely used, in the form of a 20 per cent emulsion. Its main disadvantage is that it causes temporary smarting; and in a few atopic subjects, might cause dermatitis.

(3) Sulphur ointment, containing 10 per cent sulphur in a vaseline base, is the oldest known effective remedy, although it has several disadvantages. Its greasy nature and unpleasant smell makes it a disagreeable treatment; and repeated treatments (which, indeed, should not be necessary) can cause dermatitis.

(4) Several compounds, which were successfully used in the past, fell out of use in favour of benzyl benzoate and lindane. These include: tetraethyl thiuram monosulphide ('Tetmos'), 2,7-dimethyl thianthrene ('Mitigal') and crotamiton (*N*-ethyl-*o*-crotonotoluidide, or 'Eurax'). Should the use of lindane be further restricted, these compounds could be re-examined.

(5) A novel treatment, which has shown some promise in Mexico [49] and South America [262], is thiabendazole, used as a 5–10 per cent cream. This substance was introduced as an anti-helminthic in 1962 and later shown to be fungicidal.

A second application of the chosen medicament should be made a few days after the first and in the same week. This is in the nature of a safeguard, since a very high proportion of cases are cured by one treatment.

The patient may complain of continued symptoms even after all mites have been destroyed, for the sensitized skin sometimes remains irritable for a time. Occasionally itching weals may appear periodically for a few weeks after a cure. There is much variation between different individuals and the symptoms appear to be partly nervous in origin.

After the second treatment, however, no further applications should be made, since too frequent use of these medicaments may lead to dermatitis.

Sterilization of fomites. Where facilities exist, the patient's bedding and clothing may be disinfested by heat. However, this is of minor importance compared to efficient medication. Mites are not likely to survive for long in the bedding or clothing used by the infested person and the residue of ointment applied will destroy wandering mites attracted to the skin in the day or so following application.

Secondary infections. These are medical problems strictly beyond the scope of this book. Briefly it may be remarked that mild secondary infections often clear up spontaneously as soon as the scabies is cured. Where further treatment seems necessary it is usual first to remove any scabs and crusts (e.g. by boracic fomentations or poultices) and then treat the underlying sepsis with a dilute mercurial ointment, or by an acriflavine or calamine lotion or by gentian violet. A small proportion of cases may be found refractory and these require the attention of a competent dermatologist.

Prophylaxis

Since the burrows of the ovigerous females are usually concentrated in the hands and wrists, it might be hoped that regular treatment of these regions might stamp out scabies from a community. A large experiment in which this was tried with school children, however, failed to show very much better results than in the control groups. The results indicated that regular inspection (and treatment where necessary) has a beneficial effect, and this has been confirmed elsewhere.

A similar type of prophylaxis, which has given excellent results among a closed community (in an institution), is to issue soap containing 5 per cent tetraethylthiuram monosulphide ('tetmos') in the place of ordinary soap. This cured almost all the sufferers in about 13 weeks and there were no signs of dermatitis in the 400 persons using the soap for that period.

(b) The follicle mite (Demodex folliculorum)

Like others of its group, this mite has a curious body form (Fig. 7.8); the long striated abdomen giving a somewhat worm-like appearance. It is a small mite (less than 0·4 mm long) and, though often present in human sebaceous glands, it usually escapes notice. These mites may be sometimes found if the contents of a 'blackhead' are squeezed out and examined under the microscope. There is, however, no evidence to show that blackheads are caused by the mites which appear to be quite harmless, in most cases. On rare occasions, however, it appears that the mites are associated with a rash of tiny red follicular papules on the face or scalp. In such cases, an ointment containing 10 per cent sulphur and 5 per cent balsam of Peru was found to be an effective cure.

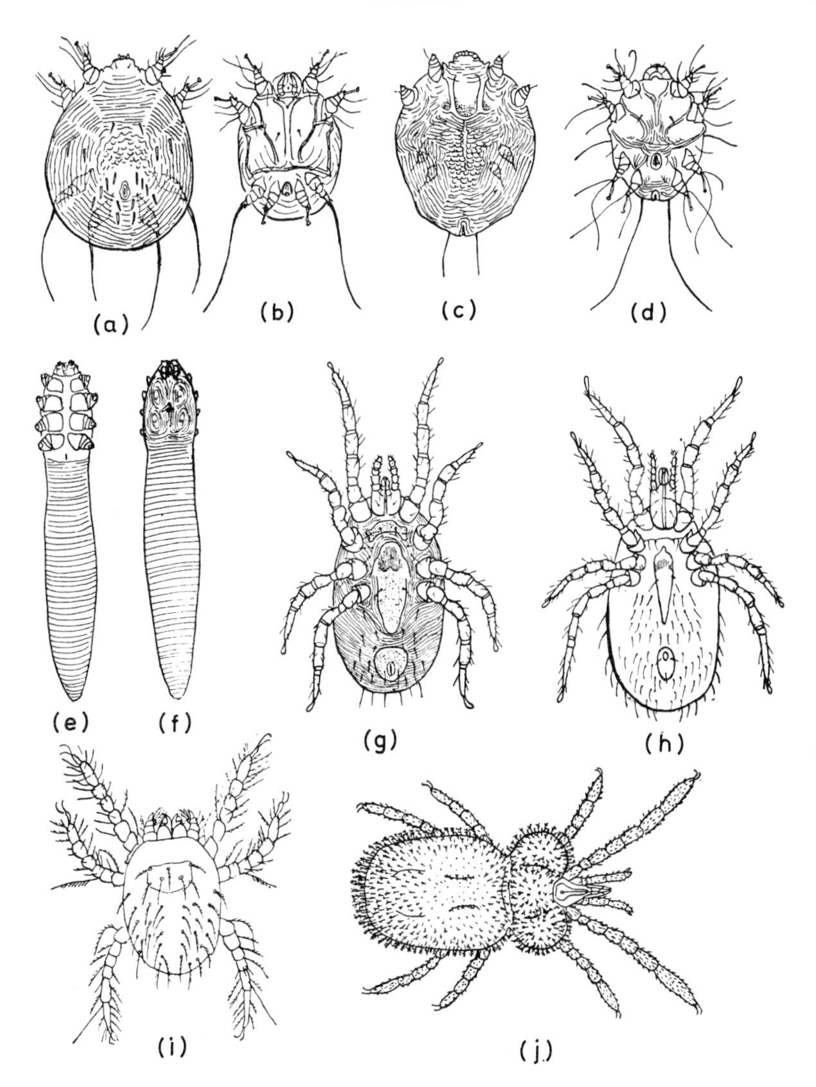

Figure 7.8 Parasitic mites. (*a*) *Notoedres cati* var. *cuniculi* (dorsal, female); (*b*) the same (ventral, male); (*c*) *Cnimidocoptes mutans* (dorsal, female); (*d*) the same (ventral, male); (*e*) *Demodex folliculorum* (ventral, female); (*f*) the same (dorsal, male); (*g*) *Dermanyssus gallinae* (ventral); (*h*) *Ornithonyssus bacoti* (ventral); (*i*) larva of *Trombicula autumnalis*; (*j*) adult of same. (*a*) to (*h*) after Hirst [279]. (*i*) (*j*) after Finnegan [178]. (*a*), (*b*) × 100; (*c*), (*d*) × 60; (*e*), (*f*) × 150; (*g*), (*h*) × 40; (*i*) × 100; (*j*) × 20.

Other species of *Demodex* occur in domestic animals, and cause manges of a more or less severe type (see page 287).

Life history. Very little is known. The eggs, which are heart-shaped, give rise to larvae with three pairs of legs. There are probably two nymphal stages, both with eight legs. These young stages vary considerably in size.

In the adults, the sexes can be distinguished by the penis of the male which protrudes dorsally at the level of the second pair of legs. In the female, the genital orifice is a ventral slit between the last pair of legs.

7.2.2 ANIMAL MITES SOMETIMES ATTACKING MAN

(a) Mange mites

Recent investigations by Hewitt and colleagues have shown that a definite proportion of cases presenting to general medical practitioners as dermatitis are due to animal parasites [12, 266]. The skin lesions produced vary from papular urticaria to bulbous and erythema-multiform reactions, and in some cases, pustular and necrotic sequlae. Because the reaction depends so much on the patient (being especially severe in atopic subjects) there is little or no clear indication of the parasite responsible. Clues must be sought from the habits and background of the patient; and it is usually found that a pet animal is responsible. Many of the cases were due to cat or dog fleas (see page 255); others arose from various mites, especially mange mites.

A considerable number of mites of the family Sarcoptidae live parasitically on the skins of various domestic animals and birds. Some of them live on the surface of the skin and cause superficial irritation; others enter the skin or its glands and cause the conditions known as mange or scab. The most important genera are *Psoroptes*, *Sarcoptes*, *Notoedres*, *Cnemidocoptes* and *Otodectes*.

As mentioned in connection with *Sarcoptes scabiei*, most of the species concerned have a number of varieties which are very similar or identical in appearance, but which are physiologically adapted to living on different hosts. Thus it is very difficult, or impossible, to infest a different kind of animal with one of these mites.

(i) Species concerned

Sarcoptes scabiei

A variety of this mite causes sarcoptic mange in dogs, a contagious infection which may give rise to transitory human infestations. It is an unpleasant disease but, fortunately, very easy to cure, in contrast to demodetic mange. Usually the disease appears first round the eyes, outside the flaps of the ears, on the elbows, hocks or abdomen; thereafter it may spread all over the dog unless checked. The affected areas become bare and covered with small spots, like fleabites. The mite causes irritation and the scratching of the dog causes numerous small scabs and sores to develop.

Sarcoptes causes a mange in rabbits; the nose being mainly attacked and becoming thickened and enlarged. From the nose, the mange may spread over the face and ears and sometimes to the body. A serious and sometimes fatal condition results if the parasite is not checked.

Sarcoptic mange in horses, though relatively rare, is an unpleasant complaint which causes intense itching. The signs are little round hairless patches and small elevations which can be felt if the hand is passed over the skin. The withers are usually affected first and the disease spreads along the back and sides.

Notoedres cati (Fig. 7.8)

The genus *Notoedres* is closely related to *Sarcoptes*, the two mites being very similar in form. This mite is responsible for a mange in cats. The attack generally starts on the neck and afterwards spreads over the head. Greyish crusts are produced and finally the whole affected skin becomes leathery.

Another variety of *N. cati* (var. *cuniculi*) cause a type of mange in rabbits; and yet another species *N. muris* attacks rats.

Psoroptes communis

This non-burrowing form occurs on the following domesticated animals: rabbits, horses, sheep, goats and cattle.

People who keep tame rabbits are sometimes troubled to find infestations of *Psoroptes communis* var. *cuniculi* which mainly attacks the ear. The presence of the mites produces a yellowish substance, consisting mainly of dead flakes of epidermis, which may practically fill up the auditory meatus. This disease does not seem to be very infectious, for it often does not spread from mother to young or from one ear to another.

Cnemidocoptes mutans (Fig. 7.8)

This mite causes 'scaley leg' in poultry; it proliferates under the scales of the legs and causes a serous exudate, which may become very thick and nodular. A related species (*Cn. laevis* var. *gallinae*) causes 'depluming itch'. It lives in the skin at the base of the feathers and causes irritation. The bird plucks at the feathers which become broken until often only the midrib is left.

Otodectes cyanotis

Mites of this species live in the ears of dogs, cats and ferrets. *O. cyanotis* var. *felis* is fairly common in cats, and the otitis caused by it is generally known as 'canker'.

Demodex

While *Demodex* does not belong to the Sarcoptidae, it is mentioned here because it produces mange in domestic animals.

Demodex canis causes a severe and highly unpleasant mange in dogs, which is very difficult to cure. It gives rise to variable clinical signs and the condition may be complicated by a bacterial infection. Usually, small patches of bare skin appear on various parts of the body; later pimples and pustules develop, and finally secondary infections are liable to produce running sores accompanied by a disgusting odour. The disease is, fortunately, not very infectious.

Miscellaneous mites

A number of other mites occur among fur or feathers, not necessarily causing any injury. Some of them are known to be predatory on harmful mites.

(ii) Control of mange mites

It will be appreciated that the whole life cycle of these mites is passed on or in the epidermis of their host. The biology of many of them is not known in detail, but it appears most unlikely that any of them can survive for a long time away from the host. Some of the infestations are of very low infestivity and it is not known how the mites spread. Most varieties are specific parasites of one type of host though transitory infections on other animals may occur; on people, for example, who continually handle infested animals.

Control measures consist in skin treatments, either local or general, according to the severity of the disease. In some cases it may be advisable to remove exudate or crusts to allow the medicaments access to the skin.

The lesions should be treated with (i) sulphur ointment (which is successful in relatively bare places), *or* (ii) benzyl benzoate ointment (page 283) applied with a brush *or* (iii) 'Tetmosol' solution or emulsion (a preparation of tetra ethyl thiuram monosulphide, made by Messrs I.C.I. Ltd).

For treatment of fowls attacked by depluming mite, dipping the whole bird in a sulphur bath (½ per cent soap, 2 per cent finely ground sulphur) is commonly used.

(b) The harvest mite or chigger (Trombicula spp.)

(i) Distinctive features

The adult mites of the family Trombidiidae may be recognized fairly easily from their velvety appearance (due to a coating of stiff serrated bristles) and the central constriction, giving them the shape of a figure 8 (Fig. 7.8). Many of them are bright red or orange in colour and are readily noticed when abundant.

Whereas the adults live a (to man) harmless predatory or scavenging life, the larvae are parasitic on vertebrates and sometimes attack humans. The six-legged larvae are only about 0·2–0·3 mm long, red to pale yellow in colour and bearing numerous feathered setae (Fig. 7.8). A considerable number of larval forms of these mites are known from different parts of the world and they are assigned to six genera. One species, *Trombicula autumnalis*, is known as the 'Harvest mite' in Britain, 'bete rouge' or 'aoutat' in France, 'Herbstgrasmilbe' in Germany, etc.

(ii) Distribution

T. autumnalis occurs over most of northern Europe. In any country its distribution is rather patchy, for unknown reasons. In Britain, for example, chalk downs are particularly liable to heavy mite populations.

T. alfreddugesi is prevalent throughout the New World, from Canada to South America. Other, less common species are *T. splendens* in the eastern U.S.A. and *T. lipovskyana* in Tennessee, Kansas, Oklahoma and Arkansas. Both prefer moist situations.

(iii) Life history

The females lay their eggs in the soil. The eggs (100–200 μm long) finally hatch to produce the six-legged larvae. These young larvae climb up herbage to the tips of leaves to await an opportunity of attaching themselves to a vertebrate host. Various rodents are commonly attacked, especially rabbits; but the mites have been found on a number of other animals and birds.

Having reached a host, the mites find their way to a suitable area and insert their mouthparts and remain attached for several days. They are commonly found in the ears and around the genitalia of small mammals; on the thighs, under the wings and around the anus of birds. On man, the mites usually crawl up until they reach a constriction of clothing (garter; waistband) and then attach themselves. The secretion of the mite's salivary glands evokes a reaction from the host resulting in histolysis with peripheral hardening; thus a tube is formed, up which the mite can suck lymph (not blood). Finally, the larvae drop off to the soil and eventually moult to produce a non-parasitic eight-legged nymph. This in turn gives rise to the adult which, like the nymph, appears to live on the eggs and young of small arthropods. Thus the mites only attack a single vertebrate host during their life cycle; this attack occurs in the larval stage.

(iv) Quantitative bionomics

The duration of the post-larval stages vary considerably according to temperature, humidity and the availability of food. The entire cycle may require from 2 to 12 months or even longer. One to three generations a year are thought to occur.

(v) Importance

In large parts of eastern Asia, mites of the genus *Trombicula* are vectors of the Rickettsia which causes scrub typhus.

In Britain the harvest mite is merely a nuisance; but it can be a severe one from the intense and prolonged irritation caused by its bites. Rural workers or picnickers are liable to be attacked; in fact, anyone who sits or lies on infested ground in summer or autumn. A line of weals is often formed where the mites have bitten along the line of a constriction of clothing.

(vi) Control

At the present stage of knowledge, prevention of the bites is much simpler than destruction of the mites. The usual routes of attack of the mites on low herbage are over socks or stockings and up the legs. Agricultural workers, picnickers, etc, may expose other parts of the body when sitting or lying on the ground. The larva is so small that it readily penetrates through many forms of fabric; through socks and stockings, for example. However, it can be easily checked by rubbing the socks or stockings several times between hands moistened with a repellent. Cuffs, necklines and other openings of the clothes may be given this 'barrier' treatment, which is quite invisible on most materials. Dimethyl phthalate or benzyl benzoate (both undiluted) are equally effective. Garments treated in this way will afford protection for about a fortnight.

(c) Bloodsucking mites

Mites of the family Dermanyssidae have a rather characteristic shape (see Fig. 7.8) with plates on the dorsal and ventral sides. They are medium-sized mites moderately covered with short bristles. Certain forms are parasitic on vertebrates and occasionally suck the blood of man.

(i) The red poultry mite, Dermanyssus gallinae (Fig. 7.8)

This mite is a common parasite of wild birds in temperate climates, but it also thrives in poultry houses, giving rise to the common name. The adult is about 1 mm long and the red colour mentioned in the name is due to the blood visible soon after a meal. Later, however, the partly digested blood turns black.

Life history

The females lay batches of up to 7 eggs, 1–2 days after a blood meal. The eggs (about 0·4 mm long) are laid in crevices in the vicinity of the host; that is, in a bird's nest or poultry roost. The eggs hatch to produce a six-legged larva, which does not feed but remains quiescent. The larva moults to give an eight-legged nymph which visits the host for a blood meal and then retires to a crevice nearby. Another moult produces the second-stage nymph which again feeds and finally moults to give the adult. The nymphs and adults generally feed at night and remains hidden in crevices during the day. The adults are very resistant to starvation and have survived without a blood meal for 4–5 months.

Speed of development

Under rather hot conditions (27–28° C) the egg hatches in 1–2 days; the larval stage takes 1 day and the nymphal stages 5–6 days. Total development, then, occupies 7–8 days; but under cooler conditions this is no doubt prolonged.

Importance
Heavy infestations will severely affect chickens, reducing growth and egg-laying and even causing death. The mites will also attack man and apparently take blood meals, although they will not breed in association with man alone. Many of the complaints of *D. gallinae* are due to infestations of wild birds' nests which have invaded adjacent rooms and annoyed the occupants.

Control
The source of infestation of houses and offices should be sought in nests in eaves or attics. Such nests should be removed and all adjacent walls sprayed with an acaricide. Infested poultry houses should be cleared of litter and sprayed. Suitable sprays would be 0·5 per cent *gamma* HCH or 1 per cent malathion. Alternatively, 2 per cent malathion dust at 1 lb to 20 ft^2 could be used in poultry houses.

(ii) Tropical rat mite, Ornithonyssus bacoti (Fig. 7.8)

This mite is associated with rats all over the world, in both tropical and temperate climates. It resembles *D. gallinae* in general, but can be distinguished by the shape of both dorsal and ventral plates, the hind end being pointed in *Ornithonyssus* and blunt in *Dermanyssus*. Like *D. gallinae*, this mite is red after a blood meal, but becomes blackish as the blood is digested.

Life history and development
The life cycle involves the same stages as *D. gallinae*: egg, six-legged non-feeding larva, two nymphal stages and adults. The feeding habits are also rather similar. In a laboratory colony maintained 'at room temperature', the total length of development varied from 11–16 days.

Importance
While *O. bacoti* is primarily a parasite of rats, people living or working in rat-infested buildings may be attacked and suffer bites.

Control
Rodent control is the best measure to protect humans from bites of the rat mite. Mites which have found their way into living quarters, offices, etc, can be destroyed by the sprays recommended for *D. gallinae*.

7.2.3 TICKS

(a) Distinctive features

Ticks are rather distinctive Acari because of their relatively large size and leathery cuticle. All of them are bloodsucking parasites of vertebrates, usually birds or mammals. Instead of a true head, there is a *capitulum* composed of the rather prominent mouthparts. These comprise (1) a pair of chelicerae, ending in two toothed digits for cutting the host's skin, (2) a central beak-like *hypostome* with recurved teeth on the under side, and (3) a pair of rather stumpy palps, bearing a humidity-detecting organ in a terminal pit.

There are two quite different types of tick: the *Ixodidae* or hard ticks and the *Argasidae* or soft ticks. They may be distinguished as follows.

(i) Anatomical differences (Fig. 7.9)

Ixodidae

The designation 'hard' ticks refers to the fact that the back of the body bears a *scutum*, or shield-like plate. This covers the entire back of the male, but only the anterior part of that of the female, the remainder being flexible and very greatly extended when the tick is fully fed. The scutum may be distinctly patterned, in which case, it is described as *decorated.* The hind end of the body in both sexes is sometimes marked with a series of radial grooves; these are described as *festoons.*

The capitulum is always prominently visible at the front of the body. The basal and terminal of the 4 palpal joints are much reduced and the whole palp is stiff and moves only from the base. The base of the capitulum in females bears a pair of oval *porose areas* of uncertain function. Eyes, when present, are situated on the sides of the scutum.

Argasidae

The scutum is absent and the sexes very similar in appearance. The cuticle is wrinkled and granulated. The capitulum in nymphs and adults is situated under the front of the body and may not be visible from above. Eyes, if present, are borne on folds above the coxae of the legs.

(ii) Biological differences

Ixodidae

Most hard ticks generally infest open country, especially fields or scrublands frequented by their hosts. They tend to be more sensitive (especially *Ixodes*) to

---

Figure 7.9 (*See opposite page.*) Ticks. (1) *Dermacentor andersoni*, male; (2) the same, female; (3) *Rhipicephalus sanguineus*, male; (4) the same, female; (5) *Amblyomma americanum*, male; (6) the same, female; (7) *Ixodes ricinus* (ventral), female; (8) the same, male; (9) *Hyalomma excavatum* (ventral) female; (10) *Ornithodorus moubata* (ventral) female; (11) *Argas persicus*, female.

Key: a, anus; e, eye; f, festoons; g, genital opening; h, hypostome; p, palp; po, porose area.

(1), (2), (5), (6), (9) after Arthur [17]; (3), (7), (8) after Smart [542]; (10), (11) after Nuttall [452].

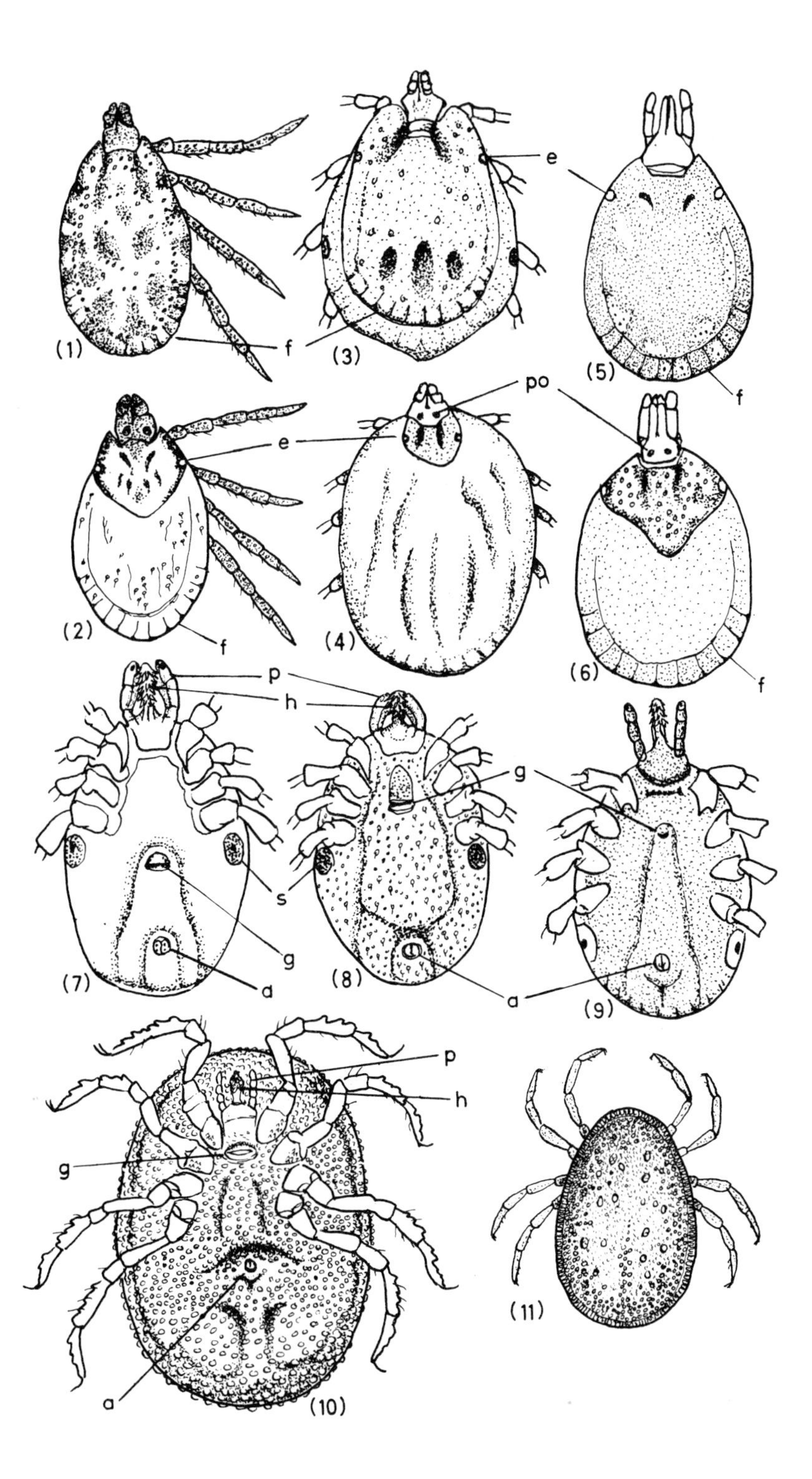

e
f
(1)
(3)
(5)
f
po
e
(2)
f
(4)
(6)
f
p
h
g
s
g
a
(7)
(8)
a
(9)
p
h
g
a
(10)
(11)

arid conditions than the soft ticks. The females feed only once, taking an enormous meal of blood, which may increase the body weight 20–150 times. Thereafter she lays one very large batch of eggs, often several thousands. After hatching, the larvae climb up vegetation, endeavouring to cling on to any passing animal. Then, some ticks (e.g. *Boophilus*) pass the rest of their lives on the same host. The larvae feed and moult into nymphs. These also feed and after moulting emerge as adults, which feed and mate on the host. After feeding, the female drops to the ground and lays her eggs. Other hard ticks (e.g. *Rhipicephalus evertsi*) require two hosts. The larval and nymphal stages are passed on the same animal, but the nymphs drop off and moult on the ground, so that the adults have to find another host. The great majority of ticks, however, feed at each stage of their lives on different hosts. Thus, ixodids can be described as one-host, two-host, or three-host ticks.

Argasidae

Soft ticks tend rather to infest the habitat of the host and, at least in the later life stages, visit the animal only to feed and then retire to a hiding place. The larvae, indeed, attach to the host for several days, or even a week or so; but the nymphs and adults take only a few minutes or an hour or two to feed. The adults feed at intervals and after each adequate meal, the females lay a batch of about 20–50 eggs. Their meals are modest compared to those of female hard ticks and only increase their weight by 2–3 times. Nevertheless, they are very resistant to starvation and can survive many months in a cool quiet environment. They are also resistant to very dry conditions (especially *Ornithodorus*).

(b) Some important hard ticks

(i) European hard ticks [489, 531]

The most important species in relation to human disease are the following. (1) Several virus vectors; *Ixodes ricinus* in Central Europe and, in the U.S.S.R., *I. persulcatus* and one or two species of *Dermacentor* and *Hyalomma*. (2) Vectors of tick typhus; *Rhipicephalus sanguineus* in the Mediterranean region and, in Siberia, *Dermacentor* spp. and *Haemaphysalis concinna*. In addition, a number of tick species are vectors of diseases of domestic animals.

Ixodes ricinus (Fig. 7.9)

There are several fairly closely related species of *Ixodes* in the North temperate region. *I. ricinus* and *I. hexagonus* are European; *I. persulcatus* extends through Russia to Siberia and *I. pacificus* occurs in Western North America.

I. ricinus feeds on a wide range of hosts, including wild and domestic animals (especially sheep) and occasionally on man. The eggs are laid on the ground among vegetable debris and covered with secretion to reduce desiccation. Even so, a high humidity (over 80 per cent R.H.) is required for incubation. After hatching, the larvae climb up vegetation and wait for passing animals or birds, trying to catch hold of one. This behaviour exposes them to water loss in hot dry weather and host-seeking is mainly characteristic of spring and, to a lesser extent, of autumn. The young larvae crawl over the host to find a suitable place and begin to feed. After engorgement, they drop to the ground and eventually moult to the nymphal stage. The nymphs climb up herbage in search of a new host, and the cycle is repeated. Eventually, the adults seek a third host. The females engorge slowly, reaching the size of a pea; but the males feed only for a few hours and then seek a mate.

The life cycle is somewhat variable, because new hosts may not be found quickly; but all stages are resistant to starvation. Unfed larvae have survived 15 months, nymphs for 13 months and adults for 21 months. Furthermore, cold winter conditions will prolong development. The following minima and maxima have been quoted (days). E, 25–400; L, (harden and attack host) 10–570 (feed) 3–6 (rest) 28–426; N, (harden and attack host) 10–540 (feed) 3–6 (rest) 56–360; A, (females; harden and attack host) 10–810 (feed) 6–14 (pre-oviposition) 4–27 [17]. In the U.S.S.R. (Karel peninsula) a 4-year life cycle has been observed [542].

Ixodes hexagonus and *I. canisuga*

I. hexagonus is fairly common in Britain and not infrequently attaches to domestic animals or man. It closely resembles *I. ricinus*, but can be distinguished as follows:

Front coxae with a long spine directed backwards; shield rounded. *I. ricinus*

Front coxae with a short spine; shield lozenge-shaped. *I. hexagonus*

I. canisuga resembles *I. hexagonus* and *I. ricinus*, but lacks the spines on the front coxae. It quite often occurs on dogs in Britain and has been called the English dog tick. In France the commonest tick pest of dogs is *Dermacentor reticulatus*, which is accordingly called *tique du chien*.

Ixodes persulcatus

This occurs throughout Russia, from the Pacific westward right to northern Germany. It has a range of wild ungulate hosts in the taiga vegetation zone; also small mammals, birds and occasionally, man. The larvae may or may not feed before hibernation, so that nymphs appear in the next year. Nymphs, again, may or may not feed before hibernation and the adults appear in the second year. Finally, the adults hibernate again before reproduction. The unfed larvae and nymphs tend to survive hibernation (at temperatures above -30° C) better than fed ones. The females feed for about 4 –12 days before dropping down to lay about 3000 eggs [489].

Rhipicephalus sanguineus (Fig. 7.9)

The brown dog tick is widely dispersed in the warmer parts of the world, between latitudes 50°N and 35°S. In urban areas, it is closely associated with dogs, but where dogs are scarce, other animals may be bitten. Humans are occasionally attacked and may thereby acquire tick typhus.

The minimum life cycle is as follows [17]. E, 20–60 (av. 38 at 20–25°C); L, (feed) 3–8 (rest) 6–9 , at 30°C; N, (feed) 4–9 (rest) 11–12; A (female, feed) 7–21. The entire cycle at 30°C takes 63 days; but commonly hibernation occurs, as larvae, nymphs, or as adults, which obviously extends development time. The usual large numbers of eggs are laid; 1400–5000 have been quoted. The ticks are resistant to starvation; larvae survived 253 days, but nymphs only 97 days.

Hyalomma marginatum and *H. excavatum* (*H. anatolicum*).
These ticks are vectors of haemorrhagic fever virus in the U.S.S.R. *H. marginatum* has a particularly large range, from Indo-China to the Balkans and North Africa. The adults of these ticks attack large animals and man, whereas their larvae and nymphs prefer small wild animals and birds as hosts. The incidence of the haemorrhagic fever in summer coincides with the maximum activity of the ticks.

Dermacentor spp.
Three species of *Dermacentor* are involved in the transmission of Siberian tick typhus as well as encephalitis virus. *D. marginatus* occurs in the Crimea–Don–Caucasus region, while *D. nuttalli* and *D. silvarum* are found further east in the general area of Lake Baikal. Their host preferences are as for the *Hyalomma* spp. [489].

(ii) North American hard ticks

Three important species which can act as disease vectors are *Dermacentor andersoni, D. variabilis* and *Amblyomma americanum*. Others of some importance are *Rhipicephalus sanguineus*, *Ixodes pacificus* and *Haemaphysalis leporis-palustris*.

Dermacentor andersoni (Fig. 7.9)
The Rocky Mountain wood tick occurs in the Rocky Mountain states and in Nevada, eastern Oregon and Washington. Its distribution coincides with recorded cases of Colorado tick fever, which is due to a virus; and the tick is also an important vector of Rocky Mountain spotted fever, a rickettsial disease.

The young stages feed on small wild animals, seeking their hosts (as described for *Ixodes ricinus*) by climbing up vegetation and spreading out their front legs to cling hold of passing creatures. Also like *I. ricinus*, they retreat down to the base of vegetation in hot dry conditions. Adults readily attack large animals and man, climbing up to the head and shoulder region before feeding. The adults are active from February to May or June; larvae are common in July and nymphs throughout the summer. All stages are able to hibernate. The minimum life cycle is as follows [17]. E, 7–10 (at 32°C) or 38 (at 22°C); L, (feed) 6–7 (rest) 10–12; N, (feed) 6–7 (rest) 14 or more; A, (female, feed) 5–15 (pre-oviposition) 11; males (feed) 3–4, then mate. About 2500–5000 eggs are laid.

Dermacentor variabilis [544]
The American dog tick occurs throughout the eastern half of North America and, in addition, along a coastal strip of California. The immature stages feed on small mammals (especially wild rodents), but the adults prefer larger wild animals of many kinds as well as dogs or man.

Adults are active from mid April, rising to a maximum in mid June, then declining in numbers to September. The larvae have a peak abundance in early spring (on wild hosts) due to host-seeking after hibernation. Nymphs are most active in July or August.

Amblyomma americanum [237]
The lone star tick is most abundant in the south-eastern parts of 'the U.S.A., but has a scattered distribution across the rest of the country, and especially in the Ozark region of Missouri. The hosts may be large wild animals and very often man is attacked. These ticks can overwinter as fed larvae, unfed nymphs or fed nymphs which moult to adults. New hosts are attacked in spring and fed females drop to the ground and lay the usual large egg batch. The eggs hatch in about 30 days and larvae climb up the vegetation to seek new hosts.

Rhipicephalus sanguineus (Fig. 7.9)
Originally an Old World species, the brown dog tick was introduced to the New World together with the domestic dog. It now occurs in eastern Canada, most of the eastern side of the U.S.A. and a few western areas. Its favoured host is the domestic dog and it frequently infests kennels and hence dwelling houses. It can be a troublesome pest, requiring eradication, although it does not bite man in the U.S.A.

Haemaphysalus leporis-palustris
This tick occurs on small wild animals, especially rabbits and hares, and occasionally on dogs and cats. It is widely distributed in the U.S.A. and Canada; and it is involved in maintenance of spotted fever and tularaemia among wild reservoirs.

(c) Some important soft ticks

Only two genera are of importance; they can be distinguished as follows.

A distinct rim separating the dorsal and ventral surfaces of the body. *Argas*

Dorsal and ventral surfaces not clearly separated by a rim. *Ornithodorus*

(i) European soft ticks

Argas persicus [17,531] (Fig. 7.9)
The fowl tick occurs widely over the world in the tropical and sub-tropical regions, reaching up to 52° N in Europe, 55° N in Siberia and 40° N in America. Though pre-eminently a pest of poultry, it will attack other birds and some animals including man. It measures about 4 mm long and is pear shaped in outline. Unengorged ticks are greyish or sandy in colour, but after a meal they become deep garnet, which changes to greyish-blue after a few days. They are involved in the transmission of avian spirochaetosis and can pass spirochaetes to man, although they are not, in practice, involved in human disease. Their bites, however, cause unpleasant symptoms.

Batches of 200–600 eggs are laid by the females, after blood meals, in crevices in poultry houses or runs. The eggs hatch in 8–11 days in warm weather, but in cooler seasons this may extend to 3 weeks or more. The larvae feed for 5–10 days, often at the base of a fowl's wing, and then drop off and seek shelter. Moulting occurs in 4 or more days. The nymphs feed twice within the course of a few hours and moult to the adult 7–14 days later. A third nymphal moult may sometimes occur. Feeding of nymphs and adults is of short duration and usually occurs at night. Fed adults can survive for up to 13 months without a meal.

Ornithodorus spp. (Fig. 7.9)
O. erraticus occurs in northern Africa and extends up into Spain and Portugal, where it sometimes transmits Spanish–North African relapsing fever. It commonly occurs in wild rodent burrows and is sometimes found in piggeries. *O. tholozoni* which occurs in southern Russia and the Near East, is another occasional vector of relapsing fever. Its hosts include wild rodents as well as dogs and occasionally, man.

(ii) North American soft ticks

Argas persicus
A. persicus occurs in many of the southern states, where it is variously known as the chicken tick, adobe tick, tampan or blue bug; the last name referring to the colour of the female a few days after engorgement.

Ornithodorus spp.
Three species of *Ornithodorus* are important in North America as vectors of relapsing fever: *O. hermsi, O. turicata* and *O. parkeri.* They are prevalent in the western states. *O. hermsi* is a serious pest in mountain cabins, even up to 8000 feet (2400 m). Often these cabins are only used in the summer vacations and in the winter small rodents invade them, bringing in the ticks. As usual, they are resistant to starvation; larvae can survive 95 days and nymphs and adults as long

as 7 months. The duration of the life cycle is 202–314 days at at 30°C and 364–602 days at 21°C [490].

The other species, though common at lower altitudes, are less frequent vectors. *O. turicata*, which has an exceptionally large range of wild hosts (including reptiles) occurs west from Oklahoma, Kansas and Texas. *O. parkeri* is a more northern species, occurring west from Montana and Wyoming. Its wild hosts are mainly birds and small mammals, especially rodents.

(d) Importance

(i) Tick bites

Ticks may become attached to dogs or human beings, usually, though not exclusively, in rural areas. The bite of the tick is not usually serious, although it may be alarming on account of the size of the tick and the difficulty of inducing them to release their hold. If the tick is torn off by force, the body may be pulled away leaving the capitulum embedded in the flesh. This may lead to irritation and perhaps to septic conditions.

Occasionally, the ticks embed themselves in the skin with exceptional firmness and the reaction of the bite stimulates the tissues to proliferate around the tick. This forms a cuplike cyst, which may practically cover the tick.

The best way of removing ticks is to dab them with chloroform or ether, after which the capitulum should be pressed inward to loosen the teeth, and then the tick is gently pulled away. If it is not possible to use a volatile solvent, the tick should be covered with a pledget of cotton wool soaked in medicinal paraffin or olive oil. In this case, removal should not be attempted for some hours after application of the dressing.

(ii) Tick paralysis [17]

In some regions, tick bites give rise to a serious paralysis. The species responsible are mainly hard ticks. In North America, a moderate number of human cases have occurred in the Rocky Mountain region of Canada and the U.S.A. (250 from Canada and 100 from U.S.A. with about 10 per cent mortality). Outbreaks among cattle are much more numerous, often with severe losses. In Europe, human cases seem to be rarer, but there are outbreaks among cattle and sheep.

The species responsible in America are *Dermacentor andersoni* in the western states and *D. variabilis, Amblyomma americanum* and *A. maculatum* in eastern and southern regions. In Europe, the trouble is mainly due to *Haemaphysalis* spp., *Ixodes ricinus*, and *Rhipicephalus* spp.

The bite of the tick is often not noticed for several days which gives the creature enough time to introduce the (unknown) venom. For this reason, it is desirable to remove any ticks as soon as possible. If the paralysis develops, in severe cases, death may ensue from respiratory paralysis. Fortunately, the symptoms generally cease when the ticks are removed.

(iii) Disease transmission

Ticks are responsible for transmission of human infection with various diseases, some of them serious. The diseases are not specific to particular ticks, nor are the ticks specific human parasites. The diseases are generally maintained in wild animal reservoirs, or in the ticks themselves, which are unharmed by them and can transmit some of them to their offspring. Human cases may result from ticks acquired in open country, by rural workers, picnickers or soldiers on manoeuvres. Domicilary infestations are unusual, although ticks may be brought into the home by dogs or (as mentioned earlier) wild animals sheltering in mountain cabins used only in summer.

Several types of pathogen are involved; spirochaetes, rickettsiae, viruses or bacteria. Although some of these can prove fatal, if untreated, modern antibiotics have reduced the dangers (except for viruses).

Diseases due to spirochaetes

Tick-borne relapsing fevers, due to *Borrelia* spp., occur in a wide band round the tropics and sub-tropics, especially in Africa. In America, the disease occurs mainly in an area south and west of a line from Idaho to Texas, the vectors being *Ornithodorus hermsi, O. turicata* and *O. parkeri*. In the Old World, the disease extends up into the Iberian peninsula (vector *O. erraticus*) and into the Caucasus (vector *O. verrucosus*).

Diseases due to Rickettsia

American tick typhus, due to *Rickettsia rickettsi*, comprises Rocky Mountain spotted fever of the western U.S.A., transmitted by *Dermacentor*, with a reservoir in various wild animals. Also eastern Rocky Mountain fever in British Columbia, Alberta and Saskatchewan, now spreading eastwards, transmitted by *Dermacentor variabilis, Rhipicephalus sanguineus* and *Amblyomma americanum*. Wild animals, again, constitute the reservoir and human cases often result from ticks brought into homes and gardens by dogs.

Tick typhus in the Mediterranean area, due to *Rickettsia conori*, is known as *fievre boutonneuse*. It extends as far as the Balkans and the Crimea. It is transmitted by *Rhipicephalus sanguineus*, with a reservoir in dogs. Siberian tick typhus is transmitted by *Dermacentor nuttalli* and *Haemaphysalis concinna*. Another rickettsial disease is Q-fever, maintained among animals (including cattle) by ticks and occasionally transmitted to man in the western U.S.A. More commonly, however, it is acquired by handling infected animals in stockyards or slaughter-houses.

Diseases due to viruses

Ticks are only second to mosquitoes as vectors of viral diseases. These include both haemorrhagic and neurotropic types. Most of the important ones occur in the Old World. Throughout the vast evergreen forests of the U.S.S.R. they are represented by Russian spring–summer encephalitis, a dangerous disease with about 25–30 per cent mortality. The vectors are *Ixodes persulcatus* and species of *Dermacentor*. In Central Europe, this is replaced by central European encephalitis, a somewhat similar but less virulent disease. Finally, in Britain, a related virus causes louping ill of sheep, which is occasionally passed to man. A virus causing haemorrhagic fever occurs in south-west Siberia. It is known as Omsk haemorrhagic fever, transmitted by *Hyalomma* spp. The natural reservoirs of these viruses are wild animals and birds.

Tularaemia

This disease, due to the bacterium *Francisella tularensis*, is prevalent among various wild animals (especially rodents) in Europe and North America. Various blood-sucking arthropods can transmit the disease among them, but ticks are probably the most important. Occasionally, they transmit it to man; but the infection can also be acquired from bites of deer flies (see page 181) or from contaminated water.

Protozoal diseases of animals

Ticks transmit a number of diseases due to protozoa (*Babesia Theileria*, *Anaplasma*) to domestic livestock, mostly in Africa, but to some extent in North America.

(e) Control

(i) Outdoors

The wide dispersion of ticks in open country makes it difficult to control (let alone eradicate) them from anything but quite limited areas. In some cases, however, it may be worth attempting to control them in the natural environment; for example, in parks, lawns and limited recreational areas in regions where ticks abound. For this purpose, DDT treatments (at 0·25 g m^{-2}) have been used in Europe against *Ixodes* ticks, which are sensitive to that compound [145]; but organochlorine compounds would now be restricted in many countries. Some success has been claimed for 0·5 per cent tetrachlorvinphos as an aqueous suspension, applied at intervals [237]. Fenthoate and fenitrothion also seem promising [145].

For people foraging, working or resting in infested areas, the best precaution is to wear protective clothing (e.g. trousers tucked into socks) treated with repellents such as deet, indalone or benzyl benzoate.

(ii) Infested dwellings and pets

Ticks do not normally infest dwellings in the temperate region although, as explained, some may be introduced by domestic animals, or (in mountain huts) wild animals. A possible exception is the brown dog tick, which may infest places frequented by dogs. To eradicate these, various organo-phosphorus insecticides may be used as a spray or dust. The dog itself may be treated with a dip containing 0·1 per cent dichlorvos, 0·2 per cent naled, 1 per cent carbaryl or 0·15 per cent dioxathion. The treatment is best done by a veterinarian. These compounds are not suitable for cats, which tend to lick their fur extensively; they (or dogs) can be dusted with a pyrethroid-containing powder.

7.3 ACCIDENTAL PARASITISM: ALLERGIES

7.3.1 INVASION OF THE HUMAN BODY BY MITES

Various mites and their eggs, either living or dead, have been found to occur in various parts of the human body, such as the alimentary canal and the urinary and respiratory tracts. It should be remembered that, in investigating such cases, great care must be taken to exclude the possibility of accidental contamination of specimens or of laboratory glassware, for the mites concerned are ubiquitous. However, many records are probably authentic. The presence of the mites has been cited as the cause of various conditions including enteritis, nocturnal enuresis, haematuria and bronchitis. The evidence, however, is not straightforward, for in many cases the mites are quite harmless.

(a) Lungs

Mites of the genus *Pneumonyssus* habitually live in the lungs of monkeys, apparently as a form of parasitism. Man does not appear to be susceptible to these mites, however. On the other hand, there are several records of various free-living mites being found in the sputum of patients suffering from asthma and eosinophilia, especially in the tropics (*Tyrophagus casei*, *Acarus siro*, *Histiogaster* sp., *Tarsonemus* sp., etc). Whether the mites were responsible for the asthmatic condition is not certain.

More recently, one particular form of mite-induced asthma has been conclusively demonstrated. This is due to mites of the genus *Dermatophagoides*, which are widely prevalent in ordinary house dust [591]. *D. pteronyssimus* is most common in the Old World, whilst *D. farinae* tends to be prevalent in America. These mites are present in nearly all houses (except, perhaps, at high altitudes) and they are especially numerous in damp conditions. The mites feed on miscellaneous organic debris, but particularly on human dander (skin debris). For this reason, they are very liable to be present in small quantities of dust in crevices of mattresses and, to some extent, in stuffed furniture.

To the great majority of people, these mites are quite harmless; but they cause asthma to atopic subjects, especially children. They are affected by inhaling the mites, their cast skins, or their faeces. Such sufferers are treated medically for the symptoms of the asthma and also, so far as possible, to desensitize them by progressive doses of the mite antigen.

Various attempts have been made to find acaricides to destroy the mites in rooms used by the sufferers. Two compounds which have shown most success are lindane and pirimiphos-methyl [260]. There are, however, considerable difficulties in eradicating the mites by chemicals, which must be used in proximity to atopic subjects.

(b) Intestine

Since mites occur so commonly on many foods, they must frequently be swallowed by man and animals; and, in fact, they are quite often found in faeces. The most common species are *Tyrophagus casei*, *Acarus siro* and *Glycyphagus domesticus*. Nearly always the mites in faeces are found to be dead, but there are records of eggs hatching after passing through the alimentary canal. It seems possible that diarrhoea with occasional bloody discharges and even intestinal ulcers have been caused by continued feeding on food heavily infested with mites. There are several records of cases, both human and of animals, which have suffered these disturbances while producing stools containing numerous mites; and the symptoms have ceased on a change of diet. Experimental studies have indicated a great variability in susceptibility of different animals to these effects.

(c) Other sites

Mites (mainly *Tarsonemus*, *Tyrophagus*, *Carpoglyphus*) have been found in urine, apparently from the urinary tract. There are also one or two records of mites found in pathological lesions; for example, in liquid from a scrotal cyst and in a carcinoma of the jaw.

7.3.2 MITES CAUSING DERMATITIS

A number of mites, which are normally free living, are liable to cause an irritation of the human skin, possibly of an allergic nature. An irritant but transitory dermatitis results, which normally clears up when association with materials infested with the mite ceases. Individuals vary considerably in their sensitivity to these mites, but a considerable proportion are likely to suffer if they have repeated intimate contact with heavily infested materials. Such men as dock labourers, porters and warehouse men are especially prone to attack by reason of their occupation; and these men may carry the mites back to their homes and cause trouble among their families.

More recently, it has been shown that a number of non-parasitic mites, occurring in the fur or feathers of domestic pets may cause dermatitis in sensitive people. In particular, atopic subjects, liable to hay fever or other allergies, are likely to suffer from such mites [267].

Another source of mites which may cause dermatitis is house dust.

(a) Effects

The effects of the attack are urticarial lesions, the individual spots ranging from a few millimeters to over an inch in diameter. Various parts of the body may be affected, but especially the hands, arms and chest. Symptoms of intense irritation are caused and scratching (as usual) may give rise to secondary septicaemia. In severe attacks, there may be pyrexia. It is not clear what causes the trouble since it is unlikely that the mites actually bite the skin; also, severe irritation can be caused by dead mites in large numbers. Possibly an allergic reaction is set up by the mites or their faeces.

(b) Mites concerned

(i) Mites in stored products

It is probable that a wide variety of mites associated with stored products can cause dermatitis in people regularly coming in contact with large numbers of them. Perhaps the most common cause is *Pyemotes ventricosus*, the grain itch mite with a curious life cycle, described on page 376. As it is parasitic on various insect larvae occurring in stored products, it has been reported in association with a wide variety of goods, including wheat, dried peas, linseed, tobacco, hay, straw, etc.

Other mites causing this type of dermatitis belong to the family Acaridae (Sarcoptiformes). Mites on cheese not infrequently cause trouble; they include *Tyrophagus casei*, *T. longior*, *Acarus siro* and *Glycyphagus domesticus*. Another source of trouble is copra, which may be infested with *Tyrophagus putrescentiae* (*T. castellanii*) or *Caloglyphus krameri*. Dried fruit, infested with *Carpoglyphus lactis*, has been implicated once or twice, and there is a record of *Suidasia nesbitti* on wheat.

Occassionally, outbreaks are sufficiently severe for a popular name to be coined to describe the disease. For example, 'copra itch' due to *T. putrescentiae*; 'grocer's itch' due to *G. domesticus*, which was formerly common on poor-quality sugar; 'vanillism' due to *T. casei* on vanilla pods.

(ii) Mites in pet animals' fur [267]

Species of *Cheyletiella* have been mainly incriminated as a source of cutaneous reactions. Although the habits of these mites in fur are not well known, they are believed to be predators on parasitic forms.

(iii) Mites from house dust [267]

The species of *Dermatophagoides* already mentioned as a source of asthma, are also liable to cause dermatitis in sensitive people.

(c) Control

Accidental dermatitis due to stored product mites clears up spontaneously when contact with infested material ceases, usually within a few days. Soothing treatments, such as calamine lotion, may be desirable.

For persons obliged to handle infested materials, mite repellents (DMP: benzyl benzoate) may be beneficial. However, in view of the possibility of effects caused by dead mites (or their skins or faeces) repellents cannot be guaranteed to protect.

For control of the mites, see page 373.

CHAPTER EIGHT

Pests of foodstuffs

A variety of different insects preys upon domestic food stores. For convenience, they can be grouped together as 'larder pests', but they are rather a motley assembly with varying degrees of dependence on human food. Two rather different kinds of pest are included. The first group, which may be described as 'visitors', live in various harbourages in the structure of buildings (or even in the ground outside) and make journeys to visit the food stores. Cockroaches, ants, crickets and bristletails belong to this class. The second type of domestic food pests will be called the 'residents' because they live and breed in the stored food. These include certain beetles, moths and a number of mites.

8.1 VISITING PESTS

8.1.1 COCKROACHES (BLATTIDAE)

(a) Distinctive features

(i) General

Cockroaches belong to one of the primitive orders of insects, being allied to crickets, grasshoppers, preying mantids and stick insects. This is a very ancient assembly of insect types, formerly grouped together in a single order, the Orthoptera. Fossils not so very different from modern cockroaches occur in the upper Carboniferous rocks, 250 million years old.

Typically, cockroaches are rather large, robust insects, with whip-like antennae and two pairs of wings, the front pair being slightly stiffened and covering the hind pair when at rest. These front wings, however, show a distinct network of 'veins', which distinguishes them from the horny forewings of beetles.

Apart from their importance as pests, the large size of some cockroaches makes them convenient for laboratory studies. Two excellent books on these insects have been published by Cornwell [124, 126].

(ii) Identification of common species

Cockroaches were originally adapted to hot climates and the numbers and variety are most prevalent there. There are, however, certain species which have become cosmopolitan and, in particular, some which have penetrated cool climates by living almost exclusively inside warm human habitations.

The pest cockroaches most likely to be encountered indoors, may be grouped as follows: common pests, nearly always associated with man, *Blatta orientalis, Blattella germanica, Periplaneta americana*; less common, but frequently associated with man, *Supella longipalpa, Periplaneta australasiae, P. brunnea, P. fuliginosa, Leucophaea maderae*; wild species which occasionally invade buildings or occur in goods, *Pycnoscelus surinamensis, Nauphoeta cinerea, Blaberus* spp. and about ten other species. A key to identify the named species is given in the Appendix, page 533.

(b) Origin and distribution

(i) Species closely associated with man

The countries indicated in the specific names of cockroaches (*B. germanica, P. americana, P. australasiae, L. maderae*) are by no means reliable guides to their original home. *B. germanica* is known as the Russian cockroach in Germany and the Prussian cockroach in Russia. An American entomologist comments 'European nationals name their cockroaches after their neighbours across the border, an honour which is always reciprocated '[371]. In fact, a survey of the evidence suggests that most of the pest cockroaches originated in Africa. They show varying degrees of dependence on a warm environment.

Blatta orientalis probably originated in north Africa and then was spread by commerce into eastern Europe and the Mediterranean region in ancient times. It reached Holland and Britain by the early 17th century and later crossed the Atlantic and spread through north America.

Blattella germanica appears to have come from east Africa, in the region of the great lakes, where several wild species of the genus occur. Its subsequent migrations were probably similar to those of *B. orientalis*; via eastern Europe and the Mediterranean to northern Europe and the U.S.A.

Periplaneta spp. probably originated in central Africa and were transported to America by commerce, or even by the slave trade. Subsequently, they have spread throughout the world, but are less common in temperate climates than the tropics. *P. americana* is probably the most troublesome species in temperate climates. It extends further north in the U.S.A. than the other species of the genus; and although not common in Britain, occurs north as far as Belfast and Glasgow. The other species flourish in southern U.S.A. even out of doors; but in the north they are mainly found in greenhouses, in which *P. brunnea* and *P. fuliginosa* have been recorded as far north as Philadelphia and Chicago, respectively.

P. australasiae has been found breeding in greenhouses in France, Italy and Britain (including the tropical houses at Kew Gardens).

Supella longipalpa is also believed to have originated in Africa. It is now distributed throughout much of the Old World tropics and sub-tropics. Introduced into the West Indies, possibly by the slave trade, it reached Florida in

1903 and then spread gradually through the U.S.A., although it remains most prevalent in the southern states. In recent years, it has become established in a few places in Britain.

(ii) Species occasionally associated with man

The four types of cockroach mentioned below are mainly outdoor forms in warm climates, but from time to time, they may invade buildings in the U.S.A. or Europe.

Leucophaea maderae, the Madeira cockroach, seems to have come from West Africa and spread north to Madeira, Morocco, Spain and Corsica; then west to central America and the southern U.S.A. and finally all round the tropics.

Pycnoscelus surinamensis, which was indeed first recorded from Surinam, is now prevalent all round the tropics. It has invaded the southern U.S.A., from Florida to Texas.

Nauphoeta cinerea, the lobster cockroach, is believed to have come from south-east Africa, but is now distributed all round the tropics. Occasionally it is introduced into temperate countries and it has been recorded in Britain and Germany in heated buildings and has become established as a pest in Florida.

Blaberus spp. These cockroaches of the American tropics live under rotting vegetation in tropical rain forests. Sometimes imported into temperate countries in tropical cargoes, they excite interest because of their large size; and for the same reason, they are cultured in some biological laboratories.

(c) Incidence of the pest species

Statements about the varied incidence of different species have often been made on rather biased evidence. For example, an advisory entomologist may rely on the proportions of different species involved in queries, or limited inspections. Thus, both A. A. Green and the present author believed that *B. germanica* was more common in Britain than *B. orientalis*, on such evidence; and the enquiries received by the Danish Pest Infestation Laboratory about the same time (1959–63) also showed a preponderence of the German cockroach [9]. However, an extensive survey conducted in Britain in 1964–66, by the staff of Rentokill Ltd, gave a different picture [124]. Inspections of some 4000 premises showed that *B. orientalis* was about four times as common as *B. germanica*, overall. There is some evidence, however, that the ratio is changing in favour of the German cockroach. Thus, more recently built buildings have relatively fewer *B. orientalis* than old, pre-war structures. Because of its higher temperature preference, one would expect *B. germanica* to be relatively more prevalent in centrally heated buildings; but according to the survey, there is no evidence of this, although both species are favoured by heating. The ratio was found to vary in different parts of the country, there being a tendency for *B. germanica* to be more common in London and the South-East. On the other hand, Glasgow and southern Scotland

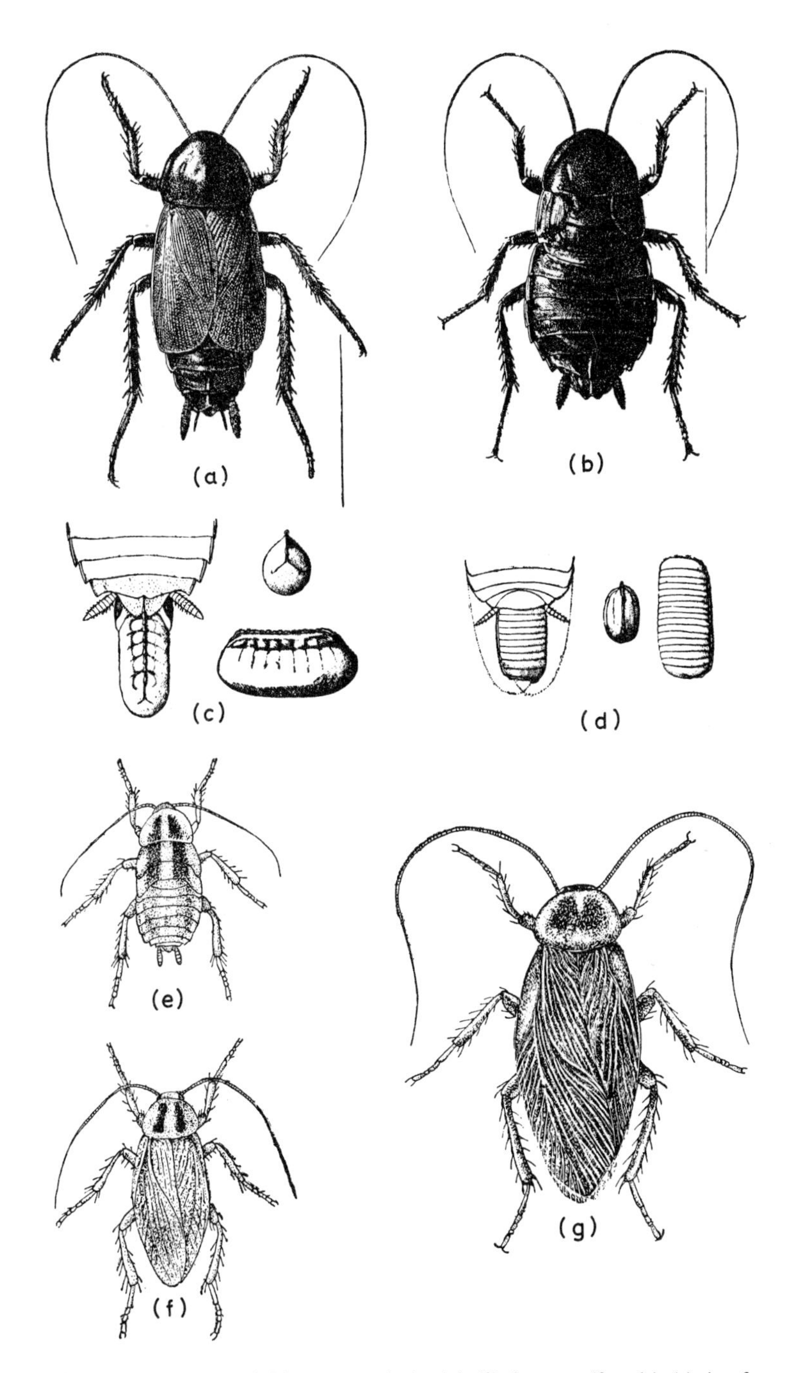

Figure 8.1 Cockroaches. (*a*) *Blatta orientalis* (male); (*b*) the same (female); (*c*) tip of abdomen of female to show the formation of egg capsule (*Blatta orientalis*); (*d*) the same (*Blatta germanica*); (*e*) *Blatella germanica* (nymph); (*f*) the same (female); (*g*) *Periplaneta americana* (male). From Laing [338]. (*a*), (*b*) × 1·5; (*e*), (*f*) × 2; (*g*) × 1.

were exceptional with relatively high *B. germanica*, while Brighton and Bournemouth were exceptions with high *B. orientalis*.

During the survey, some 16 000 records of site infestation were collected. *B. germanica* was most common in kitchens of restaurants, hotels, schools and hospitals and also in the bars of public houses and clubs. *B. orientalis* favoured basements, boiler houses, heating ducts, toilets and bathrooms.

Apart from the two common species, there were about 0.3 per cent of other types, mainly *P. americana, P. australasia* and *S. longipalpa.*

There appears to be little information on the relative incidence of cockroaches elsewhere in Europe, but there have been a few useful surveys in the south-eastern U.S.A. A 1952–3 investigation of different sites in Georgia was made by trapping [236]. The following figures give numbers per 100 traps, averaged over a year. Indoors, *B. germanica* (270) was most abundant, followed by *P. brunnea* (43) and *P. fuliginosa* (18), while *P. americana* (3) was rather rare. Outside houses, *P. fuliginosa* (23) was most common and *P. americana*, again least so. In or near privies, *P. fuliginosa* was most common (77) with the other *Periplaneta* spp. next (*c.* 20). In sewer manholes, it was *P. americana* (33) which far exceeded the others (*c.* 1–3). A more detailed analysis of the data for seasons, showed that *B. germanica* was relatively less abundant in the summer, when the *Periplaneta* spp. were at their maximum. Somewhat similar indications resulted from trapping in two Texan towns in 1955 [151]. *B. germanica* again was most prevalent in houses, followed by *P. brunnea*; while *P. americana* was most abundant in privies. A 1965 survey of different types of building in N. Carolina showed *B. germanica* to be most common [641]. It occurred in 75 per cent of houses and apartments (combined), 80 per cent of the shops and all of the 12 restaurants searched. Next most common was *B. orientalis* (in nearly half the houses) while *P. americana* and *S. longipalpa* were found in smaller numbers. In Bedford, Indiana, *B. orientalis* was very prevalent, both in houses and in sewer manholes [8].

(d) Life history [124]

(i) Oviposition

Female cockroaches lay their eggs in characteristic pod-like egg cases. These are formed in a chamber behind the egg pore which can be closed by a pair of flaps. Certain glands line the sides and back of this chamber with a viscous fluid, which is white at first but later turns brown and hardens. The eggs are laid into the pouch formed of this secretion and, as it becomes filled, the valve-like flaps relax and allow it to protrude. This occurs several times in the course of formation of the capsule, and the marks left by the retaining flaps can subsequently be distinguished as a series of ridges. In the later stages, the egg case can be seen projecting from behind the female as she runs about (Fig. 8.1). The cases are quite large compared with the size of the cockroach; those of *B. germanica* average

5·5 mm, of *B. orientalis* 10·5 mm and *P. americana* 9 mm. (Averages of six measurements from cockroaches reared at 25° C.)

There are several different types of behaviour of cockroaches in regard to oviposition. Among the pest species, three varieties can be distinguished. (1) Some species deposit the ootheca long before the eggs hatch. In such cases, the walls of the ootheca are waterproof and it contains an adequate water supply for the developing eggs. *B. orientalis* and *P. americana*, who belong to this group, generally cement the ootheca to the substrate and attempt to cover it with debris. *S. longipalpa*, another example, glues the ootheca to vertical surfaces and different females tend to deposit them together, in groups. (2) In some other species, the females carry the ootheca about with them until it is just about to hatch; *B. germanica* is an example. After completion, the seam of the ootheca which is vertical during formation, rotates 90° to one side or the other. Its anterior end which is in contact with the female, absorbs water from her during incubation. (3) Another group of cockroaches extrude the ootheca during formation, but on completion draw it back into the female's brood chamber, where it remains until the young hatch, in an apparently viviparous manner. Examples are *N. cinerea, L. maderae* and *P. surinamensis*.

(ii) Eggs

Inside the egg cases, the long narrow eggs lie in two rows, arranged alternately for compactness. The embryos are formed facing inwards, with their heads towards the aperture.

When they are fully developed, the young insects struggle upwards, burst open the closed seam and wriggle out. At the same time they undergo a moult and emerge as perfectly white, first stage nymphs. After a few hours the cuticle darkens and hardens; nevertheless, they are strong enough to run about immediately after hatching.

(iii) Nymphs

The body form and habits of the nymphs do not differ greatly from those of the adults and will not be discussed separately. As usual with insects showing partial metamorphosis, there are gradual changes to the adult form, such as increases in the number of joints of antennae and cerci and the growth of the wing pads. If a limb or antenna is damaged during the nymphal stage, it can be largely regrown in the course of subsequent moults.

Development is slow, even at high temperatures, and it is not easy to determine the exact number of moults, especially in view of the fact that the cast skin is usually eaten after moulting. In an investigation where the insects were reared separately and marked with paint to verify moulting [620], the following numbers of nymphal moults were recorded – *Blattella germanica*, 6–7; *Blatta orientalis*, 7–10; *Periplaneta americana*, 10–13. It appears that injuries needing regeneration of an appendage may cause extra moults.

(iv) Adults

Apart from their mature sexual organs, the adults differ from nymphs in possessing fully formed wings, Even in species (*B. orientalis*) with degenerate wings, the adult wing stubs can be distinguished by their distinctly marked 'veins'.

The general appearance of a cockroach is fairly well known; and although *Blatella germanica* is sometimes called a 'steam fly' its true nature is apparent if it is critically observed. The head projects downward beneath the characteristic hood formed by the dorsal shield of the first segment of the thorax. The head bears the whip-like antennae which are frequently in motion gathering sense impressions. They are often cleaned, like other appendages, by drawing them through the mouthparts. In this way a cockroach may destroy itself by its cleanly habits; for after contamination with certain stomach-position insecticides, the insect often swallows the poison cleaned off its limbs in this way.

Two large kidney-shaped eyes testify to fairly acute vision.

The mouthparts are primitive (and are commonly offered to students of zoology as a basic type). They are adapted to biting and triturating solid food and include stout mandibles and maxillae bearing hooked teeth and spines. Two pairs of palps are present.

Cockroaches can feed on practically anything edible and a large variety of substances may form their diet. Sometimes they occur plentifully in places where their source of food is problematical. When they live in warm, dry, centrally heated buildings, they need water; and they are sometimes trapped in sinks or wash-basins which they have visited in search of a drink.

The alimentary canal consists of a tube leading to a sack-like crop, where much of the digestion takes place, followed by a sort of gizzard with 'teeth' inside to grind up the food. The midgut begins with some blind pouches and continues as a simple, folded tube. At the beginning of the hindgut, a large number of Malpighian tubes are joined and they discharge their excreta into it among the food residues. The faeces vary in form according to the food, from dry pellets to dark-coloured liquid which forms smears.

The thorax bears the powerful legs and the wings, the latter being of minor importance or completely useless in the domestic species. The sturdy legs, especially the second and third pairs, are furnished with spines which give purchase on rough surfaces. If one picks up a large cockroach in a clenched hand, one is surprised at the energy with which it can thrust itself through a small crevice.

The feet of cockroaches comprise a central pad (the 'ariolium') and a pair of claws. In addition, the first four tarsal segments bear ventral pads called euplantulae. All these structures assist the insects in walking, especially on vertical or inverted surfaces. The claws, of course, are only helpful on roughened surfaces such as coarse paper. The ariolia and euplantulae, however, and especially the former, assist the cockroaches to walk on inclined polished surfaces. In *B. germanica* and *P. americana*, this ability enables them to climb up polished tiles or glass with ease. In *B. orientalis*, however, the ariolia are poorly developed and do not permit the females or some males to climb polished surfaces, although a few males are able to do so [511].

In temperate regions, none of the domestic cockroaches have been observed to fly, but in the southern states of the U.S.A., *P. americana* has been reported flying round street lamps at night like moths. This species, and *B. germanica*, may make gliding or fluttering flights in northern latitudes, but the action of the wings seems more to sustain gliding than actual flight. *B. orientalis* never uses its wings.

The cockroach abdomen is clearly segmented and bears at the end a pair of cerci, which are somewhat primitive appendages. On the dorsal side are scent glands, apparently responsible for the disgusting smell of the insect; and other glands, connected with sexual activity in some species.

Mating is initiated in *P. americana* by an odour omitted by virgin females which excites the males. In *B. germanica* no attractive odour, acting at a short distance, is involved. Direct contact with the female is necessary to induce sexual activity in the male and this usually happens by mutual stroking of the antennae. It appears that a chemical attractant is involved, since contact with male antennae or bristles will not do; yet the antennae of nymphs or of *B. orientalis* will excite males of *B. germanica*.

The copulation procedure is much the same in most species [512]. The excited male raises his wings vertically and usually vibrates them. He walks in front of the female and then moves backwards, bringing his abdomen under her body. During this stage, the female licks over his dorsal side, apparently being gratified by some secretions on it. Finally the male gropes for the female genitalia and, if successful, effects a union. Then the couples diverge until they face in opposite directions, while still connected by their sexual organs. They remain in this position for an hour or so. During this period, the spermatozoa are transferred to the female genital opening in a tiny bulb, the 'spermatophore'.

Ootheca formation begins a few days after copulation. Oothecae are also produced by unfertilized females, and in some cases there may be parthenogenetic development. The unfertilized eggs of *B. germanica* do not hatch; but some of them do in *B. orientalis* and more frequently in *P. americana*. These parthenotes can be reared to adults; they are always females. Generally speaking, however, parthenogenetic reproduction must be rare in most cockroach colonies.

(e) Resting places, movement, dissemination

(i) Resting places

During the daytime, cockroaches mainly rest in suitable harbourages. The factors that influence their choice of resting place are: concealment, temperature, access to moisture, and gregarious instincts.

Concealment

Like other evasive indoor pests, cockroaches tend to avoid daylight; also they are prone to crawl into small crevices, the narrowness of which is surprising. *B. germanica* nymphs and males can hide in cracks only 1·6 mm deep, while the gravid females need only a depth of 4·5 mm [616].

Temperature

Experiments have shown that *B. germanica* and *P. americana* chose temperatures in the range 25–33° C; while *B. orientalis* prefers a lower range, about 20–29° C [234].

Moisture

Cockroaches need regular access to water, especially if the temperature is high. *B. germanica*, with its high temperature preference, is especially dependent on moisture. If water supplies are scarce, *B. orientalis* will choose a lower temperature range, in the region 12–23° C.

Gregariousness

It is commonly found that groups of cockroaches congregate together. This may be partly due to their all finding the same favourable resting place; but it is accentuated by the existence of an aggregating pheromone which they give off. It is also present in their faeces, so that they tend to rest in sites previously contaminated by cockroaches [45].

The interaction of various factors on the main species of pest cockroaches lead them to infest the following sites. *B. germanica* is especially prevalent in kitchens, under sinks, at the backs of drawers, under or behind kitchen equipment and inside refrigerators or washing-up machines. Because of its climbing powers, it may be found at any level in a room; hot water pipes are particularly attractive. While kitchen sites are most heavily infested, *B. germanica* will spread throughout a centrally heated building into living rooms or offices, especially if people take lunchtime snacks in the latter.

B. orientalis is a poor climber and is found at floor level, often in basements, boiler rooms and cellars. It also occurs in bathrooms and privies and readily enters ducts and chases in centrally heated houses.

In the U.S.A., *P. americana* is a fairly common pest in restaurants, bakeries and food stores; but it is one of the most common cockroaches in sewers. Formerly, it was the most prevalent pest of ship's galleys, but more recently it is being replaced by *B. germanica* there.

S. longipalpa seems to be less confined to the kitchen than other domestic species and may occur in other parts of dwellings, showing a preference for high locations, such as shelves in cupboards, behind pictures and in furniture. The egg capsules may be deposited in crevices in furniture or (as already mentioned) glued to walls.

(ii) Movement

Activity, like other biological processes of insects, is greatly dependent on temperature [398]. At low temperatures, cockroaches enter a chill coma, the level of which is dependent on previous acclimatization. Thus, *B. germanica* kept at 14–17° C will remain active if the temperature is lowered, down to 2° C; but this increases to 7·5° C if the previous exposure was to 30° C, and to 9·5° C after rearing at 36° C.

Flying

Although some species have functional wings, cockroaches are not generally good at flying. In cool temperate climates, the pest species scarcely ever fly, although *B. germanica* and *P. americana* may make short semi-gliding flights. *B. orientalis* and the females of *S. longipalpa*, cannot fly at all.

In the warmer parts of the temperate region, such as the southern states of the U.S.A., flying out of doors is possible, especially in the males of *S. longipalpa* and both sexes of *L. maderae*; but even the wild species, which seldom enter houses, seldom make long flights and in many cases the females do not fly.

Walking and running

From what has been said about flying, cockroaches generally (and pest species in particular) rely on crawling when making exploratory journeys and running when disturbed or alarmed. In those circumstances, they can move quite fast, *B. germanica* males can cover 30 cm s^{-1} on paper and 11 cm s^{-1} on glass; females, 18 and 4 cm s^{-1}, respectively. This insect can also jump a few centimetres when alarmed [616].

All the domestic species of cockroach tend to be active about dusk and early night [242]. This is largely conditioned by darkness; but a diurnal rhythm becomes implanted and will persist for a day or two in constant light or darkness. German cockroaches can be seen walking over the walls of kitchens, in small numbers, even when the lights are on.

(iii) Dispersion

Active

In warm or hot climates, various species of cockroach may travel outdoors for considerable distances, sometimes in great numbers. 'Mass migrations' have sometimes been recorded in the southern U.S.A. It is likely that these are not directed migrations (as, perhaps, in some butterflies) but random wandering, stimulated by high temperature, overcrowding and/or adverse environmental conditions. Quite commonly in the same warm regions, there may be extensive movements of *P. americana* along sewers, for several hundred feet. Occasionally they escape from them and find their way into new dwellings [298].

Passive

Probably the main means of dispersion of domestic cockroaches is in transported goods or their crates or cartons; certainly this is the normal means of dissemination in cool climates. There are, however, comparatively few actual data on the details, although one example of widespread dispersion in crates used by a brewery has been documented [372].

The places actually infested by cockroaches in a building may not necessarily coincide with the mode of transportation. Cornwell [124] points out, for example, that the popular association of these pests with coke (why should gasworks be a source?) and laundry baskets, may be secondary to their invasion of a new building.

On the international scale, cockroaches have been dispersed by accidental carriage in ships and aircraft. The wide distribution of the commoner pest species seems to have been due to shipping. Early records suggest that *P. orientalis* was the common ship pest, but *B. germanica* is more prevalent; the latter was by far the most common cockroach on ships entering New Orleans (1960–63) and Miami (1957–61) [174]. Also, in ships arriving in Britain (1958–66), *B. germanica* was present in 79 shipments, *P. americana* in 59, *B. orientalis* in 7 and other species in 6 [124].

Collections of insects transported to New Zealand by aircraft in 1951 also showed predominance of *B. germanica*, followed by *P. australasiae* and *P. americana*. Subsequently, the types and numbers of aircraft travelling over international routes have enormously increased; but there do not seem to be any more relevant data.

(f) Quantitative bionomics [213, 294, 616]

Some observations of the duration of embryonic and nymphal development have been made by different workers. The incubation periods (at the same temperatures) determined in different studies were roughly similar; but there were discrepancies in the nymphal development times (in days), probably due to nutritional differences.

B. germanica. E, at 21° C, 24; at 25° C, 28; at 30° C, 15 and 17. N, at 21° C, 172; at 25° C, 103; at 30° C, 74 and 41.
B. orientalis. E, at 21° C, 81; at 25° C, 57; at 30° C, 42 and 44. N, at 25° C, 530; at 30° C, 300 and 155.
P. americana. E, at 21° C, 88; at 25° C, 57; at 30° C, 32 and 39. N, at 25° C, 519; at 30° C, 194 and 179.
S. longipalpa. E, at 23° C, 85; at 25° C, 73; at 28° C, 50 and 58. N, at 23° C, 220; at 25° C, 175; at 28° C, 104 and 90; at 30° C, 84.

At 25° C, *B. germanica* adults lived an average of 260 days and the females produced 6 oothecae (4.4 fertile). *B. orientalis* lived 140 days and produced 8 oothecae (5 fertile). *P. americana* lived 440 days and produced 53 oothecae (33 fertile). At 28° C, *S. longipalpa* lived 115 days and produced 10 oothecae (3·8 fertile).

The oothecae of *B. orientalis* and *P. americana* each averaged about 16 eggs; those of *B. germanica*, about 30; those of *S. longipalpa* about 13.

(i) Influence of food supply

It is generally known that cockroaches are omnivorous; many types of material may be ingested. These include, not only human foods in the kitchen, but the faeces of man or animals, various kind of refuse, book bindings, pastes or glues. From some of these, the insects are able to extract and digest the usual nutrients; carbohydrates, proteins and fats. Experiments have shown that the most rapid development and lowest nymphal mortality occurred on a diet with 22–24 per cent protein for both the German and Oriental species; while the American cockroach did best on 49–79 per cent. Longevity of the adults was also affected, the optimum level for *B. germanica* being 11–24 per cent, that for *B. orientalis*, 22–24 per cent, while for *P. americana* it varied widely over the range 2·5–49 per cent [254].

Some of the pest species are fairly resistant to starvation, although access to water is more essential. Survival periods (in days) of adults without food or water at 27° C were as follows (males/females): *B. germanica*, 8/12; *S. longipalpa*, 9/12; *B. orientalis*, 11/13; *P. americana*, 29/42. With water, the periods were: *B. germanica*, 10/42; *S. longipalpa*, 10/14; *B. orientalis*, 20/32; *P. americana*, 43/90 [619].

(g) Importance

A considerable amount of attention has been paid to the possibility of cockroaches acting as vectors of disease. Their habits certainly render them suspect, in that they frequent various types of septic matter in drains, privies and (especially *P. americana*) sewers, in search of food or water; and secondly they frequently contaminate human food. Furthermore, various pathogens have been identified in their gut contents or faeces. These include the following agents of enteric disease: *Salmonella* spp. (including *S. typhosa*), *Shigella* sp., *Eschericha coli*. Also pathogens of skin infections such as *Staphylococcus* spp. [513].

The actual danger of transference of these pathogens will depend very much on circumstance; in particular, whether the cockroaches in a kitchen could have access to virulent pathogens and whether uncooked food could become contaminated. It should go without saying that cockroaches in hospitals present a special risk.

Apart from the possibility of disease transference, it is generally agreed that cockroaches are disgusting pests, in appearence and from their characteristic odour. Not infrequently, they fall into foodstuffs or food packages; and this can lead to complaints and sometimes legal action resulting in fines and loss of goodwill for a business.

(h) Control [416]

(i) Inspection

A thorough inspection of the infested building should precede attempts at control, in order to define the extent and severity of the infestation. Three methods of assessment may be used.

(1) In the daytime, most cockroaches will be hiding; and although some may be discovered by examining recesses with a torch, it is best to flush them out by the use of an aerosol spray containing pyrethroid. It has been shown thay certain pyrethroids are most effective for activation (and subsequent knock-down) of cockroaches; notably, natural pyrethrins, bioallethrin and biotetramethrin. These are not necessarily the most efficient for subsequent kill.

(2) Visits during the earlier part of the night will usually reveal many cockroaches wandering about in search of food. Sudden illumination will not necessarily disturb them, but they are sent scurrying by heavy vibrations.

(3) It has been shown that the use of simple traps can provide even better evidence of the nature and distribution of an infestation. The traps can be formed of cardboard, folded into open-ended prisms (12·5 cm × 7·0 cm × 4.5 cm) with strips of plasticized gum along the inside base. Use of 23 such traps in an infested hospital, inspected weekly, caught a maximum of 253 *B. germanica* in one week and 57 *B. orientalis* in another, in single traps [27].

(ii) Use of insecticides

General

Since many infestations occur in places where food is prepared, it is important to chose reasonably safe insecticides and use them with great care. Surfaces on which food is prepared or stored should be covered with sheets of plastic and all cooking utensils removed from the area, before treatment. Certain other infestation sites, as sewers and privies, do not of course, need such strict precautions.

In considering various alternative types of insecticide, several facts must be remembered. Because of their retiring habits, no immediate spray treatment can reach all the cockroaches. Residual insecticides can sometimes maintain a steady mortality, but not all surfaces are suitable substrates. Furthermore, regular cleaning may remove the residues; or, on the other hand, dust and grease may cover them. Baits may be effective, but only where other sources of food are scarce; and they have to be protected from consumption by children or domestic animals.

In considering the main species, it must be remembered that while *B. germanica* oothecae hatch soon after deposition, those of other species may hatch up to 3 or 4 months later. Therefore, control measures must be continued until all are hatched.

As regards insecticides, dieldrin (if permitted) is highly effective against most species other than *B. germanica*, which has developed resistance to this type of compound in many places.

Aerosol treatments

Where residual treatments do not appear to be promising, regular treatments with an insecticidal aerosol may be more effective. A good combination would be an 'activating' pyrethroid combined with a killing insecticide (possibly another pyrethroid)[109].

Residual treatments

Emulsions or kerosene solutions are most suitable for applying to painted, tiled or metal surfaces. Aqueous suspensions should be used to treat concrete, brick or unpainted woodwork. Suitable insecticides are fenitrothion, bromophos, iodofenphos, propoxur or dioxacarb, used at 1 per cent, or bendiocarb at 0·3 per cent. Recent tests suggest that chlorpyrifos at 0·5 per cent gives particularly good results [516].

To obtain more lasting results on vertical surfaces, 1 per cent emulsions of these insecticides can be made up with 1·5 per cent methyl cellulose, to form a thick gel. This is applied with a paint brush, in bands 10–15 cm wide, along the base of walls, around doorways, entry points of pipes and the bases of kitchen equipment and sinks. The gel dries to a transparent deposit, which is not unsightly.

Dusts

Insecticidal powders are most suitable for treatment of confined, undisturbed places, such as ducts and wall spaces and behind drawers. For these sites (well away from food) 1 per cent dieldrin dust can be used (not for *B. germanica*) or boric acid power; or alternatively 1 per cent propoxur, which is rapidly acting, but not long lasting.

Lacquer

Insecticidal lacquer with very long lasting action, can be applied in bands along places which cockroaches are likely to cross when foraging for food. Dieldrin lacquer is highly effective for up to a year, although not against *B. germanica*, for which diazinon can be used. Lacquer treatments are best done by experienced pest control operators.

Baits

Formerly, baits containing boric acid or phosphorus were used; more recently, new synthetic insecticides have been more satisfactory. These include propoxur, trichlorphon, dichlorvos and iodofenphos. (Chlordecone, though highly effective, has been banned as dangerous.) To attract the insects, vegetable oil may be incorporated in the bait, or a gel base used (as an attractive source of water). Some evidence suggests that onion oil is attractive [3].

(iii) Repellents

It is possible that repellents might be useful to prevent cockroaches emerging from particular places (food packages, valuable apparatus, etc) especially if insecticide resistance had developed. There have been a few investigations of various chemicals for this purpose. One promising type is a series of heterocyclic amines [389].

8.1.2 ANTS (FORMICIDAE)

(a) Distinctive features

(i) General

Ants are among the most highly evolved of insects. They belong to a fairly advanced order, the Hymenoptera, characterized by biting mouthparts and two pairs of membraneous wings; the second pair being smaller than the first and hitched on to them by a row of tiny hooks. The higher members of the order (ants, bees, wasps) have a very narrow waist between thorax and abdomen. In the ants, this waist is actually part of the abdomen. It is known as the pedicel and contains one or two joints; the *petiole* if there is one, and the *petiole* and *post-petiole* (or *scale*) if there are two. Another characteristic is their antennae, which are 'elbowed' or held like a bent arm. Like all higher insects, they have complete metamorphosis.

More interesting than the specialized structure and development of ants is their complex social system and development of castes. This may be presumed to have originated as follows. From the gregarious habit (common in many insects) there developed nursing instincts to care for the helpless larvae. It probably came to pass that some of the females assumed more of the duties of hunting and foraging and at the same time became less fertile. Finally a caste of sterile females or 'workers' arose, differing in size and form from their fertile sisters. The workers are usually much smaller than the sexual forms and never develop wings. Thus there are three primary forms in the ant colony: males, females (or 'queens') and workers (which may show minor modifications). The factors which determine whether an ant will become a fertile female or a sterile worker are not fully understood. Some authorities believe that this is predetermined genetically in the egg. On the other hand, there is much evidence that nutrition supplied to the larvae can influence the production of sexual forms (as it does in the honey bee, whose queens are produced by feeding certain larvae with a special diet). It also appears that the rate of egg production in ant colonies can affect the development of sexual forms. Young colonies, with females laying sparsely, produce only workers; whereas flourishing colonies with plentiful food and egg production give rise to sexual forms.

Although worker ants are virtually sterile, they will sometimes lay eggs and occasionally these unfertilized eggs will develop (into males or workers). Generally, however, such eggs are eaten by the ants.

The complex behaviour of many ants has attracted the attention of several eminent naturalists and a voluminous literature describes their astonishing interdependence within the colony, their battles with other ants and slave-making habits, as well as the divers curious insects which live in ant nests.

(b) Classification [148, 559]

The ants, or Formicidae, comprise some 260 genera and 6000 species, about 80 per cent of which are tropical. Only about 200 species occur in Europe and less than 40 in Britain. Excluding some minor groups, there are five important sub-families, which show an interesting evolutionary gradation.

(1) The Ponerinae (17 per cent of species) are a primative, carnivorous sub-family, with castes not widely differentiated. They are mainly tropical, but are the dominant group in Australia, where the 'bulldog' ants of the genus *Myrmecia* are large, fierce stinging forms.

(2) The Dorylinae (4 per cent of species) include the driver and legionary ants of the tropics. They are likewise carnivorous, though blind, and without permanent nests.

(3) The Myrmecinae (46 per cent of species) is a large widely distributed group, the lower members being carnivorous and the higher ones omnivorous.

(4) The Dolichoderinae (5 per cent of species) are world-wide, omnivorous, but mainly vegetarian.

(5) The Formicinae (28 per cent of species) are the most advanced group, with many genera widely distributed. In this sub-family and the Dolichoderinae, the workers collect nectar and honey-dew and store it in their guts, the abdomen being capable of great distension.

This series is arranged in order from the most primative to the most highly evolved. Philosophically minded entomologists have pointed out that the social habits of ants have evolved in the same way as those of man. The stages represented are: the hunter and nomad (Ponerinae, Dorylinae), the herdsman (aphis tender), the farmer (fungus grower) and the omnivorous city dweller (the last three stages represented by various Myrmecinae, Dolichoderinae and Formicinae).

Keys to the important sub-families and species mentioned in this book are given in the Appendix, page 534.

(c) Distribution and prevalence of house-invading species

(i) European pests [148, 420] (Fig. 8.2)

There are several thousand species of ant, but only a dozen or so are liable to enter houses and become pests. The situation differs in the cooler and warmer regions of the northern temperate zone, in that some species which nest out of doors in warm climates and thence invade houses, may only breed and live indoors in the cooler areas.

In Britain and similar countries of northern Europe, comparatively few of the indigenous species which nest outside are liable to enter houses at all frequently. One very common pest which does so is *Lasius niger*, the black garden ant. This species occurs throughout Europe and Asia and there are three varieties in North America. It is said to resemble closely a fossil species from the Baltic amber, perhaps 50 million years old. Another related species which may invade dwellings is *L. brunneus*, which is also widely distributed from Britain to Japan and south to North Africa. A third invading species is *Hypoponera punctissima*, found in Western Europe and the Canary Islands. In Britain, it has most often been recorded breeding in greenhouses; but it also infests kitchens and bakehouses, possibly nesting in the walls, although it can breed outside.

Like various other insect pests, ants are frequently transported with goods (especially plants) in commerce. Some of them have become established in new countries in this way. One of the most successful is *Monomorium pharaonis*, Pharaoh's ant, which probably originated in Africa, but is now cosmopolitan. It was first recorded in Britain in 1828 and is now the most troublesome ant pest, after *Lasius niger*. Another successful colonizer is *Iridomyrmex humilis*, the Argentine and, although this has only been known in Britain since 1915 and is less widely prevalent. A third cosmopolitan species which has established itself in some localities in Britain is *Pheidole megacephala*, once known as the Madeira house ant since it was very abundant in that island. This has been recorded at scattered intervals since the mid 19th century, but it is not common.

The three introduced species just mentioned have been recorded as pests in dwellings. Other introduced species have come in with plant material and temporarily established themselves in hot houses (especially at Kew Gardens) but not elsewhere. Examples are: *Crematogaster scutellaris, Pheidole anastasii, Tetramorium guineense* and *Technomyrmex albipes*.

(ii) North American pests [158, 159, 548] (Fig. 8.3)

Probably the most troublesome ant pest in the U.S.A. is *Iridomyrmex humilis*, which is also an introduced species in North America, being first reported about 1890 in New Orleans and subsequently spreading through the U.S.A. Other imported pest species are: *Paratrichina longicornis*, the crazy ant, which is common in the Gulf Coast states and in many towns and cities of the U.S.A.; also

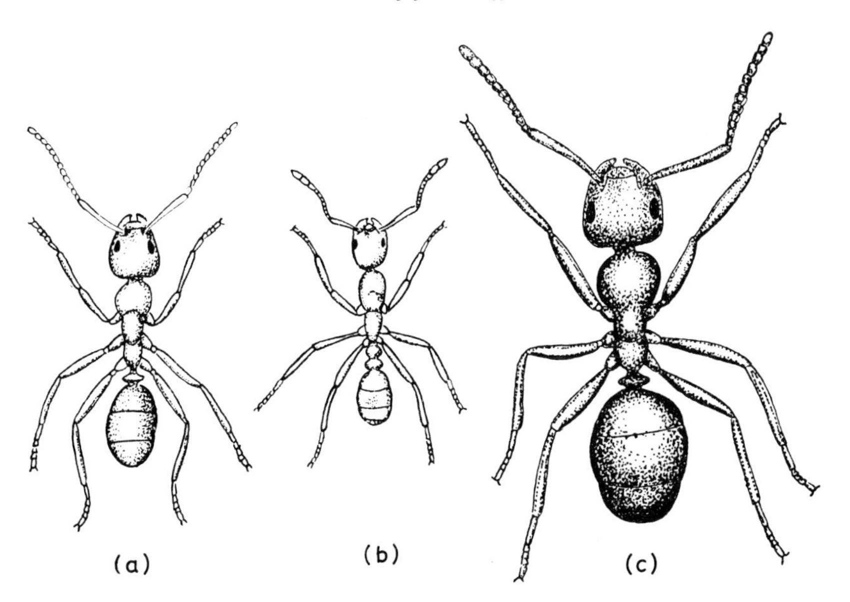

Figure 8.2 Workers of ants liable to infest houses in Britain. (*a*) *Iridomyrmex humilis*; (*b*) *Monomorium pharaonis*; (*c*) *Lasius niger*. (Partly after Gosswald [211].) × 13.

Monomorium pharaonis, which occurs throughout Canada and the U.S.A. but mainly in cities. Both these ants nest indoors in the cooler part of their range, but outdoors in the south.

Of the indigenous species, *Tapinoma sessile*, the odorous house ant, which occurs throughout the U.S.A., from sea level up to 10 000 feet, is probably the worst. Other widely dispersed species are: *Prenolepis imparis*, the small honey ant, which often replaces the Argentine ant where that has been exterminated; also *Solenopsis molesta*, the thief ant, which is related to the severely stinging fire ants, although itself rarely stings.

Other species of ant which nest out of doors, but not infrequently invade houses to become a nuisance, are: *Tetramorium caespitum*, the pavement ant, mainly troublesome in the Eastern states; and *Pheidole hyatti*, the Western big-headed ant. Somewhat less common invaders are: *Crematogaster* spp., the acrobat ants, *Dorymyrmex pyramicus*, the pyramid ant, *Lasius niger* and *L. pallitarsis*.

(d) Life histories and quantitative bionomics

The life history is complicated by the social system, which subordinates the individual to the community. It is, therefore, convenient to consider the development of a new colony or infestion. Some details of three important species will be given.

Figure 8.3 Workers of some house-invading ants of North America. (*a*) *Iridomyrmex humilis*; (*b*) *Dorymyrmex pyramicus*; (*c*) *Tapinoma sessile*; (*d*) *Solenopsis molesta*; (*e*) *Crematogaster* spp.; (*f*) *Prenolepis imparis*; (*g*) *Lasius niger*; (*h*) *Monomorium pharaonis*. (After Snelling in Ebeling[158].)

(i) The common black ant (Lasius niger) [148]

After mating, the females break off their wings and dig a cell in the earth. Here they lay their eggs and rest until they are hatched. The young are white, helpless grubs without legs. The mother feeds them with secretions of her salivary glands, until they are mature. Finally they pupate and then emerge as adult workers. During the whole of this time, the female takes no food but subsists on stores in her own body. Her fat-body and the muscles of her wings are broken down for food. When the first workers become adult, they take over the duty of feeding and tending the queen and her subsequent brood. The workers forage in all directions for food, bring it back in their crops and feed it in a semi-liquid condition to the queen and grubs. The ant 'nest' formed in this way may last for several years.

Lasius feeds on a variety of substances; it kills and devours small flies and similar insects, visits flowers for nectar and collects seeds. In addition, it has the remarkable habit (like certain other ants) of tending aphids and imbibing the sugary excretion which they produce so copiously. The workers travel considerable distances in their search for food and find their way about mainly by a sense of smell. Thus, they can follow long trials of scent drops left by themselves, or their sister ants. When they have found a good source of food, they are able to communicate the fact to their colleagues in some way (possibly by tapping them with their antennae). Consequently, when these ants nest close to a house, it often happens that a worker finds a rich harvest in the human larder; and soon there is a line of ants going each way along a trail from the nest to the food.

In addition to their foraging duties, the workers enlarge the nest and make numerous galleries and cells of earth. Also they tend and clean the grubs and the queen. If danger threatens they remove them to safety. Thus, when one digs into an ant nest one sees the workers energetically carrying away what are commonly called 'ant eggs' (goldfish food!), although the so-called eggs are nearly as large as themselves; these are the pupae.

During the late summer, sexual forms are produced in large numbers. The males are bigger than the workers and the females larger still; both sexes possess the transparent wings characteristic of the order. These sexual individuals all emerge from the nest on the same day and mate as they fly away in all directions. This phenomenon of 'swarming' often happens in many nests over a wide area on a particular day, usually in the afternoon. This 'swarming' phase occasionally gives rise to trouble inside houses (e.g. when a nest has been built in the earth underneath them). Complaints of 'flying' ants should not be serious for the nuisance is of short duration.

After mating, the males usually perish and only a very small proportion of the females survive to form a new colony. Sometimes two or three fertilized females are found in the same or adjacent earth cell but normally one dominates the others and eventually kills them. Colonies are very rarely found with more than one queen. She is the vital centre of the nest which gradually dies out if she is removed.

Quantitative bionomics

Quantitative laboratory studies of the speed of development, etc, of *L. niger* do not seem to have been made. Instead, there are scattered observations of naturalists as follows: fertilized females established in an artificial nest in March laid eggs in April. By the beginning of July pupae were present and workers began to emerge at the end of July. The following records were made of the related *L. niger americanus* in Illinois. The fertilized females enter the soil in the autumn although they may not lay their first eggs until the following May. The first workers emerge in July. The egg stage lasts 22–28 days; the larval stage 16–23 days. During the summer the entire life cycle takes about 2 months.

(ii) Pharaoh's ant (Monomorium pharaonis) [477, 480]

The biology of this ant differs in several respects from that of *Lasius* mainly because of its tropical origin. The nests are formed in hollow spaces in the walls and beneath the floors of buildings. These ants favour warm humid conditions and often make nests behind stoves or kitchen ranges and particularly in steam-heated buildings in the vicinity of hot-water pipes. These nests are usually very inaccessible.

On account of its environment *Monomorium* lives under very uniform conditions, and it is relatively unaffected by the season. Thus sexual forms are liable to occur at any time of year. In colonies with access to adequate supplies of food, this will occur at intervals which depend on the ratio of workers to queens and also on the physiological state of the latter. Excess of workers, or equally, a shortage of queens, hastens the production of the next sexual brood. The queens themselves pass through a juvenile phase of about 4 weeks, followed by a fully fertile phase, when they are fed with glandular secretions by the workers. Finally, there is a senile phase, during which the workers supply less secretion, although this is partially replaced by raw food (honey, meat). In experimental conditions, with colonies kept at 27° C, a few produced sexual forms at intervals shorter than a month and others at periods up to 7 months. (In actual infestations, intervals of 3–5 months are usual.) The period over which sexual forms were produced was usually 2–3 weeks (i.e. the time from detection of the first sexually destined larva to emergence of the last sexual adult). Of the adult sexual forms emerging, females always predominated. The numbers of females per 100 males ranged from 110 to 525.

Although sexual forms are winged, flying swarms are never observed in Britain. Mating takes place in the interstices of infested buildings. The formation of a new colony is not normally accomplished by a single female after a nuptial flight, as with many other ants. Indeed, laboratory colonies do not thrive unless there are several 'queens' present. The usual mode of formation of new colonies would appear to be undertaken by a large part of the colony, the workers carrying with them some or all of the eggs and young stages. This general migration may occur at any time when the original colony has become large and with numerous sexual forms. It is very easy to encourage this type of formation of new colonies by providing a suitable site for colonization (e.g. a tin furnished with a water-soaked cotton wool pad). New colonies formed in this way in infested buildings have been found to contain from 400 to 4800 ants (of all stages) including 5–110 females, 0–5 males and 150–1000 workers. Sub-cultures may be obtained in an analogous way in the laboratory. When two colonies co-exist in a single large container, they intermingle amicably. It is obvious that this ready formation of new colonies must facilitate the spread of the ants through suitable buildings and form numerous pockets of infestation. Clearly they are a difficult pest to eradicate.

The tiny workers of this species penetrate all over large buildings in their search for food. The 'queens' too can often be observed among the processions of workers; they are easily recognized by their larger size. This ant shows a love of warmth which may be noted in the way in which their trails will follow the course of hot-water pipes. Occasionally trails diverge from the quest for food to sinks or drains in search of water. All sorts of edible material will serve as food, including sugar, jam and honey as well as proteins (meat or cheese) and fats (butter, dripping). They display astonishing powers of finding such things in cupboards and drawers even in packets and jars. After one ant has discovered the food, a crowd of them soon collects and surrounds it, gradually removing it in small quantities.

Quantitative bionomics
The following data relate to laboratory colonies maintained at 27° C.

The duration of various stages (*days*). E, $7\frac{1}{3}$; L, 17; pre-P, 3; P, 9; total 37. This is for workers; the sexual forms require about 3–4 days more for development.

Longevity of adults. Females live longest, the maximum observed being 39 weeks; workers come next with a maximum of 9–10 weeks. The life of the males depends largely on opportunities for pairing and can range from 3 to 8 weeks.

Egg production. An average egg-yield of 3500 per female is quite possible in a thriving colony, with a maximum rate of about 30 per day.

Mortality during development was high in laboratory colonies (about 75 per cent). Most of the deaths occurred in early life, sometimes due to cannibalism.

Sex ratio. Females are more prevalent than males, approximately in the ratio 2 or 3 : 1.

(iii) The Argentine ant (Iridomyrmex humilis) [375]

This ant, also a tropical invader, is confined in Britain to warm buildings. There may be several queens to a nest and these wander about to some extent. In the warmer southern states of the U.S.A., this ant usually nests out of doors, generally on the ground in soil or refuse, but occasionally in other sites such as tree holes. They will leave their nests to forage at temperatures between 10 and

30° C. Apart from entering dwellings in the vicinity the ants can be an agricultural nuisance by fostering and protecting plant aphids, from which they take honey-dew. In houses, they avidly feed on sweet substances; but they will also eat meat, dead insects and even attack and kill live ones to feed on.

Argentine ants tend to drive out other ant species, when they become established. On the other hand, the workers will mix with others of different colonies of their own species, unlike many other ants.

Quantitative bionomics

The following data relate to an artificial colony [446].

The duration of various stages (days). E, 27 at 22° C and 20 at 27° C; L, 60 at 11° C, 50 at 17° C and 15 at 25° C; P, 25 at 14° C, 14 at 25° C, 11 at 27° C. These figures are for workers; sexual forms develop more slowly.

In California, sexual forms usually appear in May and mate in the nest. The females shed their wings and begin to lay eggs. In January next year about $\frac{3}{4}$ of these females are killed off by the workers; and soon afterwards, new workers begin to emerge. Their numbers increase until October, after which they decline until the following year [375].

(iv) The odorous house ant (Tapinoma sessile)

The common name of this species relates to the unpleasant odour emitted when it is crushed. It nests in a considerable variety of outdoor habitats and also in hollow spaces inside buildings, especially near hot-water pipes. The nests may contain anything from a few hundred to many thousand workers, with many reproductive females. Average numbers in a laboratory study were: 345 eggs, 829 larvae, 533 pupae, 2319 workers and 1–6 queens [595]. Mating usually occurs in the nest and new colonies are formed by migration of a few fertile females with attendant workers. Nuptial flights have been observed, however, and new colonies may sometimes be formed by single fertile females.

These ants will feed on small live or dead insects; they also seek and tend aphids which reward them with honey-dew secretion. The ants tend to enter houses in the autumn, when honey-dew supplies become scarce. Tests with poisoned baits, showed that sweet materials were more attractive than meat grease. The ants would take baits containing 0·1 per cent phoxim or dieldrin; but were repelled by 0·5 per cent [595].

(v) The small honey ant (Prenolepis imparis)

This ant normally nests out of doors and is characterized by activity at unusually low temperatures (2–13° C). Nests, which contain a few thousand individuals, are mainly constructed in the earth, especially moist clay or loamy soil. In winter, the colony normally contains a single queen, winged sexual forms and many workers. In spring, overwintering males and females emerge on a nuptial flight and mate. The fertilized females discard their wings and begin a new nest. When invading houses, they may take various foods, but are especially fond of sweets.

(vi) The crazy ant (*Paratrichina longicornis*)

The common name derives from the workers' habit of rapid movements with frequent change of direction, as if bewildered. In the southern states, it will nest out of doors, but in colder regions it nests inside buildings, especially hotels and greenhouses kept warm all winter. In such warm localities, it can be a pest throughout the year. The workers are omnivorous and will attack various household foodstuffs.

(vii) The thief ant (*Solenopsis molesta*)

The common name comes from the ant's habit of raiding the nests of other ants and pilfering the brood and their food stores. It is a very small ant (workers 1·3–1·8 mm long) and can travel along very minute passages to do this. Apart from outdoor nests, it will form nests in wall voids in houses. From such internal nests, it will forage for food in cupboards and food stores and it is also common around kitchen sinks. Mating occurs during nuptial flights, which usually occur in autumn.

(viii) Big-headed ants (*Pheidole* spp.)

Big-headed ants are so-called because of the large size of the head in relation to the body. They normally nest outdoors and their natural foods are live and dead insects, seeds and honey-dew. Occasionally, they enter houses and are very prone to feed on meats or fatty foods; also they may be found near kitchen sinks. They usually make their foraging journeys at night.

(e) Importance

Pharaoh's ant is a common pest in hospitals, where, in certain circumstances, it could transmit disease germs. Two of its habits suggest this possibility. In the first place, it often visits moist or liquid substances for water and has been seen at different times in drains, urinals, on faeces, wet dressings and sputum. Secondly, the foraging workers have astonishing intrusive powers and get through tiny crevices in closed containers. They have been found in 'sterile' dressings, surgical gloves and syringe dishes. By reason of its wide range, ants may travel to many parts of heated buildings and appear in such undesirable places as operating theatres, dressing stores and laundry baskets.

An investigation of the bacteria actually carried by these ants in some British hospitals revealed a number of unpleasant pathogens, including *Salmonella* spp.,

Pseudomonas aeruginosa, *Staphylococcus* spp., *Streptococcus* spp. and *Clostridium* spp. In some cases, there had been evidence of cross-contamination in the infested wards [40]. Another unpleasant habit of Pharaoh's ant in hospitals is to bite the eyelids and other parts of young infants, causing considerable irritation. Other ants, indeed, cause irritation by bites and some cause painful stings. These are not, in general the house invaders and they will be dealt with in Chapter 12.

A general objection to all kinds of invading ants is the unpleasant sight of workers clustering on food. Their actual depredations are slight, but may be sufficient to damage the appearance of food, for example, by biting holes in iced cakes.

(f) Control of house ants [420]

(i) Destruction of the nest of garden ants

Infestations of garden ants of various species which invade dwellings, generally nest within about 6 m (20 ft) of the building. The workers of most species tend to travel along well-defined trails, so that it is generally possible to trace them back to the nest. This should be dug up and quantities of boiling water poured in. If this is liable to damage valuable plants, an insecticidal dust or spray treatment should be liberally applied to the area. Chlordane emulsion, diluted to 0·2 per cent may be effective; on paths and outside walls a stronger solution (2 per cent) may be used.

(ii) Control of indoor nesting ants

It has been noted that several species of ant will often (or, in some areas, always) form nests inside buildings. *Monomorium pharaonis* is an important cosmopolitan species; others in Europe include *Iridomyrmex humilis, Hypoponera punctissima* and *Pheidole megacephala*; and in America, *Tapinoma sessile* and *Solenopsis molesta*. Occasionally it may be possible to trace the nest and eradicate it; but it is often difficult or impossible, because it is usually located in inaccessible places, such as hollow wall spaces and under floors. Moreover, large buildings infested with *M. pharaonis* may contain numerous nests.

Trap baits

Sources of food are located by the workers with astonishing efficiency and, by some means of communication, trails of workers soon form and numbers collect on the food. Some ants (e.g. *Iridomyrmex humilis*) are attracted to sweet substances; others (e.g. *Solenopsis molesta*) to meat or fatty foods. *M. pharaonis* is attracted to both, but small pieces of raw liver are especially attractive.

At one time, trap baits were used as a control measure, but they were inefficient. In one institution, ants collected on such baits were killed in boiling water and kept. After some 700 days, a total of $5\frac{1}{2}$ million workers and 6000 queens had been killed, without eradicating the colony!

Trap baits are still used, however, for locating the position of nests, which may be traced by following the trails of workers back. A survey of the extent and location of an infestation by this means should precede control measures in large buildings.

Residual treatments

Until recently, the best method of controlling indoor infestations of ants (e.g. *M. pharaonis*) was to treat the area round the nests with a residual insecticide. Spray was applied to the wall–floor junction, surrounds of radiators, cupboards and doors. Insecticides which have been fairly effective are 2 per cent chlordane or 0·3 per cent bendiocarb. Alternatively, insecticidal lacquer could be applied to all suitable surfaces, as for cockroaches (see page 319). Such treatments, if carefully applied, secure a rapid reduction of workers; but it is doubtful whether they can be relied upon to eradicate difficult infestations.

Poison baits

The principle of the poison bait is to use a slow acting poison, so that the workers will be able to return to the nest and feed it to the queens and brood. Any baiting programme should be discussed with the householder or the staff of any institution treated. The principles must be explained and their co-operation obtained. In rooms which need to be thoroughly cleaned or washed down, it may be necessary to insert the baits in small tubes, which can be taped to the walls, mouth downwards.

It appears that the most efficient bait for *M. pharaonis* is based on the insect development inhibitor, methoprene, at 0·5 per cent. This has been used to eradicate the pest from widespread infestations in hospitals [161]. The methoprene is mixed with dried powdered liver at 1 per cent and, immediately before use, further mixed with an equal weight of honey and sponge cake (1 : 1) with a trace of water to moisten it. Small pieces (about 1 g) are placed in the infested area, at the rate of about 1 per 3 m^2. The baits are renewed after 1 week and removed after a second week. Subsequently, the building is kept under observation with occasional trap baiting, to follow the decline of the ant population. Complete extermination may take 18–20 weeks. Where there is any sign of failure, a repetition (usually only needed in a limited area) should lead to complete success.

As there is always a possibility of development of resistance in pests, it may be well to mention an older bait, based on 1 per cent sodium fluoride. This must be applied in such a way (e.g. in cartons with holes for the ants) that children or domestic animals will not consume it. Such baits must be maintained, with periodic replacement, for 2–3 months to be at all successful.

8.1.3 THE FIREBRAT (THERMOBIA DOMESTICA)

(a) Distinctive features

The firebrat is related to the more common silverfish, *Lepisma sacharina*; both belong to the primitive order Thysanura. Insects of this group are wingless and moult in the adult stage; they have long, simple antennae and three tail-like processes at the back of the body (Fig. 13.1, page 478). Abdominal appendages ('styles') are present underneath the seventh to ninth abdominal segments of older insects. A few species occur in Britain; two of them can occur indoors as domestic pests, the silverfish and the firebrat. The former, which is mainly a pest in rather damp rooms, is dealt with in Chapter 13 as a nuisance. The firebrat, however, only flourishes in warm conditions and is confined to such places as kitchens and bakeries. It is greyish in colour, speckled with darker markings.

(b) Life history [354, 601]

(i) Oviposition

The female lays eggs in batches, each one in a different stage of the adult life. A separate fertilization is necessary for each batch. The insects tend to lay more eggs if they are undisturbed.

(ii) Egg

The egg is soft white and opaque when laid; weighs about 0·3 mg and measures about 1 mm × 0·8 mm.

(iii) Nymph

As with *Lepisma*, the young are hatched without scales and acquire them at the fourth stage. Three pairs of styles appear in the females in the 5th, 8th and 10th instars; the last pair seldom develop in males. Egg-laying has been observed to begin in the 14th–17th instars in different individuals.

(iv) Adult

Moulting continues in the adult life and as many as 60 stages have been recorded. The speckled markings of the adult are similar to those of the younger insects, but somewhat more distinct.

The food of the firebrat, like that of the silverfish, consists largely of carbohydrates made up with small amounts of protein. In kitchens and bakeries the firebrat finds a ready source of nourishment in crumbs and dust of flour and other food residues.

The mating procedure has been observed; it is somewhat unusual. The male performs a sort of 'love dance', turning about in short circles and repeatedly contacting the antennae, mouthparts and legs of the female. Finally he deposits the sperm bag (spermatophore) about 12 mm ($\frac{1}{2}$in.) in front of her and loses interest in the proceedings. The female now moves forward, straddles the spermatophore and receives it into her genital opening.

(c) Quantitative bionomics: ecology [565]

(i) Temperature

The eggs will not hatch at or below 22° C nor will development be completed at 25° C. Lethal high temperatures: 49° C for 1 h; 48° C for 10 h; several days at 47° C.

Over the range permitting full development, the incubation and maturation times (days) are as follows:

	27° C	29° C	32° C	37° C	42° C
Incubation	44	32	21	12	9·5
Maturation	330	247	105	92	47

Under optimum conditions (37° C and 80 per cent R.H.) the duration of an instar increases from 2 to 11 days, in the first eight instars, and thereafter remains at about this length for the rest of the insects' life.

The sexually mature females are able to mate from the 1st to the 5th day of the instar, but usually do so in the first 3 days. Males continuously exposed to females can mate for most of the time, but are most potent in the first third of the moulting cycle [601].

The adults can live for 2–2$\frac{1}{2}$ years at 32° C, or 1–1$\frac{1}{2}$ years at 37° C.

(ii) Humidity

At a favourable temperature (37° C) some eggs will hatch at humidities as low as 11 per cent R.H. and normal emergence occurs at 33 per cent R.H. and above. Development is completed at 50 per cent R.H. and above with or without a free water supply; although they do better with water available at the lower humidities.

(d) Importance

The very warm temperature preference of the firebrat restricts its prevalence to the vicinity of the oven in bakehouses and kitchens. To a certain extent it may be considered a food pest, since it presumably feeds on flour, etc, spilt in such places. Its digestive enzymes are capable of coping with fats as well as starch and protein, so that it may be regarded as omnivorous [594].

In some bakeries it has been found to nibble small holes in loaves. It also has the curious habit (like the silverfish) of biting holes in inedible fabrics, especially viscose rayon. Apart from this, the sight of these insects crawling over the walls and shelves is disagreeable; and there is always a chance that one of them may be found in the bread or rolls sold from an infested bakery.

(e) Control

The firebrat, unfortunately, usually thrives in rooms which must necessarily be kept in a condition favourable to development of the insect. Thus a bakery cannot be allowed to become cool merely to discourage the pest.

Direct attack by insecticides is difficult, in view of the pests' proximity to food. At one time, spraying the walls with an organochlorine insecticide (DDT, chlordane) was recommended; but the results were erratic. More recently [359], laboratory tests have shown some alternative insecticides to be better, especially chlorpyrifos, which was 10 times as potent as chlordane. Probably a pyrethroid insecticide would be effective, although repeated treatments might be necessary; and they should be combined with thorough cleaning measures to discourage the pest.

8.1.4 THE HOUSE CRICKET (ACHETA DOMESTICUS)

The common and scientific names of *Acheta domesticus* are both suggestive of the infestation of human dwellings and there is little doubt that, at one time, crickets were common inhabitants of warm kitchens. Since they are omnivorous, they probably lived on any scraps of food debris; perhaps they also raided the larder at night.

In recent years, however, crickets are rarely, if ever, permanent house dwellers. Living mainly on refuse tips, they sometimes tend to invade neighbouring houses during the winter. Their chief impact is as nuisances rather than food pests and accordingly they are dealt with in Chapter 13 (page 496).

8.2 'RESIDENT' PESTS

Human food stores, whether the huge accumulations in silos and warehouses or small domestic reserves, offer relatively enormous supplies to insects and mites which can live and breed in them. The only limitation is that most stored products have a rather low moisture content. For example, that of Australian wheat is generally about 11 per cent, cocoa beans and groundnuts are usually 6–7 per cent, while barley malt is only about 3 per cent. Consequently, the pests best able to thrive in stored food are those adapted to rather dry environments; and indeed some of them conserve the water produced by their own metabolism.

The great majority of insect pests of stored products belong to two orders: the Coleoptera (or beetles) and the Lepidoptera (or moths). These insects are most troublesome under warm conditions. Less restricted by low temperature but more dependent on high humidity are mites which infest stored foods.

Nearly all of the various insect pests, which will be described in the following pages, have been recorded from a wide variety of stored foods, spices and drugs, and a large proportion of them can be reared satisfactorily on many substances. Nevertheless, many pests are typically associated with a particular type of product. This probably reflects the ancestral habits of the insect, which may have been to attack whole seeds, damaged seeds, dried fruits or nuts or animal remains. Thus:

Grain pests (originally attacking seeds) include: *Sitophilus* spp., *Cryptolestes* spp., *Oryzaephilus* spp., *Trogoderma granarium*, *Rhizopertha dominica*. Also *Ephestia elutella*, *Sitotroga cerealella*.
Pulse pests (attacking larger seeds) are represented by Bruchidae.
Flour and ground cereal pests (originally attacking seeds damaged by other insects, etc) include: *Tribolium* spp., *Gnathocerus* spp., *Tenebroides mauritanicus*. Also *Ephestia kühniella*.
Dried fruit and nut pests include: *Oryzaephilus* spp., *Carpophilus hemipterus*. Also *Plodia interpunctella*, *Ephestia cautella*, *Paralipsa gularis*, *Corcyra cephalonica*.
Animal hide and bacon pests (originally attacking dried cadavers) include: *Dermestes* spp., *Necrobia rufipes*.
A miscellaneous group (possibly derived from scavengers) includes: *Stegobium paniceum* (miscellaneous cereal products and drugs), *Lasioderma serricorne* (often on tobacco), *Ptinus tectus*.

It must be stressed, once again, that these associations are not at all rigid. Thus, *E. elutella* is a common pest of cocoa and chocolate, *P. interpunctella* attacks maize seed, *Lasioderma serricorne* is a serious pest of cocoa, *Necrobia rufipes* thrives on copra.

8.2.1 BEETLES (COLEOPTERA)

(a) Distinctive features

The beetles are among the most easily recognizable insects. Their main characteristic is their stiffened forewings (devoid of any trace of 'veins') which cover the hind wings when at rest, lying above the abdomen and meeting in a straight line. Most beetles are well-armoured insects and, though many can fly, in general they are a pedestrian group. The mouthparts are always adapted for biting. Beetles undergo complete metamorphosis during development and various types of larvae are found, ranging from the active primitive type to a legless grub. The active larvae are typical of the more primitive families.

(i) Identification of beetle pests of food

A key is provided in the Appendix (page 540) to permit identification (with the aid of illustrations) of about a score of the beetles most commonly found in stored foodstuffs. Technical terms have been reduced to a minimum to facilitate the use of this key by anyone with a modicum of entomological knowledge.

(b) Occurrence and life histories of various pests [128, 319]

This section deals with some aspects of the more frequently encountered stored product beetles; the space allotted to each is roughly in accordance with their importance. They occur in six of the superfamilies of polyphagous beetles given in the revised edition of Imms' *A General Textbook of Entomology* (1977) and listed in the same way, grouped under families and genera. Short keys are given to distinguish the more important species and a few notes on general appearance are usually given to confirm identification. The common names given are those recommended by the Ministry of Agriculture, Fisheries and Food, Technical Bulletin No. 6. Prevalence and distribution are mentioned, and information given of the type of products attacked. The life history is outlined and the speed of development indicated by the following code: figures following E, L and P give the average duration of egg, larval and pupal stages (in days, except where stated) at the temperature indicated.

(i) Dermestoidea

Dermestidae (hide and carpet beetles)
The name *dermestes* is derived from the Greek and means 'skin-eater'; the common English names connote the same habit. These beetles are pests of hide, skins and woollen materials. They are dealt with more fully in the chapter on wool pests, but some mention must be made here of species that feed on food products. Thus:

Dermestes lardarius (the bacon beetle). (Fig. 10.2, p. 410). This beetle sometimes invades domestic larders and feeds on various foodstuffs with high protein content. It is also often found in corners and floor cracks. Stored bacon and ham [426], or cheese may be attacked in commercial stores. *D. lardarius*, and also *D. maculatus*, are sometimes prevalent in dog biscuit factories and the larvae may get into the biscuits.

If the infestation of such foodstuffs is light, it is sometimes possible to destroy only the affected parts; but generally, the larvae burrow right into the substrate and all has to be discarded. An infested store must be thoroughly cleared and cleaned. It may then be advisable to spray the walls and floor to eliminate the residual population of beetles.

These beetles are dealt with more fully on page 409ff.

Trogoderma granarium (the khapra beetle) (Fig. 8.5)
A little oval, dark brown beetle, with a small head. The pest originated in India and spread to Europe in foodstuffs imported during the First World War. It has become established in Britain, mainly in maltings. Under favourable conditions, the khapra beetle breeds prolifically, but it is now much rarer than formerly, owing to a change in practice. The larvae tend to crowd into crevices where they are difficult to reach with insecticides.

Life history. The eggs are laid singly either scattered among grain or sometimes in the groove of the grain. The females lay about 40–50 eggs. The larvae are hairy, like those of all dermestids, being covered with two types of yellow-brown hair; long single ones and short barbed ones. The male moults four times and the female five times, and under adverse conditions, there are additional moults. They reach a final length of about 2·3 mm. Pupation occurs inside the last larval skin.

Some of the larvae develop normally, but others enter a diapause. These latter crawl into cervices and remain without further development until fresh warm malt or other foodstuff is placed nearby. Then they emerge and continue development. Diapausing larvae *may* remain quiescent for several years. Adults live about 2 weeks at 25° C.

Speed of development. At 30° C : E, 6·5; L, 24–30; P, 4·5; adult in cocoon, 2. At 25° C: E, 9·5; L, 38–47; P, 6; adult in cocoon, 3. All at 50 per cent R.H.

(ii) Bostrychoidea

Bostrychidae
The bostrychids are mostly a family of wood-borers.

Rhizopertha dominica ('Lesser grain borer') (Fig. 8.5). *Rhizopertha* is typical in form though somewhat anomalous in habit. It is a small active beetle, nearly cylindrical in shape, with a roughened, hump-like thorax. It is a serious pest of stored grain in India and other parts of the tropics. It is imported into Britain in cereals, which it damages even more seriously than the true weevils. However, all stages are normally killed by the cold of an English winter in unheated warehouses.

Life history. The females lay 300–500 eggs at the rate of up to 25 per day. The eggs are laid either loose or attached to grains. The larvae bore into the grains and feed on them from inside, until little is left but the husk. The older grubs are fleshy and with relatively small legs; they tend to lie curved in the form of a C. After 3–5 moults, pupation occurs inside the grain. The adults also feed on the cereal after emergence.

Speed of development. At 26° C: E, 12–18; L, 53; P, $6\frac{1}{2}$. Total development (various temperatures), 24–133; average, 58.

Anobiidae

Beetles of this family have a very pronounced hood-like thorax which nearly conceals the head when viewed from above. Several species are wood-boring pests (Chapter 11).

Stegobium paniceum (The 'biscuit beetle') (Fig. 8.5). This is a small beetle, reddish-brown in colour, with a dense covering of short yellowish hairs. It is a cosmopolitan pest of cereal products (especially farinaceous materials) and of various drugs, spices and beverage concentrates. In Britain it is a very common beetle pest of food to find its way into domestic larders. The American name, 'drug store beetle', indicates its propensity for breeding in dried vegetable matter of all kinds even in poisonous substances such as strychnine, belladonna or aconite. If infested goods are left for a long time on the shelves of a store, the beetles will tend to migrate and infest other commodities.

Life history. The female lays approximately 100 eggs, over a period of about 3 weeks, either in the actual foodstuff or in crevices nearby. The first-stage larva measures about $\frac{1}{2}$ mm $\times \frac{1}{8}$ mm; it is active and wanders about, readily crawling through small openings; so that they can often penetrate into uninfested packaged foodstuff. These larvae can survive as much as 8 days' starvation while searching for food. When a suitable breeding material has been found, the larva browses in it and becomes fat, sluggish and eventually incapable of moving about. These are four moults in the larval stage and the fully grown grubs measure about 5 mm. Finally they construct a small cell of food particles cemented together with saliva and pupate inside it. The adults bite their way out and wander about, often far from the breeding site. They live for about 6–8 weeks but take no food. The sexes pair and after a few days, the females begin egg-laying.

Speed of development. At 24° C and 45 per cent R.H.: E, 9; L, 57; P, 9; adult in cocoon, 7–9; total, 83. At 19° C and 37 per cent R.H.: E, 14; L, 104; P, 15; adult in cocoon, 8–12; total, 143 [309].

Lasioderma serricorne (the 'cigarette beetle') (Fig. 8.5). The adult is somewhat like *Stegobium*, but smaller and rather more squat. It is a cosmopolitan pest and injurious wherever tobacco is grown, cured or manufactured. As well as tobacco, it infests various seeds and spices, oilcakes and locust beans, and has become, for example, a serious pest of stored cocoa in West Africa.

Life history. The females lay about 50–100 eggs, in crevices in the foodstuff. The young larva, like that of *Stegobium*, is active and can survive for 1 week without food. Later, however, it develops into a fleshy, sessile grub. After five or more moults, the larva builds a cell of fragments of food debris, in which it pupates. The adult beetles live for about 25 days at 30° C or about 45 days at 20° C.

Speed of development. Although most records of *L. serricorne* relate to tobacco, it thrives better on many other commodities. The best studies of its bionomics were made on a colony reared on wheat-feed [287]. At 20° C : E, 21; L, 69; P, 12. At 25° C : E, 10; L, 28; P, 6. At 30° C : E, 6; L, 18; P, 4. All at 70 per cent R. H. No development below 20° C or 40 per cent R.H.

Ptinidae ('spider beetles')

This group is closely related to the Anobiidae. The larvae develop into similar fleshy, curved, sessile grubs and the adults have humped thoracic hoods. But most of the beetles have a waist-like constriction at the back of the thorax which, with their rather long legs, imparts the resemblance to spiders responsible for the common name.

They are slow-moving and often sham death when disturbed. Most species have wings, but rarely use them. They are probably carried from place to place on foodstuffs or packaging. Infestations may also arise from bird's nests, which provide a common breeding ground for these beetles.

Ptinus. Adults of the two most common species are small beetles, about 2–4·5 mm long. They may be distinguished as follows.

Elytra densely clothed in brown or golden-brown hairs so that strial punctures and intervals are not distinct unless specimen is rubbed (Fig. 8.6). *P. tectus* ('Australian spider beetle')
Elytra more sparsely hairy so that striae are always distinctly visible. Prothorax with a longitudinal, feebly oblique, dense cushion of paler hairs on each side near the base. *P. fur* ('white-marked spider beetle')

Both species are now cosmopolitan and *P. fur* has been widely known since the days of Linnaeus. *P. tectus*, on the other hand, has spread from Australia (? Tasmania) within the last half-century. It reached Europe about 1900. It is now the most widespread beetle pest in warehouses in Britain.

The larvae of *Ptinus* feed on all types of dry vegetable or animal matter. They have been recorded as pests of cereals, cereal products and spices. But they often live in stores and warehouses as scavengers of miscellaneous debris. Before pupation the larvae tend to bore holes into (and thus damage) various inedible things such as cardboard boxes, books, sacks and even wood.

Life history. About 100 or more eggs are laid by the female, either singly or in small batches at intervals of a few days. Oviposition continues for about 3–4 weeks at normal room temperatures. The eggs are sticky when laid and often adhere to various objects. The larvae grow into fleshy, helpless grubs like those of anobiids. After 4 or 5 moults, they spin a cocoon cell and pupate inside it. If they are infesting food in sacks or cardboard boxes, they often bore through and form the pupal cocoon outside. After emergence the adults rest for a considerable period in the pupal cell, sometimes as much as 3 weeks. The adults avoid light and are to some extent gregarious so that large numbers may be found resting in dark places near the food. At night they wander about to a considerable extent.

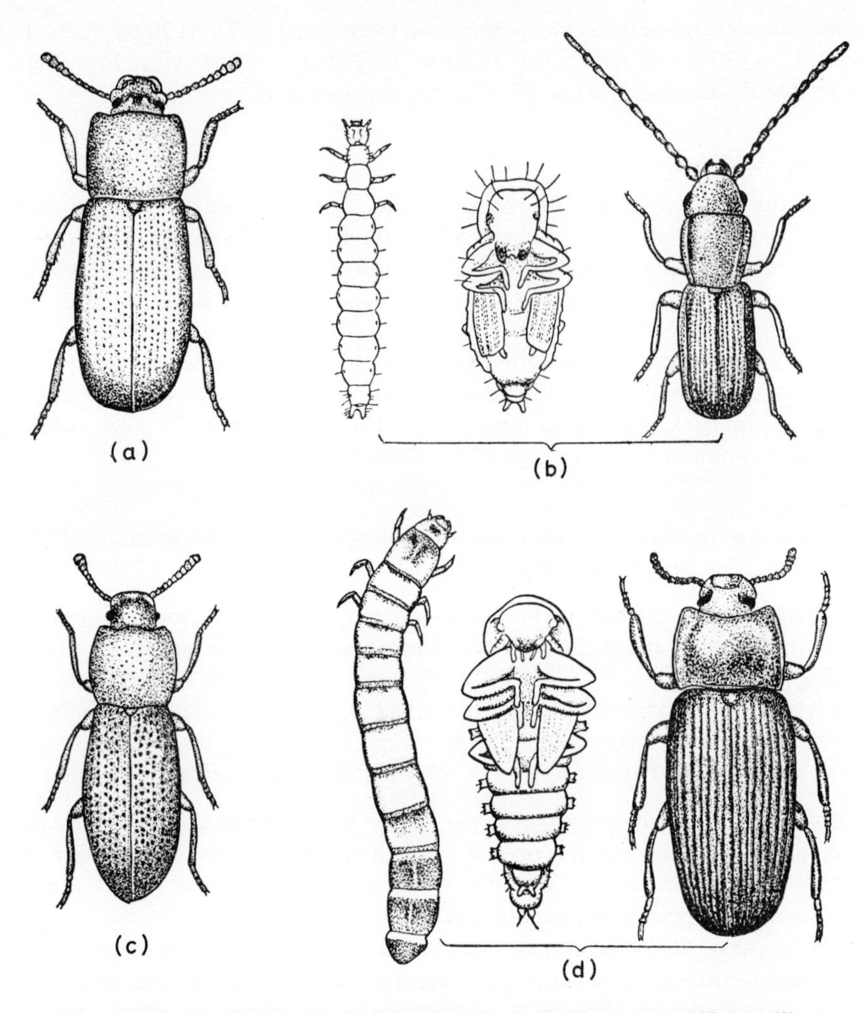

Figure 8.4 Beetle pests of stored products. (*a*) *Gnathocerus maxillosus*; (*b*) *Cryptolestes ferrugineus*; (*c*) *Palorus ratzeburgi*; (*d*) *Tenebrio molitor*; (*e*) *Oryzaephilus surinamensis*; (*f*) *Alphitobius diaperinus*; (*g*) *Trobolium confusum*; (*h*) *Latheticus oryzae*. (*a*), (*b*), (*c*), (*d*), (*f*), (*h*) after Patton [472a]; (*e*), (*g*) after Chittenden [110a]. (*a*), (*b*), (*c*) × 15; (*d*) × 3; (*e*), (*f*) × 15; (*g*) × 10; (*h*) × 15.

Speed of development (*P. tectus*). At 20–25° C: E, 3–16; L, 40 or more; P, 20–30; total, about 3½ months [289].

Trigonogenius globulus. This beetle is about the same size as *P. tectus* and is found in the same type of location, though less commonly. It is more rectangular in shape and of a grey or yellowish-brown colour, with patches of darker hairs. There are no lines of punctures on the elytra.

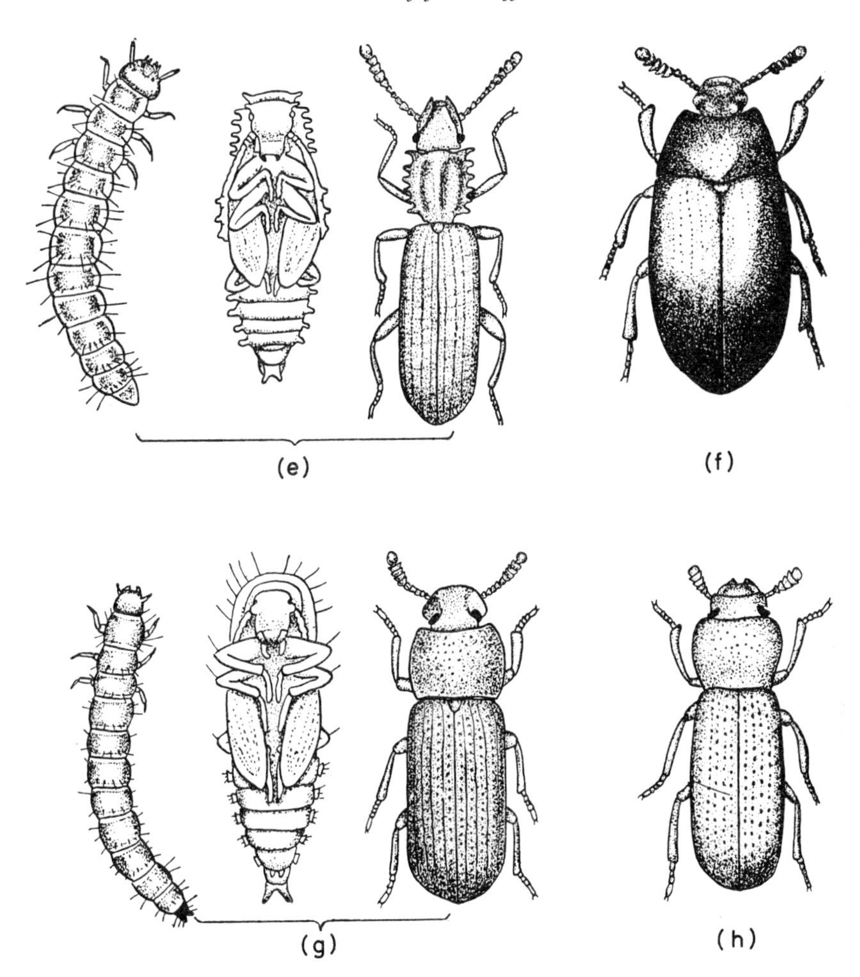

Niptus hololeucus ('golden spider beetle') (Fig. 8.6). This is a somewhat larger beetle, covered with long silky golden hairs and fine scales. It has been known in Britain since about 1836 when it was said to have been imported from Turkey. Since then it has become widely spread all over Europe.

The feeding habits resemble those of *Ptinus*, so that these beetles are not often important pests of stored food; but they sometimes damage cereals, cereal products, spices and drugs. More commonly, they feed on miscellaneous vegetable and animal remains (e.g. dead insects) and especially the debris in warehouses, cellars and ill-kept store rooms. The adults are sometimes very troublesome in houses from their habit of biting holes in various textiles (garments, carpets and bedding).

Life history. Similar to that of *Ptinus*. The adults can live as long as 250 days. They rest quietly in dark (and, preferably, moist) places during the day and

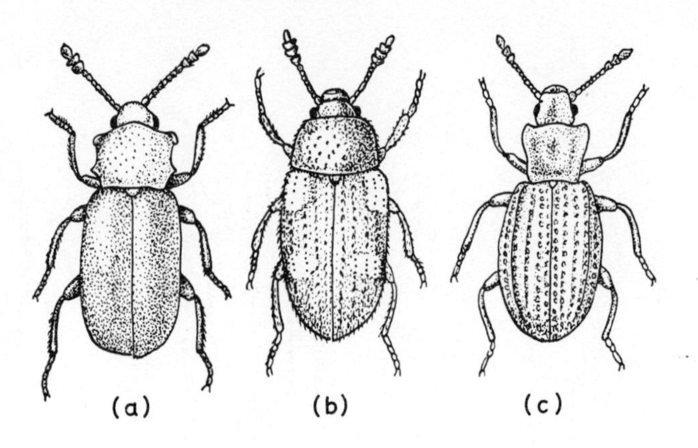

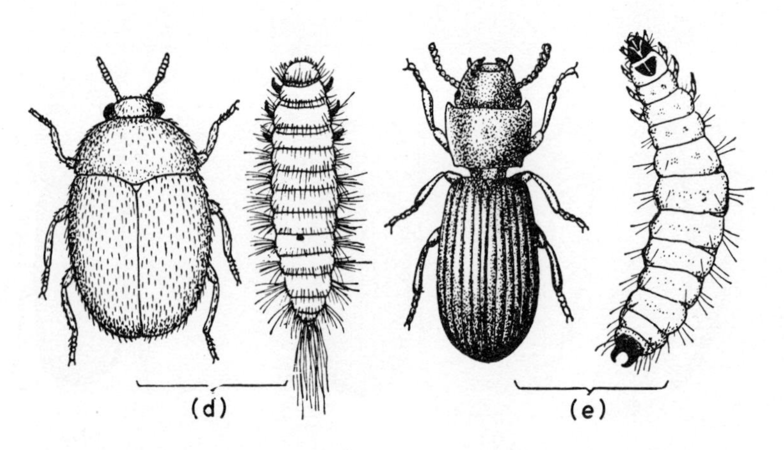

Figure 8.5 Beetle pests of stored products. (*a*) *Cryptophagus acutangulus*; (*b*) *Mycetophagus quadriguttatus*; (*c*) *Lathridius minutus*; (*d*) *Trogoderma granarium*; (*e*) *Tenebroides mauritanicus*; (*f*) *Carpophilus hemipterus*; (*g*) *Rhizopertha dominica*; (*h*) *Lasioderma serricorne*; (*i*) *Stegobium paniceum*. (*b*), (*c*) after Hinton [273]; (*e*) after Patton, [472a]; (*f*) after Kemper [318a]; (*g*) after Vayssiere and Lepesne [588a]; (*h*) after Bovingdon [59a]. Remainder original. (*a*) × 15; (*b*) × 12; (*c*) × 20; (*d*) × 15; (*e*) × 5; (*f*) × 10; (*g*) × 15; (*h*) × 10; (*i*) × 12.

wander about at night. They move about actively at temperatures down to 5·5° C.

Speed of development. At 15° C: E, 20–30; L, about 250; P, about 26. At 18–20° C: E, 11–20; L, about 150; P, 18–22.

Gibbium psylloides (Fig. 8.6). This beetle is shining and devoid of hairs or scales. It has a large rounded abdomen which gives it the appearance of a giant mite.

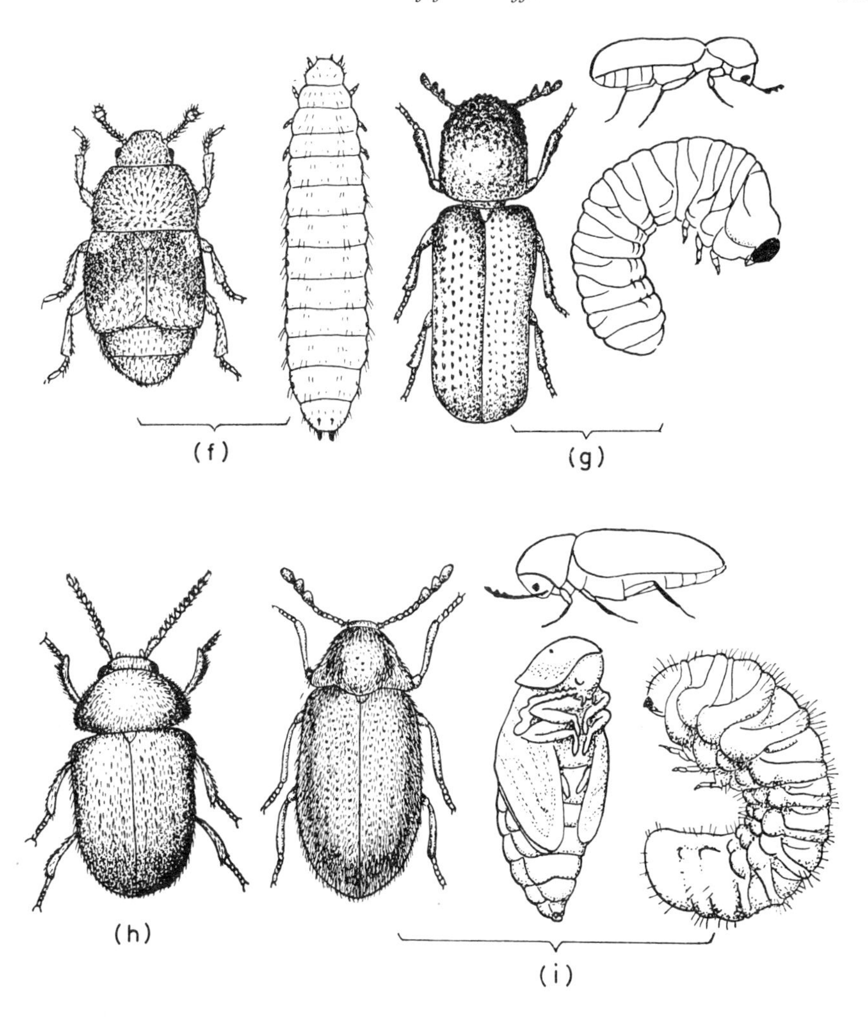

Gibbium has been recorded as a grain pest in India, but in Europe it tends, like *Ptinus* and *Niptus*, to be more of a scavenger and sometimes a nuisance. The adults have the same troublesome habit of biting holes in textiles.

Life history. Rather similar to *Ptinus* and *Niptus*. The adults are sluggish and long lived (up to $18\frac{1}{2}$ months at 25° C).

(iii) Cleroidea

Cleridae

An extensive family of mostly tropical beetles. Many examples are finely coloured and beautiful.

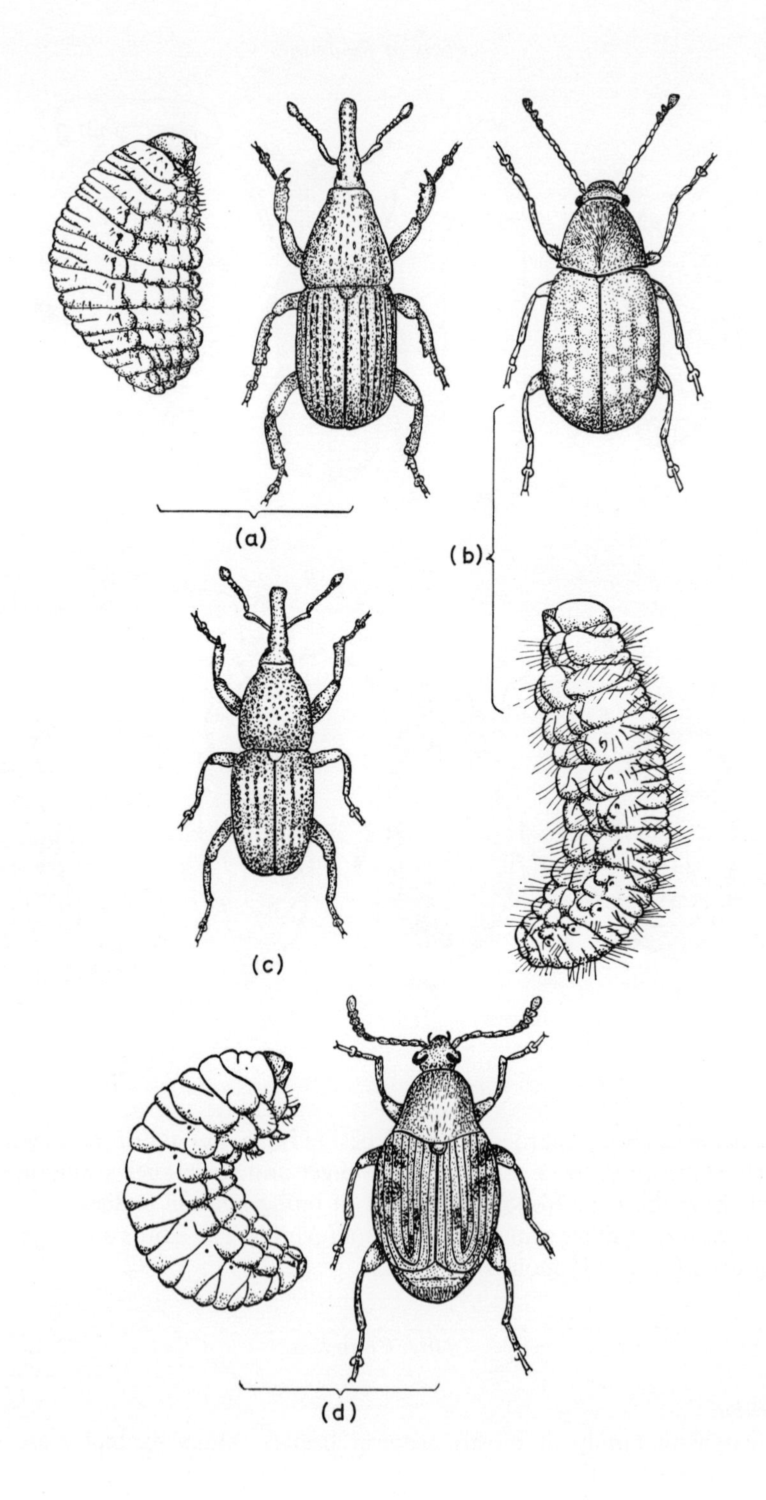

(a)
(b)
(c)
(d)

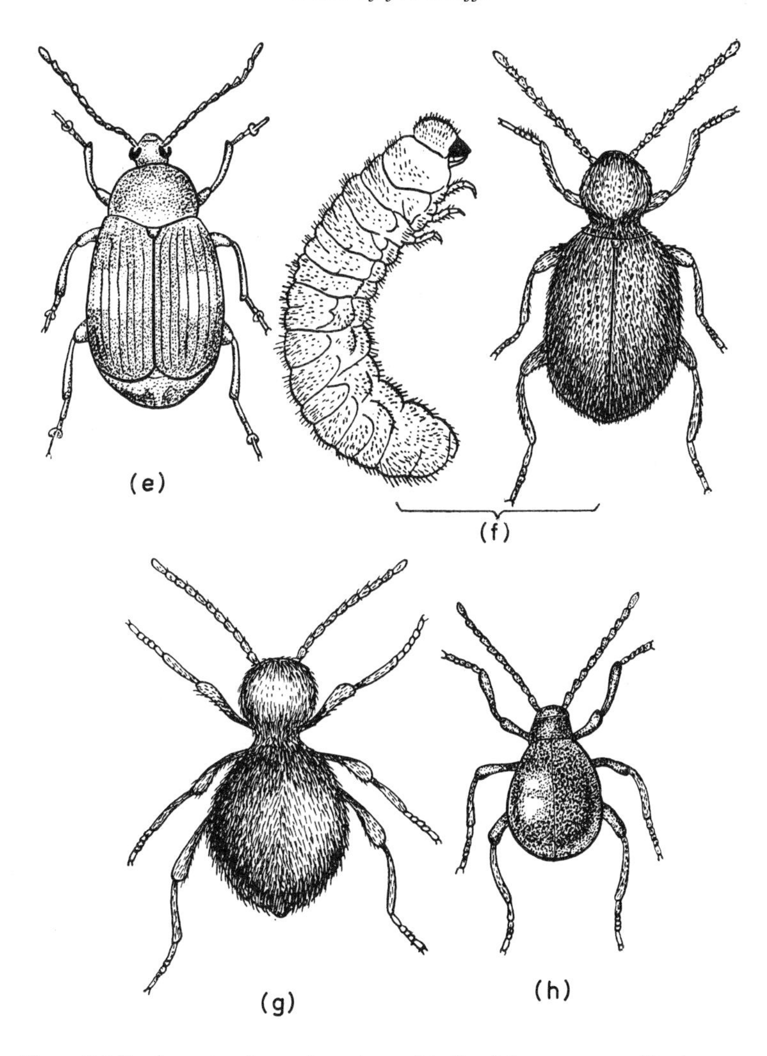

Figure 8.6 Beetle pests of stored products. (*a*) *Sitophilus granarius*; (*b*) *Araecerus fasciculatus*; (*c*) *Sitophilus oryzae*; (*d*) *Acanthoscelides obtectus*; (*e*) *Zabrotes subfasciatus*; (*f*) *Ptinus tectus*; (*g*) *Niptus hololeucus*; (*h*) *Gibbium psylloides*. (All modified after Patton, [472a]. (*a*), (*b*), (*c*), (*d*) × 10; (*e*) × 15; (*f*) × 10; (*g*), (*h*) × 7.

Necrobia rufipes ('copra beetle'). This is a distinctive, shiny, bluish-green beetle with reddish legs. Practically cosmopolitan, it is not, however, very prevalent in Britain. It thrives best in dried meats, bones, copra, palm kernels and other oilseeds.

Life history. The eggs are laid on the food material and as many as 2000 have been laid by a female, although the average is only about 300. The larvae burrow in the food, having especial preference for fatty tissues: the larvae are sometimes predaceous, e.g. on cheese skippers (*Piophila casei*). They can survive 62 days without food. They moult 2–4 times and reach a length of 10 mm. Pupation occurs in a paper-like cocoon away from the greasy food. The adults have been observed to live 14 months.

Speed of development. At 25° C: total development 6–24 weeks according to the quality (fresh protein value) of the food.

Trogositidae

Tenebroides mauritanicus ('cadelle') (Fig. 8.5). The adult is fairly easily recognized by its distinctive shape, being a black, flattened beetle with a short 'waist' between thorax and abdomen. It is a widely distributed pest, being especially troublesome in flour mills, from which it is difficult to eradicate. It has also been found attacking cereals, cereal products, spices, nuts and dried fruit.

Life history. The female lays batches of 10–40 eggs at roughly fortnightly intervals. This continues for several months and as many as 1200 eggs have been reached, although the average is nearer 500. The eggs are laid in the larval foodstuff; for example, in the superficial layers of flour. The larvae grow from a length of $1\frac{1}{2}$ mm to about 18 mm and moult at least 3 or 4 times (or up to 11 times under poor conditions). The mature larva excavates a hole in some solid material (wood, or cork) in which it pupates. This habit causes considerable damage to woodwork in grain stores, flour mills and railway wagons transporting grain. Furthermore, it provides shelter, not only for the cadelle, but for other grain pests. Many seemingly empty grain bins may actually contain a large nucleus of pests to reinfest new deliveries. Both adults and larvae have been stated to be predaceous, but this only occurs to a small extent. The adults may live as long as 21 months. They are resistant to starvation and have survived 52 days at 20° C and 4 months at 4–10° C without food.

Speed of development. At 21–22° C: E, 16; L, 232–282; P, 22–25. Under cool conditions, with a poor diet, the larval period may last up to 1248 days.

Cucujidae

These are small flattened beetles, many of which live under the bark of trees and some are predatory. A few species are grain feeders and have become pests of stored food.

Cryptolestes spp. (e.g. *C. ferrugineus*, 'rust-red grain beetle') (Fig. 8.4). The pest species are minute (about 2–3 mm long) and practically cosmopolitan. Local damp foci in grain tend to encourage their proliferation and they can spread through and spoil bulk foodstuffs. They are one of the most important pests of farm-stored grain in Britain [423].

Life history. The eggs are laid in crevices of grain or dropped loose among farinaceous materials. The larvae grow from about 0·4 mm to 4 mm in length, moulting about 4 times in the process. The larva forms itself a small cell of debris in which to pupate. Adults are recorded as living for 7 months.

Speed of development. Total, 35–75 days, according to temperature and quality of the food.

Silvanidae

This is a small family closely related to the Cucujidae.

Oryzaephilus surinamensis ('saw-toothed grain beetle') (Fig. 8.4) *O. mercaptor* ('merchant grain beetle'). *O. surinamensis* was so described by Linnaeus because his specimens had been sent from Surinam. The common name refers to lateral serrations on the thorax. *O. mercaptor* is generally similar in appearance and habits. Both species are very small, active, brown insects, cosmopolitan in distribution.

Products attacked include cereals and cereal products, dried fruit and drugs. They are liable to be serious pests of packaged food and may spread throughout a grocery store, if left unchecked.

Probably the most serious trouble from *O. surinamensis* occurs in farm-stored grain [418, 423]. The beetle damages the germ and heavy infestations cause the grain to heat. It then becomes mouldy and finally begins to sprout. Apart from direct loss of weight, serious reduction of value occurs, as the grain is discoloured and tainted.

Infestation does not take place in the field but after grain has been put into store or infested bags, which may be taken into the field on the combine harvester. Only a few insects are required to start a serious infestation. Infestations may arise from imported cereals, oilseed cake, etc; or they have been known to arise from beetles reaching farms in empty sacks, or grain from other farms or sent back after drying at a central plant.

Control measures, like those for other food pests, require cleaning and tidying of stores or barns, combined if necessary, with insecticide treatments. Beetles in lightly infested grain may be destroyed by fumigation.

Life history. The eggs are laid in crevices in or near the food material. The females lay up to 400 eggs (average about 170) under warm conditions, but cease laying in cold weather. The larvae are yellowish with flecks of brown and a brown head. They grow from about 0·9 mm to 3·0 mm and moult 2–5 times. Any dried plant materials may be suitable as food, although the larvae probably cannot attack undamaged cereal grains. Pupation occurs in a little cell constructed of bits of food and other debris. The adults are rather long-lived; they average 6–10 months, though as much as 3 years has been recorded.

Speed of development. At 20–23° C: E, 8–17; L, 28–49; P, 6–21. Total development: at 20° C, 83–108. At 31° C, 68–76. At 27° C, 22–32.

Ahasverus advena ('foreign grain beetle') is similar in general size and appearance to *Oryzaephilus* except that each side of the thorax bears only one tooth, at the anterior corner. It probably feeds mainly on moulds and refuse.

Nitidulidae

Carpophilus spp. ('dried fruit beetles') (Fig. 8.5). The common species, *C. hemipterus*, has each elytron dark brown with a large and distinct, pale (usually yellow) spot at the apex and smaller spot at the angle of the base. Other species (with elytra unicolorous) are *C. dimidiatus* and *C. ligneus*.

C. hemipterus is a cosmopolitan pest which is particularly troublesome to the dried fruit industry in California. In Britain it is usually found in imported dried fruit, which is usually completely spoilt by the mess of frass and larval skins produced. Other products less commonly attacked include nuts, damaged grain and spices.

Life history. The eggs are laid on ripe fruit on the trees or after drying. The active first-stage larva feeds on the fruit pulp and grows eventually to a length of about 9 mm. The pupa lies naked in the food or nearby.

Speed of development. At 28° C (summer in California): E, 2; L, 10; P, 7; pre-oviposition, 3; total, 22 days.

Cryptophagidae: *Cryptophagus* sp. (Fig. 8.5); Mycetophagidae: *Mycetophagus* sp. (Fig. 8.5); Lathridiidae: *Lathridius* sp. (Fig. 8.5).
These are very small brown beetles (mostly 1–2 mm long) which live among vegetable debris, either as scavengers or feeding on moulds and fungi. They are not to any large extent food pests, but often they occur in warehouses, stores and domestic larders, especially under rather damp conditions. Hence they are sometimes recorded as contaminating various foodstuffs (see also pages 486–488).

Tenebrionidae
This is a very large family, varying considerably in size and other characteristics. Most of the stored product pests are oval-oblong in shape, dark brown or black in colour and their larvae are all rather similar in appearance.

Tenebrio spp. (mealworm beetles) (Fig. 8.4). The two common species are most easily distinguished by the larvae. Those of *T. molitor* are shining yellow, while those of *T. obscurus* ('dark mealworm') are tinged with brown.

The mealworm beetles are among the largest of the beetle pests of food. The larval stages are the familiar mealworms used for feeding small animals and reptiles. Both larvae and adults feed on cereals and various cereal products; but they grow and proliferate rather slowly and, in spite of their size, do less damage than many smaller beetles. They thrive best in dark and rather damp situations and their presence is an indication of long neglect of cleaning. Both species are cosmopolitan.

Life history. The females have been observed to lay 77–576 eggs, singly or in batches, as many as 40 being deposited in a day. The larvae, as already mentioned, are a bright yellowish colour shading to a yellowish brown at the ends of the body and along the intersegmental joints. They are practically omnivorous and will devour scraps of meat and the dead bodies of insects as well as the cereals that constitute their usual diet. They are very resistant to starvation and may live 6–9 months without food or moisture. After a varying number of moults (9–20) they reach a length of up to 28 mm. When about to pupate they wander away from the food and may be discovered in various unusual places. Often they may be found in bags or packages of other products than those actually infested. A few days before pupation, they become sluggish and often lie on one side in a slightly curved position. The pupae occur naked among the food mass. The adults, which tend to avoid the light, usually live about 2–3 months.

Speed of development. At 18–20° C: E, 10–12; L, 1–1½ years; P, about 20. At 25° C: L, 6–8 months; P, 9; total development, 280–630 days.

Blaps spp. ('churchyard beetles'). These insects are not specifically food pests though they may occur in spilt or damaged food debris in sheds, warehouses and stores. The larvae should not be confused with those of the mealworm, which they greatly resemble. The last visible segment of the body bears a single upturned dark spine; whereas, in *Tenebrio* spp. there are two such spines.

Tribolium spp. ('flour beetles')

1. Black or very dark brown (5–6 mm).
T. destructor ('dark flour beetle')
Brown (3–4·5 mm) (2)

2. Antennae with a distinct, moderately compact 3-segmented club. Eyes separated ventrally by much less than 2 diameters of an eye. No ridge present above the eye. *T. castaneum* ('rust-red flour beetle')
Antennae with a loose, indistinct 5- or 6-segmented club or without a club. Eyes separated ventrally by a space equal to 3 diameters of an eye. A slight ridge evident above each eye (Fig. 8.4).
T. confusum ('confused flour beetle')

Tribolium is a cosmopolitan genus, the two most common species being *T. castaneum* and *T. confusum*; these are common and serious pests of cereal products, especially flour. They do not appear to be able to attack sound grain, but often add to the damage of primary grain pests. *Tribolium castaneum* is the most commonly intercepted insect on many kinds of imported food, being especially numerous on oilseed, oilcake and rice bran. *T. confusum* is less common and tends to be more restricted to cereal products.

Life history. The eggs are laid singly, 2–10 per day, according to temperature, over a period of many months and reaching a total of about 450. Oviposition ceases below 15° C. The eggs are sticky when laid and become coated with food particles or other debris. The larvae grow from a length of about 1 mm to 5 mm, moulting 5–11 times (usually about 7 or 8). The pupae lie naked in the food, gradually darkening before emergence of the adult. The adults feed on the same substance as the larvae. Under very warm conditions, *T. castaneum* adults readily fly; but those of *T. confusum* rarely do so. Males have been recorded as living about 600 days and females about 450. At 15° C the larvae starve to death after some 50 days and the adults in half that time.

Speed of development (*T. confusum*). At 22° C: E, 14; L, 60; P, 17; total, 91. At 27° C: E, 6; L, 22; P, 9; total, 37.

Gnathocerus cornutus ('broad-horned flour beetle'), *G. maxillosus* (Fig. 8.4) ('slender-horned flour beetle'). These beetles are slightly larger than *Tribolium*, being about 4 mm long, but generally similar in appearance. They can be easily distinguished, however, by the greatly enlarged mandibles of the males, curving outwards in front of the head (those of *G. cornutus* being broader, as the common names imply).

Cosmopolitan in distribution, these are mainly pests of flour and meal, although they also occur on grains.

Development (*G. cornutus*). 'In warm weather': E, 5; total, 8 weeks.

Palorus ratzeburgi ('small-eyed flour beetle') (Fig. 8.4). These are the smallest of the flour beetles (about 2·5 mm long). The antennae have no well-marked club. The thorax and head are wider, the legs more slender than *Tribolium*. Cosmopolitan; mainly a pest of ground cereals.

Latheticus oryzae ('long-headed flour beetle') (Fig. 8.4). The head is relatively larger than *Tribolium* and the antennae wedge-shaped. Pale yellow in colour. Length 2·3–3 mm. Cosmopolitan; mainly a pest of ground cereals.

Alphitobius diaperinus ('lesser meal-worm beetle') (Fig. 8.4), *A. laevigatus* ('black fungus beetle'). These beetles are distinctly larger than *Tribolium*, being about 6 mm long. They are black, more oval than oblong, with the back of the thorax strongly bi-sinuate.

They breed mainly in damp situations in grain or cereal products which are spoiled or are out of condition. The former is often found in large numbers in the deep litter of broiler houses.

(iv) Chrysomeloidea

Bruchidae ('pulse beetles') [419]

The beetles of this family are mostly ovoid in shape and have characteristically abbreviated elytra which leave exposed the tip of the abdomen. They breed in leguminous seeds, which many attack in the pod in the field; others will breed in dried stored seeds. The following simplified key may help to distinguish some common genera.

1. Hind tibiae with two movable spurs. *Zabrotes*
Hind tibiae with only fixed teeth. (2)
2. Lateral margin of thorax with a tooth near the middle. Hind legs always black. *Bruchus*
Lateral margin of thorax without teeth. Hind legs variable. (3)
3. Thorax conical, with straight sides. Hind femur with 2 teeth. *Callosobruchus*
Thorax with sides convex. Hind femur with only 1 tooth. *Acanthoscelides*

Species of *Bruchus* are field pests which lay their eggs on the developing pods in the fields. The larvae may, however, complete their development after harvest and emerge during storage. Peas or beans damaged by these beetles are useless for processing or canning. *Bruchus rufimanus*, the bean beetle, is indigenous to Britain and occasionally attacks broad beans or field beans. *B. pisorum*, the pea beetle, is not a native species, but often imported from abroad. *B. ervi* is a major pest of lentils.

The other genera will breed in stored peas or beans as well as in the field. *Acanthoscelides obtectus*, the American seed beetle (Fig. 8.6), *Callosobruchus maculatus* and *C. chinensis*, the cowpea beetles, and *Zabrotes subfasciatus* (Fig. 8.6) are common species.

Loss to seed depends on whether the germ or only the cotyledons are badly damaged. If the former, the seed will not germinate; if the latter, the seed may grow, but will be handicapped by lack of nutriment.

Life history (in stored beans). The eggs are laid among the dried beans. The first-stage larvae are active with well-developed legs. They bore into the beans and eat them from inside. The second- and third-stage larvae grow to fat and sessile grubs. They pupate in small oval cells excavated immediately under the seedcoat so that the adults can escape easily. The adults live about a week at 35° C or 3–4 weeks at 18° C, but do not feed.

Speed of development. At 18° C – *C. maculatus*: E, 21; L and P, 119. *Z. subfasciatus*: E, 7; L, 34; P, 12. At 35° C – *C. maculatus*: E, 4; L and P, 15. 'In Uganda' – *A. obtectus*: E, 5; L, 14–21; P, 5–6 [170].

(v) Curculionoidea

Anthribidae

This family is closely allied to the true weevils, but the adults do not have a distinct rostrum (or snout) and the antennae are not elbowed.

Araecerus fasciculatus ('coffee bean weevil') (Fig. 8.6). A small, robust dark-brown beetle, with the tip of the abdomen projecting behind the elytra. The first tarsal joint on each leg is as long as the remainder together.

Widely distributed in the tropics and sub-tropics, it is imported into warehouses and factories in Europe, but dies out in a cold winter. It occurs most commonly in nutmegs and cocoa beans.

Life history. The female lays eggs on cocoa beans (on the tree, in warm climates) and the larvae develop inside. The larvae become fleshy and legless in the older stages, like those of weevils, but they are somewhat more slender. There are 5 moults and pupation occurs in the seeds.

Speed of development. (In Java.) At 27° C: E, 6–7; L, 23–29; P, 7–8; total, 29–57.

Curculionidae ('weevils')

The weevils are a large vegetarian family, easily recognized by the snout-like prolongation of the head between the eyes, which carries the mouthparts at the tip. The antennae are 'elbowed' with a long basal segment and the last 3 or 4 segments enlarged to form an oval club. The larvae, which feed inside various parts of plants, tend to be white, fleshy, sessile grubs, often curled into the form of a C.

Sitophilus. The common species may be distinguished as follows:

Prothorax with punctures distinctly oblong or oval-oblong. Hind wings absent.
Uniform chestnut-brown colour (Fig. 8.6).
2·5–5 mm *S. granarius* ('grain weevil')
Prothorax very densely set with round or irregular punctures. Hind wings present. Elytra usually with four reddish spots (Fig. 8.6).
2–3·5 mm *S. oryzae* ('rice weevil')
3–3·5 mm *S. zea-mais* ('maize weevil')

S. zea-mais is rather larger and more shiny in appearance than *S. oryzae*; but for certain distinction, the penis of the male must be examined by an expert. For many years they were thought to be races of *S. oryzae*, but they are now considered as valid species.

Several distinct types of damage can be caused by these weevils. The feeding larvae destroy the endosperm, thus reducing weight and food value of the grain. Although the germ is not specifically attacked, so that germination is invariably prevented, the damaged grain produces a weak seedling; and it is more susceptible to fungi and moulds. Secondly, the larvae produce copious powdery frass, making the grain dusty and, if plentiful, taints it so that it becomes quite unpalatable. Thirdly, infestation by quite a small number of weevils may eventually lead to heating of the grain. If this develops into damp heating, even higher temperatures and more serious damage will result.

Sitophilus granarius [422]. This species is widely distributed in temperate and warm-temperate climates. It is imported from abroad, but can also breed freely in this country. Grain weevils attack all kinds of whole grain and some cereal products (e.g. macaroni). Being wingless, it is well adapted to life in stored food and does not attack grain in the field.

Life history. Before oviposition, the female bores a small hole in the food grain with her mouthparts, deposits the egg in it and then seals the hole with a drop of gum-like secretion. The larvae complete their development inside the grain. In a large seed like maize, several larvae can develop successfully; but only one can usually survive in a small grain of wheat or rice. The fleshy shape of the larva can be seen in Fig. 8.6. There are three larval stages, after which pupation occurs in the grain. Adult life ranges from about 160–260 days at 20° C to about 100–120 days at 27·5° C. The total number of eggs laid is not greatly dependent on temperature; but under warmer conditions they are laid more rapidly for a shorter time. Humidity, however, is important, considerably more being laid at 70–80 per cent R.H. than at low humidities. Under optimum conditions, about 200 eggs are produced at the rate of 2–3 per day.

Adverse conditions. Breeding does not occur below 13° C or above 35° C. The adults, however, can survive for over 100 days at 4·5° C and the larvae for over 70 days. The adults normally feed on the grain, but can survive starvation for 65 days at 13° C or 19 days at 30° C. Normally they avoid light and often enter grains hollowed out by larvae; but if the infested grain is disturbed, they tend to come out and walk about excitedly on the top of the grain. If handled, they 'feign death' for a short time (i.e. they lie still with their legs pressed close to the body).

Speed of development. At 18° C: E, 10·5; L, 55. At 20° C: E, 8; L, 40. At 25°C: E, 4; L, 27. At 27–28°C: P, 7·5. Total development: at 14–16° C, 113; at 23·5° C, 38; at 27° C, 29; at 30°C, 26.

Sitophilus zea-mais [194, 491], *Sitophilus oryzae.* Both species are widely distributed in the tropics and are common pests on grain imported into Britain. According to German authors, *S. zea-mais* has been common for some years on maize from the River Plate and accordingly was known as the La-Plata maize beetle. This species is more common on maize, while *S. oryzae* is most prevalent on wheat and rice. Laboratory experiments show that from a mixed colony on maize, *S. oryzae* dies out (in 17–34 months); while *S. zea-mais* dies out from a mixed colony on wheat (in 12–24 months).

The maize weevil has a preference for feeding on and laying eggs on maize rather than wheat, while the reverse is true of the rice weevil. Egg-laying preferences, however, are largely determined by the grain on which they are reared; but wheat has a stronger inducing effect than maize in both species.

Both forms can fly, in very warm weather (30° C), and the maize weevil can lay eggs on maize cobs in the field. The rice weevil is more sluggish, however, and is virtually restricted to breeding in stored grain.

Life history. The life history of both species, and many of their habits, resemble those of *S. granarius.* The number of eggs laid by the females are greater at 70 per cent R. H. than at 50 per cent R.H. Softer grains (e.g. wheat) allow more eggs to be laid than hard ones (rice). On the same grain (wheat), rice weevils laid an average of 148 eggs, as compared to the maize weevil's 217 (in 12–14 weeks at 25° C). At 20° C, about the same number of eggs were laid over 24–32 weeks.

Speed of development. Total development (of *C. oryzae*) at 16–18° C, 3 months; at 22–23° C, 2 months; at 27–28° C, 1 month. The maize weevil is slightly quicker. Thus, at 20°C, total for *S. oryzae* is 48–53 days, for *S. zea-mais* 44–48 days.

8.2.2 MOTHS (LEPIDOPTERA)

(a) Distinctive features

The moths and butterflies are the best known and most easily recognized of all insects, but the species which attack stored products are rather inconspicuous representatives of the order. They are small and devoid of striking colours, the adults being mostly buff or greyish in tone and the larvae cream-coloured with brown markings.

Moth pests of stored food, like the beetles, are responsible for losses due to absolute damage and to spoilage. In two respects, however, their problem is somewhat different. In the first place, certain larvae tend to migrate when fully grown, seeking pupation sites. Some species (especially of *Ephestia*) spin dense silken webbing over the goods. This is unsightly, traps dirt and debris, protects the larvae from insecticides and (especially in flour mills) tends to clog up machinery. For some years, therefore, control measures against these wandering larvae have included the deposition of films of insecticide in non-volatile oils. The other special character of some moth infestations is the emergence and flight of the adults in the summer. This offers an opportunity of attacking the pest in the adult stage, either by aerosols or by residual fumigants.

(i) Identification of moth pests of stored food

About 40 species of Lepidoptera have been recorded as pests of stored food, but many of them are comparatively rare. Most of the serious pests belong to the following groups of families: (Superfamily) *Pyraloidea* (dried fruit moths, flour moths), (Superfamilies) *Tinaeoidea, Gelechoidea* (clothes moths and house moths).

(ii) Identification of moth pests of food

Keys are provided (Appendix, page 535) to distinguish the dozen or so species of moths most likely to be found in domestic infestations. Since, in many cases, the larvae only will be present, a separate key for larvae is given (page 538). These keys may be found rather difficult to use in the absence of entomological training, although they have been simplified as much as possible.

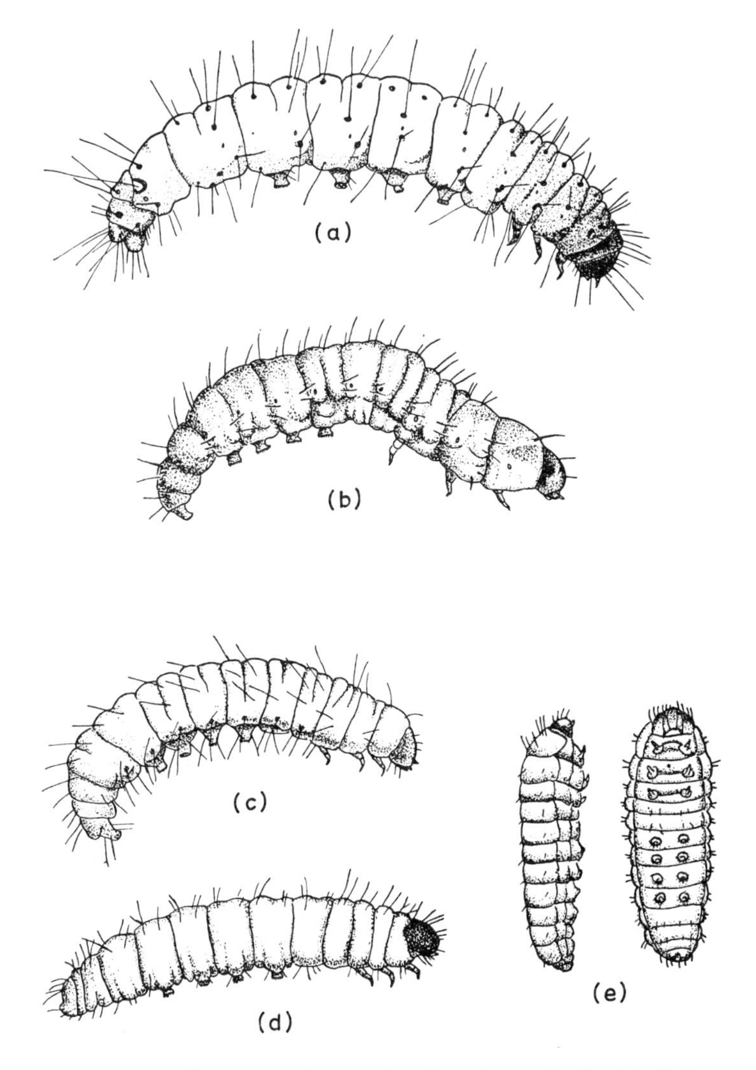

Figure 8.7 Larvae of stored product moths. (*a*) *Ephestia elutella*; (*b*) *Plodia interpunctella*; (*c*) *Hofmannophila pseudospretella*; (*d*) *Tineola bisselliella*; (*e*) *Sitotroga cerealella*. (*a*) after Bovingdon [59a]; (*b*) after Back and Cotton [21a]; (*c*), (*d*) & (*e*) after Patton [472a]. Approx. × 5.

(b) Occurrence and life histories of various pests

The details given in this section correspond to those given for the beetles (see page 336) except that, as the keys provided are more full, no descriptive matter is included.

(i) Pyraloidea

Pyralidae
This family comprises several sub-families, which were formerly considered as families. The ones with important members are the following.

Phycitinae, a large group, which includes the very important genera *Ephestia* and *Plodia*.

Galerinae, a small but widely distributed group, of which the best known species is probably *Galleria mellonella*, the honeycomb moth, a destructive pest of bee hives. The main food pests in this series are *Corcyra cephalonica* and *Paralipsis gularis*.

Pyralinae, mainly a tropical group, which includes *Pyralis farinalis*.

Ephestia elutella ('the warehouse moth') [425] (Figs. 8.7 and 14.6). This moth is widely distributed in temperate parts of the world; elsewhere its place is taken by its close ally, *Ephestia cautella*, the tropical warehouse moth.

In Britain *E. elutella* can be a serious pest of cocoa beans and chocolate confectionery, of dried fruit and nuts. Manufacturers may suffer as the result of the depredations of this pest not only directly, but also by loss of reputation if pests are discovered in the finished product by the consumer. A number of other substances liable to infestation by this pest include tobacco, wheat or other grain stored in bulk, oilseeds, oilseed products or manufactured animal feeding stuffs.

It is seldom imported except on products from other temperate areas; the insect which is found on cocoa beans on arrival in Britain is *Ephestia cautella*: during storage this dies out and is replaced by *Ephestia elutella*.

Life history. Eggs may be laid in cacao beans in warehouses in Britain, from late May to the end of July. Up to 260 eggs have been recorded as laid by a single female, in batches. The larvae gnaw their way into cacao beans and continue feeding inside, producing a mess of frass stuck together with silk threads, which often protrudes from a hole in the bean. When they are fully grown, the larvae measure about 11–12½ mm in length. They become restless, leave the food and wander about for 1–3 days. They tend to move upwards, yet they avoid the light, so that they collect in the darker corners of the walls and ceiling. Finally, they spin rather thin, greyish-white cocoons in corners and crevices. They remain in diapause as larvae until May the following year, when they pupate. The adults, like those of *E. kuehniella*, are able to take liquid food, but seldom do so. At normal room temperatures they live 2–3 weeks and at 25° C about 9 days. They fly about mainly at dusk and dawn, mate soon after emergence and the females begin to lay eggs a day later. Most of the eggs are laid in the first 4 days of adult life.

Speed of development. E at 20–24° C, 6; at 25–27° C, 4; at 28·5° C, 3; L, ?; P at 18–20° C, 16–19; at 22–24° C, 14; at 26° C, 10. Total, at ordinary room temperatures, 82–206. At 26° C (on tobacco), 60–100 (majority).

Ephestia kuehniella ('Mediterranean flour moth') (Fig. 14.6). In spite of the common name of this pest, it seems likely that its original home was Central America. It began to be widely dispersed by commerce about 1880 when roller milling was replacing grindstone milling; and it has since become cosmopolitan. The foodstuffs most frequently infested by this moth are ground cereal products and ground nuts or seeds or spices. It is a serious pest in flour mills, where its copious web interferes with milling. To some extent dried fruits are also attacked. Although it may occur in a wide variety of food substances, some of them are more favourable than others. For example, the larvae grow twice as quickly on rolled oats as they do on chocolate. They cannot develop on pure starch or gluten since they require food containing vitamins A and B (but not C).

Life history. The female lays about 200 eggs on average although as many as 500 has been recorded. The elliptical white eggs measure 0·6 mm × 0·3 mm; they are dirty white in colour and often stuck to various objects by a sticky secretion.

The newly hatched larvae measure 1–1½ mm and grow to a length of 15–19 mm after undergoing 3–5 moults. They begin spinning silk at once and cover much food with webbing in the course of their lives. When fully grown, they become restless and wander away from the food. Finally they choose a dark corner for pupation which a proportion of them undergo without spinning a cocoon. The normal cocoon is formed of a thick outer layer and a thin inner one with a head cover over the emergence hole.

The adults rest in shady corners in the daytime and fly about at dusk or at night. If they are disturbed during the day, they fly off with a characteristic zig-zag flight.

Pairing can occur immediately after emergence and it lasts a long time (usually 12–15 h). At normal room temperatures, the females begin egg-laying about half a day later and continue for about 11 days.

Speed of development. At 17° C: E, 8; L, 128; P, 16. At 18–20° C: E, 11; L, 56–70; P, 17–20. At 30° C: E, 3; L, 29; P, 7–10. At 18–20° C there may be 4 generations per year, but in unheated store rooms there are not more than 3.

Ephestia cautella (the 'tropical warehouse moth') (Fig. 14.6). This moth, widely distributed outside temperate areas, attacks almost any stored commodity except tobacco and animal products. It is the second most common insect found on imported food cargoes. It is an important pest of dried fruit, nuts, oilseeds and oilcakes. It is an important problem of the dried fruit industry in Australia, California and the Mediterranean. In the first two countries it is perhaps less serious than *Plodia*.

Life history. The female lays eggs on fruit which is somewhat drier than that chosen by *Plodia*. Figs and dates are frequently attacked and also sultanas and other fruit. The course of the life history has not been carefully studied, but is probably similar to that of *Ephestia*. The larvae pupate close to the food and the cocoon is very similar to that of *Plodia*. *E. cautella* seems unable to survive the winter in unheated rooms in Britain.

Plodia interpunctella ('Indian-meal moth') (Figs. 8.7 and 14.6). This important pest is cosmopolitan but it is primarily a nuisance in areas where there is a large fruit-drying industry (e.g. California, Mediterranean, Australia). Dried fruit and nuts are principally attacked while maize, chocolates, cereal products and various seeds and drugs also suffer. Grubs which are overlooked can spoil a whole packet of dried fruit. Apart from the actual damage by feeding, the frass and webbing are unsightly.

Life history. The females exercise some choice in the type of fruit upon which they oviposit. Dry and shrivelled as well as very sticky fruits are avoided. Also, certain treatments and handling processes in the industry render the fruit more or less attractive. The number of eggs laid depends on the food of the mother in her larval life; the maximum observed was over 500. The larva readily feeds on fruit tissues and grows to a length of 12 mm and a breadth of about 1·75 mm (i.e. relatively slimmer than *Ephestia*). There are about 4–7 moults. The pupa, which measures about 7 mm, is formed in a fairly thick white cocoon. The adults emerge and mate and, in large fruit stores, continue to proliferate the infestation.

Speed of development. E at 20° C, 8; at 25° C, 4; at 30° C, 2; L (depends on food as well as temperature); P, 12–43 according to temperature. Shortest total development (at 25° C), 35. Normally the larval stage is greatly prolonged by the winter in unheated store rooms. There are 1 or 2 generations per year.

Pyralis farinalis (the meal moth) (Fig. 14.6). This relatively large moth is a minor pest of grain products and other stored vegetable matter (for example, hay, especially from clover and lucerne). It is cosmopolitan.

Life history. The eggs are laid in small clusters on the food material to a total of 120–160. The larvae live in long silken tubes of their own silk, sometimes several together. They reach a final length of 12–14 mm and pupate in a cocoon which incorporates particles of the food debris.

Speed of development. There are usually two generations per year.

Corcyra cephalonica (the 'rice moth') (Fig. 14.6). Primarily a tropical insect, this is often imported into temperate regions with foodstuffs; but in northern countries it can only survive the winter in heated store rooms. The larvae are pests of grain (especially rice), grain products, oilseeds and oilseed products, beverage concentrates, nuts, dried fruit and spices.

Life history. About 120–160 eggs are laid in or near the larval food. The larvae meander over the food spinning a copious and strong webbing. The pupal cocoon is thick, strong and dense white and easily distinguished from the more transparent greyish cocoons of *Ephestia elutella*.

Speed of development. Under 'favourable conditions': L, 15–20; P, 7–10. Usually, however, only one generation a year in northern Europe.

Paralipsa gularis (Fig. 14.6). This moth has become a pest in Europe since about 1930. It breeds mainly in nuts and dried fruit and may cause considerable losses.

Life history. About 150–250 eggs are laid. The larva grows to a length of 20–30 mm and pupates in a rather dense, strong cocoon. The cocoons are formed in corners or crevices, or else many may be spun close together.

Speed of development. At 30° C: E, 4–5; L, ?; P, 20–115.

(ii) Tineoidea

Tineidae

These small moths breed in a wide variety of dry animal and vegetable matter. They are principally important for damaging wollen garments, furs, stuffed furniture and skins and are dealt with more fully as 'fabric pests' in Chapter 10. Several species will, however, breed quite successfully in food products, notably farinaceous materials and dried meats and spices.

Nemapogon granella (the 'corn moth') (Fig. 14.6). This is one of the most common tineids to occur in grain and grain products. It is widely distributed in Europe and has spread as far as North America and Japan.

Life history. Up to about 100 minute white eggs are laid by the females, often among grains of stored corn. The larvae wander over the food, feeding and spinning silken trails. They grow to a length of about 7–10 mm and, just before pupation, tend to wander away from the food and pupate in some dry crevice (e.g. between floor-boards). Sometimes, however, they pupate inside corn grains which have been eaten away till they are hollow. The adults, like the larvae, avoid bright light and mainly fly at dusk and at night.

Speed of development. Under 'normal conditions': E, 10–14; L, 2–4 months; P, 14–21. In northern Europe usually one generation per year (overwintering as mature larvae).

Gelechiidae

Sitotroga cerealella (the 'Angoumois grain moth') (Figs. 8.7 and 14.6). This pest is possibly of European origin, for a severe attack in France was recorded as long ago as 1671; or it may have been introduced from America. Its common name is associated with the province of Angoumois in France, but the pest is now widespread, although the pest cannot develop in the British climate. It is an important pest of stored grain of various kinds.

Life history. The eggs are laid, preferably soon after harvesting, on rye, maize, wheat or oats. The larvae burrow into the grains and develop inside. They become rather squat with very short legs, but always remain small, so that several may develop in a single large grain. They pupate inside the grain under a very thin cover which the adult pushes aside when it emerges. The adults do not feed, but mate and the females soon begin egg-laying.

Speed of development. E, 3–14. L at 20° C, 16; at 30° C, 10. P, at 15° C, 40; at 30° C, 7. Total, at 14° C, 115–118; at 21–26° C, about 42.

Oecophoridae

Endrosis sarcitrella ('white shouldered house moth') (Fig. 14.6); *Hofmannophila pseudospretella* ('brown house moth') (Figs. 8.7 and 14.6). These two moths are cosmopolitan and in Britain they breed out of doors as well as in houses and sheds. They will infest stored grains (especially if damaged) and also various grain products. They are often quite troublesome in outhouses where poultry food is kept, especially if it gets somewhat damp. In rooms of dwelling houses, where cleaning measures are somewhat neglected, *Hofmannophila* will also breed in debris under linoleum or carpets and may cause damage to the latter.

The life histories and bionomics are described in Chapter 10 (page 400).

8.2.3 IMPORTANCE OF BEETLE AND MOTH PESTS AND THEIR CONTROL

(a) Importance

(i) Types of damage caused by infestation [191, 192]

Insect pests in stored food consume it and, in heavy infestations, their depredations are detectable as actual loss of weight. Some pests tend to feed on the more nutritious part of the food (e.g. *Ephestia elutella* larvae select the germ of the wheat) and the loss of nutritive value due to infestation is therefore greater than the actual loss of weight.

A very troublesome consequence of infestation of bulk grain is heating. Insects, though cold-blooded animals, produce a certain amount of heat as a result of their respiration. Normally, this rapidly dissipates; but grain is such a poor conductor of heat that pockets of infestation in a large store tend to become warm. This warmth accelerates the respiration and proliferation of the insects, so that the heating of such 'hot spots' is a kind of chain reaction.

Grain heating begins in small spots but, if unchecked, may spread throughout the whole bulk. Insect grubs which bore in the grains (*Sitophilus, Rhizopertha, Sitotroga*) are eventually killed by the heat produced; but other, more mobile larvae (*Cryptolestes, Oryzaephilus*) gradually move outwards as the temperature rises.

Associated with this heating is a movement of moisture, from the infestation centres to the top of the grain, or to other cooler regions. This is due to warm air at the heating centre, absorbing moisture from the grain, rising by convection and depositing the moisture by condensation. The damp regions thus caused are readily attacked by moulds and become caked by them. In some cases the grain begins to sprout.

The damage to grain by heating may comprise loss of germinative capacity in grain for seed or malting and spoiling of the baking quality of wheat. The encouragement of moulds will cause mustiness.

Apart from heating of grain, other special troubles may arise in infestations of bulk foodstuffs. Thus, moth infestations of flour mills cause clogging of machinery by their webbing. Several beetle pests tend to bore holes in woodwork structures or in various packages.

On a small scale, problems of pest infestation are mainly consumer reactions of disgust at the tainting and spoilt appearance of packaged food.

(ii) Losses due to infestation of stored food [468]

Enormous quantities of stored foodstuff are destroyed by insects throughout the world; estimates range from 5 to 10 per cent, quantities greater than the total amount involved in actual trade. Most of these losses, however, occur in tropical countries, where food is stored under primitive conditions, and half the crop may be destroyed by pests before the next harvest. It would be quite wrong, therefore, to expect the overall world figure of losses to apply, with any accuracy, to Britain. On the other hand, the standards of hygiene in this country are such that comparatively minor contamination of food can cause complaints or rejection of goods. Consequently, the financial losses are out of proportion to the degree of infestation. This situation is maintained by two kinds of purchaser. (1) The rights of the private consumer in Britain are protected by the Food and Drugs Act and the Food Hygiene Regulations and sale of infested food may result in prosecution by a local authority. Generally speaking, however, reputable firms selling branded goods are most anxious to avoid loss of reputation by such an occurrence, irrespective of legal action. (2) Other important purchasers, who maintain high standards, are meticulous foreign markets for our manufactured foods. The American Food, Drug and Cosmetics Act, for example, imposes stringent regulations on imported foods, so that the presence of even parts of dead insects can cause refusal.

(iii) Nature of the problem in Britain and similar countries [191, 193]

Stored food infestation in Britain and other industrialized countries in cool climates has certain distinctive characteristics, due to the relatively few indigenous pests and the enormous importation of food. There are two main aspects of our problems: (1) the continuous stream of insect pests, imported with various products. Infestations at the start of a long voyage (perhaps through the tropics) can proliferate extensively, and may spread to other goods in the cargo. (2) The consequent infestation of the structure of ships and the storage warehouses, on arrival. From such residual insect populations, new cargoes (even uninfested) can become contaminated.

Infestation in imported foodstuffs [192]
One of the most serious problems of this type in the past, has been the shipments of grain and cereals, especially wheat. At one time, about 80 per cent of cargoes from some countries were infested with grain pests. Strong representations to the exporting countries, however, resulted in stringent measures being taken at the countries of origin. As a result, imports of wheat from the Argentine, Australia, Canada and the U.S.A. have declined to low levels, over the period 1960–70, largely by treatments with malathion. At present, there remain serious infestation problems with other cereals (rice, sago, tapioca, etc) from the tropics; also maize from South Africa and the Argentine. Khapra beetles are prevalent in oilseed and oilcake and lepidopterous pests in dried fruit.

Infestations in the structure of ships and warehouses
Numerous crevices exist in most ship holds and in warehouses, in which spilled grain or other substances can provide breeding sites for residual pest populations. An important aspect of this is the possible introduction of new exotic pests into the receiving country.

The permanence of such infestations depends on the adaptability of the insects concerned, one of the most important characteristics being cold-hardiness, to survive the winter in a cool climate. As regards Britain, some of the pests mentioned can be grouped as follows.

Pests surviving the British winter without special protection:
Beetles: *Sitophilus granarius, Oryzaephilus surinamensis, Cryptolestes ferrugineus, Ptinus tectus, Stegobium paniceum, Tenebrioides mauritanicus, Tenebrio* spp. and *Dermestes* spp.
Moths: *Ephestia elutella, E. kuehniella, Plodia interpunctella.*
Pests surviving mild winters:
Beetles: *Sitophilus oryzae, S. zea-mais, Lasioderma serricorne.*
Moths: *Ephestia cautella.*
Pests not surviving the winter, unless protected or warmed in some way:
Beetles: *Tribolium* spp., *Rhizopertha dominica* and *Carpophilus hemipterus, Araecerus fasiculatus, Necrobia rufipes.*
Moths: *Sitotroga cerealella.*

(b) Control [288]

The numerous problems in the control of insect pests of food in bulk storage are beyond the scope of this book. A great deal of effort is expended in endeavouring to provide the consumer with products free from insects. In Britain, this work is supervised by the inspectorate of the Ministry of Agriculture, Fisheries and Food and the research problems investigated by the Pest Infestation Control Laboratory.

The subject, which can only be outlined here, must be considered in relation to the magnitude of the food stores (ships and warehouses; farm stores; retail stores) and also in regard to the control objectives (rapid disinfestation by fumigation; design and treatment of storage spaces; long term protection of food in store).

(i) Infestations in ships, barges, warehouses, mills and factories [193]

On arrival, food cargoes may be fumigated in the hold of the ship, although this presents considerable difficulties. Alternatively, the cargo may be fumigated in barges or on shore. Bagged or boxed goods can be treated in large steel fumigation chambers or, in some cases, under gas-proof sheets. Grain, or oilseed in bulk, is most satisfactorily fumigated in specially equipped silos. In most cases where large bulk goods are to be treated, it is most efficient to use rather toxic fumigants; notably methyl bromide, phosphine or occasionally ethylene oxide; and such operations would only be done by experienced servicing companies.

The use of these materials is regulated under the Health and Safety at Work Act, 1974; and, in some cases, by the Factory Act, 1961. Some valuable official leaflets have been prepared concerning the fumigants mentioned (the properties of which are outlined on page 80). These leaflets emphasize the dangers of the respective fumigants, outline the responsibilities of the users and describe the precautions for using them, including protective equipment and methods of protection. The following are available.

Fumigation using Methyl Bromide was issued by the Health and Safety Executive in 1976. The liquefied gas is available in steel gas cylinders or disposable cans. Instructions are given for fumigation of (1) buildings, (2) commodities under gas-proof sheets, (3) ships and barges, and (4) fumigation chambers. The next three leaflets were all issued by the Ministry of Agriculture Fisheries and Food.

Fumigation with Aluminium Phosphide Preparations (1973) describes the two forms commercially available: *Detia gas ExB* which consists of green paper packets, and *Phostoxin* tablets (see page 80). For treating bulk grain, these preparations can either be laid over the surface, or pushed down into the grain. Instructions are given for fumigating grain in silo bins, or in bulk on the floor, in barges, or under gas-proof sheets, or as individual bags. The fumigation is slow, requiring several days; and precautions for airing are necessary. Details of detection are given and notes on the danger of the gas and on first aid treatment.

Fumigation with Ethylene Oxide (1969) deals with an old established fumigant, the use of which is now declining. In addition to toxic hazards, this gas presents dangers of explosion and fire, for which reason it is usually mixed with carbon dioxide.

Fumigation with Liquid fumigants, Carbon Tetrachloride, Ethylene Dichloride and Ethylene Dibromide (1973) deals with somewhat less hazardous fumigants, the first two being most commonly used in Britain, often in combination. In small quantities, these may be used by competent persons, not necessarily professional fumigators. Nevertheless, these compounds can be dangerous and instructions in the leaflet must be carefully followed. Details of the compounds are given and notes on residues in food, methods of detection, and protective equipment. The instructions include treatment of grain in deep silo bins, small bins, grain in bulk on the floor and of bagged cereals and of empty sacks.

Several Ministry leaflets recommend a mixture of ethylene dichloride and carbon tetrachloride in equal parts, for bulk grain over 2 m (6 ft) deep, using 1 litre tonne^{-1} (1 gall ton^{-1}). For bagged grain, bulk grain less than 1 m deep, and empty used sacks, a 3:1 mixture of ethylene dichloride and carbon tetrachloride is recommended. The bagged grain is treated in metal bins at 5 litre tonne^{-1} (6 pint ton^{-1}); while the sacks are dosed at 1 litre per 50 kg of sacks (2 pint cwt^{-1}).

Prevention or elimination of residual pest populations [288]

Good storage conditions. The nuisance of residual pests is much less in well-built, clean storage premises. Buildings intended for bulk food storage should be well ventilated and adequately insulated, to keep cool. It has been pointed out that a Nigerian hut store, with thick mud walls and a thatched roof, can remain at 25° C with an outside temperature of 28° C; while a metal and brick building with a skylight in England, can rise up to 30° C on a sunny day in May.

Humidity is also important; and the nature of the stored product can affect this, as it often accounts for 60 per cent of the interior volume. Temperature and humidity are, of course, interdependent. A fluctuating temperature or a sharp temperature gradient can load to deposition of moisture and spoilage of grain or similar foodstuffs.

Unfortunately, some old and imperfect warehouses have to be used for food storage; but attempts should be made to reduce the harbourages for pests, by sealing crevices, removing flaking paint or paper from the walls and excluding lumber. When the warehouse is empty, it should be thoroughly cleaned and any spilled food, which could provide breeding material for pests, removed.

Insecticide treatment [422]. At this stage, it may be desirable to spray or dust the walls and floor with suitable insecticides. About a score of insecticides have been cleared for use in food stores in the U.K., but most commonly recommended are the following.

Sprays (all to be used at about 5 litre/100 m^2; 1 gall/1000 ft^2). 1·5 per cent malathion; 1 per cent pirimiphos-methyl; 1 per cent fenitrothion; 1 per cent iodophenphos; 1 per cent lindane; 2 per cent bromophos.

Dusts. 2 per cent malathion; 2 per cent pirimiphos-methyl; 2 per cent fenitrothion.

Prevention and control of infestation

Prophylactic treatment of grain in the process of storage can prevent infestations developing for a year. Low toxicity insecticides in liquid or powder form are mixed with the grain, most conveniently by application to the grain on a conveyer belt. The following application rates (per tonne) have been recommended, where the conveying rate is less than 5 tonne h^{-1}.

1·5 litre (3 pint) malathion emulsion at 0·6 per cent; or pirimiphos-methyl emulsion at 0·3 per cent.

500 g (18 oz) malathion dust at 2 per cent; or 200 g (7 oz) pirimiphos-methyl dust at 2 per cent.

If the conveying rate is greater than 5 tonne h^{-1}, the emulsion concentrates should be doubled and the application rates halved. The water present in the highest application rate will only amount to 0·2 per cent on the grain and should not affect it. Dust treatments are not altered at different conveyer rates.

Regular inspection of stored foodstuffs is necessary, preferably once a week. Therefore, storage must allow access for this (e.g. bags should be kept away from walls and adequate inspection passages left).

The obvious place to look for insect pests is on the surface of grain or other produce in bins or sacks. In addition, samples from bulk grain can be taken with special 'spears'. Another means of detecting infestations in bulk foodstuffs (especially grain) is by the rise in temperature caused by their metabolism. Hence, a thermometer inserted in the produce should be inspected regularly for suspicious rises.

Control of internal pests can only be achieved by fumigation, which has already been described.

The control of lepidopterous pests offers special opportunities in that the fully grown caterpillars tend to migrate extensively over the outside of the foodstuff, prior to pupation. These can be attacked by spraying a mist of pyrethroids in white oil, which settles on all surfaces forming an insecticidal film. A suitable spray should contain about 1·3 per cent natural pyrethrins, or an equivalent concentration of synthetic pyrethroid, or a synergized mixture. As a rough guide to dosage, about 2·5–5 litres should be applied to 250 m^2 (2500 ft^2) of surface [425].

In addition to the attack on the wandering larvae, adult moths are vulnerable to treatment by aerosols when they are flying about (especially at dusk) in the summer months. An aerosol containing pyrethroid applied by generators combining heat and mechanical dispersion should emit a dose of 0·5 litre per 1200 m^3, or double this dose with a mechanical generator. An alternative for attacking the adult moths is to use slow-release dichlorvos strips. The numbers used should be adjusted to the amount of free air space. Thus, if one unit is recommended per 30 m^3 (1000 ft^3) of empty room, this can be adjusted by the amount of the bulk of storage present. The units are hung up in early summer and should last 12–14 weeks. Ventilation during the day may reduce the air concentration, but this builds up when the warehouse is closed at night. The vapour does not penetrate into bagged materials.

(ii) Infestation of farm-stored grain [423]

Grain is often stored on farms before sale for malting, for animal fodder on the farm, or for seed; and during the period of storage may suffer from attack by insects. Among the more common beetle pests on British farms are *Sitophilus granarius, Oryzaephilus surinamensis* and *Cryptolestes ferrugineus*. To prevent or reduce losses from this source, the first essential is good storage conditions in a damp-proof building. Regular inspection and cleaning are desirable and this entails storage in a manner allowing access. Where there is a danger of the barn being infested from previous storage, the walls and floor should be sprayed or dusted with insecticide, as described earlier (page 364).

Grain brought from outside sources should not be stored alongside farm-grown grain; this will reduce the risk of cross-infestation. Where there is any suspicion of infestation present in grain brought in for storage, it can be treated by addition of non-toxic insecticide as described earlier (page 365). So far as possible, all stocks should be used in rotation, to prevent the accumulation of very old stocks.

When infestations have developed, fumigation is convenient to destroy the pests. The mixture of ethylene dichloride and carbon tetrachloride should be used, as described earlier (page 363).

(iii) Infestation in retail shops and domestic larders [421]

To a large extent, the manufacturers or wholesale merchants will have eliminated insect infestation from foodstuffs before they reach the shop or home. But, inevitably, there are occasionally a few insects which slip through into bags or packets of the finished product despite precautions. Perhaps the most common offenders among the beetles are *Stegobium paniceum* and *Ptinus* sp. Among the moths, *Ephestia elutella* turns up from time to time in chocolate, while *Plodia interpunctella* or less often *Ephestia cautella*, are found in dried fruit. When there is rapid sale and consumption of the product, these insects may escape notice, unless they are well grown. It cannot be said that these insects are actually harmful, if eaten, other than psychologically. But if they are detected, complaints and legal action may follow. In general, it will be admitted that insect infestations tend to indicate poor hygienic standards.

If infested food is kept for long in storage, the pests will multiply and probably spread, eventually, throughout the store. Contamination of clean packages can be prevented by adequate wrapping; but this is more difficult to achieve than might appear. Cardboard and metal foil wrapping is not usually effective, because crevices are left through which tiny insects can penetrate. Sealed wrappings of wax paper or plastic are adequate for mechanical exclusion of feeble pests. But some insects are easily able to bore through paper, card or thin plastic. For protection against such intrusive forms, insecticidal treatment (based on pyrethrins) may be desirable.

In order to prevent insect damage in food stores, it is important, therefore, to be alert for the first signs of a pest. Adults or larvae may be seen wandering about on shelves or on adjacent walls or windows. Even suspicious-looking holes in packets or bags should give the warning.

The first essential is to trace all the breeding sites and to segregate the infested products. When a store has been badly neglected and there is a very widespread and general infestation, it may perhaps be necessary to resort to fumigation. Normally, however, it should be sufficient to remove and destroy very badly infested materials. Less badly infested products can be salvaged, either for human or animal consumption. Dry foods (cereals, beverage concentrates and spices) can be simply disinfested by heat treatment (see page 47). This is not always feasible with moist sugary or fatty foods (dried fruit or dried meats) but if they are only lightly infested they may be cleaned by discriminative elimination.

The possibility of allowing a permanent infestation to develop, in spillage and debris, exists even in small stores and larders. Such pests as *Stegobium* and *Ptinus* thrive in almost any dry organic matter. Therefore, the eradication of an infestation should conclude with a thorough cleansing of the whole store and the application of an inoffensive insecticidal spray. Finally, good ventilation is most important, for many pests can only thrive under damp conditions.

8.2.4 BLOWFLIES ETC

A few kinds of fly are liable to breed in various foodstuffs; but these have been dealt with in Chapter 6 as 'houseflies and blowflies'. They include:

(a) Blowflies, which breed in raw or uncooked meat or fish, e.g. *Calliphora Lucilia, Sarcophaga* and *Phormia* (pages 219–224).
(b) Fruit flies, which breed in over-ripe fruit, decaying vegetables and milk curds, e.g. *Drosophila* (page 226).
(c) The cheese skipper *Piophila casei*, which occurs in cheese and smoked meats (page 228).

8.2.5 BOOKLICE (PSOCOPTERA)

These small, flattened, yellowish, frequently wingless insects may occur in various places in dwelling houses. They feed on minute moulds and fungi and are therefore only liable to thrive under slightly damp conditions. They are dealt with more fully as 'nuisances' in Chapter 13; but occasionally they may be found infesting various kinds of foodstuff, especially packaged cereals, sugar, etc. It is quite possible that this may often be accidental contamination due to the foodstuff being kept in a rather damp store infested with psocids. Lightly infested foods can easily be disinfested by heat treatment (page 47).

8.2.6 MITE PESTS OF STORED FOOD [550]

Various mites may occur in stored foodstuffs, particularly under conditions of high humidity. Most abundant are the herbivorous mites, which feed on the food, belonging to the family Acaridae. In addition, there may be predaceous mites present, which are beneficial inasmuch as they destroy the harmful species. These predaceous mites occur in diverse branches of the Acarina.

(a) Distinctive characters

The mites concerned are about 0·5 mm long; that is, about the size of the eggs of many stored food insects. To the naked eye, they appear as small white dots, which move slowly away from a source of illumination. When very numerous, these mites form a characteristic buff-coloured dust, made up of the mites themselves and their cast skins. Also there is a peculiar 'minty' smell, easily recognized, if one crushes some of the mites. Under the microscope, the mites are pearly white, often with tan-coloured legs, and usually with numerous outstanding long bristles. Their general appearance can be seen from the examples in Fig. 8.8.

The acarid mites of stored food belong to two sub-families distinguished as follows:

Anterior part of the body separated by a transverse suture. Tyroglyphinae
(e.g. genera: *Acarus*, *Tyrophagus*, *Caloglyphus*, *Rhizoglyphus*, *Suidasia*, *Thyreophagus*)

No transverse suture separating anterior part of body. Glycyphaginae
(e.g. genera: *Glycyphagus*, *Carpoglyphus*, *Gohieria*)

(b) Occurrence and biology of various mite pests [424, 292]

(i) The flour mite (Acarus siro) [327, 550] (Fig. 8.8)

This is the commonest of the mite pests of stored food.

Appearance

The body is colourless, the appendages varying from pale yellow to reddish brown; the hind end being smoothly rounded in the male (0·32–0·42 mm) and slightly indented in the female (0·35–0·65 mm). Two pairs of bristles trail from the hind end of the body.

Occurrence

In addition to flour, *A. siro* occurs on all kinds of farinaceous products, on cheese, grain, hay, etc.

Life history

Under optimum conditions, the females can lay up to 500 eggs, during a lifetime of about 40 days. The eggs, laid loosely in the food, measure about 0·15 mm × 0·08 mm. The larvae which hatch from the eggs are about 0·15 mm long and have three pairs of legs. After feeding for some days, the larva becomes inert and then moults to produce the first nymphal stage, which, like all later stages, has four pairs of legs. Subsequently, the nymph moults to produce a second nymphal stage, which finally moults again to produce the adult. In some mites a relatively (or completely) inactive stage called a 'hypopus' is produced, instead of the second nymph; this is resistant to low humidity and other adverse conditions. Formerly it was thought that hypopi commonly occurred in strains of *A. siro*; but it is now recognized that they are seldom formed in this species, most of the earlier reports relating to a closely related species (*A. farris* Oud.) which does not generally occur in stored food.

Speed of development. At 18–22° C: E, 3–4; L, 4–5; NI, 6–8; total about 17. At 10–15° C: total about 28. At 5° C: total about 140.

Ecology

Flour mites require a moderately high humidity in order to proliferate. They will not thrive on grain or flour with a water content below 13 per cent (in equilibrium with air at 65 per cent R.H.). A slight rise above this level (say, to 14 per cent water content; 70 per cent R.H.) will allow ready breeding. At 15–18 per cent water content (75–85 per cent R.H.), infestations increase to serious proportions. The mites can distinguish air at favourable humidity and choose regions with 75–85 per cent R. H. if offered a choice (higher, as well as lower, humidities being avoided). Quite often, an infestation is confined to the superficial layer of flour or grain, which has, in some way, become damp.

The optimum temperature range for the mite is rather low (about 18–25° C). This is affected by the humidity, the maximum proliferation rate at 80 per cent R.H. being at 25° C, while at 90 per cent R.H. it is at 22° C. As well as the optimum, the overall temperature range of flour mites is also rather low, as compared to most stored food insects. Thus the upper limit, at which they can breed and survive, is just over 30° C; whereas they can increase (slowly) at low temperatures, such as 5° C and persist for a long time at temperatures around freezing.

The flour mite is often preyed upon by *Cheyletus eruditus* (Fig. 8.8, and page 376). Environmental conditions act differently upon the two mites, *Cheyletus* being more dependent on warmth and *Acarus* on moisture. In mixed populations, therefore, *Acarus* is more abundant in winter and *Cheyletus* gains the upper hand in summer.

Mites are sometimes accused of spreading moulds and fungi in cereals. However, it seems more likely that they appear together because the same moist conditions favour both of them. In fact, there is evidence that the mites feed on the fungi and reduce their prevalence.

(ii) The cheese mite (Tyrophagus casei) (Fig. 8.8)

Appearance
Mites of the genus *Tyrophagus* have at least 4 pairs of long bristles trailing from the hind end of the body. *T. casei* is one of the larger, more clumsy species, with well-tanned legs. Males measure 0·45–0·55 mm; females 0·50–0·70 mm. Other species of *Tyrophagus* not uncommon are *T. putrescentiae* and *T. longior*. The former prefers foods with high fat or protein content (linseed, dried egg, cheese); the latter occurs on cheese but also on home-produced grain, straw and hay.

Occurrence
As well as on cheese, *T. casei* is found on various other stored foods, on grain, damp flour, old honeycombs, etc. *T. casei* has been maintained in culture for addition to the particular brand of cheese known as Altenburger, to which it is supposed to impart a piquant taste. When 'ripe', the surface of the cheese is covered with a greyish powder consisting of enormous numbers of the mites (living and dead) together with their cast skins and faeces. Other cheeses may sometimes become infested with *T. casei* or other mites, but in Britain this is generally regarded as distasteful. The mites eat out small holes and cause accumulations of brownish dust.

Biology
The life history resembles that of *A. siro* in many respects. The life cycle takes 15–18 days at 23° C and 87 per cent R.H.

(iii) The dried fruit mite (Carpoglyphus lactis) (Fig. 8.8)

Appearance
As in *A. siro* there are two pairs of bristles trailing from the hind end of the body; but there is no suture dividing the body into two. The appendages are slightly pinkish. Males, 0·38–0·40 mm; females, 0·38–0·42 mm.

Occurrence
This mite is commonly associated with dried fruits and jam, though it is by no means the only species found in such products. (Another mite favouring this type of food is *Thyreophagus entomophagus*, which may also occur in flour stored for long periods.)

Biology
The females have been recorded as laying 25–72 eggs in a week. The mites themselves can survive for 40–50 days.

(iv) Other mites infesting stored food [550]

Glyciphagus destructor, resembling the house mite *G. domesticus* (Fig. 8.8) (see page 492), is common on many kinds of stored food, often in company with *Acarus siro* and the predator *Cheyletus eruditus*.

A mite with notably different appearance from other tyroglyphid pests is *Gohieria fusca* (Fig. 8.8). The body is covered with a pinkish-brown cuticle, giving it a superficial resemblance to the beetle mites (page 508). It occurs in flour, rice, corn, bran, etc.

Certain mites seem to require exceptionally moist medium to thrive and are found in particularly damp grain, nuts, bulbs, etc. Examples: *Rhizoglyphus echinopus* (Fig. 8.8) and *Caloglyphus berlesei* (Fig. 8.8).

(c) Importance of mites in foodstuffs

(i) Harm to foodstuffs

Several of the species of mites referred to above may cause deterioration in stored food, but the most common source of trouble is *Acarus siro*. This cosmopolitan mite is troublesome not only in grain stores, mills and bakeries, but also in larders of private dwellings. It is primarily a pest of cereals and cereal products. In stored grain it tunnels into the germ and consumes the best parts of the seed; and in flour too it probably attacks the more valuable food constituents, selectively. In addition to reducing the food value, a heavy mite infestation causes tainting. Bad tainting occurs if there are over 500 mites per 100 g food. The cereal acquires a characteristic musty or sour smell which is usually described as 'minty'. (This is due to a secretion of the mites, apparently arising from two dorsal glands.) The taint may be largely removed from grain in the cleaning and washing processes before milling; but in grain products, such as flour, it is a serious problem. Mitey flour used in mixtures such as pudding powder, custard powder, etc., usually results in unpalatable cooked articles. If it is used for bread making, the bread has a sour taste, poor odour and may not rise adequately.

(ii) Dermatitis

Various acarid (and other) mites have caused dermatitis in people handling infested products. Probably almost any species could cause skin reactions in sensitive people coming into contact with large numbers. The following have actually been recorded as causing this trouble: *Tyrophagus casei*, *T. longior*, *T. putrescentiae*, *Caloglyphus krameri*, *Suidasia nesbitti*, *Glyciphagus domesticus* and *Carpoglyphus lactis* (see page 304).

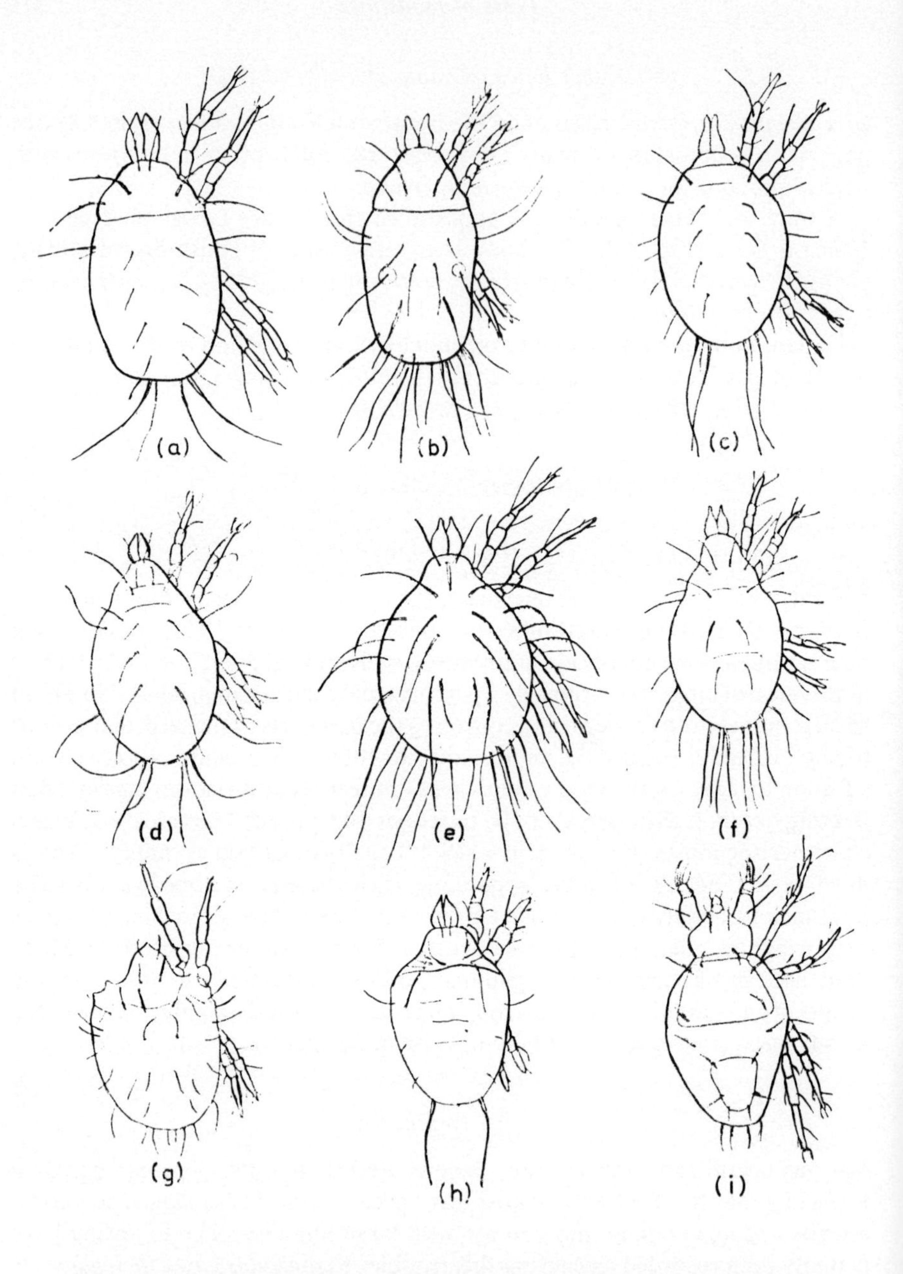

Figure 8.8 Mites occurring in stored food. (*a*) *Acarus siro* (male); (*b*) *Tyrophagus casei* (male); (*c*) *Carpoglyphus lactis* (female); (*d*) *Caloglyphus berlesei* (male); (*e*) *Glyciphagus domesticus* (female); (*f*) *Rhizoglyphus callae* (male) (rather similar to *R. echinopus*); (*g*) *Gohieria fusca* (female); (*h*) *Suidasia nesbitti* (male); (*i*) *Cheyletus eruditus* (female). Magnifications: (*f*) × 40; (*a*), (*d*), (*e*), (*g*) × 55; (*b*) × 80; (*c*), (*h*), (*i*) × 90. (After Hughes [292].)

(d) Control of mites in foodstuffs

(i) Prevention of infestation [424]

Physical measures
The simplest method of mite prevention is to keep cereal products sufficiently dry, so that their water content remains below 12–13 per cent. Stores and larders should be reasonably well ventilated and care taken that food in bags is not allowed to come in contact with condensed moisture on stone or concrete walls or floors.

Packages of foodstuffs which are free of mites can be protected from them by efficient sealing, with wax-paper, foil or plastic sheeting. The sealing must be complete, however, because packaging which allows the mites entry may actually encourage them by retaining moisture. Cheeses may be protected by careful dipping in wax to provide a complete coating.

Chemical measures
Contact acaricides can be used to treat the walls of empty stores, in the same way as insecticides are used to eliminate residual insect populations (see page 364). Also, certain low toxicity compounds can be mixed with grain as a prophylactic measure. Not all those recommended for use against insects are effective against mites; thus, malathion and fenitrothion have only limited value, as they will not control all the species commonly found. The most effective compounds for use against mites in food stores are pirimiphos-methyl and bioresmethrin; they can be thoroughly mixed with grain entering a store, as described for insect pests (page 365).

(ii) Control of infestations

Physical measures
Stored grain and oilseed can be disinfested by re-drying to a safe moisture content. If the infestation has been severe, however, the taint will not be removed and the grain may be unfit for consumption.

Vacuum cleaning or brushing the surface of the infested goods may give some temporary relief, especially with cheeses.

Chemical measures
If the infestation is confined to the surface of a heap or bin of grain, acaricidal dusts (pirimiphos-methyl or bioresmethrin) can be applied to the surface, at the rate of 140 g m^{-2} and mixed into the top 0·3 m of the bulk.

Radical treatment of infestations requires fumigation, but since mites are rather tolerant (especially in the egg stage) this should only be done with serious and urgent cases. For large stores, methyl bromide or phosphine treatments by

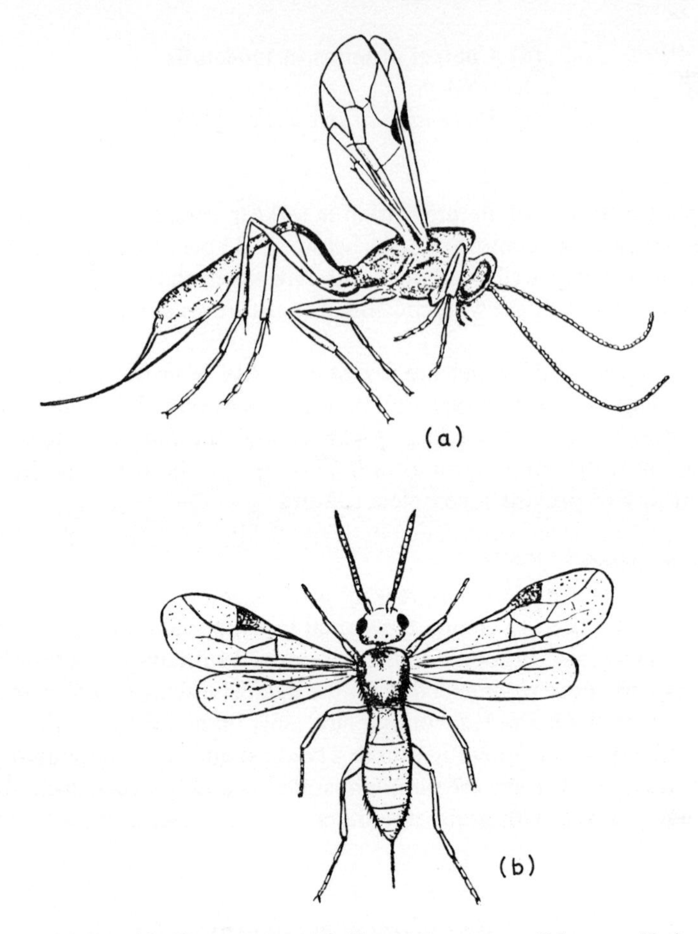

Figure 8.9 Parasitic Hymenoptera. (*a*) *Nemeritis canescens* (Ichneumonidae); (*b*) *Microbracon hebetor* (Braconidae). (*a*) after Freeman and Turtle [193]; (*b*) after Richards and Herford [501a]. (*a*) × 7½; (*b*) × 15.

servicing companies may be needed; and doses or exposures 50 per cent greater than those for insects will be needed.

Liquid fumigants (ethylene dichloride plus carbon tetrachloride) may be used in smaller stores, such as farm barns, as described earlier (page 363).

8.2.7 BENEFICIAL INSECTS AND MITES

The following organisms, which may be found in or near stored food products, may be described as beneficial since they destroy a certain proportion of pests, which they attack as parasites or predators. Generally, however, they are not very important since they are only able to flourish when the pest species is thriving.

(a) Insects

(i) Hemiptera

Two species of bug may be encountered in food stores, especially large ones such as warehouses and granaries. In their young stages, they resemble the nymphs of bed bugs and they therefore sometimes give occasions for surprise and concern. However, they are quite harmless to man and even beneficial, since they live by sucking the juices of the beetles and moths or sometimes of mites. These two species may be distinguished as follows:

Second segment of the antennae considerably longer than the third. About 3·5–4 mm long in the adult. Back of thorax dark brown to nearly black. Forewings and legs yellowish brown. *Lyctocoris campestris*
Second segment of the antenna about the same length as the third. Adult about 2 mm long, more reddish than the above and with the wings stunted. *Xylocoris flavipes*

(ii) Diptera

Scenopinus fenestralis (the 'window fly')
The larvae of this small fly prey upon the larvae of various other insects. For example, it may occasionally be found among grain, destroying moth caterpillars and beetle grubs: and it has also been said to attack larvae of carpet beetles and clothes moths.

The larva of *Scenopinus fenestralis* is fairly easily recognized. It has a long, narrow, segmented body, yellowish white in colour, with a distinct head and a few, rather prominent bristles. The abdominal segments are sub-divided, so that the insect appears to have 17 abdominal and 3 thoracic segments. The adult fly is a small, rather hump-backed fly, with greyish-yellow bristles on the thorax and a dark flattened abdomen.

(iii) Hymenoptera

Ichneumonidae (Ichneumon flies); Braconidae (Braconid wasps) (Fig. 8.9)
A large number of rather specialized members of the order Hymenoptera have larvae which develop parasitically inside other insects. Those species which attack pests of stored products are small or minute black insects which look rather like tiny flies but are easily distinguished by possessing two pairs of wings. The wings, especially the hind pair, are rather sparsely veined and there is often a dark spot about the middle of the anterior edge of the front pair. The Ichneumonidae are characterized by very narrow, extended 'wasp-waists'.

Although these parasitic insects are beneficial in so far as they attack pest insects, they do not occur in large numbers unless the host population is already very high. Therefore, their presence should be taken as an indication that control measures against the pest are long overdue.

Some typical parasitic Hymenoptera are shown in Fig. 8.9. The life histories of these parasites are similar in general form. The adult female injects an egg into the body of the host caterpillar or grub with her long ovipositor. The egg hatches inside and the parasite larva develops as a white, legless maggot. This grows gradually as it consumes the body of the host which eventually dies. The parasite then emerges and spins a small, yellowish, oval cocoon and pupates inside. Eventually the adult emerges, the sexes pair and the females seek fresh prey for their offspring.

(b) Mites

(i) Pyemotidae

Pyemotes ventricosus (the grain itch mite)
This mite is curious in several respects. It is theoretically, beneficial in that it is parasitic on a variety of pest insects in their larval stage. (Biologists have suffered when laboratory cultures of certain pests have been destroyed by the mite!) But the mite itself can be a nuisance to people handling infested foodstuff for long periods, since it causes an irritating, though transitory, dermatitis (see page 304).

The life history is unusual. The young fertilized females, which are normal-looking mites, seek out suitable hosts, insert their mouth-parts and become permanently attached. The hind part of the body becomes gradually swollen into a relatively large white globe about 1 mm diameter. The eggs hatch and the entire development of the young takes place inside the body of the mother. After about 10–11 days (at 25° C) the progeny begin to emerge and they may continue to be produced over a period of 9–33 days. As many as 36 births have been observed in a single day, the largest total number being 242.

Only about 3–4 per cent of the offspring of fertilized females are males. These remain for 20–30 days on the body of the mother, apparently living parasitically on her, for if they are removed they die in 20 h. The males remain close to the genital aperture of the mother and assist in the emergence of their sisters with whom they immediately mate. Each male mates with about 30 females altogether. The fertilized females scatter to seek new hosts to parasitize and the cycle begins again. If they are unsuccessful, they die after about 48 h.

Females which are not fertilized sometimes produce young parthenogenetically, in which case the progeny are all males.

(ii) Cheyletidae

Cheyletus eruditus (Fig. 8.8)
This mite occurs fairly commonly among grain and grain products where it lives

by feeding on other mites, chiefly *Acarus siro*. Under conditions which are unfavourable to the latter, the predator may completely suppress it.

The life history is normal except that reproduction is parthenogenetic; there are one larval and two nymphal stages before the adult. The younger stages may feed on the eggs rather than the active stages of other mites. The mites seize their prey with their powerful palps, insert their beak-like mouthparts and suck it dry. The whole process takes about $\frac{1}{4}$ h.

There is an interesting ecological balance between *Cheyletus*, its chief victim *Acarus siro* and the environment. *Acarus* is very dependent on water content of the food, and its numbers rise enormously as this increases. *Cheyletus* populations increase too with the rise in numbers of their prey, but less rapidly. On the other hand *Cheyletus* can occur at lower water contents than *Acarus* and under these conditions it may completely suppress the grain mite. When the latter is absent, *Cheyletus* feeds on other mites or, for a short time, may continue existence by cannibalism.

As already mentioned, *Cheyletus* tends to be dominant in the summer and *Acarus* in the moister, cooler conditions of winter.

CHAPTER NINE

Insect pests in waste products

The waste products of a modern community are eliminated in two ways: as relatively dry refuse and as water-borne sewage. It is perhaps worth noting that, in most cases, valuable minerals and organic matter, which could form useful fertilizer, are wastefully discarded. This matter will be briefly discussed later, under the headings of reclamation and composting. More relevant to this book is the fact that in most disposal systems opportunities arise for excessive breeding of various flies and of crickets, which may cause serious annoyance. These troubles will be considered in this chapter. Other nuisances are also possible, from the encouragement of rodents, pollution of water from effluent, foul odours and smoke from open burning.

9.1 REFUSE [184]

9.1.1 NORMAL DOMESTIC REFUSE

(a) Nature and amount

The average composition of domestic refuse from urban districts in Britain has been analysed in various ways at different times. The following are typical figures obtained in summer months [428]. *Non-combustible matter*: fine dust 16–20 per cent; metal, 7 per cent; glass, 6–9 per cent. *Combustible matter*: small cinders, 9–10 per cent; large cinders, 6–9 per cent; paper, 18–21 per cent; rags, 1–2 per cent. *Vegetable and other putrescible matter*: 25–26 per cent. *Miscellaneous*: 2–4 per cent. A comparison of these data with earlier assays suggest that the proportion of fine dust has declined (possibly owing to the decline in coal fires) while proportions of metal and glass have risen.

The overall figures give an average of about 0·8 kg per person per day; but the figures for winter months may be about 25 per cent above and in the summer 25 per cent below the mean. The composition as well as the total amount varies in different seasons, for the simple reason that in winter there are more domestic fires burning, with consequent increases in ashes and cinders. In summer, on the other hand, there are rather more fruits and vegetables available. Thus, in December to February cinders may constitute 32 per cent and putrescible matter only 9 per cent of the total; whereas in June–August the cinders may amount to only 15 per cent and the putrescible matter rises to 24 per cent.

Data from other industrialized countries are not greatly different from the foregoing. A 1956 W.H.O. report gives an estimate of 0·3–2 kg (0·7–4·5 lb) *per capita*, per day of mixed refuse [212]. This will probably contain 0·09–0·4 kg (0·2–0·9 lb) of putrescible garbage, 0·1–1·3 kg (0·25–3 lb) combustible rubbish and 0·1–0·3 kg (0·3–0·8 lb) non-combustible materials (excluding ash and cinders). A more recent American survey indicates a continual growth in *per capita* refuse production; from about 2·7 lb in 1920 to about 4·4 lb in 1964 [52].

(b) Fly breeding in cities and city dumps

(i) Inside cities

There have been a few surveys of the actual sources of flies and blowflies in towns and cities. In Savannah, Georgia, a city of about 130 000 inhabitants, fly breeding was found in about 60 per cent of refuse bins [492] and a somewhat similar figure reported from a London district [221].

The necessity of covering refuse bins is emphasized by the finding in Concord, California, that an average of 1128 maggots would migrate from uncovered bins to pupate, before the refuse was collected each week [101]. Moreover, the cover must be adequate, since it has been shown that a *Lucilia* adult can enter an opening as small as 0·125 in. (3·2 mm) diameter [540].

Investigations in some smaller American towns of about 8000 population, showed that, while blowflies predominated in garbage bins, houseflies were more prevalent in scattered refuse [523]. In all the American surveys, dog faeces were found to be a considerable source, mainly of *Sarcophaga* spp. Surprisingly, the average number of flies developing from a single specimen was as high as 144. Houseflies and also lesser houseflies will also breed readily in chicken dung.

(ii) In refuse dumps

It has been estimated that, even with reasonable sanitary standards, refuse collections from a British district of 10 000 houses will deposit more than 20 000 maggots on a refuse tip each week, during the summer months. In warmer climates, the number will certainly be larger. The majority of the larvae will be of blowflies; but there will be a proportion of housefly maggots. Furthermore, houseflies have the habit of following refuse vehicles to the tip.

On emergence, adult blowflies do not generally stay at the dump, but migrate and infest the neighbourhood for several miles around. In the Savannah investigation, large numbers of a mutant strain of *Cochliomyia macellaria* were released at the refuse tip and were recovered at distances up to 1·5 miles within 7 h [492].

Houseflies are somewhat less liable to migrate readily from a refuse dump; and if suitable material is exposed, they will breed in it. Because of the warmth of fermentation, fly breeding on refuse tips can continue, even in winter.

Another nuisance insect which can breed continuously in refuse is the common cricket, the biology and control of which is dealt with on pages 496–501. These too can overwinter in the tip, but they often migrate into adjoining houses in the warm days of late summer.

(c) Hygienic collection and disposal of refuse

(i) Collection

The dustbin area should be clean and tidy. Where a number of dustbins are collected (e.g. behind restaurants or hotels) it is very desirable that they should stand on a smooth concrete surface, sloping to a drain, so that the area can be hosed down regularly. The dustbins themselves should be in good condition with well-fitting lids, which are kept in place. Too often one sees a collection of battered bins and sometimes odd drums and wooden boxes used to store refuse. Old bins sometimes accumulate a layer of filth in the bottom which clings to the bin when the refuse is collected. This can be avoided by wrapping wet refuse in paper, or at least lining the bottom of the bin with paper.

To supplement the hygienic measures, especially where many bins or especially attractive refuse is present (e.g. from fishmongers, butchers, etc) the outsides of the bins and the adjoining walls should be sprayed weekly with a suitable insecticide.

Collection of refuse by the local authority should be made at least weekly and preferably twice weekly in the summer.

One of the most promising new developments, which unfortunately has not yet been widely adopted, is the use of paper sacks to replace dustbins. Their somewhat greater cost is offset by reduction in labour of collection and they present less opportunity for insect breeding, since they are renewed at each collection.

(ii) Disposal

The method of disposal of domestic refuse can be chosen by the local authorities responsible for its collection, in accordance with the facilities available and its own financial policy. Some boroughs endeavour to recover part or all of the costs of refuse elimination by various utilization schemes (see page 382).

Utilization schemes, however, are not very common, either because the financial benefit is considered uncertain or because the smaller boroughs do not have the special plant necessary. Therefore a very common method of refuse disposal is by simple dumping. Apart from disposal at sea, this involves forming a large accumulation of refuse in some isolated waste ground. (Choice of a suitable site is difficult in many areas.) Wherever possible, it is desirable to utilize the refuse for land reclamation. Marshy lands, sandy wastes, ravines, disused quarries and pits may be eventually improved by dumping. Furthermore, in most of these cases, the refuse will be used for filling in cavities and depressions instead of forming unsightly mounds. (Two large dumps at South Hornchurch, formed of refuse from the city of London, were estimated in 1928 to be $\frac{3}{4}$ mile long, nearly $\frac{1}{3}$ mile wide and up to 90 ft high.) Pits and depressions containing standing water, however, cannot always be used; and in any case they require special consideration, because of the probable pollution of the water by percolate from the refuse [428].

Refuse dumps can be highly unpleasant, both in themselves and in their effects on their surroundings. The organic materials present may provide food for multitudes of insects and for rats; fires may start and spread smoke and an unpleasant effluvia; light rubbish (paper and rags) tends to get blown about. Most of these defects are due to a large exposure of recently tipped refuse, and they can be substantially reduced by proper systematic tipping.

(iii) Controlled tipping [306]

The principle of this method is to keep the tipping face small, so that fresh refuse is rapidly covered up by further loads. The dump is therefore formed in long strips, which fill up the site gradually. These strips should be about 6 ft deep and of a width depending on the number of vehicles constantly discharging. When the floor of a large depression (e.g. an old quarry) has been filled in, a further layer may be commenced, provided that sufficient time has been allowed for subsidence of the first layer of strips. As the face of the tip advances, the top and the exposed side (or sides) should be covered with earth, flue dust or fine ashes, if possible, to a depth of 9 in. For large tips, the use of bulldozers or earth-lifting machines will be found exceedingly helpful [184]. The track taken by the tipping vehicles must, of course, move forward with the tipping face. Often it is necessary to lay sleepers on the top of soft rubbish, to carry the vehicles. The frequent passage of loaded vehicles over the dump has a beneficial effect in consolidating the tipped refuse. Where possible, it is desirable to flatten tins and drums and to fill tin baths, etc., with small refuse. Otherwise these hollow articles will collapse after some months and cause unequal subsidence in the completed dump. There will, in any case, be some general subsidence owing to the decomposition of organic matter in the refuse.

Under the conditions described, the heat generated by decomposition of organic materials is retained and the whole mass of refuse becomes hot for several weeks after tipping. The degree and persistence of heating depends partly on the nature of the refuse and partly on the way it is tipped. Where the rubbish is spread out (indiscriminate tipping) heat is soon lost; when controlled tipping is practised, the large volume of the refuse with the outside covered or, at least, well compacted, correspondingly retains its heat.

Liability to insect infestation is greatly reduced by the practice of controlled tipping for the following reasons:

(1) The use of earth or other material to cover the refuse and the reduction of the surface area of attractive fresh refuse, prevents extensive oviposition of flies after tipping and therefore reduces additional breeding in it. For this reason, the

cover must be applied as soon as possible and certainly within 24 h of tipping. The cover also prevents access of crickets to the food and, where they are present, will imprison and destroy many of them. To be effective, the covering material must be finely grained; coarse cinders and clinker are useless.

(2) The retention of refuse in a large mass and the compaction of the outside, conserves the heat of fermentation so that the bulk of the dump is too hot for insect breeding. Some of those present may actually be destroyed by the high temperature.

(3) A cover of earth prevents the emergence of a proportion of fly adults, even if the maggots were present before tipping; but to retain a high percentage, a layer of at least 23 cm (9 in.) deep is required. This is seldom practicable.

(iv) Use of insecticides

Where an insect infestation develops to large proportions in spite of good tipping practice, recourse must be made to insecticides. A simple measure is to apply powder insecticide to the surface of the tip, at the rate of about 1·25 kg per 100 m^2 (1 cwt $acre^{-1}$). Formerly, 0·5–1 per cent *gamma* HCH was found effective; but in many cases, resistance to organochlorines has developed. Malathion dust is an alternative, but this may also be vitiated by resistance. Other organophosphorus compounds might be tried; or alternatively, synthetic pyrethroids.

The object of these treatments is to kill flies emerging or attempting to oviposit; and applications may have to be made weekly.

(v) Reclamation and composting

It has been pointed out that 'Unless an objectionable dump is nearby, the average citizen's interest in waste disposal is limited to having his refuse collected regularly. Lack of public concern is a real handicap to local officials responsible for collecting the funds necessary to operate adequate refuse collection and disposal systems and often prevents the planning and construction of needed facilities' [52]. Furthermore, the trends of modern capitalist civilization have encouraged obsolescence and replacement, rather than saving and mending. With rising labour costs, it is uneconomical to clean and re-use paper, bottles or metal and the volume of these discarded materials steadily grows. Perhaps more important than the discarding of glass and metal is the loss of nitrogen and useful minerals which can be used as fertilizers and the reckless sluicing away of phosphorus in sewage.

Although it must be recognized that reclamation and composting schemes are seldom economic in the short term, the germinating philosophy of reclamation and modern respect for ecology should encourage a careful consideration of their possibilities. A considerable number of such operations are already working in Europe and North America. The stages in the process can only be briefly outlined here. First, refuse must be sorted; and while ferrous metals can be extracted magnetically, large bottles and other objects may need manual removal. Alternatively, if recovery is to be disregarded, the organic matter for composting can be separated by grinding and sieving and the remainder burnt or buried. Details of composting, which may involve admixture with sewage sludge, are given in various publications [212, 488]. It is sufficient to note here that, properly carried out, composting avoids proliferation of insect pests and produces usable fertilizer.

9.1.2 REFUSE IN RECREATIONAL AREAS

(a) Introduction

The great expansion in motor car ownership, combined perhaps with the high cost of hotel accomodation, has led to very large numbers of people taking holidays camping or in caravans. Most of them visit well-known beauty spots and a further tendency to aggregation is caused by the restricted number of permitted sites and the facilities at permitted camping sites. Sanitary and efficient disposal of refuse, particularly garbage, presents special problems, since the limited period of usage may not justify the expensive equipment necessary. It has been estimated that the combined refuse from campers averages about 630 g or 0·5 m^3 (1·4 lb, 0·18 ft^3) *per capita* per day. With moderate or large camps, this would clearly result in nuisance from rodents and insects, unless dealt with properly.

(b) Field investigations

Some experiments in sanitary disposal in the U.S.A. have produced some interesting results. Each camp site was provided with (1) a refuse bin, (2) a garbage pit, and (3) pit privies [381].

Refuse bins consisted of metal drums (55 U.S. gall.: 200 litre.) with self-closing lids. They were used for dry refuse, such as paper cartons and tins and were emptied 3 times weekly. The contents were taken to a central disposal site where a combination of burning and sanitary land-fill was employed.

Wet garbage pits for food refuse were made by digging cylindrical holes 1·7 m ($5\frac{1}{2}$ ft) deep and lining them with sections of concrete pipe or a pair of metal drums welded together and open at both ends. The liners projected 15 cm (6 in.) above the ground and soil was packed round, sloping away to prevent water soaking into the pits. The pits were covered with a concrete slab, with 20 cm × 20 cm (8 in. × 8 in.) openings, fitted with self-closing lids.

Pit latrines were constructed in the same way as the garbage pits, with wooden risers fitted with toilet seats and lids. Each was enclosed in a wooden hut, with small screened windows and a spring-closing door.

(c) Fly production

Breeding in the garbage pits and latrines was assessed by trapping emerging flies throughout the summer months. The three-week running averages for 10 garbage pits ranged from 1404 to 3155. A total of 20 804 flies were collected, of which 33 per cent were Psychodidae, 48 per cent *Lucilia* spp., 6 per cent *Callitroga macellaria*, 3 per cent *Phormia regina* and 2 per cent *Fannia* spp.

The collections from 5 privies ranged from 38 to 273 flies, which comprised 54 per cent *Fannia* spp., 19 per cent *Ophyra leucostoma*, 14 per cent Psychodidae and 5 per cent *Hermetia illucens*.

(d) Control experiments

Two types of control were tried in a group of wet garbage pits and in 5 privies: spray treatments and dichlorvos strips. Monthly spraying with 1–2 per cent diazinon or dimethoate, used at 1 litre to treat walls and contents, failed to reduce fly breeding appreciably. Dichlorvos strips (0·5 cm × 6 cm × 25 cm), containing 20 per cent toxicant, were used at 1 or half a strip per pit or privy. These gave 95 per cent control or better for 8 weeks in the pits and about 90 per cent for a further 4 weeks. In the privies, one unit each gave over 70 per cent reduction for 4 weeks of *Fannia* and *Ophyra*, but were ineffective against the other flies. The poorer results may have been due to greater ventilation.

9.1.3 REFUSE FROM SLAUGHTER HOUSES [219]

There are some 500 licensed slaughter houses in England and Wales, ranging from very large municipal types to small affairs, little more than an outhouse to a butcher's shop. Almost all suffer from blowfly infestation to a greater or lesser degree. The flies concerned are *Calliphora* (mainly *C. vicina* with some *C. vomitoria*), *Lucilia* (about 95 per cent *L. sericata*, with some *L. caesar*, etc) and *Phormia terrae-novae*. Their biology is described in Chapter 8 (pages 219–224).

In basic design, the slaughter house comprises (1) the *lairage*, where animals awaiting slaughter are kept, (2) the *slaughter room*, where the killing and separation of edible and inedible materials occurs and (3) the *hanging room*, where meat and offal are hung to cool and inspection and grading takes place.

(a) Nature of the refuse

The refuse from slaughter houses comprises the following:

(i) Condemned meat and offal

This is usally stored in an outbuilding but often heaped in an open yeard. The period of storage ranges from a day at most large slaughter houses to 3–4 days at small rural establishments.

(ii) Inedible gut

This is usually kept in open metal drums and the period of storage varies as much as for the condemned meat and offal.

(iii) Blood

As much blood as can be saved is stored in tanks or metal drums awaiting collection.

(iv) Refuse

This consists of dung from the gut of slaughtered beasts, sweepings from the yards and floor of the slaughtering bays, waste pig hair, wool and hide and sometimes congealed blood from traps in the drainage system. This mixture is usually tipped into a pit or heaped in an open yard where it may remain any period from 2 h to 8 days. It forms an ideal breeding material for blowflies and because of the heat of putrefaction, promotes their rapid development.

(b) Hygienic control measures

A great deal can be achieved in preventing blowfly nuisance by a high standard of cleanliness. This is much easier to achieve at modern, well-designed slaughter houses than at old-fashioned inefficient premises, where the staff tend to become frustrated and eventually resigned to hordes of blowflies. The following measures are recommended.

(i) The midden

Blowfly eggs can develop to migrating larvae within 3 days so that refuse should be removed within this period, and preferably daily. To facilitate this, in a large establishment, rotational storage and collection should be employed. The midden yard is divided by brick walls into bays, which are filled and emptied in turn, so that no refuse remains in place more than 2 days. As each bay is emptied, it is thoroughly cleared and hosed down.

This arrangement will prevent a local breeding nucleus in the dangerous neighbourhood of the slaughter house. Furthermore, many of the larvae in the refuse will be killed by heat due to storage in depth during and after transportation to the disposal site.

(ii) Blood, inedible meat and offal

These should be stored in outbuildings. The blood and offal should be kept in heavy, galvanized-iron bins with strong lids, not in the open oil-drums often used. All bins should be emptied every second day at the maximum and hosed out before re-use.

(iii) Destructible waste

Destructible waste should be burnt, daily if possible. This includes discarded aprons and other clothing, sacking and rags used for swabbing, blood-soaked sawdust, etc.

(c) Chemical control measures

The hygienic measures described are aimed mainly at larvae; chemical control is directed against adults, to prevent them ovipositing. Experiements have shown that the simplest and most effective insecticides are dusts containing *gamma* HCH or a synthetic pyrethroid. The dust should be liberally applied daily to the surface of all exposed refuse, to adjacent walls and vegetation. A convenient time for dusting is late afternoon, when all refuse has been deposited for the day. Every 6 weeks, a dusting of vegetation and walls within about 50 yd of the midden is also advisable. This practice has been found to give good control, when combined with hygiene, and no resistance was developed in 3–4 years of regular use.

9.2 SEWAGE

9.2.1 COLLECTION AND TREATMENT OF SEWAGE

In nearly all towns and cities in Britain, water-borne waste is collected by a sewage system. This removes water closet excreta and dirty water from dwelling houses and rainwater from gutters and road drains. (The rainwater is often conducted along separate sewers.) In addition, some manufacturing effluents can be discharged into the drains. Sewage is collected in a network of conduits, so constructed that there is normally a gentle flow which changes to a rapid, self-scouring stream at the daily periods of peak load. In this way, the effluent is carried to the disposal works.

Sewage purification consists, essentially, in the removal of the grosser solids, the coagulation and precipitation of finer suspended solids and the oxidation of most of the remaining impurities. These objects are attained in stages, as follows:

(1) Coarse screening (to remove debris).
(2) Partial sedimentation (to remove grit and sand).

(3) Further sedimentation in temporary storage tanks (to allow settlement of 'sludge').
(4) Oxidation (i.e. largely, aerobic bacterial decomposition) by (*a*) irrigation of land (a slow and rather obsolete process), or (*b*) aeration, by bubbling air through 'activated sludge' tanks, or (*c*) sprinkling the effluent over percolating filters (see below).
(5) Final sedimentation.
(6) Discharge into a river.

9.2.2 THE FAUNA OF PERCOLATING FILTERS [582]

Only one of the above processes is important in regard to insect nuisances; that is the percolating filter. This type of filter is efficient and simple and it is very widely used. It consists of a bed of a coarse filtering medium (pebbles or clinker) either circular or rectangular in form and 1–3 m (3–10 ft) deep. The sewage is sprinkled over this from either stationary or, better, from travelling jets. Over the pieces of broken solids, which constitute the filter, there grows a thick film of fungi and other organisms. In the upper layers, there is a leathery growth of the fungus *Phormidium* with a slight admixture of *Ulothrix* and other algae. In the depths, there is a slimy zoogloea of bacteria, fungi, protozoa and other organisms. This 'biological film' is important in the purification of the effluent trickling over it. Living in the film, and feeding on it, are a number of small animals, chiefly insect larvae and worms. The number of types present is not very large and a few species predominate. The characteristic fauna includes the following groups [356]: (1) Oligochaete worms, especially *Lumbricillus lineatus*. (2) Springtails (Collembola) especially *Hypogastrura viatica* and *Tomocerus minor*. (3) Small flies, belonging to the families Psychodidae, Chironomidae and Anisopodidae. (4) Spiders of the family Linyphiidae.

Some of these organisms are beneficial, in that they prevent the fungal mat from growing so dense as to clog the filter. This may happen in the winter, when the rate of reproduction of insects is low, so that there is a danger of 'ponding'. In the spring, the activity of the scouring organisms accelerates and much of the old film is destroyed and broken away; this is known as 'off-loading'. Subsequently, however, a new and more vigorous film regenerates.

During the summer, in certain seasons, there may be enormous proliferation of some of the flies, resulting in swarms which constitute a nuisance.

(a) Species liable to cause a nuisance

The flies most frequently reported as causing nuisance in the vicinity of sewage works are the following:

(1) The psychodids, especially *Psychoda alternate* and *P. severini*.
(2) *Anisopus fenestralis*.
(3) Various chironomids, such as *Spaniotoma, Metriocnemus* and *Chironomus*.

(The adults can be distinguished with the help of the key on page 528 and Fig. 6.2; the larvae are shown in Fig. 6.3.)

Most irritating are the minute psychodids which get into the eyes, mouth and nostrils of workmen in the sewage works and people in the vicinity. All the species tend to enter houses (see page 219). The chironomids may breed in large numbers in places other than sewage works.

(b) Biological data [356, 584]

(i) Psychoda spp. Moth flies

Life history

The eggs are laid singly or in groups on the biological film at various depths in the filter bed. These minute white, oval eggs are about 0·2 mm long. The larvae which hatch from them are slow-moving, legless grubs, with a distinct head bearing powerful mandibles. These grubs feed on the algae, fungi, bacteria and sludge which made up the film and break it down into small faecal pellets which are easily washed away. They breathe atmospheric oxygen through two spiracles on a 'siphon' at the end of the body; usually this siphon is the only part of the body to project from the biological film in which the larva lives.

The larva grows gradually from about 1 mm to 9 mm in length, moulting three times in the process. Then it turns into a pupa, with two respiratory horns at the anterior end. Finally the pupal skin splits and the adult fly emerges. The flies cluster together on the drier parts of the filter, on the filter wall, etc. Here mating takes place, except in one species (*P. severini*) in which males have never been discovered in Britain, and in which the females lay fertile eggs without mating. The females lay up to about 100 eggs each.

Speed of development

The incubation of the eggs takes 1–6 days and larval life 10–50 days according to the temperature. There are apparently about eight generations in the course of the year. There are usually several peaks of abundance of the adults corresponding to the emergence of various generations.

(ii) Anisopus fenestralis [253]. *The window gnat*

Life history

The female lays its eggs in a mucous ribbon which is traversed by a network of refringent strands. This ribbon forms a grey spherical mass, containing upwards of 150 eggs; and these masses may be found on the film or the wet filtering medium. After a few days the gelatinous material breaks down and the young larvae emerge. Like the larvae of *Psychoda*, they are legless grubs, but they are brownish in colour and considerably more active. They, too, feed on the biological film and remain partly immersed in it, with the posterior siphon projecting. The larvae grow to a length of about 20 mm and then pupate. The pupa is capable of limited movement which enables it to move to a relatively dry place suitable for emergence of the adult fly. The adults are relatively inactive and are often seen at rest upon the walls of the filter in which they breed; they may, however, be wind-borne for distances up to a mile from the filters.

Speed of development
At 20° C the life cycle is completed in 35 days (4 days in the egg; 20 days in the larval stage; 8 days as a pupa and 3 days pre-oviposition period). At 10·5° C the complete cycle takes 88 days, of which 50 are spent as larvae. At summer temperatures, the life of the adult fly is about 7 days.

(iii) Chironomids. Midges

Flies of the genera *Metriocnemus* and *Spaniotoma* breed in the biological film of the filters. *Spaniotoma minima* and the less common *S. perennis* are distributed throughout the filter bed; but *Metriocnemus longitarsus* and *M. hirticollis* larvae tend to climb up into the upper layers. *Metriocnemus* spp. are strongly predaceous, feeding on eggs and pupae of other species and, perhaps, their own. The eggs are laid in gelatinous masses and hatch to give active creamy-white worm-like larvae. These feed not only on the film but also on eggs, larvae and pupae of their own and other species. Both genera form cocoon-like cases in which to pupate. The adults emerge and form 'dancing' swarms in the air before mating. *Metriocnemus longitarsus*, however, can mate in confined spaces, unlike the other species.

The larvae of *Chironomus* spp. do not occur in the actual filters but often breed in drains and effluent channels. They contain haemoglobin dissolved in the blood and are commonly known as 'blood worms' from their red colour. The larvae make tubular shelters which adhere to the humus, sludge, etc, on which they feed. The adult fly emerges from the pupa at the surface of the water.

(c) Factors influencing the insect populations

Nuisances from insects in the filter beds develop when, owing to a combination of favourable climatic and other factors, vast numbers of flies develop over a relatively short period. The fluctuation of numbers are difficult to predict on account of the numerous factors involved and the complexity of their interactions, which have formed the basis of several ecological studies. The following are some of the critical factors.

(i) Temperature

Several of the filter breeding flies have rather low temperature thresholds, a few degrees above freezing, and can breed successfully at 5–10° C. Nevertheless, development and proliferation are slow during the winter and greatly accelerated by warm spells in summer.

In hot weather, on the other hand, drying of the upper layers of the beds may cause downward migration of the predaceous *M. longitarsus* larvae and death of pupae and eggs. This will reduce all fly population.

(ii) Loading of filters

By the 'loading' of a filter is meant the product of the concentration of organic impurities times the rate of application. The growth of the biological film in the surface layers of the filter and also the rate of proliferation of *Psychoda* increases with the load. On the other hand, most of the chironomids breed for preference in clean, lightly loaded filters. *Anisopus fenestralis* occurs in both heavily and lightly loaded filters.

(iii) Characteristics of filters

In general, fewer flies emerge from finer than from coarser grades of filtering medium. Also, in proportion to the amount of sewage treated, fewer flies emerge from alternating double filters than from continuous action single filters.

Table 9.1 Total development (in days) of different flies breeding in sewage filter beds.

Temperature		Species				
(° F)	(° C)	*Psychoda alternata*	*severini*	*Metriocnemus longitarsus*	*Spaniotoma minima*	*Anisopus fenestralis*
37	3	[a]	90	243	260	—
45	7	140	60	153[b]	103	—
50	10	80	36	94	80	88
59	15	35	26	49[c]	43	—
68	20	22	[a]	26	29	35
77	25	16		—	[a]	—

[a] Did not complete development; [b] 6° C; [c] 16° C.

(iv) Competition

Some of the interspecific competition is due to actual predation. Thus *Metriocnemus longitarsus* is a powerful predator in the upper layers of the filters. In filters where *Metriocnemus* and other chironomids are absent (perhaps because of long rests of the filters) *Psychoda* larvae tend to be more numerous.

Other competition is merely for the available food and may be accidental, as in the dislodging of biological film by activity of the worm *Lumbricillus*.

(v) Time of year

The prevalence of various species at different times of the year is dependent on all of the foregoing factors. The most serious nuisance is, perhaps, *Psychoda alternata*, outbreaks of which occur mainly from June to August. Trouble from *Anisopus fenestralis* or *Psychoda severini* usually occurs earlier, from April to July. The chironomids are most abundant, as a rule, in the autumn, from August to October.

(d) Control of filter fly outbreaks [583]

Unfortunately there is no simple way of preventing sewage fly outbreaks by suitable management of the filters. Usually it is impossible without a large expansion to alter the load in works liable to outbreaks. It is, of course, possible to alter the size of the filtering medium and use a small granular material. But during the winter, when the insects are not very active, the growth of fungus tends to clog fine filtering medium and lead to 'ponding'.

A simple control method can be employed using insecticides. After a considerable amount of experimentation, involving large-scale trials, it has been decided that the most satisfactory treatment is *gamma* HCH applied at the rate of 1 lb $acre^{-1}$ (2·75 kg ha^{-1}). The insecticide, in the form of a dispersible powder, is mixed with water and introduced into the sewage line running to the filter sprinklers. After application, the filter is rested for as long as possible (at least an hour or two). Weekly applications may be necessary during the fly outbreak seasons. Care must be taken that the effluent is not toxic to fish; but at this rate of dosage the risk is exceedingly small.

CHAPTER TEN

Clothes moths and carpet beetles

10.1 THE ORIGIN OF FABRIC PESTS

Several insect pests with biting mouthparts, which occur in houses, will occasionally bite holes in fabrics; they include cockroaches, crickets and bristletails. The reason for this destructive habit is unknown, for the fabrics attacked do not serve as nourishment. Accordingly, any type of fibre, whether natural or artificial, may be damaged.

This type of sporadic indiscriminate attack is quite distinct from the actual feeding on woollen cloth (or fur, or feathers) by the larvae of certain insects. This seems to have originated as follows.

Before the advent of man, certain small moths and beetles had found a useful role in the economy of nature by acting as scavengers. They utilized as food the more indigestible portions of animal cadavers; the fur and feathers which were neglected by other animals. Another natural breeding site was the neglected nests of birds or lairs of animals, where miscellaneous organic waste, such as fur and feathers, food debris and excrement, would be present. The insects become a nuisance to man when they turn their attention to hides, skins, furs and feathers used in clothes, furnishings and other human belongings.

The insects concerned are the clothes moths (Tineidae), house moths (Oecophoridae) and carpet beetles (Dermestidae). The species concerned possess the unusual ability of digesting keratin, the chief constituent of fur, wool and feathers [528, 597]. Keratin is a scleroprotein, consisting of helical polypeptide chains, linked together by cystine groups. Before such molecules can be attacked by proteinases, the —S—S— bonds of the cross-linkages must be broken and the chains set free. This is achieved in the gut of the clothes moth by highly alkaline conditions combined with an extremely low oxidation–reduction potential (maintained by a sparse tracheal system which keeps down the oxygen supply). Under these conditions the chain links are broken, thus:

$$X—S—S—X \rightarrow X—SH\ HS—X$$

(This digestive process forms the basis of a method of mothproofing, see page 408.)

While clothes moths, house moths and carpet beetles are exceptional in their ability to live on wool, they are by no means restricted to this diet. They have been recorded as breeding on a wide variety of dry materials of high protein content, both of animal origin (fish meal, albumen) or of vegetable nature (e.g. cereals). Furthermore, laboratory experiments have shown that they can thrive on many such substances better than on wool, fur or feathers. One must suppose that their association with animal fibres is the exploitation of an ecological niche where there is little competition and that this choice is maintained by the habits of the adults.

One more comment on this type of diet may be added. Pure keratin is very deficient in certain important amino acids. Furthermore, keratin fibres contain little or none of the vitamin B complex, which are necessary for insect development. In natural breeding sites (bird nests; cadavers) these may be supplemented by miscellaneous contamination. In contrast, the scoured wool or cleaned fur used by man, forms a very sparse diet. It is notable that the parts of garments most liable to successful attack by clothes moth larvae are places stained with sweat (or even more favourable, urine). These contaminants may supply some of the missing food requirements. In artificial (laboratory) colonies of clothes moths, addition of yeast to woollen cloth will very greatly favour development, for the same reason.

10.2 CLOTHES MOTHS AND HOUSE MOTHS

10.2.1 DISTINCTIVE FEATURES

The Lepidoptera which are liable to damage woollen materials in Britain belong to two families, the Tineidae and Oecophoridae. They are small moths with narrow wings bordered with long hair fringes. Their colouring is shining buff or mottled brown and cream. The genera and species concerned are:

Tineidae.	*Tineola bisselliella* (the common clothes moth)
	Tinea pellionella (the case-bearing clothes moth)
	Trichophaga tapetzella (the tapestry moth)
Oecophoridae.	*Hofmannophila pseudospretella* (the brown house moth)
	Endrosis sarcitrella (the white-shouldered house moth)

These may be distinguished by the keys in the Appendix (page 535).

10.2.2 LIFE HISTORIES

(a) The common clothes moth (Tineola bisselliella) [443, 581, 608]

(i) Oviposition

Eggs are laid by both mated and unmated females, but none of the unfertilized eggs hatch. The female may begin oviposition within a few minutes after copulation. She displays characteristic flickering movements of the ovipositor and occasional jerks of the abdomen, interspersed by a few steps forward now and then. The ovipositor is rubbed against the ground, possibly testing it for suitable places to lay the eggs. When fabric is present the moth will push the eggs in among the fibres. In an otherwise bare container, the eggs will be thrust up against small objects; but in the absence of these, they are scattered at random. There is no evidence to show that any insecticide will repel clothes moths to the extent of preventing them ovipositing on cloth.

(ii) Eggs

When they are laid, the eggs are damp and slightly sticky and will adhere to any surface, but they are not tightly fastened and are easily dislodged by shaking or brushing.

The egg is roughly oval (about 0·53 mm × 0·3 mm, weight 0·26 mg) and covered with a reticular pattern of ridges visible under good magnification.

Inside the egg, the larva develops as a U-shaped embryo. Hatching is achieved by biting an irregular hole in the shell, the whole process of escape taking 10–30 min.

(iii) Larva

The newly hatched larva measures about 1 mm long and 0·2 mm wide. Most of the body is creamy white, sparsely beset with bristles. The hardened head capsule is a light golden brown. The general form is shown in Fig. 10.1. In the course of development, the larva grows to a length of about 10 mm and the head becomes darker in colour; otherwise there are no great changes in appearance.

The larva bears short antennae on the head but no eyes (ocelli); nevertheless, they are able to distinguish the general direction of a source of light. They crawl away from the direction of illumination and, if placed on top of a piece of cloth, will tend to crawl underneath it.

The mouthparts are modified for biting the food and for spinning silk. The mandibles are strong, ridged, cutting organs. On the inside of the labrum and of the maxillae are spines to assist in grasping the food. When, as usually, the larva is feeding on fur or woollen fibres, its usual procedure is to bite off a fibre at a convenient point and reject the loose end. After a few bites at the stump, the larva often moves on to another fibre, thus leaving a trail of damaged hairs and destroying more than it consumes.

Apart from fur and wool, many substances can form larval diets, some of them being highly suitable; examples are meat- or fish-meal, egg albumen and various cereals.

Well-grown larvae may bite holes in many fabrics which they cannot digest and sometimes small quantities of inedible materials (cork, cotton wool, beeswax) may be actually swallowed. Nevertheless, the larvae are able to discriminate between animal and vegetable fibres; in a mixed textile only woollen fabrics are eaten and cotton is left.

Some kinds of animal fibre are too coarse to be attacked by young larvae whose mandibles only span about 90 μm. Pig bristles and horse hair are too wide to be bitten whereas wool fibres (10–100 μm) are readily attacked.

Digestion is evidently slow, for the food moves slowly through the intestine. (Starved larvae expel no faeces for 2 days after being re-fed.) This slow rate of digestion is probably due to the difficulty of coping with keratin, which is sometimes excreted partly unchanged.

The midgut is alkaline but the hindgut is acid owing to the presence of acid ammonium urate, which appears to be the main vehicle for elimination of nitrogen. Sulphur (from cystine) is excreted in the form of sulphates, and melanin passes through unchanged [528].

The larval silk glands are located in the head and thorax and the silk emerges from a tube or 'spinneret' beneath the head. The diameter of the silk varies with the size of the larva, being of the order of one-tenth to one-twentieth the diameter of white wool. The average amount of silk spun by 500 larvae in a week was found to average 0·1 mg each, which was 1·7 per cent of the mean body weight.

The larvae of *Tineola* may crawl about naked and unprotected, but very often they spin themselves frail tubes of silk, open at both ends, in which they can move freely back and forth. Or the silk may be spun as a flattened tent under which they can crawl about. Small pieces of fibre and other debris are usually woven into the webbing.

The larvae also make tough cylindrical cases in which they retire to moult or for resting periods. These cases are quite different from the straggling galleries or the smooth, portable cases made by *Tinea pellionella*. They have an untidy look owing to the incorporation of a great deal of miscellaneous debris (particles of food material, excrement and cast skins) which makes excellent camouflage. But inside they are smoothly lined with silk.

There are the usual three pairs of legs ending in claws on the thoracic segments and prolegs on four abdominal segments, each proleg bearing a circle of hooklets. On smooth surfaces the larvae are dependent on spinning a silken web on the surface to provide a grip for these claws and hooklets.

The minimum number of moults in the larval period is five; but various circumstances may increase the number and as many as 40 have been observed. Pupation occurs in one of the tough cocoon-like moulting cases and the last larval skin may be found at the bottom of the case.

(iv) Pupa

The pupae of *Tineola* vary very much in size, ranging from about 4–7 mm in length (3–10 mg weight). In form the pupa differs slightly from most Lepidoptera in that the appendages, though fused to the body, are free at the tips; and more than three of the abdominal segments are mobile. On the posterior dorsal margins of the abdominal segments IV to IX are comb-like rows of hooklets which assist the larva to emerge.

Before emergence, the adult in the pupal cuticle wriggles forward and, bursting its cocoon, projects out of it. Afterwards the pupal skin is left sticking out of the empty cocoon. The wings of the newly emerged adult take 10–30 min to expand and afterwards they are held vertically upwards for a further 5–20 min. Meanwhile, the moth excretes a few drops of accumulated faeces.

(v) *Adult*

The size of the adult, like that of the pupa, is very variable and depends on the amount of food taken in larval life. Normally the moths measure about 10–15 mm wing span and 6–8 mm long when at rest with the wings folded. The colour is a shining golden hue and distinguished from *Tinea* as described in the key (page 537).

The body is entirely and densely covered with scales of which there are about half a dozen different kinds. The size of the head is exaggerated by erect scales. The antennae are thread-like, of an indefinite number of joints. The mouthparts are greatly reduced and are functionless; tiny mandibles and maxillae are evident but there is no coiled sucking tube formed from the maxillae as in most Lepidoptera.

The thoracic region is covered with fine scales. The scales on the legs give them a characteristic appearance; at each joint there is a sleeve-like projection of scales which tends to become belled or ruffled by walking. The feet bear claws and there are also spines on the mid and hind legs, presumably to aid in walking.

The borders of the wings are beset with fringes of long hairs which increase their effective surface area. The ratio of the wing surface to fringe surface is about 2 or 3 to 1. The wings are coupled together by bristles (6–8 in the male, 2 in female) overlapping one strong dark seta.

The adult moths have three methods of progression: flight, running and jumping.

Flight

Only males and dwarf females (resulting from inadequate larval food) fly voluntarily. The speed of flight is about 50–70 cm s^{-1}. If normal females are launched into the air, they flutter down to the earth. Observations on a number of moths show that the larger ones have proportionally less wing area (doubling of the body weight only resulted in 30 per cent wing increase). Thus although a few females occasionally fly, it is the larvae and running females which are mainly responsible for spreading infestation. In one experiment, a piece of clean wool was suspended above a heavy moth infestation in a chest and it remained uninfested for a whole season.

Running

Moths very frequently make use of this means of locomotion, travelling at the rate of about 10–30 mm s^{-1}.

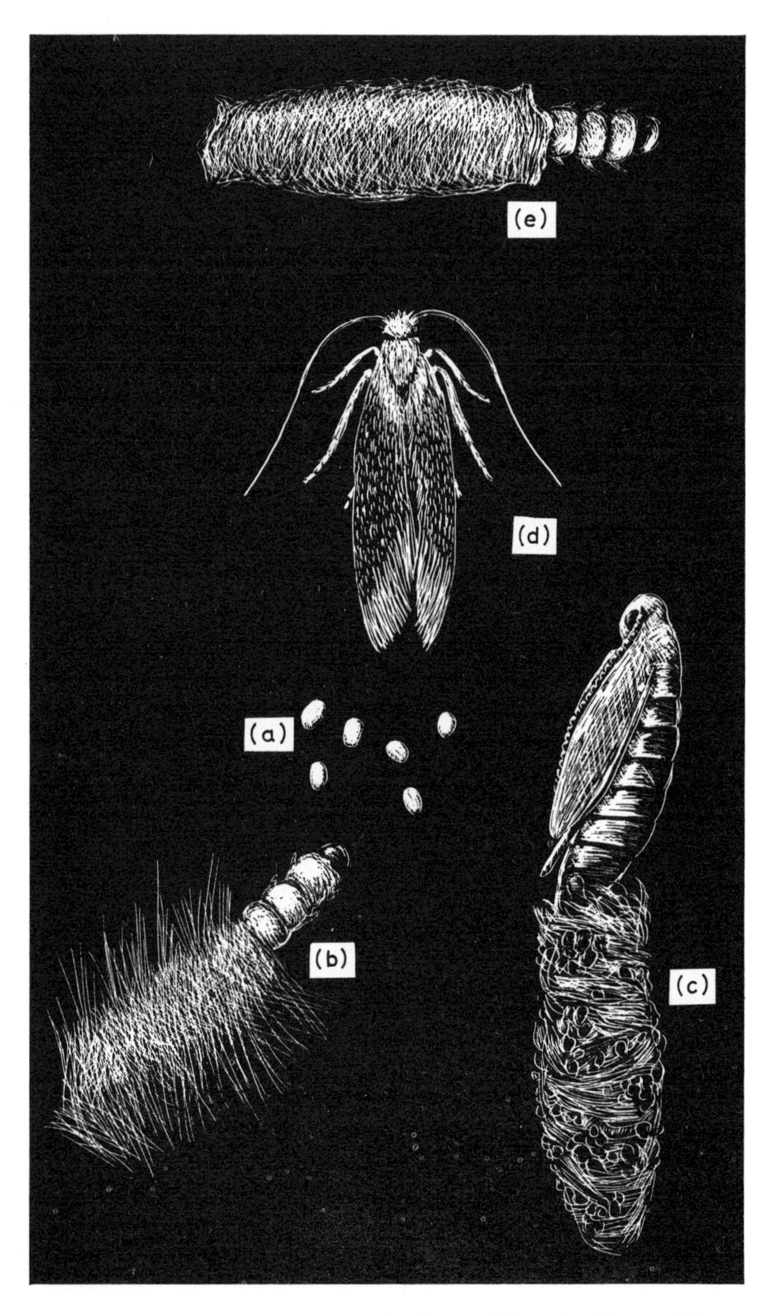

Figure 10.1 Life history of the clothes moths. (*a*) eggs; (*b*) larva in silk tunnel; (*c*) pupa wriggling out of the pupa case before emergence of (*d*) the adult. All of *Tineola bisselliella*; (*e*) larva of *Tinea pellionella* in its characteristic case. (Original.)

Jumping

This is characteristic of males and is rarely observed among females. The spring is given by the hind legs and the wings give one or two beats.

Decapitation does not prevent normal running, jumping or flying, but headless moths will not initiate these movements spontaneously.

Copulation

The males flutter round the females which remain largely passive, except for occasional protrusion and retraction of the ovipositor. The sexual excitement of the males is enhanced by scent which is given off from the abdomen of the female. A single female will elicit response at a distance of only 1·5 cm, but several females can be perceived by the males at rather longer distances.

Copulation is effected by the male twisting his abdomen round so as to make contact with the tip of that of the female. When union has been effected, the male usually moves round to the linear position (heads pointing in opposite directions). Copulation lasts about 15 min. Sometimes the pair are unable to separate and remain connected until they die.

(b) The case-bearing clothes moth (Tinea pellionella) [110]

This moth is less common than *T. bisselliella* and it has received less attention from biologists. In general, its life history is not very different.

(i) Oviposition

Females begin to lay eggs 1–6 days after emergence. The eggs are laid singly on, or inside, the folds of woollen fabrics, where these are available.

(ii) Eggs

The egg of *T. pellionella*, though about the same size as that of *Tineola bisselliella*, can be distinguished by its longitudinal ridges (visible on magnification); whereas that of *Tineola* has a reticular pattern.

(iii) Larvae

The newly hatched larva crawls about on the food material for 24 h; after this, it settles down to make the case, which gives it its popular name. The case (Fig. 10.1) is shaped somewhat like a pillowcase, open at both ends; and the larva can turn round and feed from either end. The case is constructed from pieces of fibre, bitten off by the larva, and bound together with silk. Sometimes the external appearance is rough, but it is smooth inside. The case is somewhat flattened and never fastened to the surface; from all these points it is easily distinguished from both the meandering galleries and the rough oval pupa cases made by *Tineola*.

The fore-part of the larva emerges from the case, which is carried about with the larva as it moves; the fore-part is stretched out and then the rear, covered by the case, is pulled up afterwards. In spite of this, the larva can move about as fast as the larva of *Tineola*, that is, about 2–3 cm min^{-1}.

Normally, larvae of *T. pellionella* never abandon their cases and, if the case is fastened down, the larva will struggle to pull it along for about $\frac{1}{2}$ h before abandoning it; but a new case can be spun overnight if necessary.

As the larva grows, it enlarges its case by inserting a V-shaped patch into a longitudinal incision. The progress of enlargement during larval life can be illustrated by placing them successively on fibres of contrasting colours which are then woven into succeeding areas of the case.

As with the common clothes moth, there are a minimum of five larval instars; but the number can be increased by unfavourable conditions.

(iv) Pupae

Before pupation, the larva wanders away from the food material and, finding a suitable niche, fastens its case down to the substrate, by silken threads. One end of the case is also closed and the larva rests inside and finally casts its skin and pupates. The pupae are pale yellow at first, darkening to brown; about 4–5 min long.

At the time of emergence, the pupa wriggles partly out of the case and, when the adult has emerged, the pupal shell remains protruding from the old case.

(v) Adult

The colour of the wings is darker than that of *Tineola* and they are marked by three rather indistinct dark spots on the forewings. In general the moth is not unlike *Tineola* in appearance and size, the distinguishing characteristics being given in the key (page 537).

The mouthparts of the adult are fairly primitive in that, like those of *Tineola*, they include vestigial mandibles. There is a trace of the coiled sucking tube (formed by processes of the maxillae) which is the normal feeding mechanism of most Lepidoptera; however, it is non-functional and the adult does not feed, but lives on reserves accumulated during larval life.

(c) The tapestry moth (Trichophaga tapetzella) (Fig. 14.6)

This is a larger moth than the two clothes moths. The larva is about 14 mm long when fully grown and the adult has a wing span of about 22–25 mm. The larvae form tunnels in the infested material lined with silk and incorporating many bitten off fibres. In making such galleries they damage a considerable amount of the infested fabric. They feed on such things as carpets, horse blankets, tapestries, felting, furs and skins and the larvae are able to attack rather coarser fibres than those damaged by the more common clothes moths. Upholstered furnishing may also be consumed.

The larva forms itself a very rough pupa case (like *Tineola*) and the adult has the same habit of partly wriggling out in its pupa case before emergence so that empty cases may be found with the pupal skins still protruding from them.

The adult holds its wings at a roof-like angle over the back like the other Tineids and in contrast to the Oecophorids which hold their wings folded flat over the back.

(d) The house moths (Hofmannophila pseudospretella and Endrosis sarcitrella) [118, 337, 624] (Fig. 14.6)

The house moths are probably species which originated as feeders on dry vegetable matter and have become adapted to dry animal remains, including hair or wool. As pests, they can infest and damage stored food (see page 360) or, in some cases, they will feed on wool clothing and carpets. *Endrosis* tends to be more of a food pest and *Hofmannophila* more of a scavenger and more likely to attack fabrics. Mixed infestations not infrequently occur. Their biology has been studied on cereals; but both species have been reared on wool dusted with yeast, so that they can probably digest keratin. They often occur in bird nests [626].

(i) Oviposition

The eggs of *Hofmannophila* are laid singly, quite freely, though with some preference for rough surfaces. *Endrosis* females lay reluctantly in captivity and prefer to insert the eggs into fine crevices.

(ii) Eggs

The eggs of *Hofmannophila* are oval, hard and shiny, not sticky when laid, about 0·5–0·6 mm long. The eggs of *Endrosis* are dull white, sticky and adhere together, about 0·55 mm long.

(iii) Larvae

Larvae may be distinguished by the key in the Appendix (page 543); in addition the following points may be noted [275].

The larvae of *Hofmannophila* are white and glossy and, when fully grown, reach a length above 16 mm (weight 50–108 mg, Fig. 8.7). Those of *Endrosis* are dull white and seldom exceed 14 mm (10–20 mg). Both species burrow in the food material, forming a silk tunnel. They emerge and wander away to pupate.

(iv) Pupae

Both species form rough cocoons incorporating debris. If these cocoons are torn open, both are found to be rather tough. But that of *Hofmannophila* is brittle and tears like brown paper exposing the pupa lying loosely inside; whereas that of *Endrosis* tears like cotton wool and the pupa is closely invested with the innermost layers.

(v) *Adults*

Adults of the two species are quite distinct (see Appendix key, page 537). In mixed cultures they have been seen copulating together but without producing progeny. Both forms (but especially *Hofmannophila*) are occasionally seen in houses, sheds and attics.

Table 10.1 Rate of development of clothes moths and house moths at different temperatures.

		Period in days				
Insect	Stage	13° C	15° C	20° C	25° C	29–30° C
Tineola	Egg	37	24	10	7	6
bisselliella	Larva	?	186–195	123–135	72–89	62–72
	Pupa	52	35	18	12	10
Tinea	Egg		—	4·5*	5	7
pellionella	Larva		—	46*	33	48
	Pupa	—	—	18·5*	10	10
Hofmannophila	Egg	42	25	14	9·8	10
pseudospretella	Larva	145	126	78	71	
	Pupa	56	48	25	15·5	13
Endrosis	Egg	23	15	8·6	6·3	6·8
sarcitrella	Larva	102	73	42	38	
	Pupa	31	25	15	10·4	7·4

* at 21·5° C.
Humidity: 70 per cent R.H. for *Tineola*, 90 per cent for other species. Foods: Wool impregnated with infusion of horse dung for *Tineola*; wool plus yeast for *Tinea*; wheat 'middlings' for other species. Data from Titschack [581], Cheema [110], and Woodroffe [624].

It should be noted that the diet for *Tineola* was rather unfavourable. On a good diet, larvae can develop in 45–50 days at 25° C.

10.2.3 QUANTITATIVE BIONOMICS [581, 624]

In assessing the effects of environmental factors on developing stages of clothes moths and house moths, a favourable influence will be indicated by short development and a high percentage emergence as adults. A favourable influence on the adults should promote long life and fertility.

(a) Temperature (Table 10.1)

For two of the developmental stages, the egg and the pupa, temperature is virtually the only important environmental factor.

Eggs of *Tineola*, *Hofmannophila* and *Endrosis*, hatched at the lowest temperatures tested (12·8° C, and 10° C). On the other hand, *Tinea* eggs failed to hatch at 14° C. At the upper limit *Tinea* hatched at 32·5° C and *Tineola* at 30° C, but 29° C was the limit for the house moths. Between the extremes, temperature had the usual accelerating effect, except near the upper limit, where a slight prolongation was evidence of a harmful influence.

The effects of temperature on larval and pupal development of these moths was largely similar to that on the egg stage, except that the upper limit for larval development was a little lower.

Larvae of *Hofmannophila*, however, tend to enter a diapause when fully grown, especially if they have been reared under warm (summer) conditions. This diapause prior to pupation adds a considerable, indefinite, period to the figure for larval development for this moth in Table 10.1. It appears that the favourable temperature range of the clothes moths is slightly higher than that of the house moths. Within the range of possible development, it appears that the slower development at cool temperature produces heavier adults.

Adult life is curtailed by high temperature in all species, thus (in days): *Tineola*, 10–13 at 20° C, 5–6 at 30° C; *Tinea*, 6·6 at 22° C, 5·2 at 26·5° C; *Hofmannophila*, 19 at 15° C, 12 at 25° C; *Endrosis*, 16 at 15° C, 5 at 25° C.

(b) Humidity [228]

The larval stages of the clothes moths, both *Tineola* and *Tinea*, can develop at humidities down to 20–30 per cent R.H. On the other hand, the rate of development is much prolonged at low humidities. The optimum for *Tineola* is about 75 per cent R.H. The two house moths will only complete larval development at humidities above 80 per cent R.H.; and 90 per cent R.H. appears to be optimum.

Low humidity somewhat curtails adult life. Thus at 25° C (77° F) in days: *Tineola* (males) 32 at 75 per cent and 23 at 20 per cent R.H., (females) 20 and 14 days respectively. *Hofmannophila* (females) 12 at 70 per cent, 9·4 at 20 per cent. *Endrosis* females live longer if allowed access to water.

(c) Food of larvae [581]

A wide range of larval food materials has been investigated, especially with *Tineola*. The results are not very easy to interpret briefly, because the two criteria of a good diet (rapid development and high percentage adult emergence) do not always agree. On a sparse diet, development is greatly prolonged and the number of moults increases [443].

Woollen cloth sprinkled with yeast was found to be a very satisfactory diet for the clothes moths and also reasonably satisfactory for the two house moths. Apparently all species are able to digest keratin. Nevertheless, supplementary vitamins are highly desirable, because wool (raw or scoured), fur and feathers by themselves were unfavourable even for the clothes moths. Fish meal formed a satisfactory breeding material and larvae were also able to develop on dead insect remains. Cereal products (e.g. flour) were satisfactory if not extracted or if vitamins (yeast, etc) were added.

(d) Longevity and fertility of adults

As already noted, high temperature and low humidity curtail the life of the adult moths. But under any given conditions, their longevity and egg production depend to a large extent on their size, which in turn depends on larval life. Thus, with *Tineola*, females about 6 mg weight averaged 106 eggs each, as compared to 55 eggs from 4 mg females, or 27 eggs from 2 mg females. Analogous results are recorded for *Tinea*, *Hofmannophila* and *Endrosis*.

10.2.4 NATURAL ENEMIES [179]

(a) Predators

Probably the most important predator of *Tineola* is the larva of the window fly *Scenopinus fenestralis*, which attacks the larvae. The mite *Pyemotes ventricosus* attacks newly emerged adults as well as early stages.

Laboratory culture of *Hofmannophila* have been troubled by the mite *Cheyletus eruditus* (Fig. 8.8 and page 376), which destroys eggs and newly emerged larvae.

(b) Parasites

Various hymenopterous parasites have been observed to attack clothes moth larvae, including *Apantales carpatus*, *Meteorus cespitator*, etc.

(c) Micro-organisms

The common clothes moth *Tineola* suffers from two types of disease: polyhedrosis, due to a virus, and protozoan infections.

Polyhedrosis may occur to different degrees in laboratory colonies. The signs are loss of appetite, sluggishness and a dull white appearance. Finally the larvae become swollen, the skin rupturing at a touch and exuding a cloudy liquid. The disease can exterminate colonies.

The protozoan infections are rather less prevalent and less serious. Two protozoan parasites have been recorded: *Adelina masnili*, a coccidian, and *Nosema* sp., a microsporidian.

10.2.5 IMPORTANCE

A survey of losses due to moth damage was conducted in 1948. About 26 per cent of ordinary households were found to have suffered some moth damage in the previous 15 months. At a fairly conservative estimate the total loss in the country worked out at £1·5 million (a figure which would need to be increased to correspond to present-day prices) [467].

The most important of the clothes moths, in regard to damage of domestic clothing and furnishings, is *Tineola bisselliella* followed by *Tinea pellionella* and *Hofmannophila pseudospretella*.

Materials damaged by the clothes moths include many kinds of fur, any woollen clothing or furnishing (but especially carpets and blankets), fabrics made with other animal hair (alpaca, camel), baize, felt used for insulation or piano hammers, feather stuffings, skins of animals and other museum specimens. Clothes moth larvae damage far more than they actually consume, owing to their wasteful feeding habits and to utilization of fibres in making cocoons. Moreover, the mere severing of threads in a fabric may cause weak patches or even holes and in the same way the pile of carpets or of velvet upholstery may be loosened and fall out leaving small 'bald' patches.

As already mentioned, clothes moths only thrive on woollen fabrics which are somewhat soiled and also they require undisturbed conditions, preferably in partial darkness. Where an infestation exists in some corner of a house, it represents a constant threat to all other susceptible articles. Thus eggs laid on perfectly clean fabric produce larvae which may find it difficult to grow and develop. Even a small and unsuccessful colonization will cause holes which seriously spoil whole garments.

The tapestry moth, *Trichophaga tapetzella*, is less successful in human dwellings, but it is able to breed in coarser materials than the clothes moths. Among other places it is encountered in stables and harness rooms, where it breeds in soiled horse blankets and formerly often damaged coach upholstery.

The house moths (which have been mentioned as food pests, page 360) have been described as rubbish feeders; they breed in dried animal or vegetable matter. Both species are very common in the nests of various species of wild birds. Sheds, barns and other outhouses are another favourite breeding ground from which the adults often fly into human dwellings; and upon occasion they may cause damage to carpets or upholstery.

10.2.6 CONTROL

There are two aspects to the control of clothes moths: eradication and prevention. The destruction of an existing infestation (often after much damage has been done) is comparatively easy. Prevention, however, which is much more desirable, is considerably more troublesome.

(a) Destruction of clothes moth infestations

Fumigation is a satisfactory way to eradicate an infestation of clothes moths. Small articles (clothing, furs, etc) may be simply treated in any convenient container such as a suitcase, trunk or chest. A variety of organic liquids can be used; the chosen fluid is poured over the clothing fairly liberally (say $\frac{1}{2}$ pint per 10 ft^3) and allowed to give an exposure for about 6 h, or overnight [263].

Where a widespread infestation of clothes moths has developed, for example among a lot of stored upholstered furniture and carpets, it may be necessary to resort to van fumigation. Hydrogen cyanide, as used against bed bugs is the most efficient treatment. Or possibly the cheaper and simpler, but less reliable, sulphur fumigation might be tried.

(b) Preventive measures

Both adults and larvae of clothes moths avoid the light and the eggs are easily dislodged from furs and fabrics by shaking or brushing. Therefore infestations are mainly prone to develop in materials left undisturbed in dark and shady places. Clothes lying in drawers or cupboards (especially furs or woollen winter garments put away for the summer), furniture in storage, or in seldom-used and rarely cleaned rooms, are especially liable to attack.

The first line of defence against the moth is constant vigilance for signs of attack. Unless reliable precautions have been taken, stored furs and woollen fabrics should be periodically examined for signs of damage. It is also very important to take special care of susceptible articles that are to be put in storage. The precautions which may be taken are given below.

(i) Cleansing

Furs and garments to be put in storage should be thoroughly brushed and shaken to remove any moth eggs which may have been laid on them. Unless other precautions have been taken, this shaking and brushing should be repeated periodically during the period of storage.

Fabrics and clothing generally should be dry-cleaned if possible, since the presence of grease stains is favourable to the development of an infestation.

Regular household cleaning with brushing of furniture and carpets will tend to discourage moths attacking furniture. Areas of carpet inaccessible to these cleansing measures (e.g. under low cupboards, etc) are liable to attack.

(ii) Mechanical barriers

One simple method of protecting garments from moths is to make sure that they are uninfested and to store them in large bags of stout paper or plastic. If the bag is unbroken and properly sealed by folding, it will prevent access of the moths [77].

Fairly efficient protection of such things as blankets can be achieved by folding them in large sheets of paper. Good quality newspaper can be used, though the idea that the newsprint has some additional mothproofing value is, unfortunately, fictitious.

(iii) Insecticides

Vaporizing solid insecticides made up into 'moth balls' have been in use for many years to prevent clothes moth damage. One of the earliest substances employed for this purpose was camphor; more recently naphthalene and paradichlorbenzene have been used, the latter being the more efficient [263].

These odoriferous substances are often described as moth 'repellents' but there is no good evidence that the mere odour of any substance will drive away moths [1]. Insecticides of the moth-ball type act merely as fumigants and therefore they can only be relied upon in nearly air-tight containers. However, they may be found useful in suitcases, trunks and chests used at the rate of 46 g ft^3 naphthalene or 12 g ft^3 paradichlorbenzene. It must be realized, however, that unless hermatically sealed up, these substances will evaporate away, in proportion to their volatility. (Paradichlorbenzene, though more potent, is considerably more volatile than naphthalene.) In recent years, the highly effective modern insecticides *gamma* HCH and heptachlor have been used in the same way, in crystalline form. They are much more effective and, at 5 g ft^{-3} (0·14 g m^{-3}), have given as much as 5 years' protection of crated furniture, as compared with 1 year's persistence of naphthalene (at 46 g ft^{-3}; 1·3 g m^{-3}). Since it may be undesirable to scatter these insecticides over furnishings in substantially pure form, they may be used adhering to cardboard sheets. The crystals are simply scattered over the cardboard after applying a layer of quick-drying adhesive. The treated cards, placed among furnishings (to give 180 g crystals per m^3) have given good protection from carpet beetle larvae, which are likely to be more tolerant than clothes moth grubs [76].

(iv) Cold storage [22]

For the protection of valuable furs in the summer months, it is the practice in many large towns to put them into large commercial refrigeration vaults which are maintained at a cool temperature (about 5° C). This does not destroy any eggs or moth grubs that may be present, but it prevents the eggs from hatching and the grubs from feeding. Therefore no damage can occur during cold storage. The method is quite reliable but rather expensive.

(v) Mothproofing [199, 248]

The mothproofing of furs or woollen fabrics involves treating them in some way so that clothes moth larvae cannot feed on them. The ideal treatment should be permanent, imperceptible, resistant to washing and dry-cleaning and compatible with dyeing and bleaching and other processes of textile manufacture. The credit for the earliest conception of making fibres immune to moth damage seems to belong to two Americans who, in 1887, took out a patent for moth-proofing curled hair. In subsequent decades the subject has received much attention from technologists; and there are thousands of patents for various processes.

The various methods may be grouped into three categories: impregnations of stomach poison insecticides; DDT impregnation; and treatments rendering wool indigestible to clothes moth larvae.

Impregnations of stomach poisons

The principle involved in this method is simply to add a poison which kills the moth grubs when they begin to feed on the treated fibres. (The amount of damage they can do before being killed is negligible, with an efficient treatment.) As early as 1905 it was noted that eosin was toxic to clothes moth larvae and subsequently a number of dyestuffs were examined for this property by the German firm of Bayer. However, the only dye found to be promising in early years was Martius yellow (2,4-dinitro-α-naphthol) which is still used for certain purposes. More progress was made with colourless compounds, especially fluorides and silico fluorides. A series of these was marketed by another German chemical company, the I.G. Farbenindustrie, in the 1920s, under the general name 'Eulan' (Eulan W, M, NK, NKF, etc) [251]. These were usually applied at various stages in the dyeing processes, but they were never entirely successful since they were leached out by ordinary laundering. Other treatments of the same order of efficiency depend upon pentachlorphenol and on formaldehyde applied in an acid bath. Other insecticides, soluble in organic liquids, can also be used; for example cinchona derivatives (fatty acid salts of the alkaloids) or such compounds as lauryl pentachlorphenol.

A very considerable advance was made in the following decade when compounds were employed which, though colourless, behaved like good dyes in having a strong affinity for the wool fibres. They were therefore fast to washing and a few were found to be sufficiently toxic to moth larvae to act as proofing agents. In particular the later forms of Eulan (Eulan N and CN) and Mitin FF should be mentioned (see page 60).

Impregnation of contact poisons

In the early 1950s, mothproofing of garments was offered as a regular service by many dry cleaners. Impregnation was achieved by dipping in dry-cleaning liquid containing DDT (or lauryl pentachlorphenate). 0·2 per cent DDT on the garment weight gave good results; but the effect was slowly lost after washing or immediately after ordinary dry cleaning. By the 1960s it appears that this type of moth proofing had become obsolete.

At present, the only impregnation of this type, which is extensively employed and apparently safe and reliable, is the impregnation of dieldrin by the Dielmoth process. By this method, the dieldrin is applied under standard conditions in a hot emulsion and remains bound in the wool fibres more firmly than is possible by impregnation from a volatile solvent. The wool takes up about 0·1–0·25 per cent by weight. This treatment is largely employed for carpets and outer garments and not for undergarments or children's wear. Tests with some more modern insecticides, showed that dermestid beetle attack could be prevented by impregnation with 0·03–0·1 per cent iodophenphos [78]. This would probably be satisfactory for protection against clothes moths.

On the retail market, it is possible to buy pressurized aerosol packs containing insecticides for application to woollen garments for mothproofing. Such treatments would be unreliable unless the whole garment was carefully treated. Attempts to protect clothing by casually puffing spray into a cupboard full of clothes are likely to lead to disappointment.

Rendering wool inedible to clothes moths
It has been mentioned earlier that the digestion of wool by insects relies on splitting the disulphide cross-linkages in the keratin molecules. This can be largely prevented by the following ingenious chemical treatment. The disulphide linkages are broken *in vitro* (by reduction to sulphydryl groups) and then reunited by intercalation of a short hydrocarbon chain, thus:

$$X\text{—}S\text{—}S\text{—}X$$
$$X\text{—}SH \quad HS\text{—}X + (CH_2)_n Br_z$$
$$X\text{—}S\text{—}(CH_2)_n\text{—}S\text{—}X + 2HBr$$

Molecules with these new cross-linkages are more difficult for the insect to digest and the physical properties of the wool are not greatly affected.

Moth-proofing tests. In order to introduce standard comparisons between estimates of moth-proofing value based on work in different countries, the International Wool Textile Organization published a standard test method in 1956 [119]. This depends on measurement of damage by moth larvae based on weight and adjusted by comparison with proofed material (treated with dinitro-α-naphthol). This method is useful, but somewhat narrow in application (since a particular kind of serge must be used). Other methods are employed in commercial investigations.

10.3 HIDE BEETLES (DERMESTIDAE)

10.3.1 DISTINCTIVE FEATURES

About 700 species of dermestid beetles are known. They are compact, oval or nearly round insects, mostly very convex, and usually rather small (2–4 mm; though a few reach as much as 12 mm in length). The body is nearly always covered with hairs or scales, often of various colours, making up distinctive patterns.

The head is often partly sunk in the thorax and bears a pair of antennae which can be laid back to rest in grooves in the front (or sometimes on the underside) of the first thoracic segment.

The larvae of dermestids are characteristic of the family. They are densely covered with hairs of various lengths and sometimes of different types. Thus the larvae of *Anthrenus* bear bunches of segmented hairs, with arrow-like heads. All the larvae have moderately well developed, five-segmented legs.

10.3.2 RECOGNITION, OCCURRENCE AND LIFE HISTORIES OF VARIOUS PESTS [274]

(a) Dermestes

Over fifty species are known but only two or three are at all common as domestic pests in Britain. The *Dermestes* spp. are rather large members of the family, being 5·5–12 mm in length. They are oblong in shape and densely covered with hairs which are always round, never flattened and scale-like. *Dermestes* can be distinguished from the other genera by the absence of the median ocellus on the head. The two commonest species may be distinguished as follows:

(i) Adults

Anterior half of each elytron with a small undulated black mark in the centre of an extensive pale area (Fig. 10.2) *D. lardarius* (bacon beetle)

Elytra uniformly coloured, black (or, when immature, brown). Apex of each elytron produced backwards into a fine point (Fig. 10.2) *D. maculatus* (= *vulpinus*) (the leather beetle)

(ii) Larvae

Conical processes (spines) on the penultimate segment of the abdomen, nearly at right-angles to the body, curved and placed close together as in a V, the actual tips directed backward; no pronounced median bristle *D. lardarius*

Spines sloping to the rear, moderately close together, with the actual tips directed forward; a pronounced median bristle placed slightly behind them (Fig. 10.2) *D. maculatus*

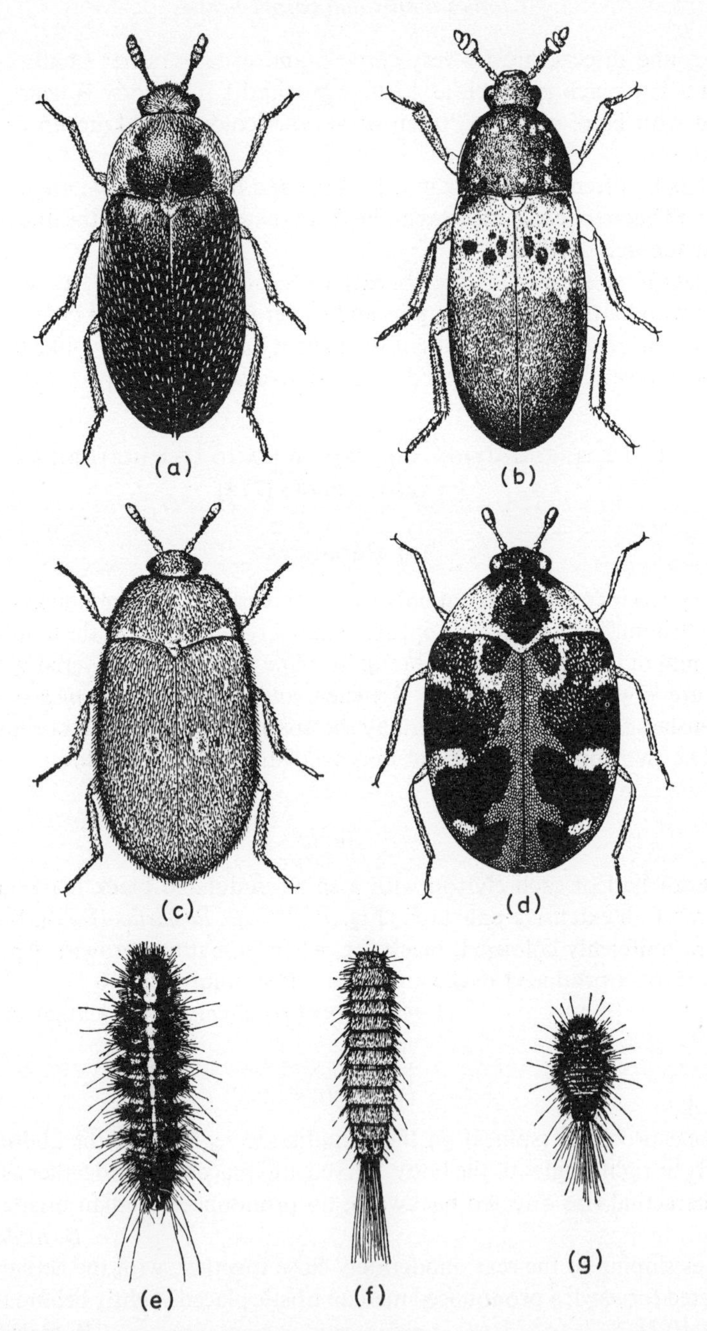

Figure 10.2 Dermestidae (carpet beetles). Adults: (*a*) *Dermestes maculatus*; (*b*) *Dermestes lardarius*; (*c*) *Attagenus pellio*; (*d*) *Anthrenus scrophulariae*. Larvae: (*e*) *Dermestes maculatus*; (*f*) *Attagenus piceus*; (*g*) *Anthrenus scrophulariae*. (After Hinton [174].) (*a*), (*b*) × 6; (*c*) × 10; (*d*) × 15; (*e*), (*f*), (*g*) × 3.

(iii) Occurrence

The *Dermestes* beetles breed in dry animal proteins and are common in hide and skin warehouses, bone factories and dog biscuit stores. They sometimes attack dried meats and fish meal in domestic stores but rarely breed in woollen clothes or furnishings unless these are badly contaminated with animal matter. When specimens are often encountered in dwelling houses, the larder should first be examined for the breeding site. Alternatively they may be feeding on a dead mouse or bird in an attic or floor space.

(iv) Life history

The females lay eggs on material suitable as food for the larvae, often in crevices, in skins, hides, etc. The eggs are about 2 mm long, white in colour.

The larvae are whitish when first hatched, but become darker in a few hours, when exposed to light, or in 2 days, if kept in darkness. They will feed on almost any animal matter which is dry or in a state of decomposition; and vegetable matter is also sometimes consumed. Feeding is continuous and sometimes their faecal pellets emerge joined together in a bead-like chain.

Throughout development the larvae avoid light. They are very active but sometimes if disturbed they will become suddenly immobile, partly curled up, apparently feigning death. Normally there are 5 or 6 moults; but, under rather indefinite conditions, this number may increase to 12. When fully grown, they reach a length of 10–15 mm.

The mature larvae cease feeding and seek a place for pupating. Often they wander away from the foodstuff and excavate holes in quite hard inedible materials. These shelters may be merely a little longer than the larva's body, or they may extend up to a foot in length. The mouth of the burrow becomes choked with debris and finally with the last larval skin, which is cast off and acts as a protective plug. But if the larvae are forced to pupate in the open, they will usually do so inside the last larval skin.

The boring habits of the mature larvae cause quite a serious economic problem. One of the earliest records of this kind of injury is that referred to in 'The Last Voyage of Thom. Cavendish' (*Hakluyt's Voyages*; ed. Goldsmid, Edinburgh, 1890), where there is an account dating from 1593 of a ship carrying a cargo of dead penguins which sank because of the honeycombing of sides and bottom by dermestid larvae.

In recent times, there are numerous references of damage to crates and boxes and to walls, floors and roofs of factories and stores and sometimes of ships and barges due to these pests.

The adult beetles feed on substances eaten by the larvae, this food being necessary to the female for the maturation of her eggs. The sexes will copulate at temperatures above 16–18° C, usually for periods of 3–5 min. A single mating is sufficient to allow the female to produce fertile eggs for the rest of her life; but normally she will mate a number of times. After a pre-oviposition period of 10–15 days, egg-laying begins and is continuous for 2–3 months. The total number of eggs laid varies considerably; 200–800 have been counted. The adults may live for over 3 months.

(v) Speed of development

D. maculatus. E at 26–27° C, 3 days; at 23–24° C, 5 days. L at 23° C, 44 days; P (winter), 35 days; 20–25° C, 14 days; 27° C, 8–14 days. Total, at 28–30° C, 42–46 days; at 23° C, 55 days; but may increase up to several years with very unfavourable conditions.

D. lardarius. E at 17° C, 9 days; at 24° C, 3½ days; at 25–28° C, 2½ days; P, 8–15 days. Total, at 18–25° C, 2–3 months.

(b) Anthrenus

An obvious feature of the *Anthrenus* beetles is their dense covering of scales of different colours, giving the body a variegated pattern. They are rather small beetles (1·5–4 mm), roughly oval in outline and strongly convex. The antennae lie back in deep recesses on the thorax when at rest. Some of the commoner species may be distinguished as follows:

(i) Adults

1. Antennae 8-segmented with a club of 2 segments (2–2·8 mm).
A. museorum ('museum beetle')
Antennae 11-segmented with a 3-segmented club. (2)
2. Eyes smoothly rounded. Antennal club with nearly parallel sides.
A. verbasci ('varied carpet beetle')
Eyes indented on inner side. Antennal club oval.
A. scrophulariae ('common carpet beetle') (Fig. 10.2)
and *A. vorax* ('furniture carpet beetle')

(N.B. *verbasci*, *vorax* and *scrophulariae* show a number of varieties of colour pattern and the last two are difficult to distinguish from each other and from *A. pimpinellae*.)

(ii) Larvae

1. Abdomen with first 8 sternites distinctly sclerotized. Arrow-headed hairs of caudal tufts with heads strongly produced and filiform. (2)
Abdomen with first 8 sternites entirely membraneous. Arrow-headed hairs with heads not produced apically, never filiform. (3)

2. Heads of arrow-headed hairs about 0·17 mm long; basal struts about one-ninth as long as the complete head (Fig. 10.2). *A. scrophulariae*
Heads of arrow-headed hairs about 0·10 mm long; basal struts about one-sixth as long as complete head. *A. vorax*
3. Heads of arrow-headed hairs as long as combined length of 4 or 5 preceding segments. *A. museorum*
Heads of arrow-headed hairs as long as combined length of 7 or 8 preceding segments. *A. verbasci*

(iii) Occurrence

Anthrenus larvae will feed on woollen materials, hair, furs, bristles, leather and skins and on insect specimens. (*A. verbasci* and *A. museorum* are the worst pests of dried insect specimens.) They are not serious pests of carpets except where these are kept tacked down and undisturbed for long periods.

(iv) Life history

The eggs are laid on larval foodstuffs, sometimes on furs and woollen fabrics and very readily on dried animal remains (especially dead insects). The eggs are often thrust into crevices of the breeding material, but even if not, they are apparently stuck to it by a secretion which resists dislodgment by shaking.

The larvae emerge from the eggs and begin to feed. They are rather squat, brown, hairy grubs, bearing three pairs of bunches of characteristic golden hairs on the posterior segments of the abdomen. When these hairs are examined under the microscope, they are found to be segmented and to bear small arrow-like heads. When the larvae are disturbed, they often curl themselves up and spread out these curious tufts of hair fanwise, giving the appearance of tiny golden, hairy balls.

Throughout life, the larvae tend to avoid the light and to burrow into their feeding material. The normal diet contains keratin and all types of keratin are suitable. However, like the clothes moths, they require additional nutrients to complete development. Some extensive investigations with *A. vorax* were made to determine the effects of the additional nutrients. (An infusion of horse dung was used, at different strengths, added to clean wool.) As the quality of the food decreased (more dilute solutions used to impregnate the wool), the larval mortality increased and the growth rate fell off rapidly. Another effect of a sparser diet on *Anthrenus* (which also occurs with other insects) is the prolongation of each stage, of the total development and an increase in the number of moults (from the minimum up to about 30). On a very inadequate diet, death finally sets a limit to this prolonged development with numerous moults.

Well-fed larvae of *Anthrenus* are very resistant to starvation. Under rather cool conditions, larvae of *A. scrophulariae* have survived 10 months without food.

Temperature has its usual accelerating effect on the larvae, providing they have an adequate diet. One species (*A. vorax*) requires a fairly high temperature (25° C) for pupation, and it will not complete its development unless fairly warm conditions prevail for this part of the life history.

The effects of moisture have only been roughly studied. A high relative humidity (90–100 per cent) was found to be more favourable than a rather low one (30–40 per cent) and development was shorter under most conditions.

The fully grown larva measures about 4–5 mm in length. Pupation occurs inside the last larval skin, usually without leaving the breeding site. The newly formed adults also remain resting inside the old larval skin for a period ranging from 4 or 5 days to a month, according to the temperature.

Finally, the adults emerge and tend to seek the light, so that they usually fly to the windows and eventually escape out of doors. Mating can take place at once and the females of some species can lay eggs without taking any food. Usually, however, the adults congregate on various flowers and feed upon pollen and nectar. Thus the beetles may often be found in the garden during the summer months.

In late summer and autumn, the females tend to re-enter houses to lay their eggs. The numbers of eggs laid range from about a score to a hundred or so.

(v) Life cycle

In temperate climates there is usually only one generation per year. Eggs are laid in the summer. The larvae feed until winter when they hibernate if conditions become cold. Feeding resumes in the spring and pupation occurs in February or March. The adults appear from the end of March and occur throughout the summer.

(vi) Speed of development

A. verbasci. E at 18° C, 31 days; at 23–24° C, 14 days; at 29° C, 11 days; L (room temperature New York), 222–323 days; P, 10–30 days. Total (New York; Germany), 7–14 months.

A. vorax. E at 18° C, 32 days; at 24–25° C, 16–17 days; at 30° C, 10 days; L at 20° C, 370–406 days; at 25° C, 118–135 days; at 30° C, 77–95 days; P at 25° C, 13 days; at 30° C, 9–10 days. Total at 20° C, P at 25° C, 420–450 days.

A. scrophulariae. P at 18–20° C, 18–19 days; at 25° C, 14 days; at 27° C, 10–11 days.

A. museorum. P at 20–22° C, 9–10 days. Total at 18° C, 10–11 months.

(c) Attagenus

Beetles of this species are intermediate in size between *Dermestes* and *Anthrenus*, being of the range 3·4–6 mm in length. The two commonest species are oval-oblong beetles, densely coated with hairs which are mostly, or entirely, brownish black. They may be distinguished as follows:

(i) Adults

Whitish or yellowish spot about the middle of each elytron (Fig. 10.2). *A. pellio*
Elytra unicolorous. *A. piceus* ('black carpet beetle')

(ii) Larvae

Numerous flat, striated, broadly lanceolate scales on back of thorax and abdomen. *A. pellio*
Body without flattened lanceolate scale (Fig. 10.2). *A. piceus*

Distribution
A. pellio: Europe, Asia, Africa, North America. *A. piceus*: Cosmopolitan. Indigenous in Oriental region.

(iii) Occurrence

These beetles are fairly common in warehouses and in dwelling houses, where they may breed in furs, skins and woollen fabrics (especially carpets). They are also found among stored grains and cereals, partly feeding on the remains of other grain pests. As a domestic nuisance, *A. pellio* is more prevalent in Europe and *A. piceus* in North America.

(iv) Life history

The life history of *Attagenus* resembles that of *Anthrenus* in outline. The females enter houses in the summer months to lay their eggs on materials suitable for larval food. The larvae will feed on woollen fabrics (especially carpets), furs and other dry substances of animal origin. They are also found among dry vegetable products which they undoubtedly consume, but they may also subsist on the dead bodies of other insect pests.

The larvae, which consistently tend to avoid the light, are of the usual hairy type characteristic of the family and they have a distinctive tuft of very long hairs at the end of the body. They moult from 6 to 20 times or possibly even more under very adverse conditions. As with other pests of this type, inadequate diet, as well as low temperature, prolongs the larval period and increases the number of moults.

Pupation finally occurs inside the last larval skin and the adult rests inside for a period (of 3–20 days) before emerging to live an active life.

The adults tend to fly out of houses and congregate on flowers (especially

Spiraea) where they feed on pollen and nectar and mate with each other. The life of the adult at 29° C is about 15–25 days; at 18° C it is about 35–40 days if mated and 60–75 days if unmated. The females lay about 50–100 eggs, averaging about 75.

(v) Life cycle

In temperate regions the life cycle shows considerable variation, from about 6 months to 3 years, according to diet and other unknown circumstances. A one-year cycle is probably most common. In this cycle, the adults occur from late April to August and the eggs are laid from late May to late August. The larvae hatch in June to September and pupate in the following spring between April and June.

Speed of development. E at 18° C, 22 days; at 24° C, 10 days; at 30° C, 6 days. L at 25–30° C, 65–184 days (depends upon food rather than temperature within this range). P at 18° C, 18 days; at 24° C, 9 days; at 30° C, $5\frac{1}{2}$ days.

(d) Trogoderma

Trogoderma beetles are rather small (2–5 mm), egg-shaped, convex beetles, covered with brownish hairs, frequently with bands or patches of paler hairs. Two species are considered in this book:

Eyes indented on the inner side	*T. versicolor*
Eyes gently rounded on inner side (Fig. 8.5)	*T. granarium*

(i) Occurrence

T. granarium (the 'khapra beetle') is essentially a grain pest and is dealt with on page 337.

T. versicolor is a well-known pest of insect collections and it is also liable to develop in woollen fabrics and furs. This beetle is also found among stored food, partly feeding on it and also devouring remains of other grain pests. Out of doors, *T. versicolor* is a common scavenger of dead insects in various situations.

(ii) Life history

The eggs are laid on a variety of dry organic substances, although the larvae thrive best on animal remains. The speed of development and the number of moults vary according to the quality and quantity of the food and the temperature. As usual, sparse food induces prolonged development with numerous moults. The larvae show astonishing powers of resistance to starvation. (They have survived as long as five years without food, and at any time before death, they are able to recuperate if offered suitable food.) During starvation they moult and *decrease* in size so that they may actually reduce below their original size on hatching from the egg.

Pupation occurs inside the last larval skin, and adult life begins with a quiescent period before they emerge to live an active life. These beetles require no food or water to attain full fecundity. They tend to avoid the light for most of their life except towards the end.

The number of eggs laid by each female varies according to the diet during larval development. The usual average is about 50–100.

(iii) Life cycle

In North America and in Russia, two generations per year have been reported.

Speed of development. E, 8–12 days; L, about 5 months; P, 11–17 days. Total, (under 'optimum conditions') 2 months; (usually) about 6 months.

10.3.3 IMPORTANCE

The family name dermestid ('skin feeder') indicates the principal type of damage caused by this group. Great losses are incurred by the hide and fur trades by attacks of these beetles, which are especially prone to destroy raw skins and hides, although finished garments or furnishings of wool, fur or silk are also liable to attack. In addition, the beetles may damage various stored food products. Dried materials of animal origin are mainly infested (e.g. dried or smoked fish or meat, cheese, dried milk, etc), although some forms will also feed on cereals, cereal products and seeds.

In order to understand the destructive propensities of various dermestids, it is desirable to review their feeding preferences. Almost all the serious damage is due to the larvae. The adults of several genera (e.g. *Anthrenus* and *Trogoderma*) do not eat the larval food, but normally feed out of doors on the pollen or nectar of flowers. Other forms require little or no food in the adult stage; only the adults of *Dermestes* do more than insignificant damage to commodities, but even in this genus the larvae are much more serious pests.

Nests of birds and rodents are two important reservoirs likely to provide sources of these domestic pests. A considerable number of species has been recorded from these situations, where the larvae apparently feed upon cast feathers or hairs, fragments of food and dead insects. The larvae have even been found attacking weak nestlings and feeding on the wings.

Various dermestids have the habit of breeding near spiders' webs and feeding on captured insects or even on the eggs or young of the spiders. Others live in the nests of wasps and bees, where again they feed on dead or dying insects and also, to some extent, on honey or pollen.

The original natural diet of the dermestid larvae appears to have been dry proteinaceous materials of animal origin. Most members of the group are unable to complete development unless they obtain some food of this type; but others have adapted themselves to a vegetarian diet and can develop successfully on it, though they normally prefer animal matter. Only one species (*Trogoderma granarium*) habitually breeds in vegetable matter (see page 337).

Among the forms restricted to animal diet, there are some which usually or always confine their feeding to more or less raw products such as dried or smoked meat, skins and hides, museum specimens, silkworm eggs and pupae. These include all species of *Dermestes*, some *Attagenus* and some *Anthrenus*. Other forms (*Attagenus piceus*, *Anthrenus vorax*, *A. verbasci* and *A. scrophulariae*), as well as attacking these raw products, are also serious pests of highly processed commodities containing animal matter, such as carpets, woollen clothes, furs, leather or silk. To some extent, this ability to develop on processed animal products is due to a capacity for feeding on food with a very low water content. (Whereas fish meal might be expected to contain 45 per cent moisture, a woollen fabric under the same conditions would contain only 11–12 per cent.) Apart from this, the nutritive value of sterols present in the raw products (and absent from scoured wool) may be essential for some species.

In addition to the hides, skins and other animal products which dermestid larvae attack for food, a very considerable amount of damage is caused by a habit of the fully grown larvae. After completing larval development they have a strong tendency to excavate cavities in quite hard, inedible materials to form shelters in which to pupate. The woodwork of boxes, buildings, ships and barges may be attacked and also many other products stored in the vicinity of infested materials. Even lead covers of electrical cables have been badly damaged by dermestid larvae with resulting short-circuiting.

Finally, two unusual but quite troublesome nuisances due to dermestid larvae may be cited; damage to insect collections and to silkworm pupae. Dead insects form a natural food of many dermestid larvae and it is therefore easy to see why they are about the most serious pest of insect (and other) specimens in museums. The species mainly responsible are *Anthrenus verbasei*, *A. museorum*, *A. scrophulariae* and *Trogoderma versicolor*.

In the silkworm industry, dermestids cause trouble by devouring the helpless pupae and eggs. Cocoons which are damaged in the attack on pupae are completely spoilt for silk production.

(a) Dermestids and disease

Dermestids play only a very minor role in disease transmission. There have been records of mechanical transmission of anthrax germs by beetles from infected carcases, but this must be a rare occurrence.

The hairs dislodged from dermestid larvae have been observed to cause skin irritation and conjunctivitis to workmen unloading a badly infested cargo of skins.

(b) Uses of dermestids

One ingenious use of these insects is to allow them to clean up vertebrate skeletons for exhibition purposes. They are said to remove all unwanted tissues from even the most delicate structures.

10.3.4 CONTROL OF DERMESTID BEETLES [469]

Two kinds of infestation must be considered. In hide and skin warehouses, in boneyards and other large accumulations of animal matter of this type, there may be heavy and obvious infestations. Infestations in dwelling houses, on the other hand, are usually small, but they may be annoying and the breeding sites are often hard to trace.

(a) Warehouse infestations

When large quantities of hides or skins are infested (usually with *Dermestes* spp.) it may be necessary to dip them in an arsenical solution. A $2\frac{1}{2}$ per cent solution of sodium arsenite is effective and the arsenic does not affect the leather in any stage of its manufacture. The men carrying out the treatment must wear protective rubber gloves.

Alternatively, under suitable conditions (reasonably gas-tight buildings) hydrogen cyanide fumigation may be employed. The dermestid beetles are rather resistant to fumigants so that a good concentration of the gas should be ensured. Fumigants other than hydrogen cyanide should not be used unless there is good evidence of their effectiveness against these pests.

As a protective measure, skins and raw hides are usually baled up with layers of flake naphthalene before despatch on ships or other long journeys.

(b) Domestic infestations

Very often the site of breeding in a dwelling house is obscure and the presence of the insects becomes noticed by repeated appearance of adults or, sometimes, of wandering larvae. The first essential is to trace the focus of the infestation. When the pest has been identified, some indication of the possible breeding ground will be gathered from the food preferences of the various species, as already described. (See also the section on the importance of the group.) The larder should be examined for neglected pieces of dried meat or other dry protein. An alternative source of dried animal matter may be a dead rodent under the floor or a dead bird in the attic. Neglected woollen garments or furnishings (especially carpets) should be examined, particularly any which have been allowed to become dirty. Woollen felt used in lagging pipes or for other insulating purposes should not be forgotten. Finally, it should be remembered that dermestids sometimes breed in bird nests and that they may be entering the house from such places under the eaves.

When the breeding focus has been traced, it should be thoroughly extirpated, if possible, by burning the infested material. If an article of value is attacked, it may be necessary to use spray insecticides; these should be applied very thoroughly to kill all the larvae.

For protecting woollen fabrics or furs from possible attacks of dermestid beetles, the same measures may be employed as against the clothes moths (see pages 404–408). Recent trials of impregnation against *Attagenus megatoma* showed best results with 0·03–0·1 per cent iodophenphos [78]. Fumigation with dichlorvos in various forms (wax sticks, resinous granules, etc) in a closed 42 m^3 chamber, gave concentrations of 0·46–3·3 $\mu g\,l^{-1}$. This killed the beetle larvae in 32 h [58].

CHAPTER ELEVEN

Wood-boring insects

11.1 BEETLES

11.1.1 LIABILITY OF WOOD TO ATTACK

In order to understand the potentialities of different beetles for damaging different kinds of woodwork, it is necessary to consider some of the factors which render wood liable to attack.

(a) Water content

Wood which is prepared for use as structural timber or furniture is obviously much drier than the wood of living trees or even the wood of logs lying on the forest floor. It is, indeed, a very dry foodstuff; and this fact probably excludes many of those wood-feeding insects which attack trees in the forest. The domestic timber pests (like many pests of stored foodstuff or fabrics) are well adapted to a dry diet. The wood which they swallow needs to be soaked in digestive juices to be assimilated; but all possible water is extracted by the hind part of the gut, so that they pass dry and powdery faeces ('frass').

As wood dries from fresh green timber it approaches an optimum for the domestic wood beetles and further drying renders it unsuitable as food. Thus the *Lyctus* or powder-post beetles will not attack newly felled timber which has a water content of 30 per cent or over. As it dries it becomes more and more liable to damage, the optimum being about 12–15 per cent. Below this level, the lack of moisture becomes deleterious and prevents attack below 8 per cent water content [466].

(b) Chemical constituents of wood [373]

Apart from its lack of moisture, dry wood does not appear to be very promising as a diet. Until recent years there was very little knowledge about the nutrition of the wood-boring insects. The main constituents of wood are: cellulose, 40–62 per cent dry weight; lignin, 18–38 per cent dry weight; hemicelluloses, 8–37 per cent dry weight; starch, up to 6 per cent dry weight; sugar, up to 6 per cent dry weight; proteins, 1–2 per cent dry weight.

It has been shown that different wood-feeding insects consume different constituents. The lignin is only attacked to a minor extent, but several groups of insects are able to digest cellulose. Some (Anobiidae) can do this directly, by the action of their own digestive enzymes. Others, like certain termites, rely upon the preliminary digestion of the cellulose by micro-organisms (protozoa) in their intestines.

One important group, the Lyctidae, cannot digest lignin, cellulose or hemicellulose. They depend entirely upon starch and small amounts of sugars in the wood. *Lyctus* beetles are therefore only able to develop in wood which contains starch.

It is probable that both the anobiids and the lyctids assimilate small quantities of protein from the wood cells; but their nitrogen requirements are not great.

(c) Condition of the wood [53]

It has been known for some time that structural timber or portions of it which have become decayed by continual exposure to rather damp conditions, is especially liable to attack by the death-watch beetle. Laboratory experiments have shown that this beetle has difficulty in establishing itself in sound timber and that its development in sound, or nearly sound, wood is very slow indeed. The length of the larval period and the larval mortality are both decreased to a large extent if the wood has been subjected to fungal attack. The type of fungus is not particularly important since the most important benefit of decay to the insect is the softening of the wood, allowing more rapid larval penetration. The fungi which were in fact investigated were brown and white rots (basidiomycetes); these attack both cellulose and lignin and other constiuents equally, so that their only overall chemical effect is a slight concentration of nitrogen, which may be somewhat beneficial. More recently, similar investigations with furniture beetle revealed the same facilitation of attack in wood attacked by these fungi and also by soft rots [54]. It is also noteworthy that wood-boring weevils and the wharf borer are only prone to infest wood which has been attacked by fungi.

(d) Plywood

The cheaper grades of plywood of the birch-alderwood type, made up with blood-casein glue, are very frequently attacked, and it has been shown that these glues accelerate growth of woodworm. However, plywood bonded with synthetic glue, such as urea-formaldehyde, seems to be immune to furniture beetles, though it may be attacked by powder-post beetles.

11.1.2 THE DOMESTIC WOOD-BORING BEETLES [269]

(a) Distinctive features

The principal wood-eating insects are the termites, various members of certain families of beetles and the wood-wasps. Of these, the termites fortunately do not occur in Britain; and the wood-wasps and many of the beetles are pests of the forest and do not attack wood in human dwellings. Certain beetle pests of the

Table 11.1 Signs of attack by the commoner wood-boring beetles.

Insect genus	Wood attacked: Hardwoods Softwoods	Sound Decayed	Heartwood Sapwood	Bore dust	Exit holes
Anobium	H and S	S and D	Usually S	Ellipsoidal pellets	Round 1·5 mm
Xestobium	Old H rarely S	S and D	H and S	Coarse bun-shaped pellets	Round 3 mm
Lyctus	H	S	Usually S	Fine soft powder	Round 1·5 mm
Hylotrupes	S	S	Usually S	Large cylindrical pellets and powder	Oval 3 mm × 6 mm
Euophryum	H and S	D	—	Tiny ellipsoidal pellets	Irregular oval 0·8 mm × 1·5 mm

forest may produce holes in wood which may be subsequently confused with those caused by domestic timber pests. These are the so-called pin-hole or shot-hole borers, belonging to the families Platypodidae or Scolytidae [185]. They attack trunks and branches of living trees or, in some cases, freshly felled trees. When the wood dries sufficiently for use as structural timber or furniture, the beetle larvae die, so that they cannot be responsible for infestations in houses. Their burrows, however, remain and may be visible as holes in cut timber or plywood. These tunnels are round, ranging between 0·5 and 3 mm diameter (hence 'pin-hole' or 'shot-hole'). It is necessary to distinguish these harmless relics from the signs of attack by beetles which can damage seasoned wood. This is easily done by two criteria. (1) The wood surrounding the galleries of the pin-hole or shot-hole borers are always stained, usually a purplish brown. This staining is a residue of the fungi which the beetles cultivate in them for food, for which reason they are sometimes called 'Ambrosia beetles'. (2) The galleries are empty. In contrast, the galleries of domestic wood-boring beetles are unstained and are usually full of frass (powdery faeces).

The beetles which do damage furniture or structural timber are forms which, in nature, breed in dead branches of trees or old logs. These beetles, together with certain fungi, perform a useful function in nature by breaking down the waste lumber of the forest. They become troublesome to man when they attack the particular timber which he uses for his own purposes.

(i) Classification

The most 'domestic' of wood-boring beetles belong to the Bostrychoidea group, which includes the Anobiidae (furniture beetle, death-watch beetle), Lyctidae (powder-post beetles) and the Bostrychidae (false powder-post beetles). Another family in the group is the Ptinidae or spider beetles, which do not attack wood, but include some stored product pests.

Other wood feeding beetles include many species breeding in woodland timber, one or two of which may be troublesome in houses. The Cerambycidae, or longhorn beetles, with some 15 000 species are distributed world wide; whereas the Buprestidae, a family of similar size, are mainly tropical, and the Oedemeridae (600 species) are mainly temperate. The Cerambycidae and the Buprestidae include some large and beautiful forms. Two weevils (Curculionidae) are included; they are anomalous members of this enormous family with over 35 000 species.

(ii) Notes on the British species

The following are liable to damage woodwork and timber in Britain.

Anobiidae: *Anobium punctatum, Xestobium rufo-villosum, Ptilinus pectinicornis, Ernobius mollis.*
Lyctidae: *Lyctus linearis, L. brunneus, L. planicollis, L. cavicollis, Trogoxylon parallelopipidum.*
Bostrychidae: a few species occasionally imported.
Cerambycidae: *Hylotrupes bajulus.*
Curculionidae: *Euophryum confine, Pentarthrum huttoni.*
Oedemeridae: *Nacerdes melanura.*

The adults of the more important genera (*Anobium, Xestobium, Lyctus, Hylotrupes, Euophryum, Pentarthrum*) may be distinguished by the main beetle key on page 540. The two other anobiids, which are rather uncommon, resemble *Anobium punctatum* but can be distinguished as follows.

Ptilinus pectinicornis is slightly larger (3–6 mm) and more cylindrical. The males are easily recognized by their branched antennae (Fig. 11.1); those of the females are merely toothed like a saw.

Ernobius mollis is also larger (up to 6 mm) and when freshly emerged is covered with golden hairs (though these tend to become rubbed off). The wing cases are less horny than those of *Anobium.*

The more important larvae may be identified by another key (page 544).

The two common species of *Lyctus* are *L. brunneus*, an American species, which has now become much more common than the native British species *L. linearis.* The adults may be distinguished as follows:

Key to Lyctid species in Britain (*based on Hickin* [269])

1. Pubescence on elytra general and confused. Anterior angles of prothorax prominent and acute. *Trogoxylon parallelopipidum*
Pubescence on elytra confined to longitudinal ridges. (2)
2. Y-shaped depression on pronotum. 5 rows of punctures between elytral suture and first ridge. *Lyctus brunneus*
Depression on pronotum not Y-shaped. Fewer than 5 rows of punctures between elytral suture and first ridge. (3)

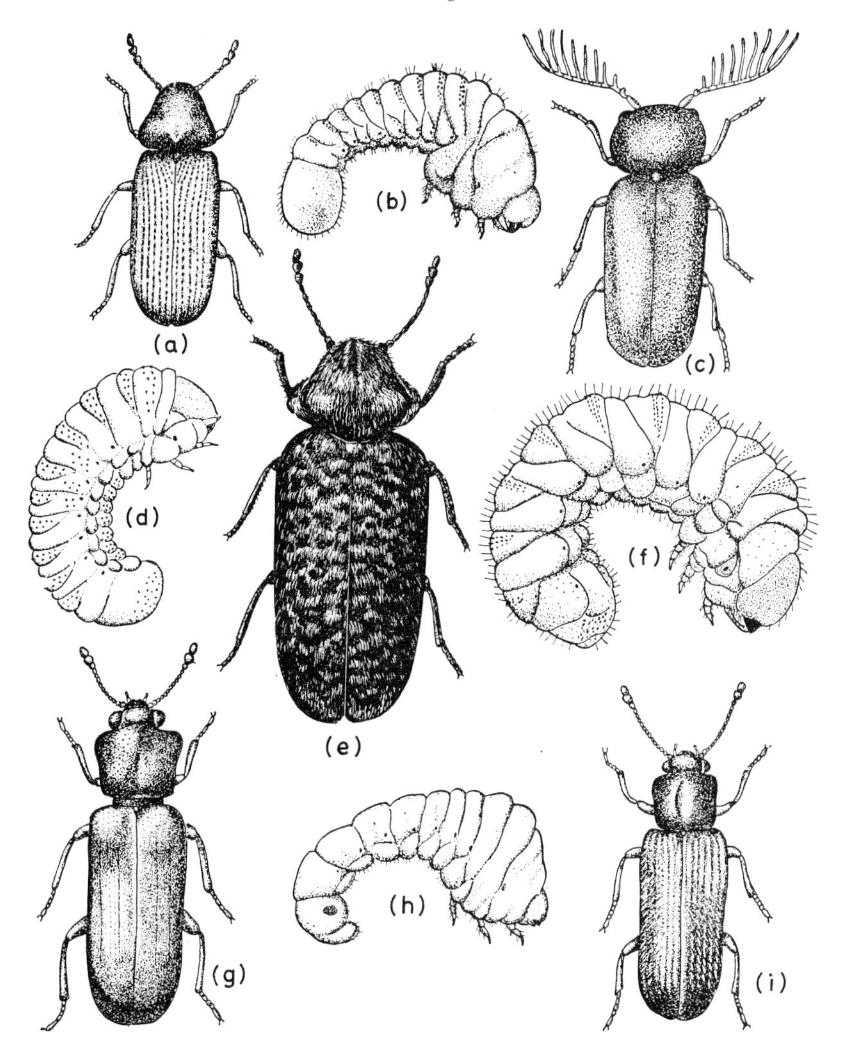

Figure 11.1 Wood-boring beetles and their larvae. (*a*) adult and (*b*) larva of *Anobium punctatum*; (*c*) adult and (*d*) larva of *Ptilinus pectinicornis*; (*e*) adult and (*f*) larva of *Xestobium rufovillosum;* (*g*) adult and (*h*) larva of *Lyctus brunneus*; (*i*) adult of *Lyctus linearis*; (*e*) original; (*f*) after Parkin [464]. Remainder after Gahan and Laing [196]. All × 7½.

3. Punctuation on median third of elytra not in distinct rows. *Lyctus sinensis*

Punctuation on median third of elytra in distinct rows. (4)

4. Punctuation on elytra in single rows, except between suture and first ridge. *Lyctus linearis*

Punctuation on elytra in double rows between ridges. (5)

5. 10th antennal segment wider than long. *Lyctus planicollis*
10th antennal segment not wider than long. *Lyctus cavicollis*

(iii) Notes on North American species (Fig. 11.3, page 434)

The most important domestic wood boring beetles in North America are as follows.

Anobiidae: There are about 260 species in North America which feed on dead wood. Apart from the furniture beetle and the death-watch beetle, a few also attack furniture and domestic timber. *Anobium punctatum, Xestobium rufo-villosum, Hemicoelus carinatus, H. gibbicollis, Ptilinus ruficornis, Xyletinus peltatus, Enobius mollis*

Lyctidae: Ten species occur in the U.S.A. and 6 of them are pests. *Lyctus planicollis, L. cavicollis, L. brunneus, L. linearis, Trogoxylon parallelopipidum, Lyctoxylon japonicum*

Bostrychidae: *Polycaon stoutii, Scobicia declivis, Xylobiops basilaris*

Cerambycidae: *Arhopalus productus, Hylotrupes bajulus;* Oedemeridae: *Nacerdes melanura.* Most of the important genera can be recognized from the key mentioned or from Fig. 11.1. The Bostrychidae, though related to the Lyctidae, are easily distinguished by their serrated antennae, the hood formed by the prothorax over the head and the spines on the front of the prothorax and at the end of the elytra. *Polycaon stoutii*, however, is an anomalous species; the prothorax is not hoodlike and there are no spines. It is a cylindrical black beetle, with coarse punctures on the head and pronotum and fine ones on the elytra. *Arhopalus productus* has the typical long antennae of the Cerambycidae. It is a narrow black beetle 2–3 cm long.

(b) Life histories

(i) Anobiidae

The females chose suitable places to deposit their eggs, the condition of the wood surface being apparently more important than the kind of timber. They are able to extend their ovipositors telescopically, to a length nearly half that of the whole body, and the tips of these organs are used to probe the surface to discover small holes and crevices into which the eggs can be wedged as they are laid. In addition an adhesive secretion on the surface tends to bind them to the wood.

The eggs are oval or lemon-shaped, white in colour and about 0·4 mm long. Under the microscope each egg shows a fine honeycomb-like sculpturing which extends for one-third of its length from one end, the remainder being smooth.

The larvae emerge from the egg shell surface in contact with the wood and begin to tunnel into the timber with the capsule still in place [316]. The first-stage larvae are very small and straight; they acquire the characteristic curled shape of this type of larvae as they grow older. The appearance of the older grubs is shown in Fig. 11.1. It will be seen that they bear rows of small brown spines ('spinules') on the more prominent ridges of some of the dorsal segments; these probably assist in gripping the sides of the burrow. The larvae possess powerful mandibles with which they dig into the wood. Part of the wood is swallowed, digested and extruded as faeces in the form of small oval pellets. The mixture of wood fragments and faeces which fills up the burrow as the larvae move forward, is known as 'frass'.

The burrows in the wood usually move along in the direction of grain. The larvae gradually grow in size and increase from a body weight of 0·005 mg on hatching to about 13 mg when mature [42]. The fully grown larva reaches a length of about 5 mm ($\frac{1}{5}$ in.). As the larva grows, the size of the burrow naturally increases, till it reaches a diameter of about 2 mm ($\frac{1}{2}$ in.). Very badly attacked wood may contain 2 or 3 larva per cubic centimetre.

When it is fully grown, the larva prepares to pupate. The course of the burrow is directed towards the surface of the wood and, just below the surface, a slightly enlarged chamber is excavated. In this cell pupation takes place. The pupa is of the usual beetle type with the adult limbs lying close to the body but not adhering to it.

When the adult beetle is ready to emerge, it waits until its jaws and body have hardened, and then bites an emergence hole through the thin layer separating it from the free air. The appearance of these beetles is shown in Fig. 11.1. They are 2·5–5·0 mm long and of a reddish-brown colour, modified somewhat by a covering of short yellow-grey hairs. The beetles may be seen crawling about on walls, ceilings or windows in the summer months. On warm days, they fly readily and thus may spread infestation to new sites; in flight they may be mistaken for small flies. Like many other insects, they have the habit of 'shamming death' when suddenly disturbed; the appendages are drawn close to the body and they remain motionless for some time.

In Germany *Anobium* is given the name 'Totenuhr' (death watch) which is reserved for *Xestobium* in Britain, because of the tapping noise which it is said to produce. There seem to be no British records of the production of audible sound by *Anobium*.

The sexes are about equal in numbers and pairing takes place freely on warm days. The females may mate several times in the course of their lives. A day or so after copulation the females begin egg-laying without requiring any food in the meantime. At 20–22° C and 80–90 per cent R.H., the adults were mostly observed to live for 20 days and a few for over 30 days [41].

Natural enemies

The larvae of *Anobium* are preyed upon by the larvae of certain other beetles of the family Cleridae, which seek them out in their burrows. *Anobium* larvae are also attacked by the mite *Pyemotes* (see page 376) and by various tiny parasitic wasps of the family Braconidae.

Life cycle
The duration of the life cycle extends apparently for about 3 years under normal room conditions. Wood may be found containing larvae of all sizes corresponding to different ages; those of 7 mg and over may be presumed to be in their third year [41].

The adult beetles emerge, mate and lay their eggs throughout the summer (June–August).

Speed of development
E at 15° C, 59 days; at 20° C, 19–20 days; at 28° C, 15–16 days; L, (?) 3 years; P at 20° C, 14 days; at 28° C, 10 days.

Death-watch beetle (Xestobium rufo-villosum) (Fig. 11.1)[182, 189]
Egg-laying seems to occur mainly in the daytime. The pregnant females walk slowly over woodwork, exploring possible egg sites, first with their antennae and then by probing with their ovipositors. Into suitable holes or crevices the ovipositor is extended and an egg laid. Sometimes groups of 2 or 3 up to more than 100 may be found in favourable sites. There is no evidence that the females are able to choose the decayed portions of wood which are most favourable for larval development.

The eggs are lemon-shaped, white in colour and about 0·6–0·7 mm long.

The young larvae emerge and they may crawl over the wood to choose a site for boring. Larval life is spent tunnelling through wood. The burrowing is able to progress much more freely in wood which has been weakened by fungal decay and the larvae thrive better in such wood.

As with *Anobium*, pupal cells are formed near to the surface of the timber. After completing pupation, the newly formed adults remain inside the pupal cells until the following spring before biting their way out. The beetles resemble *Anobium* in shape, but they are larger (7–8 mm long) and the elytra are mottled by patches of short yellowish-grey hairs.

The activity of the adult beetles depends upon the temperature. In natural infestations out of doors, the beetles are usually found resting under bark except in hot weather. On warm sunny days, or inside warm buildings, they tend to walk about over the surface of the wood. The beetles rarely, if ever, fly voluntarily although they will flutter their wings sometimes if launched into the air. The frequent infestation of wooden beams in the roof suggests that they may sometimes fly unobserved. However, they certainly depend a great deal on walking and may very often be carried into an uninfested building on a piece of infested timber. Intermittently they produce the characteristic tapping sound which is responsible for the common name of death-watch beetle. This is a series of 7 or 8 clicks in quick succession, caused by the beetle rapping its head against the woodwork. The sound is, no doubt, somewhat eerie and foreboding in a large, quiet, old oak-panelled or oak-timbered building. Both sexes will emit the noise; it is apparently a sexual call and can be elicited in a captured specimen by similar noises made artificially. The beetles mate quite freely at a temperature of 20° C, the male mounting on the back of the female and remaining in copulation for about $\frac{1}{2}$–$1\frac{3}{4}$ h.

Life cycle. The total life cycle depends not only on the temperature but on the suitability of the infested timber. Experiments at the Forest Products Research Laboratory [182] indicate a life cycle of 1–6 years indoors at a temperature of 22–25° C (71–77° F). Outdoors in a place sheltered from sun and rain, the life cycle extended for 3–6 years.

Speed of development. E, at 15° C, 36–50 days; at 20° C, 20–24 days; at 25° C, 12–16 days; L, (in decayed wood) 10–17 months (in undecayed wood) up to 10 years; P, at 18° C, 20–28 days, at 23° C, 18 days.

Ptilinus pectinicornis

The mating and life cycle of this beetle has been studied in Germany [135]. In undamaged wood, after copulation, the female makes a short bore hole at right angles to the grain. Secondary attacks make use of emergence holes and old pupal chambers, from which a new brood chamber is excavated. Often females mate soon after emergence and return to their own emergence holes.

The long thin eggs (1·5 mm × 0·075 mm) are pushed into the lumen of wood vessels; and the females remain in the brood chamber until death. The egg laying habit resembles that of the Lyctidae and this beetle, though an anobiid, is related to that family.

(ii) Powder-post beetles; Lyctidae [188]

Oviposition occurs mainly at the period of maximum activity which is at dusk. As already mentioned, the larvae of *Lyctus* depend for their nourishment on starch present in the wood. The egg-laying females 'taste' the wood before depositing their eggs and are thus able to distinguish suitable breeding material for their progeny. On a suitable exposed piece of wood, the females seek the transverse or longitudinal cut surfaces and lay their eggs in the pores; that is, in the exposed openings of the wood vessels. The ovipositor is inserted into the mouth of the pore and one or more eggs inserted, so that they may lie some distance from the surface.

The eggs are long and narrow, with a stalk-like prolongation at the top. If they are thrust into rather narrow vessels, they may be distorted by pressure so that they become even longer and narrower than usual.

The mature first-stage larvae begin by feeding on the residual yolk mass in the egg and eat their way forward and out of the shell. Then they may consume a few particles of the walls or contents of the wood vessel before settling down to moult. The second-stage larvae begin to bore into the surrounding wood (and often turn through one or two right angles to avoid reaching the surface of the wood).

The larvae gnaw their way through the wood swallowing a portion of it and filling their burrows up with 'frass' consisting of powdery faeces and wood fragments. There are no salivary glands to digest the swallowed wood, but the digestive juices (enzymes capable of hydrolysing starch, sugars and protein) are secreted by the midgut. The skeletal substance of the wood passes through the gut unchanged, the nourishment being derived from the cell contents. To a large extent the larvae rely on starch and they cannot develop in starch-free wood. For this reason they are confined to the sapwood; and where (as often happens) the starch is irregular in distribution, the larval attack will often be concentrated in patches or bands of high starch content.

The larvae gradually grow to a length of about 6 mm ($\frac{1}{4}$in.) in length. They resemble those of *Anobium* in general appearance but differ in the absence of spinules and the enlarged eighth abdominal spiracle which shows up very distinctly as a brown spot.

When they are mature, the larvae direct their burrows towards the exterior and pupate just below the surface. After the pupal period, the adults remain inside the pupal chamber for 3–4 days before biting an exit hole and emerging. The adults are reddish-brown to dark brown in colour and vary considerably in size (3–6 mm in length).

The activity and reaction of the adults are closely related to temperature and light conditions prevailing. They are very active at temperatures above 20° C when they run about on the surface of the wood. The activity is, however, mainly in the evening; during the day they often crawl into crevices between boards or hide in old exit holes. But at dusk they become active and frequently fly about. Copulation can take place immediately after emergence; it is most frequently observed at dusk. Both sexes usually live about 3–6 weeks, the longest observed life in the laboratory being 68 days.

Life cycle

The normal life cycle in Britain, under favourable conditions, is an annual one; but in wood with a low starch content, this may be extended to 2 years. The adults emerge, mate and reproduce in the spring and summer and the next generation larvae overwinter as half-grown or nearly mature larvae.

Trogoxylon parallelopipidum development took 3–4 months; *L. planicollis*, 9–12 months [204].

Speed of development

(*L. bruneus*) E at 20–23° C, 8–8½ days; P, about 3 weeks. Total at 23° C in sapwood of English oak, 6–12 months [465].

(*L. planicollis*) at 31° C and 85 per cent R.H., in red-oak chips and wood flour. E, 8·2 days; L, *c.* 40 days; P, 8·1 days; A, lived 33–35 days. Eggs per female, av. 51 [640].

(iii) Bostrychidae

The lead-cable beetle; Scobicia declivis [79]
Unlike most anobiids and lyctids, bostrychid beetles always bore into wood to form their egg laying chamber. The lead-cable beetle occasionally mistakes lead-covered electric or telephone wires for wood and burrows into them, with the possibility of causing a short-circuit. Normally, the female begins boring into wood at a knot or some other irregularity which gives her purchase. At a depth of about 8 mm, the burrow turns at right angles and continues parallel to the surface for some 6 cm. Eggs are then deposited along it.

The larvae mine along the grain of the wood, packing their borings behind them. There six larval stages. By the time of pupation, the larval burrow is about 50–60 cm long. After pupation, the adults eat their way out to the surface.

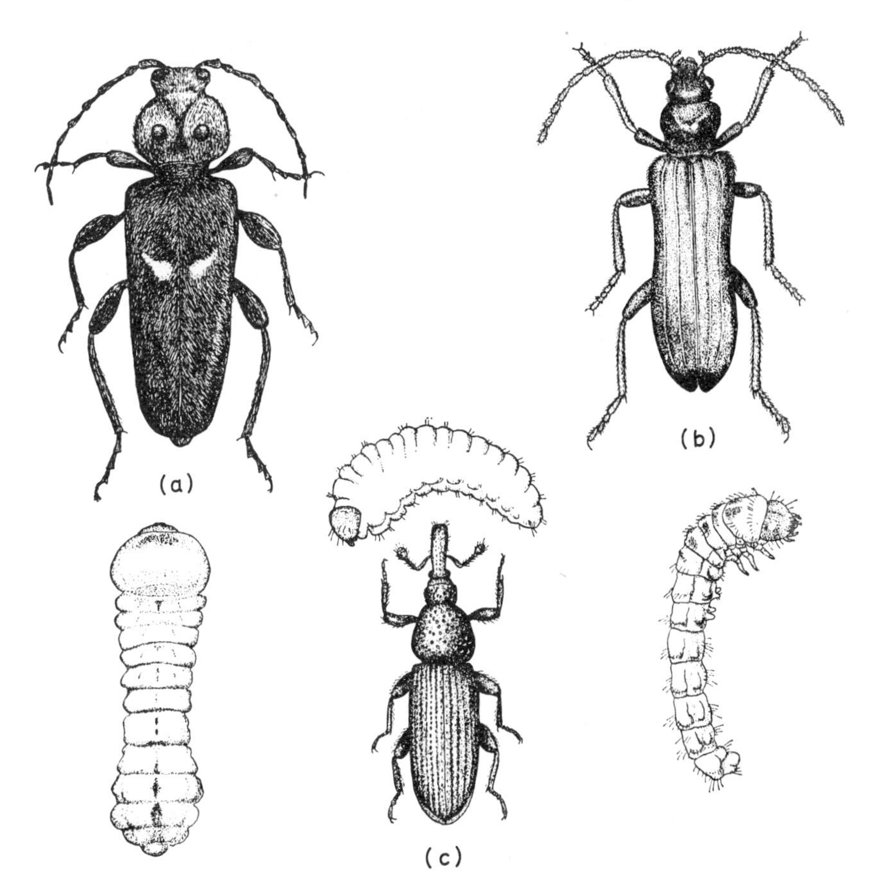

Figure 11.2 Further wood-boring beetles and their larvae. (*a*) *Hylotrupes bajulus*; (*b*) *Nacerdes melanura*; (*c*) *Euophryum confine*. Magnifications: (*a*) × 1·5; (*b*) × 5; (*c*) × 10. (*a*) after Forest Products Research Laboratory [187]; (*b*) (*c*) after Britten [64].

Life cycle and speed of development. In California, there is usually 1 generation a year, the adults being prevalent in the summer months. E, 3 weeks; L, 9 months; P, 2 weeks; A, live 10–40 days.

(iv) Longhorn beetles: Cerambycidae

House Longhorn beetle (*Hylotrupes bajulus* [42, 554]) (Fig. 11.2)
The females oviposit several times (1–7 times) and lay a total of about 100 eggs, although as many as 500 have been recorded. They are attracted to soft woods by the odour of their characteristic resins and lay eggs in small crevices.

The eggs are spindle-shaped, about 2 mm long.

The larva is a straight-bodied fleshy white grub, divided by deep transverse folds into rings or segments. The head is sunk into the prothorax, so that only the dark brown jaws are visible. They burrow into the wood forming a tunnel which becomes crammed full of powdery frass. If the walls of the tunnel are carefully examined, the bite marks of the larvae can be observed, even with the naked eye. These marks, said to resemble ripples in the sand left by the tide, are apparently characteristic of the pest. The larval tunnels are often parallel to the surface and separated by a this layer, which may bulge outward, like a blister.

The larvae (Fig. 11.2) grow to a length of about 30 mm and then pupate, near the surface as usual.

The adults (10–20 mm) bite their way out of the wood, leaving their characteristically large emergence holes. Since relatively few larvae can destroy a large amount of wood, the emergence holes are relatively rare. The adults do not live long (males 1–2 weeks, females 2–3 weeks). In the laboratory they do not feed, but probably do so in the field.

Life cycle. In Europe, the life cycle ranges from 3 to 11 years, being affected by the nature of the timber as well as environmental conditions. The adults emerge during the summer months. In the southern U.S.A. the cycle is 3–5 years; northward from Washington D.C., 5–8 years [557].

Speed of development. E at 16·6° C, 48 days; at 31° C, 6 days. L and P (in South Africa), 1·75–5 (av. 3) years. In U.S.A., 2–10 (usually 3–5) years.

(v) Weevils: Curculionidae

Euophyrum confine [277] (Fig. 11.2)
Little is known of the life history. The larvae tunnel in the wood forming a series or inter-communicating, more or less parallel galleries, thinly divided from each other and partly choked with frass. Many galleries are close to the surface and only covered by a thin layer of wood.

Life cycle. Adult beetles (3–5 mm) have been found from March to December, which seems to indicate that there may be two overlapping generations in a year.

Pentarthrum huttoni [240]
The life cycle has been studied at 25° C and 95–100 per cent R.H. About 4 days after mating, the female lays her eggs, singly in holes excavated by the mandibles or in crevices in the wood; the holes are then sealed with glue exuded by the ovipositor.

The larvae, which are legless, tunnel in the wood, forming burrows parallel to the surface. Pupation is in a cell which becomes lined with fungal hyphae. Adults emerge at tunnels at an angle of 45° to the surface.

Speed of development. E, 16 days; L, 6–8 months; P, 16 days.

(vi) Oedemeridae

Wharf borer (Narcerdes melanura) (Fig. 11.2)
The life history has not been carefully studied. The larvae tunnel mainly in rotten wood, although they may penetrate for an inch or so into sound wood. Water-sodden wood is chiefly attacked, especially wharf piles a foot or so above water level, or timber sunk in the ground. The pest is practically confined to wood infested with fungus, either dry rot (*Merulius lachrymans*) or one of the other fungi which attack wet wood, such as *Coniophora cerebella.*

The adults (6–12 mm) sometimes emerge in large numbers, especially in early summer, and invade dwellings and office buildings.

(c) Importance [269]

Nearly all the 17 million private dwellings in Britain are of partly timber-framed construction. Part of the wood is structural, as in walls, floors and roof; part is in fittings, such as doors, window frames, skirtings and cupboards. Over 90 per cent of the timber used in building is imported from North America or Europe. A very high proportion derives from *Pinus sylvestris* (Scots pine, Baltic deal, European redwood). Of secondary importance is *Picea abies* (Norway spruce, European whitewood) and, more recently, *Tsuga heterophylla* (Western hemlock). Various other softwoods have also been used.

During the period 1960–65, a wide survey of damage by wood-boring beetle pests was conducted by a commercial firm concerned with remedial treatment [269]. Some 142 000 buildings were inspected, including all types; but private dwellings constituted 85 per cent of them. About half were constructed before 1919 and only 5 per cent after 1940. The survey revealed that the biggest pest was *Anobium punctatum*, responsible for 75 per cent of cases, while *Xestobium rufo-villosum* and the wood boring weevils each contributed about 5 per cent. Surprisingly, *Lyctus* damage amounted to only about 1 per cent and *Hylotrupes bajulus* about 0·3 per cent.

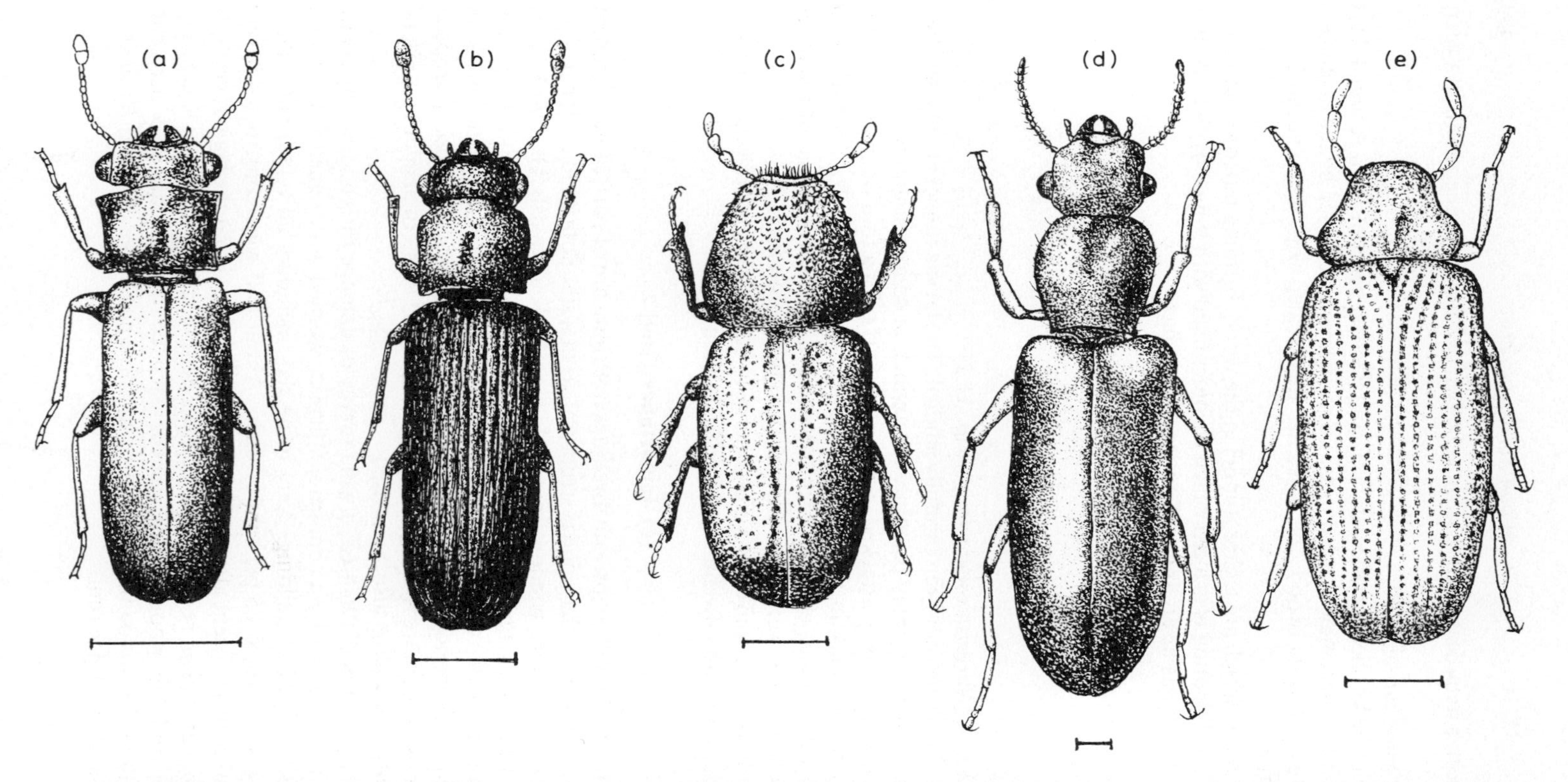

Figure 11.3 Some North American domestic wood-boring beetle pests. (*a*)*Trogoxylon parallelopipidum*; (*b*) *Lyctus planicollis*; (*c*) *Scobicia declivis*; (*d*) *Polycaon stouti*; (*e*) *Hemicoelus carinatus*. (*a*), (*b*) after Gerberg [204]; (*c*) after Burke *et al.* [79]; (*d*) after Doane *et al.* [146]; (*e*) after White [607].

The various types of wood boring pests have different spheres of activity, as given below.

(i) Anobiidae

Anobium punctatum [186]
Anobium breeds out of doors in dead or dying parts of trees and in fence and gate posts. Inside buildings, the beetle may breed in most forms of wooden furniture (including plywood and wickerwork) as well as in flooring, rafters and other structural timber. The presence of emergence holes spoils the appearance of the woodwork and the damage caused by the larvae may seriously weaken articles of furniture. In large wooden beams, however, *Anobium* does not usually penetrate sufficiently to endanger their mechanical strength. Almost any type of wood, whether from a deciduous or coniferous tree, is liable to attack; but softwoods are more usually damaged. Attack is mainly confined to sapwood (except in timber where the heart-wood is not clearly distinct, such as beech, birch and spruce). Accordingly, floorboards with a high proportion of sapwood may be seriously weakened. Very old wood is less frequently infested than wood only a few years old.

Optimum conditions for development of *Anobium* are a temperature of 22–23° C and a relative humidity of 80–90 per cent. Temperatures above 28° C and relative humidities below 40 per cent prevent development; infestations do not flourish in very warm dry rooms, such as those in centrally heated buildings. On the other hand, the beetle can develop satisfactorily, though slowly, in unheated places such as cellars and store rooms.

Xestobium rufo-villosum
The natural habitat of this beetle is in decayed or dying trunks or large branches of hardwoods, chiefly oak and willow; it has also been found in other hardwoods, and once or twice in conifers. As a pest in buildings, *Xestobium* is most troublesome for its attacks on structural timbers, which may actually endanger the stability of timber-framed buildings. Large pieces of furniture or fittings, such as church pews and screens or oak panelling, are also liable to attack. The beetle is essentially a pest of old seasoned wood and it has seriously damaged the roofing timbers of many old and historically important buildings.

The wood most frequently infested is oak, but other hardwoods may be involved and, occasionally, pinewood.

Other European species
Ptilinus pectinicornis is not very often responsible for damage to domestic woodwork, being more restricted in its choice of wood. Generally speaking, only hardwoods are attacked, most commonly beech, maple and sycamore.

Ernobius mollis lays its eggs only on softwoods which retain the bark and the borings are superficial. This is not a serious pest and its main importance is a possible confusion with *Anobium punctatum*. Since it only attacks wood with some bark attached, it is more likely to be found in rough sheds or in rustic furniture than in dwellings.

Other North American species

The eastern death-watch beetle, *Hemicoelus carinatus*, is widely distributed in the eastern states; it attacks sills, joists, beams and flooring of various hardwoods. The damage resembles that done by lyctus beetles, but the emergence holes are somewhat larger and the frass coarser.

H. gibbicollis, the Californian death-watch beetle, breeds in well-seasoned Douglas fir timber, along the Pacific coast, causing serious damage.

Priobium sericum, in eastern U.S.A. and *P. punctatum* in the west, cause considerable damage to wooden house fittings.

Xyletinus peltatus is one of the most common anobiids in the southeastern U.S.A. It is particularly damaging in moist conditions with associated fungi, and it will attack both hard and soft woods.

(ii) Lyctidae

In complete contrast to *Xestobium*, the *Lyctus* beetles are mainly injurious to new or recently prepared timber. Planks and boards or partly manufactured articles are very liable to become infested while being stored in sheds by the timber merchant or manufacturer. Very often the infestation is not discovered until the wood has been made into furniture or employed for panelling, flooring or other structural uses. Complaints are made and the suppliers have to replace the damaged wood and moreover suffer from loss of reputation.

Lyctus is fortunately more restricted in the type of timber which it will infest. It will only breed in the starch-containing sapwood of deciduous trees. Furthermore, it has the habit of laying eggs in the pores of the wood (as described in the life history) and certain kinds of wood with very narrow pores are not infested (e.g. horse chestnut). Others (cherry) have rather narrow pores which only just permit attack, and these are not very often infested. In spite of these restrictions, a very large number of important timbers are susceptible to this pest, including oak, walnut, ash, hickory, sycamore, sweet chestnut, elm and African mahogany; also imported tropical hardwoods like agba and obeche.

Lyctus brunneus is by far the most common powder-post beetle in Britain and it is widespread in the U.S.A., as are *L. planicollis* and *L. cavicollis*. *L. linearis* is mainly prevalent in the eastern states. Another important Nearctic species is *Trogoxylon parallelopipidum*.

(iii) Bostrychidae

Scobicia declivis

According to Ebeling [158] ‘Lead cable borers are common along the Pacific

Coast, particularly in California and southern Oregon, occurring at all elevations up to 1600 m (5300 ft). Outdoors, they normally infest dead and seasoned oak . . . , but they will also infest acacia, eucalyptus, maple . . . and other hardwoods. They can reinfest the wood from which they emerge. They occasionally attack newly painted houses, apparently being attracted to them, and have been known to infest hardwood panneling and floors. They are also attracted to corks in bottles containing alcoholic beverages, or to hard liquors or wine in oak casks. . . . '

Polycaon stoutii
This species occurs in the Pacific Coast area and in Arizona. It attacks any softwood and several hardwoods. It often damages furniture, especially if veneered; also seasoned hardwoods in lumberyards and buildings.

(iv) *Cerambycidae*

Hylotrupes bajulus
Longhorn beetles (family Cerambycidae) are essentially forest insects, which breed in the bark and wood of trees and logs. *H. bajulus*, however, is exceptional, since it can breed in seasoned wood (although it will develop even more quickly in recently felled timber). It attacks the sapwood of softwoods only.

During the present century, the status of the beetle in Europe has changed from that of a rather uncommon insect to a widespread pest. It has been introduced into North and South America and also South Africa, where it is a severe nuisance. Though long resident in Britain, it has only become a pest in the past 25 years. Although there are records throughout England it is only serious in the south, especially in Surrey. Damage in houses is usually found in structural timbers in the roof space; but doors, window frames, etc, may be attacked. It has been recorded breeding out of doors (in tree-stumps and telegraph posts) but this is not very common. Infestations may be spread by adults flying from house to house or by movement of infested timber.

In the U.S.A., this beetle is known as the old house borer, because it can infest buildings after construction, in contrast to the new house borer, *Arhopalus productus*. *H. bajulus* is currently confined to the eastern states and mainly breeds in houses, especially attic framing, along the Atlantic Coast.

Arhopalus productus
This beetle occurs throughout most of western U.S.A. and Canada. It breeds in dead or dying trees, especially after forest fires. When infested timber is used for house construction, larvae may continue developing and emerge within a few months. However, as the wood dries out, the younger larvae are unable to complete development; nor are fresh infestations possible.

(v) Curculionidae

Euophryum confine and *Pentarthrum huttoni*
Both these weevils attack only wood which has been (or is being) attacked by fungi; for example, *Coniophora cerebella*; one of the less serious dry rots. As a rule, then, they are most liable to occur in wood which has become damp; near leaking pipes in the vicinity of sinks, lavatories and bathrooms. They also occur in damp situations in breweries, wine vaults and beer cellars. *Euophryum confine* has been introduced into Britain from New Zealand (where, apparently, it is not a pest). The first record in Britain dates from 1937, and subsequently it has been found breeding out of doors in south-east England. A number of infestations have been reported in houses in the London area.

(vi) Oedemeridae

Nacerdes melanura
This beetle is thought to be introduced from the Great Lakes region of North America. In England it has become quite common in wharves of many estuaries in the south, from the Thames round the south coast to Cornwall. It occurs not only in piling and timber supporting the river bank, but also in wooden barges. It has also been found in sodden wood largely buried in the ground and also in badly maintained houses (e.g. in leaking lavatories). On one occasion, however, it was found in a London church, 35 ft above ground.

(d) Control

(i) Detection of infestations

The most obvious sign of attack by wood-boring beetles are the exit holes made by the emerging adults. It is sometimes possible to distinguish emergence holes which have been recently made. Their outline is sharp and the wood within fresh and bright; dull and weathered holes are probably relics of an earlier attack. The presence of the holes proves that a piece of timber has, at some time, been infested; it is less easy to determine whether living grubs are still present. One method is to place some clean paper below the suspected wood for a day or so. If an extensive infestation is present, the activity of the grubs is liable to dislodge frass from the ramifying galleries and some of this tends to be thrown out of the exit holes on to the ground below. Frass may also be thrown out by the activity of the predatory enemies of the wood-boring beetle larvae.

During the summer months the actual beetles may be found. Adults of *Anobium* are often seen crawling on walls, ceiling or windows; or they may be seen in flight. The death-watch beetles are less active and often fall from an infested ceiling to the floor below. Since they do not wander far, their numbers and distribution may give a useful indication of the extent of activity above. Their absence from the floor, however, does not mean that the timber above is free from attack.

(ii) Preventive measures

Anobiidae

Attacks of *Anobium* in furniture are less likely if the wood is kept in a warm well-ventilated room. Structural timber which is kept in good condition and free from decay, is not liable to damage by *Xestobium*. Therefore, especial care in house construction should be taken to prevent damp conditions through leakage and bad ventilation. Some sites very liable to attack are: wall-plates, wall-posts, the feet of the principal rafters and hammer beams and timbers near the ridge.

Care should be taken not to introduce beetle larvae in timber used for repairs or in second-hand furniture, which should be examined carefully and treated if necessary. Old or disused articles of furniture, especially ones containing plywood (as in backs of picture frames) and wickerwork, should not be stored in attics or cellars without regular examination. They are a frequent source of infestation of structural timbers.

Certain insecticidal treatments may be made to timber to prevent insect attack. For rough structural timbers where discoloration is unimportant, creosote or creosote derivatives are effective, especially if applied by impregnation methods. For the treatment of decorative woodwork, impregnation with certain proprietary proofing agents may be carried out at the time of manufacture.

The modern contact insecticide treatments found effective against *Lyctus* have shown promise in protecting timber against attacks of other beetles, such as *Anobium*, *Xestobium* or *Hylotrupes*; on the other hand, it is not certain for how long the protection can be relied upon. Thorough application by dipping or spraying are most effective. The use of smoke generators cannot be relied upon to leave deposits on vertical or inverted surfaces; therefore they have little protective value, although they are likely to kill any beetles present and may prevent successful oviposition for a short time.

Lyctidae

The powder-post beetles present a rather different problem from the Anobiidae. Their dependence on the presence of starch in the wood has suggested the possibility of preventing attack by eliminating the starch before the wood is exposed to attack. The various methods depend on the natural metabolism of starch by the timber if the wood cells are not killed too quickly. One way is to season logs with the bark on for a few months. The seasoning may be accelerated by cutting into planks and heating gently for a week or so [316]. The most practical method appears to be seasoning with the bark intact, taking precautions against rot from too damp conditions.

The optimum temperature is about 40° C; stronger heat, say 45° C or above, kills the cells and conserves the starch.

Modern contact insecticides provide a useful method of preventing *Lyctus* damage; for example in sawn timber awaiting use. The best treatment is actual immersion, for 10 s in an emulsion containing 2 per cent DDT or 0·5 per cent HCH or 0·5 per cent dieldrin. This should give protection for 3 years. Where dipping is not feasible, some protection can be obtained by thorough spraying of the above-mentioned insecticide emulsions, preferably in March–April before the emergence of adult beetles begins. This should be repeated annually.

These insecticides are also useful for addition to synthetic glues used during plywood manufacture.

(iii) Eradication of infestations

The measures to be adopted in dealing with wood-boring beetles will, of course, depend on the extent and severity of the infestation. Where there is likelihood of extensive damage to structural timber (e.g. by *Xestobium*) in Britain expert advice should be obtained from the Building Research Establishment.

The big British servicing company which undertook the survey mentioned earlier (page 433) has a standardized method of dealing with wood beetles in dwellings. These involve specific methods of treatment for (1) roof voids (attics), (2) floor boards, (3) joinery (skirting boards, architraves, door lintels, etc), (4) staircases, and (5) external woodwork on half-timbered houses.

Heat

Relatively small articles can be disinfected by heat in a temperature- and humidity-controlled kiln. The exposures necessary for the penetration of an effective temperature into planks of different thicknesses are given in the Building Research Establishment Leaflet no. 43. Care is, of course, necessary to prevent damage to varnish and glue in polished furniture, which should not be appreciably affected at temperatures up to 55° C with humidities up to 80 per cent R.H. (wax finishes are always affected; but the wood can be re-polished).

At 55° C and 80 per cent R.H., the exposure periods necessary to kill *Lyctus* larvae range from $2\frac{1}{2}$ h for 1 in. thickness of wood, to 4 h for 2 in. or $6\frac{1}{2}$ h for 3 in. (see also page 48).

Insecticides

A suitable insecticide should be effective, harmless to the treated article and not dangerous to man. The main difficulty is to penetrate the wood and perhaps the most suitable agent is a liquid with a toxic vapour. *Ortho*-dichlorbenzene has been found very satisfactory but it should not be used too freely in confined spaces, because continued exposure to the vapour is liable to cause chronic poisoning of the liver. Alternatively, various metallic naphthalenates or solutions of pentachlorphenol or *para*-dichlorbenzene in benzene have been used with success.

Relatively large articles can be treated by brushing the liquid over the unvarnished surfaces. Small and valuable articles may be treated by injecting *ortho*-dichlorbenzene into the exit holes and other crevices with a syringe.

Fumigation with methyl bromide is a method of destroying wood worm in valuable woodwork which might be damaged by liquid insecticides [243]. Laboratory experiments at 15° C indicate that a lethal concentration time products (mg/litre × h) for eggs of the common wood beetles is about 70, for larvae 200–300 and for adults 100–200. Small objects could be readily treated in a fumigation chamber. Also, field trials have shown good success in controlling *Xestobium* in H.M.S. *Victory* and the Round Tower, Windsor Castle. House fumigation in Denmark against *Hylotrupes bajulus* employs 1 kg per 10 m^3 [496].

The use of dichlorvos fumigation strips in attic spaces can only provide relief from attack by the adult beetles and should be done in the summer months. The larvae in the wood are unaffected, so that dichlorvos can only be recommended as an adjunct to other measures. Similar remarks could be made about the use of smoke generators to disperse *gamma* HCH. Both methods are, however, relatively simple and instructions on their use are provided by the Building Research Establishment.

Excision and fungicides

Certain wood-borers (the two wood-boring weevils and the wharf borer) only attack sodden wood, badly attacked by fungi. It is therefore essential to cut out all the infested wood and destroy it. The surrounding timber should then be treated with a fungicide as a precautionary measure. The oil-soluble pentachlorphenol can be recommended for this purpose, as it is toxic to insect larvae as well as being a fungicide.

Other methods [54]

Certain new physical methods of destroying woodworm have been investigated in recent years. Irradication by X-rays will kill the larvae; but very high doses (about 10 000 R) are necessary and in view of the difficulties of ensuring safe treatment, the method does not seem practical.

The use of infra-red radiation has been investigated but does not seem very promising.

11.2 TERMITES (ISOPTERA)

11.2.1 DISTINCTIVE FEATURES [247, 268]

Termites are insects which combine some rather primitive anatomical characters with complex social organizations. All termites damage wood and most are able to digest it, either directly by enzymes, or with the aid of intestinal protozoa.

They do not inherit these protozoa, but acquire them by anal feeding from other termites; and the stock also has to be replenished after moulting. Some termites feed on wood partly decayed by fungi; and certain higher termites cultivate special fungus 'gardens'.

As with all social insects, mutual contacts, caressing with antennae and the use of pheromones are important aspects of communal activity. Trail-marking pheromones are used to guide workers to food sources. These apparently depend on rather simple compounds, as quite ordinary chemicals (e.g. diethylene glycol monoethyl ether) are effective in deluding many species.

In their colonial organization and in their appearance, termites resemble ants, which accounts for the popular name 'white ants'. However, they can be easily distinguished on careful examination. The two pairs of wings of the sexual migrants are similar, covered with a network of veins, and project about 25–33 per cent beyond the abdomen at rest. This contrasts with the unequal wings of winged ants, with their sparse vennation. Again, both the sexual forms and the workers and soldiers lack the narrow waist of ants; and their antennae are straight and thread-like, not 'elbowed' like those of ants. Colony formation, in both sorts of insect normally begins with the migration of winged sexual forms, which discard their wings after the nuptial flight. In termites, the males remain to assist in formation of the colony, which grows only slowly and may require several years before flourishing. The queen becomes devoted to egg laying and her abdomen swells with the enlarged ovaries (to an enormous size in some species) and often she lives for years.

An important biological difference which, to some extent, affects the social organization, is that termites undergo only partial metamorphosis. Because the young termites are active nymphs and not helpless grubs, they can act as workers and are not fully committed to sterility, except in some higher termites. Nymphal forms can be of both sexes; they can begin to differentiate after a few moults into working forms or to soldiers. Workers have heads rather similar to the winged adults (though sometimes without eyes), but their bodies are pale and soft. Soldiers generally have large heads and jaws, or else have a poison ejecting apparatus well developed. The winged forms range from about 5 to 25–30 mm long, excluding their wings. Soldiers are generally in the range 3–20 mm long and workers a little smaller.

11.2.2 CLASSIFICATION AND DISTRIBUTION

(a) Classification

There are nearly 2000 species of termites, in 141 genera, falling into six families. Only four of these occur to any extent in the northern temperate region and one (the very large Termitidae) is of virtually no importance in this zone. This leaves three families: the Termopsidae, or damp wood termites; the Kalotermitidae, or dry wood termites and the Rhinotermitidae, or subterranean termites. These families can be distinguished from the winged sexual forms or from the soldiers; but the majority of workers are too similar for simple identification. The characters to be used include the *fontanelle*, a pit in the frontal region of the head, which is particularly strongly developed in the soldiers (and may be used for squirting repugnant liquid). Keys for these three families may be found in the Appendix (page 545).

(b) Distribution

Termites are most numerous in the tropics. In the Old World, three species only occur in Europe; one of them (*Reticulotermes lucifugus*) as far north as La Rochelle in France and Dnepropetrovsk, 100 miles north of the Black Sea [246]. In the New World, two species of *Reticulotermes* and two of *Zootermopsis* are found as far north as the Canadian border. In the southern U.S.A., several species are economic pests.

11.2.3 BIOLOGY

(a) Termopsidae: damp-wood termites

The Termopsidae are a primitive family, with comparatively few genera and species, each with limited and well-defined areas, just inside or well outside the tropics. Colonies are small and there is no definite worker caste. However, the nymphs in the early stages of colony development do not develop into reproductive forms until later (about 4 years in *Zootermopsis*, when about 450 individuals may be present).

They attack wood with a high moisture content and are often found in fallen coniferous trees. Wooden buildings in areas with a high water table, or near the shore, are especially liable to attack. They tend to work upwards, from the foundations to the roof rafters.

(i) Important species

Zootermopsis angusticollis (Fig. 11.4), the Pacific dampwood termite, occurs along the western part of North America, from British Columbia to southern California. It causes a considerable amount of damage, even in urban areas. Specimens have been found in timber imported into England from North America. *Zootermopsis nevadensis* occurs from British Columbia to north California and eastwards to Nevada and Montana. It can be equally destructive, though less prevalent in urban areas. *Z. laticeps* is found in Arizona and New Mexico.

(b) Kalotermitidae: dry-wood termites

This family is also rather primitive and does not possess a definitive worker

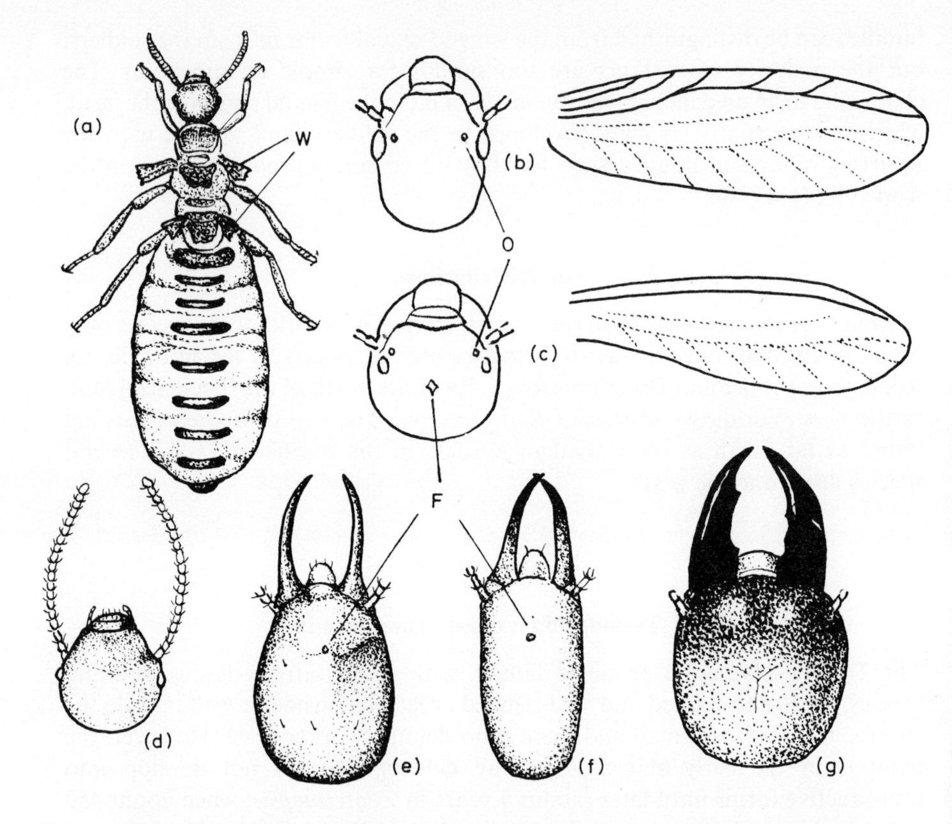

Figure 11.4 Termites. (*a*) De-alate queen, *Reticulotermes flavipes*; (*b*) head and wing Kalotermitidae; (*c*) head and wing, Rhinotermitidae; (*d*) worker and (*e*) soldier, *Prorhinotermes simplex*; (*f*) soldier, *Reticulotermes lucifugus*; (*g*) soldier, *Zootermopsis angusticollis*. Key: W, wing stumps; O, ocelli; F, fontanelle. (*a*), (*d*), (*e*) after Banks and Snyder [28]; (*b*), (*c*), (*f*), (*g*) after Harris [247].

caste. Its members live inside galleries excavated in non-decayed, relatively dry wood, on which they feed. They burrow indescriminately across or with the grain, making large pockets or chambers, connected by tunnels. They quite often infest dead branches of trees and from such sites near houses, may infest structural timbers, as well as poles, posts and lumber. Because relatively small pieces of wood can be infested, these termites are readily transported in crates and sometimes, articles of furniture.

(i) Important species

Kalotermes flavicollis is the only member of the family to occur in Europe. It lives in communities of up to 1000 (more often about 500) individuals. It often damages vines, especially after these have been partly attacked by boring beetles.

Colonies develop slowly. At 25° C, eggs hatch in 50–60 days. The first two nymphal stages take 24–31 days, but later ones are very variable. After a year, the colony will only contain 15–55 members and winged adults will not appear for at least another year.

Incisitermes (= Kalotermes) minor, the western drywood termite is common in California, where it causes considerable damage. New colonies develop slowly. Eggs take 24–90 days to hatch and further development, which may involve as many as 7 nymphal stages, 6–14 months at 21° C. One year after a nuptial pair have invaded wood, they will have produced one soldier and a dozen or so nymphs. By 15 years, besides the original queen and king, there may be a few supplementary reproductives, 120 soldiers and 2600 nymph-workers.

Incisitermes snyderi, the southeastern drywood termite, ranges from South Carolina to Florida and west to Texas.

Marginitermes hubbardi, the desert drywood termite, is found in the arid deserts of southeastern California and Mexico.

Cryptotermes brevis, the powder-post termite, was introduced into the U.S.A. from the tropics, but is now well established in Hawaii, Florida and parts of Louisiana. It develops in man-made structures, but not in natural habitats in the U.S.A.

(c) Rhinotermitidae: subterranean termites

The rhinotermitidae constitute a somewhat more advanced family, the colonies of which are nearly always found in the ground. This habit is not, however, limited to the family; many Termitidae are subterranean mound builders. The earth nests provide protection from surface temperatures and, more important, from the desiccation, to which the members are sensitive. To reach wooden structures above the ground, the working nymphs construct shelter tubes, from particles of sand, earth or wood, and sometimes their own faeces, cementing them together with a glue-like secretion. Various types of shelter tube are constructed. Some are broad flattened passages, running up the side of stone or concrete foundations. These are regularly used for communication between the working galleries in the wood and the nest. Other irregularly cylindrical tubes rise up from the earth for exploratory purposes. Still others are built downwards from the timber towards the ground. These are largely constructed of particles of wood and are lighter in colour than the others.

The subterranean termites attack mainly the soft spring growth of fairly damp wood, generally tunnelling parallel to the grain. New colonies may be formed by the migration of winged adults; but also by budding off from an existing colony and the sexual ripening of substitute reproductives. Colony development from winged adults is slow. Eggs hatch after 20–25 days (*Reticulotermes lucifugus*) or 50 days (*R. hesperus*). The early instars last a week or two and the later ones 10 days to a month. Nymphs assume active duties earlier in young colonies, but no normally before the third instar (6–7 months). Winged reproductives are no formed until the third or fourth year. A fully developed colony may comprise thousands of individuals, with extensive foraging powers.

Species of *Reticulotermes* penetrate furthest north in the temperate zone, occurring up to latitudes 30° or 40° N. *R. lucifugus* ranges from countries bordering the Mediterranean to those above the Black Sea. (This species has also become established in Massachusetts, U.S.A.). After a gap due to the cold high plateaux of Turkestan and Tibet, *R. speratus* appears in China and Japan. In North America, the genus is represented by two important species. *R. hesperus*, the western subterranean termite, ranges from British Columbia to Mexico, with eastward limits in Idaho and Nevada. *R. flavipes* (Fig. 11.4), the eastern subterranean termite, extends along the Atlantic coast, from southern Canada to the Gulf of Mexico. This species has been introduced into Europe at various times and has become established in buildings in La Rochelle and Rochefort (France) and Hamburg (Germany).

Other American species are *R. tibialis*, *R. arenicloa* and *R. hageni*. An introduced species, *Coptotermes formosanus*, which has long been destructive in Hawaii, has more recently invaded the Gulf Coast states.

11.2.4 IMPORTANCE

In the economy of nature, termites play a useful role in destroying lumber from dead trees; but to the extent that man uses timber in building construction, they can be serious pests where they are prevalent. It has been estimated that the cost of damage to buildings in the U.S.A. was about $500 million in 1968. This figure may need to be expanded (apart from inflation), because of the gradual expansion of termites in a northerly direction. Hickin [268] gave details of costs of preventive and remedial treatments in different parts of the world in 1971.

11.2.5 CONTROL

(a) Detection of termite damage

Because of their cryptic behaviour, termites are not often seen except at the times of the migration of the winged forms. In the warmer parts of temperate North America, these migrations are most likely to occur at the end of summer or in early autumn. Apart from this, it is important to keep a watch for other signs of attack in areas where they are prevalent.

(i) Drywood termites

These termites form small persistent colonies in timber, gradually enlarging the galleries as extra accommodation is required. The interior is thus gradually hollowed out, leaving a fragile shell, with some remains of the harder portions of wood core inside. The timber may then collapse at any time, under extra strain.

In order to dispose of their faeces, termite workers make small holes and discharge the pellets to the exterior. Small piles of frass may accumulate in this way, giving warning of termite action. The faeces are larger and different in shape from beetle frass. The identity of the termites responsible may sometimes be indicated by the shape of the pellets, which are about 1 mm long or slightly less.

(ii) Dampwood termites

Attacks of dampwood termites should be suspected in areas with a high water table or near the shore. These termites also discharge their frass through holes and the pellets tend to stick together when the wood is, as usual, moist.

(iii) Subterranean termites

Many colonies of subterranean termites consist of tens of thousands of insects, which can travel considerable distances foraging for wood. For this reason, their attacks on buildings are often more spectacular than those of drywood species. A careful examination of the foundations of houses should be made periodically, to look for shelter tubes from the soil to the woodwork. In the interior, dark blisters on the wood may also give indications of attack. Finally, many termites can be heard grinding away at the wood in quiet conditions. Sometimes, a sharp knock on the woodwork will provoke the soldier termites to respond by tapping their heads against the gallery walls.

(b) Prevention and control

The subject of prevention and control of termite attack is a very large one and cannot be adequately discussed here. Several excellent accounts are available [157, 268]. It can be said that in most cases, professional assistance is essential to effect proper measures. A few general principles may be stated.

(i) Prevention

New buildings should be sited with adequate drainage and ventilation assured by raising them on piers, rather than dwarf walls. Facilities for easy inspection should be provided. In construction, all woody debris must be removed, because expansion from existing colonies in wooden lumber is a more serious threat than initiation of new colonies by winged adults.

Soil poisons should provide a chemical barrier and metal shields or correctly placed concrete slabs may give mechanical ones. All the timber used should be either naturally immune to termites or be adequately treated with insecticide after seasoning.

(ii) Treatment

This requires skilled operators, especially with subterranean termites, since it is necessary to find and destroy the colonies liable to cause the damage. It is necessary to remove all wooden debris and all infested wood or supports and to insert mechanical or chemical barriers (as appropriate) to prevent further attack.

Various chemicals have been used against termites during the past 100 years. Creosote and sodium arsenite were among the earliest, followed by pentachlorphenol, sodium pentachlorphenate and copper naphthenate. More recently, modern organochlorine insecticides, such as DDT, HCH and dieldrin have been used. Restrictions for fear of environmental contamination may now limit their use, although studies have shown that most of the chemical remains *in situ*.

An interesting way of using stomach poison insecticides against termites is to employ blocks of wood as baits [38]. Blocks treated with poison are placed where foraging termites will find them and the poison is taken by workers and fed to other termites and the young in the nest.

Fumigation is also used, although it gives no protection against reinfestation. Methyl bromide is generally used, at 0·91 kg per 28 m^3 (2 lb per 1000 ft^3); or sulphuryl fluoride at half that rate. Sometimes whole buildings are covered with tarpaulines and fumigated.

CHAPTER TWELVE

Stinging, biting and urticating arthropods

12.1 INSECT STINGS

12.1.1 INTRODUCTION

The stinging insects belong to the order Hymenoptera, a highly evolved group of insects, some of which display very complex patterns of behaviour. Relatively few species are pests and many of them are beneficial, since they parasitize other, more harmful insects. The characteristic of the order which concerns us here is a tendency to elaborate the egg-laying apparatus, or ovipositor, to form a tool or a weapon. The more primitive branch of the order (the sawflies and wood-wasps) have ovipositors modified for sawing or drilling into the plant tissues in which they lay their eggs. The more advanced branch contains the wasps, bees, ants, gall wasps and insect parasites (Ichneumondae and Braconidae; see page 375). These higher types which may be recognized by their narrow wasp-waists, usually have the ovipositor modified to form an offensive weapon, the sting. The most normal use of this organ is that of the parasitic forms which employ it to puncture the host and lay an egg inside.

Wasps, bees and ants possess stings which are mainly used against other insects but can be turned against larger animals or man, with unpleasant results. The effect of the sting of a solitary form may be unpleasant, but, as is well known, some of these insects are social and the concerted attack of a large colony is naturally more serious than individual stings.

The insect colonies may be formed annually, or they may be of more permanent duration. They may be founded in two distinct ways. The more primitive method (which characterizes most tropical forms) is the departure of a group ('swarm') of workers with one or more fertile females. Alternatively, the sexual forms may be produced in large numbers over a limited time; and these fly off, mate in the air and the fertilized females begin new colonies independently. Thus, there is a tendency towards colonies containing only one fertile female (the 'queen').

12.1.2 WASP AND BEE STINGS

(a) Wasp biology

(i) Distinctive characters: classification

Certain non-stinging insects are sometimes mistaken for wasps or hornets; hoverflies (Syrphidae) for the former and giant wood wasps (*Urocerus gigas*) for the latter. Hoverflies can be distinguished by their alternate darting and hovering flight. The giant wood wasp lacks the narrow waist characteristic of the social wasps and the female has a very long rigid ovipositor.

The true wasps belong to the family Vespidae, with rather uniform structure. The eyes are kidney-shaped, the mouthparts relatively primitive and adapted for chewing and licking. The wings can be folded longitudinally, like a fan.

The sub-family vespinae comprises the common 'wasps' known to everyone, in the genera *Vespula*, and the hornet *Vespa crabro*. Although *Vespula* is divided into a few sub-genera, the species are rather similar in appearance and habits. The different types can be distinguished by proportions of facial sclerites and by the black and yellow patterns on the body (Fig. 12.1).

North European species [555]
Vespula vulgaris, the common wasp, *V. germanica*, the German wasp and *V. rufa* the red wasp (though not very red) are all prevalent, soil-nesting forms. *V. sylvestris*, the tree wasp, forms nests in trees and *V. norvegica* in gooseberry bushes or hawthorn. *V. austriaca*, the cuckoo wasp, is parasitic, using nests of other species, especially *V. rufa*; and it does not need to have a worker caste. *Vespa crabro* the hornet, usually chooses hollow trees to nest in.

North America: northern species [158]
Vespula vulgaris, *V. arenaria* and *V. maculata* occur in the northern states of the U.S.A. and southern Canada, across the continent, extending down only to high altitudes in the south. *V. vulgaris* tends to be associated with coniferous trees, nesting in rotted stumps, logs or moist ground. *V. arenaria* forms aerial nests, often under house eaves. *V. maculata* is known as the baldfaced hornet, distinguished by black and white markings; it resembles the true hornet in size and habits. The parasitic cuckoo wasp, *V. austriaca*, is widely distributed in Canada and northern U.S.A.

North America: eastern species
Vespula germanica has become well established in the eastern states. In America, it forms rather large nests, commonly in roof voids. *V. squamosa* is the commonest ground-nesting species in the southern states; while *V. vidua* is prevalent in north-eastern areas, occasionally nesting in buildings. *V. maculif-*

rons is found in central Atlantic coast states, nesting in the ground, usually in deciduous woodland.

Vespa crabro, which has been introduced from Europe, occurs sparingly in the Atlantic seaboard states.

North America: western species [56]
Vespula pensylvanica is the most common and troublesome wasp in the western states, from Canada to Mexico. Although mainly nesting in the earth, it sometimes uses roof voids and attics.

V. atropilosa, with a similar distribution, is less abundant, as is *V. sulphurea*, extending from Oregon to Arizona. Both are ground-nesting forms. Two other western species, *V. norvegicoides* and *V. albida*, usually build aerial nests; but sometimes in litter on the ground.

Paper wasps or umbrella wasps: sub-family Polistinae
The genus *Polistes* contains numerous species, many of which are social, but the nests contain much fewer individuals than those of the Vespinae. They may be distinguished by their spindle-shaped abdomens (narrowed at both ends) in contrast to the anteriorally truncated abdomens of ordinary wasps.

Palearctic species. Polistes gallicus is a prevalent species in southern France and Italy. *P. smelleni* and *P. jadwigae* occur in Japan.

North American species. Several species of *Polistes* can be troublesome pests, stinging agricultural workers if they are disturbed. *P. fuscatus* is most common, throughout Canada and the U.S.A. *P. apachus* occurs extensively in Texas and California. *P. exclamens* has sub-species in the south-west and south-east.

(ii) Life history: colony formation

The life history of most wasps approximates to that of other social insects, the young being helpless grubs tended by worker adults. In the Polistinae, these are potentially fertile females acting as workers, but capable of taking on the reproductive function if the colony-founding queen dies. In most Vespinae, however, there is a sterile worker caste, usually smaller and to some extent different in appearance from the queen. Unlike the social bees, wasp colonies are formed annually. Therefore, the life cycle must be considered in relation to colony formation, which begins each spring with the activity of a single fecundated female emerging from hibernation. This female begins to construct her nest of wasp paper, which she makes by chewing up wood scraped from weather-worn palings, fences, dead trees, etc. The wood fragments bound together with an adhesive saliva, form a thin but strong paper-like material when dry. The female begins her nest with a few cells hung upside down on a kind of stalk or pedicel. Over this is built an umbrella-like cover. The female lays her eggs in these cells; and when the grubs hatch, she feeds them with fragments of captured insects. Finally, the next generation emerges as sterile workers, who take over most of the work of provisioning and building. New cells are formed round the periphery of the original clump, to form a horizontal layer, or comb. When this reaches a certain size, another similar layer is constructed below it and suspended from the first by short stalks or columns. Eventually, 6 or 7 such combs may be formed, the whole surrounded by the cover, which is enlarged

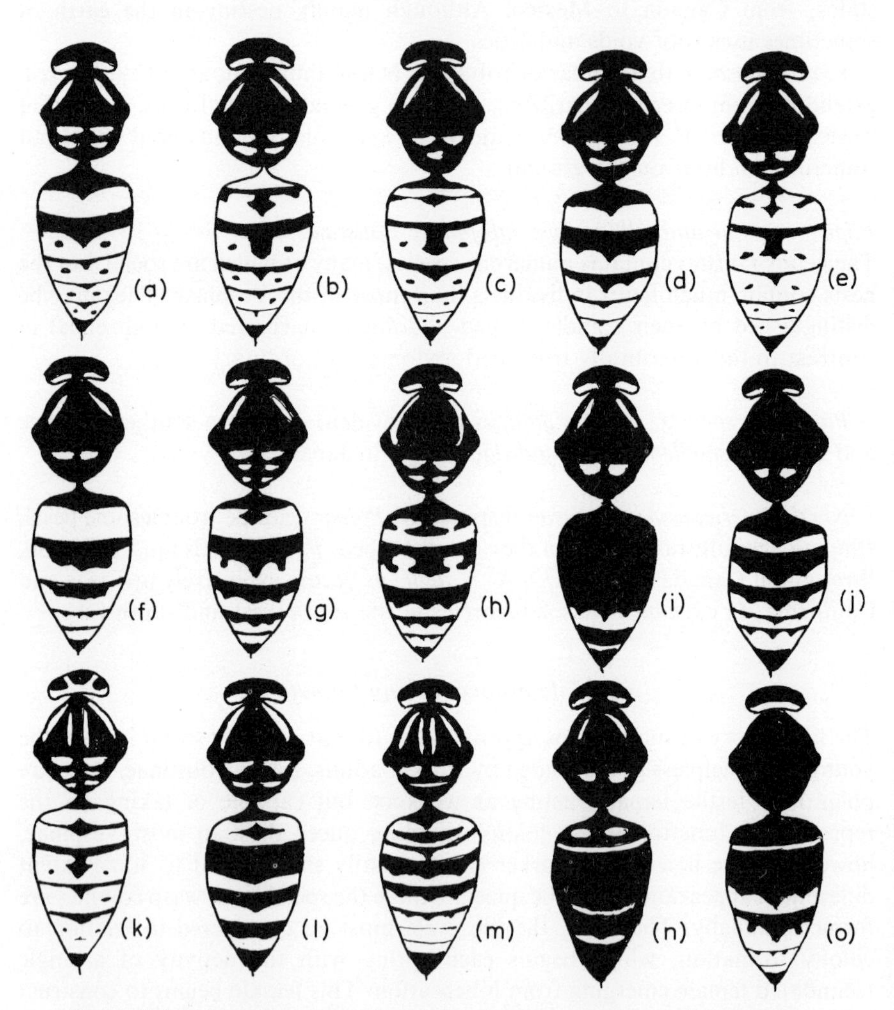

Figure 12.1 Body patterns of some common wasps. (*a*) *Vespula vulgaris*; (*b*) *V. germanica*; (*c*) *V. rufa*; (*d*) *V. sylvestris*; (*e*) *V. austriaca*; (*f*) *V. norvegica*; (*g*) *V. pensylvanica*; (*h*) *V. atropilosa*; (*i*) *V. maculata*; (*j*) *V. arenaria*; (*k*) *V. sulphurea*; (*l*) *V. maculifrons*; (*m*) *V. squamosa*; (*n*) *V. vidua*; (*o*) *Vespa crabro*. (Semi-diagrammatic; not to scale.) After photographs; (*a*) to (*f*) in Step [555]; (*g*) to (*n*) in Ebeling [158]; (*o*) original.

periodically. The cells of the combs may be used for producing another young wasp two or three times; and eventually a nest of this type may produce some 25–30 thousand wasps during a season.

As already indicated, wasp nests may be formed hanging from the branches of a tree or shrub; they may be located inside a shed or attic, or inside a wall space; but in many cases, they are formed in a hole in the ground, excavated first by the founding queen and later enlarged steadily by workers.

Later in the year, special large cells are constructed to provide the queens for the subsequent season; the young in these cells are particularly well fed and they develop into fertile females. Towards the end of the summer the female begins to lay unfertilized eggs (and some workers may also reproduce without fertilization). All these unfertilized eggs develop into males, which mate with the young queens. The fertilized young queens fly away to find a resting place in which to hibernate; the rest of the colony dies out in the autumn.

Foraging worker wasps seek food for the larvae and for themselves. The grubs require protein, for growth, and so the workers collect insects or fresh and decaying meat, including fish. The activity of the workers is promoted by a sticky secretion produced by the grubs, in exchange for their nutriment. For themselves, workers require sugar for energy; and they obtain this from the nectar of various flowers (e.g. cotoneaster, ivy) or from fresh or processed fruits. This explains their attraction to jam factories or fruit canneries.

Nests of polistine wasps are invariably formed above ground, in trees, or commonly, in barns, attics or under the eaves of houses. They are much smaller than those of vespine wasps and consist of a cluster of cells suspended by a single large pedicel. The shape of the nest resembles an inverted umbrella, from which Ebeling has suggested the name umbrella wasps [158].

(b) Bee biology [231]

(i) Distinctive characters and classification

As a group, bees may be considered as wasps which have given up the carnivorous habit and turned to feeding on the pollen and nectar of flowers. As a consequence, their bodies have become modified for collecting honey and pollen. The honey is taken up by a tongue-like process of the labium and stored in the crop; from which, however, it can readily be regurgitated for social purposes. The pollen is collected in a basket-like arrangement of hairs on the specially flattened hind legs.

Flowers, however, do not lose from the bees robbing them of their nectar and pollen, for the bodies of bees are covered with a dense coating of hairs and this assists in cross pollination, which is also of great benefit to man. Another peculiarity of social bees is their ability to secrete wax from glands under plates of the abdomen. This wax is used to build the cells and combs for rearing the young, either in a pure state or mixed with earth.

The varieties of bees are more numerous even than those of wasps and, perhaps surprisingly, about 95 per cent are solitary forms. Most of the social bees belong to the family Apidae, the best known being *Apis mellifera*, the honey bee. Another familiar type in the northern temperate region, is the humble or bumble bees, which are medium to large bees, robust and densely coated with hairs.

(ii) Life cycle; colony formation [555]

Bumble bees

The social organization of bumble bees resembles that of the wasps, rather than that of hive bees, in that new colonies are formed annually by hibernating fertilized females. After emergence from hibernation (which may be long – about 9 months in Britain) the female begins to form a nest in the earth, sometimes using an abandoned mouse hole. She lines it with dry vegetation and then makes little round cells of wax, which she furnishes with pollen for larvae which will hatch from her eggs and others with honey for herself. From the eggs, workers develop, which are smaller than the queen, but not clearly distinguished from her. New cells are now constructed, in each of which about 12 eggs are laid. At full development, a colony may comprise 100–500 bees. Late in the season, sexual forms are produced. The males are short lived, but the fertilized females remain to complete the cycle.

As with the wasps, some bumble bees have become parasitic, with the cuckoo habit. Members of the genus *Psithyrus* resemble the bumble bees (*Bombus*) although lacking the proper pollen collecting baskets. They enter the bumble nests, kill the queen and lay eggs in her cells, which are then tended by the *Bombus* workers.

Honey bees

While there are some wild species of *Apis* in India, the honey bee, *A. mellifera*, is rarely if ever found in the wild, although it has been introduced into almost every country in the world. The fertilized queen, with her long abdomen, is quite distinct from the workers, which are sterile females. She takes no part in nest building or food gathering, but confines her duties to egg laying. Over her life span of several years, she may produce $1\frac{1}{2}$ million eggs, most of which develop into workers. These produce the elaborate wax comb, with its two layers of cells, back to back. Since the comb hangs vertically, the cells lie in a horizontal position. Eggs laid in most cells develop into workers; larger ones are prepared for males and still bigger ones for prospective new queens. Still other cells are used to store honey.

New colonies are formed by the migration of the old queen, attended by a number of workers. A new virgin queen emerges and after a nuptial flight with the males, she returns to the hive to continue the original colony. The males (drones) are expelled from the hive in autumn and die. A fully developed colony may contain 50–80 thousand bees.

It sometimes happens that a migrating colony of bees will invade some part of a dwelling, and begin to infest a hollow wall space of an attic. This may be undesirable, because of the close proximity of the bees to people, especially children. An experienced bee-keeper may be able to entice the colony away into a new hive; but if this is not possible, it may be desirable to exterminate the bees.

(c) Nature of wasp and bee stings

(i) Anatomical features [522, 585] (Fig. 12.2).

The sting of a wasp or bee is carried in a cavity at the end of the body. The opening of this chamber is guarded by the last dorsal and ventral plates of the abdomen, which are separated only by a narrow curved slit, like a fish's mouth. Often the point of the sting can be seen protruding from this orifice.

On removing the lower plates, the sting can be seen lying horizontally between a pair of unjointed 'sting palps'. If it is examined more closely, the sting is found to consist of three portions, which together form the ovipositors of non-stinging Hymenoptera. There is a quill-like sting-sheath, with a basal bulb, hollowed underneath like an inverted gutter. Below this lie two long stylets, running on grooves underneath the sting-sheath and closing its trough-like hollow. The tips of these stylets, which project beyond the end of the sting-sheath, are pointed and barbed.

The bulb-like base of the sting-sheath diverges into two curved arms and the lower portions of the stylets curve outwards along them. These serve as anchoring rods and are articulated with three sets of plates. The sting is brought into use by internal abdominal pressure, which extrudes it; then muscles play on the anchoring plates and rods which rotate it downwards, to be driven into the victim.

The poison injected by the sting is secreted by two long glands inside the abdomen, which discharge into a reservoir, the poison sac. (Another gland, called Dufour's or the alkaline gland, was formerly believed to be concerned in the action of the sting; but its use appears to be connected with the normal sexual function of the ovipositor.) From the poison sac the venom enters a bulb-like enlargement at the base of the sting-sheath and it runs down, between this organ and the stylets, into the wound. If the stinging insect is rapidly brushed away by a human victim, there is a tendency (especially with bees) for the whole stinging apparatus, together with the poison sac, to be torn out of the insect's body and remain attached to the sting in the wound.

(ii) Nature of the venom [37]

Venom secretion in the honey bee begins just prior to emergence and reaches a maximum (about 0·3 mg) after about 2 weeks. Some protein is required for

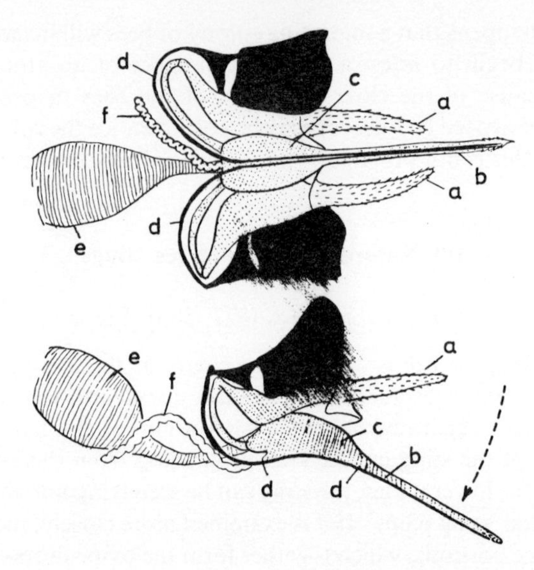

Figure 12.2 The sting of a honey bee, semi-diagrammatic. Above: ventral view. Below: lateral view with sting depressed for use. *a*, sting palps; *b*, sting-sheath; *c*, its bulb; *d*, stylets; *e*, poison sac; *f*, alkaline gland. (After various authors.)

maximum production and bees fed only on sugar secrete only a quarter of those taking pollen.

Recent studies have revealed the remarkable chemical complexity of bee venom. Histamine is present and the venom causes release of more histamine from histidine in the tissues. Histamine, however, is not the only or even the main toxin. Electrophoresis splits the venom into 8 fractions, 5 of which are bases precipitated by picric acid. Of the remaining 3, 2 are notably active. One of them, 'melittin', is responsible for some local pain and inflammation and also general toxicity. The other component contains at least two enzymes – hyaluronidase and phospholipase A – which are responsible for supplementing and spreading the effects of melittin; both are present in snake venoms. Wasp venom also contains histamine and hyaluronidase as well as other distinctive components such as 5-hydroxytryptamine. Hornet venom contains a surprisingly high proportion of acetylcholine.

The venom has four characteristic toxic effects: (1) a histamine effect, responsible for redness, flare and weal seen in the skin, (2) a haemolytic effect, (3) a haemorrhagic effect (which can be responsible for uterine bleeding), and (4) a neurotoxic effect, which tends to cause paralysis.

(d) Importance

The effects of stings are two-fold; the direct toxic action and the anaphylactic shock which may develop in people who become sensitized to it. The direct effects are well known to be painful, the severity increasing from the honey bee to the bumble bee, wasp and hornet; but they are not dangerous to man, except in large numbers. The chances of this are clearly much greater with large colonies. While numerous small colonies may sometimes prove a serious nuisance, (e.g. umbrella bees in citrus orchards [282]), the main danger comes from ordinary wasps and hive bees. Bees are especially liable to cause trouble, because of their association with man. On rare occasions, people have received over 500 stings, causing dangerous effects. The symptoms (faintness, respiratory difficulty, vomiting and diarrhoea) usually pass off in 24 h; but an attack of urticaria may follow in about a week.

The main danger from wasp or bee stings is the anaphylactic reaction, which may be provoked after a few stings or even a single one. Studies on the allergenic properties of the stings of various species of bees and wasps revealed 4–6 antigenic fractions in 5 species. Probably 2 fractions were shared by all species, while the remainder were specific. Individuals could become sensitized either to the generalized antigens or to the specific ones, or to both.

The severe allergic shock which is experienced by some people may cause death. Indeed, during the period 1950–4, deaths from stings of hymenopterous insects in the U.S.A. exceeded those from poisonous snakes (86 to 71); those from bees alone were about equal to those from rattlesnakes [472]. The symptoms arise within 20 min, reaching a maximum within 30 min and usually abating within 3–4 h. Symptoms include respiratory distress, faintness with partial or complete loss of consciousness, followed by an itching rash. The skin may be flushed, or in some cases become pale. Swelling of the face is common, even when the sting has been on a limb. Sometimes there is vomiting with abdominal pain, or cramp or diarrhoea. Death is commonly due to respiratory failure brought on by vasomotor collapse. This may be accelerated in people stung on the mouth or throat, by swelling of the mucous membranes and occlusion of the air passages. The rapid onset of these symptoms (as contrasted with most snake bites) explains their danger.

(e) Treatment [374]

When the sting can be found, it should be removed. It appears as a tiny black shaft with the white poison sac attached to its free end. It should never be pulled out by grasping this sac, or more poison will be squeezed into the wound; it is best to scrape away the sting by a knife-blade or the fingernail.

Various simple local treatments, such as the application of vinegar or washing soda, have been suggested; apparently these 'remedies' reflect the erroneous idea that bee or wasp venoms are simple acids or alkalis which can be neutralized. Probably, however, any cool, damp application will have a temporary analgesic effect and give some relief; moreover, the use of a definite and specific treatment may have a reassuring psychological effect. More scientific is the employment of an antihistamine. Phenindamine, in the form of a 5 per cent ointment has been said to cause local relief. Anti-histamine drugs (e.g. diphenhydramine) may be taken by mouth also, sometimes with good effect; but the absorption is slow.

Treatment of cases of anaphylactic shock is more serious and may require medical help. Stings on mouth or throat causing occlusion of the air passage may require tracheotomy. The best drug for treatment of acute systemic symptoms is adrenaline. An intravenous dose of 0·3–1 ml of 1 : 1000 adrenaline should be injected slowly.

Sensitivity to bee stings is at best inconvenient and possibly dangerous. Some people have found it advisable to undergo desensitization, under medical advice. The process consists of exposure to increasing doses of venom, beginning at very low levels. Bee venom is available commercially for this purpose and wasp venom could be prepared and used in the same way, if necessary.

(f) Control of wasps and bees [417]

(i) Collection of bee swarms

During the summer months colonies of bees may settle in places where they cause some alarm. The bees are quite mild in the swarming phase and may be quite easily caught by an experienced bee-keeper. Wearing a veil and with cuffs and trouser legs tied tightly, no harm should be expected. If the swarm has settled on the branch of a tree, it may be dislodged by a sharp jerk and caught in a cardboard box held below (one should be sure that the queen is not left behind on the branch). The box is sealed and returned to a suitable hive.

(ii) Destruction of colonies

Bees and wasps may forage half a mile or more, so that it may not be possible to trace their nests. In some cases, however, the source of a local nuisance may be found; for example, in the eaves of a house or some other undesirable place. Chemicals may be used to destroy such a colony. For a nest in the ground, carbon tetrachloride may be recommended, as it is rapidly effective against all stages. About $\frac{1}{4}$ - $\frac{1}{2}$ pint (150–300 ml) should be poured (or syringed) into the entrance hole, preferably at dusk, when most of the foragers will have returned. Alternatively a liberal application of 10 per cent DDT dust could be applied (if it is available). *Gamma* HCH is not recommended for this purpose, as it is somewhat repellent and the bees may avoid it. Pyrethroids, too are not always satisfactory (except, perhaps resmethrin) as they tend to excite the bees and wasps and may induce stinging. Other insecticides which have shown promise in liquid or powder form, are dichlorvos (1 per cent), propoxur (2 per cent) and carbaryl (5 per cent). Calcium cyanide, which was commonly a common recommendation, is no better than these insecticides and is more dangerous to handle.

(iii) Wasps at picnic grounds or parks

Wasps can be extremely troublesome at the seaside, at picnic sites and in zoological gardens or parks, especially where food debris from picnickers is available. The first essential is to provide waste-bins and make sure that they are used. It will be found that wasps (also flies and blowflies) tend to congregate in and around these. Good control of wasps (and other insects) can be achieved by spraying the inside of these waste-bins weekly, after emptying them, with 0·75 per cent dichlorvos emulsion. The surfaces should be heavily sprayed, with special attention to the rim [593].

(iv) Wasps in buildings

It may happen that elimination of a source of wasps is not feasible, yet some control is necessary; e.g. in a fruit-processing factory. Screening of windows and ventilators may be possible (see page 40) using wire mesh of a size not greater than $\frac{1}{8}$ in. Even if it is not possible to protect the entire building, it may be worth screening particular sites which are very attractive to the insects.

Wasps in buildings may be destroyed in large numbers by bait traps. A bait is prepared from jam, syrup, molasses, fermenting fruit or beer; a selection of the most attractive medium in the local circumstances can be made by small trials. Enough water is added so that the wasps will readily drown in the bait; the addition of a wetting agent (e.g. a domestic detergent) at about a teaspoonful per gallon will make the trapped wasps sink quickly. The prepared bait is poured into wide-mouthed jars or tins, which are placed where the wasps are likely to be active (e.g. near windows). The jars should be inspected periodically and the bait renewed when necessary.

12.1.3 ANT STINGS

(a) Kinds of ant involved [148, 559]

A general introduction to ant classification has been given earlier (page 321). In the northern temperate region, stinging ants belong mostly to the sub-families myrmecinae and formicinae.

(b) Nature of ant stings [106]

(i) Anatomical characters

The workers of ponerine, doryline and myrmecine ants all possess functional stings at the end of the abdomen. The general form of the sting is similar to that of wasps and bees; i.e. a pair of poison glands leading to a vesicle and an accessory gland discharging near to the base of the sting. In the dolichoderinae, the stinging apparatus is greatly reduced; but there is a pair of glands leading to two vescicles discharging just above the anus (the poison glands and sting lie below it). These glands produce a strong-smelling repugnatorial liquid which can be squirted out. In the formicine ants, the poison glands unite and form a coiled mass inside the vesicle. The sting is much reduced, but the ants are able to discharge an acid liquid from the orifice, projecting it several feet.

In most cases there is a tendency to combine the action of stinging or venom projection with biting. The ant may grip the victim's skin with its jaws, curl the abdomen forward and, using its head as a pivot, sting in several places. Formicine ants make wounds with their jaws and squirt poison on them.

Not all ants will use their sting or spraying apparatus when interfered with; some run away and others feign death. Certain forms tend to be more pugnacious when their colonies are large and well established. Furthermore, the readiness with which formicine ants will eject poison may depend on the weather; thus, *Lasius niger* is more prone to squirt the acid under damp conditions.

(ii) Nature of the poison

In the more primitive sub-families of ants, the sting venoms contain proteins like those of wasps or bees. The venom of *Solenopsis invicta* (Myrmecinae) contains five unique alkaloids, including solipsin A, an alkylated piperidine with necrotic, haemolytic and other properties [360]. That of *Iridomyrmex humilis* contains iridomyrmecin ($C_{10}H_{15}O_2$) to the extent of 1 per cent of its body weight [474]. It is used in defence and offence against other insects, in which it produces effects like those of DDT.

In the formicine ants, the fluid they eject contains formic acid, the quality and strength of which depends on the conditions under which the ant has been reared. In the wood ant (*Formica rufa*) the venom may contain 20–70 per cent formic acid, amounting to 12–18 per cent of the body weight.

(c) Important species

(i) Myrmecinae

Fire ants of North America [548]
Species of *Solenopsis* in the U.S.A. are renowned for their severe stings, which cause a burning sensation; hence the common name. Within minutes after stinging, a weal 4–8 mm diameter appears and the stinging sensation subsides. However, next day a small pustule appears at the site of each sting. In 3–8 days, the purulent material is absorbed or sloughed off leaving a pink area, which persists for several weeks before a scar is formed.

Solenopsis geminata, the fire ant, ranges from Texas to South Carolina and Florida, and southwards through Central America. *S. xyloni*, the southern fire ant, extends further westward into southern California. Both species mainly nest in the ground, but will sometimes colonize hollow spaces in dwellings. The slow moving trails of workers are practically omnivorous. They will take seeds from seed beds, and damage various kinds of fruit or sprouting plant. They tend aphids for honey-dew and will enter kitchens to steal various kinds of human food.

Solenopsis invicta, the imported fire ant, was apparently introduced into the U.S.A. from Brazil, and has now spread through the southern states as far as Texas. It forms large, hard earth mounds, as much as 3 ft high and 5 ft diameter; in extreme cases as many as 1000, but usually about 20–30 per acre. These interfere with agricultural operations. Each mound may house a very large colony (as many as 100 000 ants) which, however, remains distinct, apparently observing territorial boundaries.

Harvester ants of North America [132]
The large dark brown workers of the genus *Pogonomyrmex* are known as harvester ants, because they gather seeds and grasses to their nests. Most of the American species occur west of the Mississippi river. They include some of the most aggressive ants in the U.S.A. The reaction to their stings is not localized, but spreads along the lymph vessels, so that great discomfort may be experienced in the lymph nodes in the axilla or groin, long after the original pain of the sting subsides.

Pogonomyrmex occidentalis, the western harvester ant, occurs at high altitudes in the Middle West, from Idaho to New Mexico. It is a large red ant, sometimes damaging to grassland.

P. californicus, troublesome in southern California, is only active in hot weather. Like the honey bee, but unlike most ants, its sting is readily torn off and remains in the wound.

(ii) Formicinae

Ants of the genus *Formica*, often known as field ants, are widely distributed in the northern temperate zone. They are medium to large ants, brown, black or reddish in colour. Their nests are usually formed of mounds of twigs and leaf fragments. Most species are diurnal and omnivorous (predatory, scavenging and tending aphids for honey-dew). A few species will enter houses in rural areas, but they are mainly troublesome to campers and picnickers. Though lacking functional stings, they tend to bite and jet formic acid at the wound.

European species [148]
Formica rufa, the wood ant, ranges over North and Central Europe, from Britain to the Caucasas and Siberia. In southern Europe, it is found at high altitudes. It nests in or near woods, especially of fir trees. Although it avoids dwellings, it may enter gardens foraging. This ant secretes plentiful formic acid, which can be

squirted considerable distances. The fumes from workers in a disturbed nest can be so pungent as to cause coughing.

F. pratensis, the meadow ant, with a similar distribution, prefers to nest in a more open site. Its habits resemble those of *F. rufa*, but it is somewhat less aggressive.

F. exsecta, the narrow headed ant, is also found in North and Central Europe. It nests in open places, though sometimes on the edges of, or in clearings in, woodland; and it avoids human dwellings.

F. sanguinea, the blood-red ant, occurs throughout the Palearctic region and is represented by several sub-species in North America. It is a restless, aggressive ant, which attacks intruders by biting fiercely and ejecting acid. It practices slavery on other ants, especially *F. fusca*, the Negro or silky ant; i.e. it captures pupae of workers of the smaller species and uses them to perform domestic duties in the nest.

North American species [132, 548]
Formica exsectoides, the Allegheny mound ant, is a common eastern species, ranging from Nova Scotia down to Georgia; and inland, from Ontario to Iowa. As the name implies, it is noted for building large mound nests, up to a yard (90 cm) high. It avoids dwellings, but may be a pest in recreational areas.

F. occidua, the Californian red-and-black field ant, *F. obscuripes* (western thatcher ant) and *F. pilicornis* are all western species; *F. obscuripes* rather to the north and *F. pilicornis* to the southern Pacific region. *F. occidua* and *F. pilicornis* can occur in urban areas and may invade houses. *F. obscuripes* prevails at high altitudes and may become a pest in mountain cabins.

(d) Effects of ant stings, etc

The venoms of ant stings produce effects analogous to those of wasp and bee stings; however, since the quantity of venom is generally smaller, the reactions are less. Nevertheless, repeated stings from fire ants can cause severe pain. Generally people avoid being stung many times; but this has been recorded for some intoxicated men who inadvertantly disturbed a nest of fire ants; and there is another case where a stoic sufferer from rheumatism inserted his arm in an ant's nest in the belief that this might effect a cure. Few British ants cause severe stings; but even the tiny *Monomorium pharaonis* can rupture the delicate skins of infants in hospitals.

Ants which eject formic acid cause the symptoms to be expected from small splashes of that chemical. Drops in the eye can cause severe smarting; on areas of delicate skin, they cause milder smarting. The skin of the hands is not affected except after prolonged exposure, as of an entomologist investigating a nest. Ants which bite and inject formic acid into the wound, cause a pricking sensation followed by a considerable stinging feeling.

12.1.4 SCORPION STINGS

(a) Distinctive features of scorpions

(i) Anatomy and classification [115]

The general appearance of scorpions is familiar to most people; and although the species differ in size, they all show a marked resemblance to each other. The two lobster-like claws are the pedipalps, not the chelicerae, which are quite small. The cephalothorax bears the 8 walking legs and the abdomen comprises a broad basal part and a mobile tail-like portion, bearing the sting.

There are six families, the most important being the Buthidae, with about 600 species, including the most dangerous ones. Scorpions are mainly tropical or sub-tropical, although some species occur in southern Europe and the U.S.A. south of latitude 40°N.

(ii) Biology

Scorpions are solitary creatures, showing no trace of social behaviour. They are carnivorous, feeding mainly on other arthropods. Like ticks and spiders, they can withstand starvation for several months. Because of their aggressive habits, the males (like those of spiders) perform courtship rituals, sometimes long and bizarre, to signal their intentions to the females. The young develop in the female's body, in pouches analogous to uteri, and they are born alive. They commonly climb on to their mother's back and remain there, without feeding for 10–16 days, when they moult. A day or two later, they disperse, though not widely, and live independent lives. The number of moults before maturity appears to be 7 or 8.

(iii) The sting and venom

The external appearance of the sting is generally obvious. It comes into action as the 'tail' is arched over the body, while the victim (if small enough) is gripped by the chelate pedipalps.

The venom has not been studied extensively, but it appears that there are two kinds. One is local in effect and comparatively harmless to man. The other is neurotoxic and haemolytic, resembling some kinds of snake venom, and this can be dangerous. The sting of *Androctonus australis*, the fat-tailed scorpion of North Africa, is said to have venom as dangerous as that of a cobra.

(b) Some important species

(i) European species [115, 278]

A few species occur in southern Europe; *Buthus occitanus* is the most harmful. It grows to 5·5–6 cm long and is tawny with dark ridges, living among stones in arid places. Its sting contains about 8 mg venom.

Several species of *Euscorpius* occur in the region, generally smaller as well as less dangerous. *E. flavicaudis* (3·6 cm) and *E. carpathicus* (2·7 cm) in France, *E. italicus* (4·7 cm) in Italy and *E. germanus* in the Tyrol. *Belisarius xambeui* in the Eastern Pyranees, is another less dangerous form.

(ii) North American species [553]

The most dangerous scorpion is the yellow *Centuroides sculpturatus* (Fig. 12.3), which reaches about 7 cm in length when fully grown. It occurs mainly in Arizona

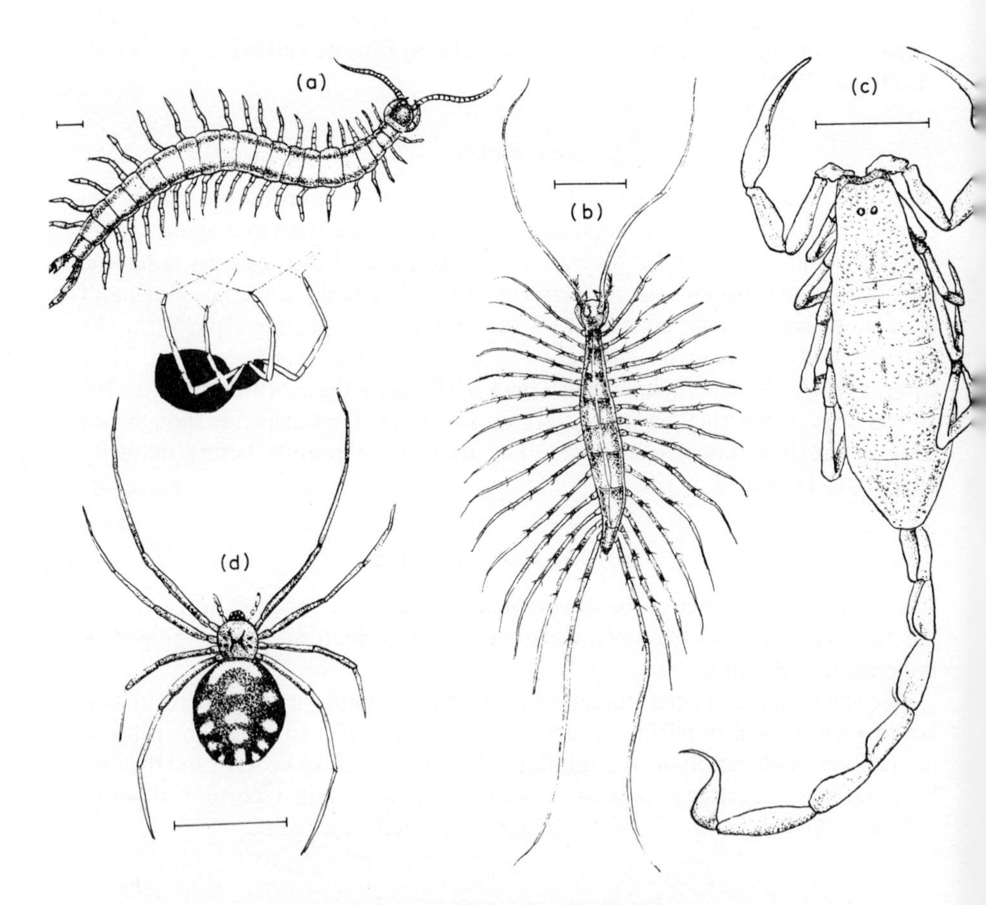

Figure 12.3 Some stinging or biting arthropods. (*a*) *Scolopendra obscura*, a European centipede; (*b*) *Scutigera coleoptrata*, the house centipede; (*c*) *Centuroides sculpturatus*, the sculptured scorpion; (*d*) *Latrodectus tredecimguttatus*, the European 'widow' spider, or *malmignathe* (and above, the attitude of widow spiders in their webs). (*a*), (*b*) after Back [20]; (*c*) after Ebeling [158]; (*d*) after Planet [487]. Scale bars represent 1 cm.

and has been responsible for 75 deaths (mainly of children) over a 40 year period (1926–65). This is twice as many deaths as from rattlesnakes and coral snakes in that state.

Another species, *C. vittatus*, is much less dangerous, the venom producing mainly local effects. This species ranges from New Mexico to South Carolina. Some other types of scorpion in North America appear to be colour variants of these two species (*C. gertschi* of *C. sculpturatus; C. pantheriensis* and *C. chisosarius* of *C. vittatus*).

(c) Control of scorpion stings

(i) Prevention

Human contacts with scorpions near dwellings may occur when firewood is handled or lumber moved. Indoors, they are liable to occur in attics or in the crawl-space under floors of the house. If such places are often infested, they could be treated with a residual insecticide.

(ii) Effects of stings; treatment

Scorpion stings with systemic effects cause numbness in the injured area, followed by spreading partial paralysis and twitching of muscles, which may lead to respiratory difficulty. Profuse perspiration and copious secretion from mouth and nose may lead to sneezing. In grave cases, there may be convulsions leading to death. Otherwise, recovery may begin within a few hours, leaving a temporary sensitivity around the stung area for several days.

If possible a physician should be summoned. The patient should be kept warm and rested and reassured. Anti-venin is available from Poison Centers in the U.S.A.

12.1.5 SPIDER BITES

(a) Distinctive features of spiders [62, 63]

Spiders belong to the Arachnida, like mites and ticks, but are far better known than the latter. The body is divided into a cephalothorax, which bears the four pairs of legs, and an abdomen, separated from it by a distinct waist. The mouthparts consist of a pair of leg-like palps and two-jointed chelicerae, which contain poison glands. The second joints of the chelicerae are fang-like, with an orifice near the top for injecting the poison. A small number of spiders have very toxic venom and can harm people with their bites; but the great majority are harmless, either because their fangs are too small and fragile to penetrate the skin, or the venom is too weak. All spiders are carnivorous and their combined attacks must take a considerable toll of insects. It has been estimated that the weight of insects thus destroyed in England may well exceed the weight of the human population.

The classification of spiders is somewhat complex and need not be considered in detail. There are two main divisions. In the Mygalomorpha, which contains the large hairy spiders of the tropics, the piercing fangs articulate downwards, closing towards the underside of the cephalothorax. In the other main group, the Araneomorpha, the fangs project forward and somewhat downward, but close together, inward. This is a very large assembly of families with different habits. These include some groups with common names. The Cribellate, or lace-web spiders have specially modified silk producers, as well as normal spinnarettes. The crab-spiders shuffle sideways, like crabs (Sparassidae; Thomisidae); the jumping spiders (Salticidae); the wolf-spiders (Lycosidae; Pisauridae); Lynx-spiders (Oxyopidae); money-spiders (Linyphiidae); the comb-footed spiders (Theridiidae and others).

(b) Important biting species

(i) Latrodectus spp.

These belong to the comb-footed family (Theridiidae) with rows of stout bristles on the tarsi of their hind legs, which they use to draw out strands of silk to fling over their prey. They live in scaffolding webs of various sizes.

European species [278, 487]
Latrodectus tredecimguttatus (12.3) occurs in Spain, France, Italy, Asia Minor and southern Russia. It makes a large irregular web, strong enough to catch quite large insects. The female, which is 12–14 mm long, has a large globular abdomen and hangs upsidedown in the web. The colour is black or dark brown and the abdomen bears a number of white and vivid red marks. This spider, known as the *malmignathe* in France, can inflict unpleasant bites, though perhaps not so serious as commonly believed (Fig. 12.3).

North American species [310]
In North America there are five related species of *Latrodectus* [349]. *L. mactans*, in the southern U.S.A., is known as the black widow spider, a name also given to *L. hesperus* in the western states and to *L. variolus* in the north-east U.S.A. and Canada. Two other species occur in Florida: *L. geometricus*, the brown widow and *L. bishopi*, the red widow. All these spiders are generally similar, with males 3–6 mm and females 8–15 mm long. Black or dark in colour, the abdomina bear various markings on the back, but underneath they nearly all carry an hour-glass shaped red mark. They live under piles of debris and logs, in rodent burrows and in shrubbery. They will also invade outdoor privies, sheds, cellars and garages; and these sites clearly offer opportunities for contacts with man.

(ii) Loxosceles spp.

Loxosceles reclusa, the brown recluse spider of the U.S.A., can cause serious bites, though less frequently than the black widow group. This species occurs in the mid-west, from Ohio to Kansas and south to the Gulf of Mexico. The adult females are about 7–12 mm long, brown, with a dark mark like a violin (base forward) on the cephalothorax. Several related species occur elsewhere. *L. unicolor* and *L. arizonica* are found in the south western U.S.A. (Kansas to southern California); *L. devia* in southern Texas and *L. rufeseens* in the eastern and south-eastern states. *L. laeta* has been introduced from South America into one or two areas.

Loxosceles spiders have been found in rocks, wood piles and miscellaneous debris; also in houses. Indoors, they often hide in lumber, empty boxes and unused garments; and human bites may result from careless handling of such articles.

(iii) Tarantula spp.

There are various species of *Tarantula* (family Lycosidae) including some small and quite harmless ones in England. The bite of the notorious *T. tarantula* of southern Italy has given rise to folklore, including much exaggeration. It was suggested that the bite causes a morbid lethargy, from which the sufferer can only be roused by vigorous dancing (the tarantella). In fact, there is considerable doubt whether the bites of these spiders are particularly severe; and in some cases, the bites of the *malmignathe* may be responsible.

(iv) Mygalomorph spiders

This group includes the large 'bird-eating' spiders, covered with hairs. One of them occurs in the southern U.S.A.; the genus *Aphonopelma*. In Britain, such spiders are sometimes encountered in crates of fruit imported from the tropics. Because of their menacing appearance and by vague association with folklore, they are often wrongly called tarantulas. In fact, their bites are no more serious than pin pricks, although their hairs may have an urticating property.

(c) Spider biology [62, 115]

(i) Life cycle

Male spiders are usually considerably smaller than females; and it is generally known that they risk their lives in mating, because of the agressive predatory habits of their prospective mates. Accordingly, some species perform various mating rituals to signal their intentions to the females; others clamp the female chelicerae with their own, for safety. Sperms from the male orifice are transferred to the female by means of curious pouches on the ends of the palps. The fertilized females spin different kinds of silken sacs (bag-shaped, bun-shaped, etc, 5–20 mm diameter) into which they lay their eggs. After hatching, the young spiders rest for a few days and moult before emerging. In some cases, there is a degree of maternal care before the spiderlings disperse.

(ii) Quantitative bionomics

Laboratory studies have shown that *Latrodectus mactans* females produce up to 8 egg sacs, with an average of 250 eggs each. *L. hesperus* produces as many as 21 egg sacs with about 200 eggs each. Incubation in both species is about 14 days at 25° C; but rather more in nature. Males pass through 4–7 instars, taking 54–88 days (*L. mactans*) or 62–151 days (*L. hesperus*). Females have 6–8 or 9 instars, taking 112–140 days (*L. mactans*) or 137–242 days (*L. hesperus*).

Loxoscelus reclusa produces 1–5 egg sacs, with about 50 eggs each. The young spiders emerge after 25–39 days. There are 8 instars, taking an average of 336 days to reach maturity. Males live about 540 days and females 630 days.

(d) Spider bite venom and its effects

(i) Nature of the venoms

The venoms of American *Latrodectus* spp. contain various constituents, including low-molecular weight compounds, polypeptides and enzymes. It is not clear which are the harmful components, but probably the peptides are responsible, while the other ingredients facilitate their distribution and action. The venom of *L. 13-guttatus* contains two fractions effective against insects and one against mammals. This last is intensely potent; 1 mg contains 230 LD50 doses per kg animal. The signs of intoxication are typical of the *Latrodectus* group [195].

Spiders of *Loxosceles* spp. carry about 0·33 mg venom; usually this is not all injected at once, but the spider often bites twice in succession. Electrophoresis studies revealed 3 antigenic fractions, which differed slightly in different species. The effects were necrotic and haemolytic, but not neurologic [546].

(ii) Bites and their effects

The quantity of venom injected is minute; and since spiders are solitary creatures, multiple bites are extremely unlikely. These facts are fortunate, since the venom of some species is excessively powerful, being about 15 times as potent as rattlesnake venom. Because of the tiny dose, however, only about 1 per cent of black widow bites are fatal as compared to 15–25 per cent of untreated rattlesnake bites.

The bite of *Latrodectus* is not always felt immediately; but pain soon develops and reaches a maximum in 2–3 h and then gradually subsides. Local pain is accompanied by sweating, difficulty in breathing, muscular rigidity (hard abdomen) vomiting and, in extreme cases, convulsions and prostration.

The effects of *Loxosceles* venom are somewhat different, being primarily local rather than systemic, although in some cases, death may result. Generally, however, the worst effect is a painful necrotic lesion, which eats away the flesh at the site of the bite. This heals slowly, leaving a depressed scar after a month or two.

(e) Control of spider bites

(i) Prevention

Spiders are not agressive to large vertebrates, so people are bitten mainly when they damage them. Not infrequently, this can happen when spiders are roughly handled in shifting old lumber, or putting on footwear which has been undisturbed for some time, or crushing spiders in bed. 'Control' therefore consists of avoiding such encounters and by preventing the accumulation of boxes, papers and lumber in areas where the spiders occur. If a spider should fall on to exposed skin, it should be brushed off gently, not crushed.

(ii) Treatment [515]

First aid treatment for *Latrodectus* bites is somewhat rudimentary. The patient should rest, be reassured and cold compresses applied to soothe the pain. Medical aid should be obtained. Anti-venin for *L. mactans* is available in parts of the U.S.A. but it is important to make sure of the species responsible for the bite. Young or elderly patients may need hospital treatment.

Treatment for bites of *Loxosceles* is somewhat empirical. In the early stages, the physician may decide on local excision of the bite area. Later, steroid therapy may be tried.

12.1.6 CENTIPEDE BITES

(a) Distinctive features of centipedes [115]

(i) Anatomy and classification

The general form of centipedes is well known and the only possible confusion is with millipedes. The latter, which are harmless, except as vegetable pests, are more cylindrical and have two legs on each body segment. Centipedes are flatter in cross-section, and each body segment bears only one pair of legs.

The jointed poison claws are not part of the mouthparts, but are borne on the first segment of the body behind the head. The tips of their strong piercing terminal segments have the orifices of the poison glands. A number of species produce painful and unpleasant bites; but there is only one recorded case of a fatality (that of a 7 year old Phillipino child).

There are five sub-orders, three of which include temperate species capable of biting man. These are the: Scolopendromorpha, with up to 23 pairs of legs (mostly 21), including some giant tropical species; Lithobiomorphia, mainly rather small, with only 15 leg-bearing segments; Scutigeromorpha, remarkable for the great length of their 15 pairs of walking legs, very agile and unique in having compound eyes.

(ii) Biology

Centipedes lack fully waterproof cuticles and accordingly tend to be nocturnal, hiding by day in damp places, under stones or debris. They are nearly all carnivorous and occasionally cannibalistic. They prey on small arthropods, including many insect pests; and they will also eat small worms and slugs.

The sexes are rather similar in appearance. Mating is not well understood. Eggs are laid in moist soil or vegetation. The females of Scolopendromorpha show a degree of maternal care for the young. Development is slow, taking about 3 years to reach maturity. There are 9 or 10 instars, during the first few of which the number of legs gradually increases. Adults may live a further 2–3 years.

(b) Biting species in the temperate zone

(i) European species [115]

Scolopendra cingulata is a common species around the Mediterranean and in the Near East. *S. morsitans* is an Asian species, but has been widely dispersed by commerce; a specimen was found in the South of France. Both species can reach considerable lengths in some areas (up to 18 cm) but the European ones tend to be rather small. They can inflict severe bites, causing discomfort and oedema.

Scutigera coleoptrata (Fig. 12.3) also occurs in southern Europe, often invading houses. There is a record of it being introduced into a paper mill in Aberdeen and establishing itself there. This can give bites causing pain similar to a bee sting.

Lithobius forficatus, not uncommon in England, will bite, causing a prick, with little sign of injury.

(ii) North American species [20]

Scutigera coleoptrata has been introduced into the U.S.A. apparently via Mexico, and is widely distributed there.

Scolopendra polymorpha, occurring in south-western states, reaches a length of about 15 cm. Its bite causes considerable pain and swelling. *S. heros*, another rather large species, occurs only in southern California. In addition to their bites, these species have claws which make tiny punctures if they crawl over bare skin. When alarmed, they secrete a vesicating liquid, which enters the wounds and these later become intensely irritating and inflamed.

(c) Control

Damp harbourages near dwellings, such as piles of rubbish, lumber or vegetable refuse, should be removed. To eradicate the *Scutigera coleoptrata* from houses, insecticidal sprays of dusts should be effective.

12.2 INSECT BITES

12.2.1 MECHANISM OF INSECT BITES

Used in this sense, 'biting' insects is a misleading description. The so-called 'bites' are really punctures made by the mouthparts of bloodsucking insects; whereas the true biting insects, such as food pests, are those with chewing mandibles.

Various bloodsucking insects and acarines have been mentioned in the pages of this book. The habit has apparently developed several times, independently, in different orders, so that the mouthparts have been modified in various ways to pierce skin and suck out blood.

(a) Formation of the wound [140, 209, 210, 342]

Relatively shallow wounds are made by some insects (and also ticks), the mouthparts making lacerating movements to create a pool of blood, which is then sucked up. This laceration is done by snipping movements of the mandibles in blackflies and biting midges, by stabbing of the mouthparts of horseflies; while the stable fly rasps a hole by rapid movements of toothed lobes at the end of the proboscis. Insects with long fine stylets are able to bend these about in exploratory movements inside the host's tissue. Mosquitoes, bed bugs and, apparently, fleas are able in this way to find a capillary under the skin, from which the blood is tapped. Mosquitoes are also able to feed on the pool of blood leaking from damaged tissues, under the skin.

(b) Injection of saliva and bloodsucking

The mouthparts of all the bloodsucking insects are shaped in such a way that they form two tubes – e.g. by the apposition of grooves on two opposite elements. One tube is usually formed by the apposition of the mandibles or maxillae and is relatively wide; the other (often a hollow along the hypopharnyx) is relatively narrow. Saliva is injected into the wound through the narrow tube, and the blood, mixed with saliva, is sucked up the wider tube by a 'pump' in the pharynx.

12.2.2 HUMAN REACTION TO BITES

Some insect bites cause a sharp stinging pain (e.g. horseflies, stable flies) but the attacks of many bloodsuckers cause little or no sensation. Nearly all the unpleasant effects develop subsequently and are mainly due to the insect's saliva which is injected with the bite.

The human reactions vary very greatly in different individuals and this is the probable reason for many stories of laymen about selective biting insect parasites. If two people sleep in the same room and one *suffers* badly from insect bites, he will imagine that all the insects have chosen to bite him; whereas his companion may be equally bitten but not show any effects.

Three types of reaction may be observed:

(1) *Haemorrhagic maculae.* These are small red marks surrounding the site of the bite, which may develop without any symptoms of irritation. In the course of several days, these marks become darker and finally brownish and they disappear only slowly.

(2) *Delayed-reaction papules.* A delayed reaction may be observed from a few hours up to as much as 14 days after the bite. Characteristically there is a red raised spot or patch ('papule') with inflammation and extensive swelling, usually accompanied by intense irritation. This reaction may persist for several days.

(3) *Immediate-reaction weals.* These weals appear within a few minutes of the bite, but do not last long, usually less than an hour. They are small whitish raised patches surrounded by an inflamed area and they cause moderate irritation.

The first of these effects described above is practically innocuous. The red colour is due to rupture of capillaries and diffusion of blood into the tissues. Liability of the patient to develop them seems to be related to proneness to bleeding (possibly associated with a low platelet count). The marks may be caused by bites of lice, fleas ('purpura pulicosa'), stable flies or ticks.

The two other effects of insect bites (which are the important ones) are allergic responses to the introduction of foreign proteins into the human body. These act an antigens, to which the body develops antibodies [390]. When the antigen–antibody reaction occurs in the tissues, there may be local production of a histamine-like substance, causing irritant skin reaction.

The foreign proteins introduced by insect bites are far from simple. In addition to saliva of the insect, there may be contamination from regurgitated fluid, so that it is probably better to describe the liquid as 'oral secretion'. Furthermore, insect saliva itself probably contains more than one component, as a rule. Thus, a variety of different chemical entities have been shown to exist in different regions of a mosquito's salivary glands. A further complication concerns differences in the sensitizing substances produced by different insects. Thus those in mosquito oral secretions are probably proteins, with a minimum molecular weight somewhat over 10 000. On the other hand, the sensitizing compounds associated with flea bites are dialysable and consequently of relatively low molecular weight; they are probably haptens which combine with proteins in the host to form antigens [43].

The responses to insect bites are often, like other allergic reactions, rather specific; so that a person sensitized to one species of mosquito [5] (or bed bug [250]) may scarcely react to the bite of another one. On the other hand, there may be varying degrees of cross-sensitization. Thus people sensitized to bites of human fleas may also react to fox fleas, to which they could never have been exposed [291]. The degree of specificity of bite reaction seems to be highest in the early stages of sensitization.

The rate at which the tissues respond in production of antibody differs in individuals and also according to the type of insect bite. The first bites of some insects usually do not elicit any reaction of this type. For example, no reactions to primary louse infestations are observed for 7–10 days. On the other hand, many insects can cause a delayed reaction after their first attack. This delay may be 1–7 days with *Cimex* [250] or 3–34 days with *Phlebotomus* [574]. If the insect bites are repeated at intervals, it is found that the time lag decreases and the severity of the reaction increases up to a maximum. Subsequently the severity may decline and even disappear; but if the insect bites cease for a few weeks, this acquired immunity is partially lost. The time required for development of maximum sensitivity appears to be about 1–2 months with *Cimex, Phlebotomus* and *Aedes*. In investigations with *Pediculus* and with infestations of *Sarcoptes*, a severe reaction developed in about 1 and 3 months, respectively; but the reaction was so unpleasant and severe that it was not feasible to continue heavy infestation of the volunteers to determine whether a subsequent desensitization could be obtained. [478].

An interesting development of this form of allergic response is the reactivation of old sites. The places bitten a few weeks previously may swell up again and become irritant when the same type of insect has bitten freely elsewhere on the body. This is interpreted as the result of local antibody production at the old bite sites. The reaction is caused by minute traces of antigen circulating in the blood from the new bites.

Immediate-reaction weals occur at a later stage of sensitization than the delayed-reaction papules; usually when the latter have reached their maximum or are beginning to decline in severity. Both types may successively appear in the same person; or the immediate weal may occur without any further reaction. It is not clear whether the weal is a reaction to a different antigen; or a changed response to the same one.

The progressive changes in reactions to bites of a particular insect seem usually to pass through various stages, as follows [257, 404].

State	Immediate reaction	Delayed reaction
I	–	–
II	–	+
III	+	+
IV	+	–
V	–	–

Individuals may not pass through all stages of this scheme. For example, they may begin in stage II, by showing a delayed reaction to the first bite. (This may, perhaps, be due to a cross-reaction from a related insect bite.) In many cases, people do not succeed in reaching stage V, of complete immunity; and some may remain in stage III.

12.2.3 TREATMENT OF BITES

As was remarked in relation wasp and bee stings, it is unlikely that any simple substances applied externally to the site will be radically effective, because of the complexity of the toxins involved. Treatment must be symptomatic, using analgesic applications to allay irritation and inflammation.

Various antihistamines have attracted attention as possible methods of relieving reactions to mosquito bites. When taken orally, they may relieve the itching, although they do not reduce the visible reaction. Local application in creams cannot be relied upon.

Scratching may, of course, lead to sepsis which would require appropriate treatment.

(a) Desensitization

Scientific evidence supports the popular impression that continual exposure to biting insects eventually results in more or less immunity. This would correspond to an individual passing into stage IV or V, in which the unpleasant delayed reaction is lost. Attempts have been made to accelerate this process artificially, by graded injections of suitable insect extracts containing the appropriate antigen. There is a good deal of disagreement in the results, probably due to investigators depending on subjective assessments by patients. There is a further difficulty in determining the previous history of bites, especially since the insects responsible are rarely caught (or even seen) and even when specimens are provided they may be 'anything from grain beetles to small weed seeds' [291]. Clinical trials may also be complicated by the changing prevalence of various biting insects at different seasons or from year to year.

Taking everything into account, it may be said that there is some evidence of successful desensitization to fleas [43], horseflies (*Chrysops*) and mosquitoes [390]. The delayed reaction (which is the more unpleasant) appears to be easier to prevent than the immediate weal reaction. Possibly further success may follow research on improved methods of preparing desensitizing inoculants.

12.3 URTICATING INSECTS

12.3.1 BLISTER BEETLES

Quite severe irritation of the skin may follow handling or crushing certain beetles of the families Meloidae or Staphylinidae. The latter are the common beetles with the wings folded up under very short elytra; two European blister beetles in this family are *Paederus limnophilus* and *P. gemellus*. Better known than these is the Spanish fly of southern Europe, *Lytta vesicatoria* (Meloidae). This contains the vesicant substance, cantharidin, formerly used as an aphrodisiac. It occurs mainly in the elytra and has the formula $C_{10}H_{12}O_4$, with a structure somewhat like dieldrin.

The blister beetles of the U.S.A. *Epicauta* spp. also contain cantharidin in their body fluid. When disturbed, they inflate their tracheal spaces and thus exert hydrostatic pressure, so that if roughly handled, their legs break off and exude the vesicating liquid. This causes severe blistering of exposed skin.

12.3.2 CATERPILLARS WITH IRRITATING HAIRS

There are also some moth larvae bearing urticating hairs on their bodies (a subject well reviewed by Southcott [551a]). Sometimes these are hollow and contain irritant liquid which is liberated like nettle poison if the hairs penetrate the skin and break off. Other urticating hairs seem to contain no poison and their effects may be mechanical, although there is some evidence of allergenic action.

(a) European species [320]

Probably the most common offenders are members of the family Lymantriidae, the caterpillars of which have long tufts of hairs along the body. Troublesome species are: *Euproctis chrysorrhoea*, the golden tuft moth, *Lymantria dispar*, the gypsy moth and *Nygmia phaecorrhoea*, the brown tail moth. Others are; *L. monacha*, the black archer, *Dasychira pudibunda*, the pale tussock moth; also *Epicnaptera quercifolia*, a lappet moth (Lasiocampidae) and the garden tiger, *Arctia caja* (Arctiidae).

(b) North American species [529]

Two of the lymantrid family have been introduced from Europe and constitute pest species, apart from their urticating hairs; these are *L. dispar* and *N. phaecorrhoea*. Other members of the family with urticating hairs are the tussock moths, *Hemerocampa* spp. Perhaps the worst offender of this kind in the U.S.A.

is the puss moth, *Megalopyge opercularis* (Megalopygidae). Others are the saddleback caterpillar, *Sibine stimulea* (Limacodidae) and the io moth, *Automeris io* (Saturnidae).

(c) Effects and treatment

The effects of many vesicant insects can be felt as stinging or burning sensations within a few minutes or $\frac{1}{2}$ h of contact. Oedema and erythema develop and, as the pain subsides, there may be pruritus for several hours. Agricultural workers who are most likely to contact these insects, may get the hairs lodged in their clothing; so that the trouble may resume when the clothing is next worn.

There is no obvious specific treatment; but soothing lotions or compresses may be used to allay the symptoms.

CHAPTER THIRTEEN

Nuisances

The various arthropods mentioned in this chapter as nuisances are grouped in the following ways:

Damp room pests. A number of pests are very sensitive to desiccation and will only thrive in damp places, such as basements, cellars, sculleries and bathrooms. Some of these pests also may be troublesome in newly built houses, the plaster of which retains moisture for a long time.

I The silverfish (*Lepisma sacharina*); II booklice (Psocoptera); III plaster beetles (Lathiridiidae and Cryptophagidae); IV spiders (Araneae); V the furniture mite (*Glyciphagus domesticus*).

Garden invaders. A great variety of different kinds of insect will occasionally enter houses and cause some dismay to the inhabitants, who may think them harmful. Among the specimens sent to the writer in the course of advisory work, were the following: soldier beetles (Telephoridae), chafers (*Melolontha*), stag beetles (*Lucanus*), water beetles (*Dytiscus*), bark beetles (Scolytidae), raspberry beetles (*Byturus*), longhorn beetles (Cerambycidae), rat-tailed maggots (*Eristalis*), ichneumon flies, leather jackets (*Tipula*), lacewings (*Chrysopidae*), water boatmen (*Notonecta*), aphids, poplar hawk moths (*Laothoe*), mason wasps (*Odernus*), thrips and ant pupae.

Obviously it is not feasible to describe all the casual insect invaders from the garden; but certain kinds of arthropod are repeatedly troublesome from their habit of invading dwellings in large numbers, usually for shelter (e.g. hibernation). The following examples do not include flying insects.

I Springtails (*Collembola*); II the cricket (*Acheta domesticus*); III earwigs (Dermaptera); IV ground beetles (Carabidae); V bagworm moths (*Psychidae*); VI the clover mite (*Bryobia praetiosa*); VII beetle mites (Oribatei); VIII woodlice (Isopoda).

Outdoor swarms. Certain outdoor swarms of insects may come to the attention of Public Health Departments as nuisances. These include: I seaweed flies (Coelopidae); II St Mark's fly and the fever fly (Bibionidae).

Flies breeding in sewage works, which could come under this heading, have been dealt with in Chapter 9.

13.1 DAMP ROOM PESTS

13.1.1 SILVERFISH (LEPISMA SACHARINA)

(a) Distinctive features

The silverfish belongs to the order Thysanura or bristletails, which is a very ancient group of insects. These primitive insects are widely distributed but no longer numerous. They are able to persist among their more efficient rivals by leading concealed lives, in the soil, under stones, leaves or tree bark.

These insects are descended from the primeval insect stock before wings were evolved (see Fig. 1.1, page 2). Consequently, there is no impediment to their moulting in the adult stage and they continue to do so throughout life. Their characteristic body form is carrot-shaped, with two long antennae at the anterior (blunt) end and three tail-like appendages from the tip of the abdomen (Fig. 13.1). A careful examination of the underside of the abdomen of older bristletails reveals two or three pairs of abdominal appendages, called 'styles'. These are probably relics of the numerous abdominal appendages of the ancestral form.

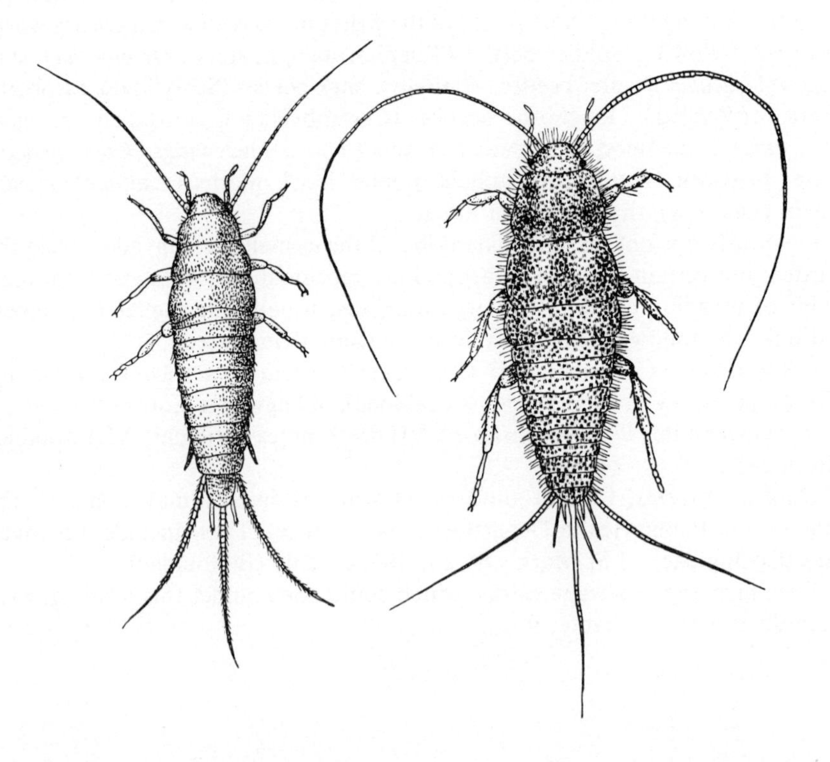

Figure 13.1 Thysanura (bristletails). Left: *Lepisma saccharina* (silverfish). Right: *Thermobia domestica* (firebrat). (After Anon. [65a].)

About 23 species (out of a total of some 350) occur in Britain; two of them occur indoors, as pests: the firebrat and the silverfish. The former demands a very warm environment and is mainly restricted to old-fashioned bakehouses. It is dealt with in Chapter 8 (page 332).

Two other species of silverfish occur in the U.S.A.; both of them resemble *L. saccharina*, although they grow somewhat larger (15 mm, as compared to 12 mm). *Ctenolepisma lineata*, which has four dark lines down the length of its back, occurs in the eastern U.S.A. and in California [547]. *Ct. longicaudata*, which is uniform grey, is the common silverfish in Australia, and has been introduced into a few eastern and southern states and to California and Hawaii [566]. The biology and habits of these species resemble those of *L. saccharina*.

The silverfish owes its name to its silvery appearance, combined with rapid darting movements and undulating turns. It is by no means confined to dwellings, being quite common in nests of birds, especially pigeons; and it can survive quite rigorous winter weather [625].

(b) Life history [500, 563–565]

(i) Oviposition

The eggs are laid singly or in groups of two or three. Only a few are laid on any one day but the total eventually reaches about 100. Usually the eggs are thrust into crevices or hidden under objects; but sometimes they may be dropped quite haphazardly.

(ii) Egg

The eggs are broadly oval, about $1\frac{1}{2}$ mm × 1 mm in dimensions. At first they are smooth and white, but within a few hours they darken to a brownish colour and become somewhat shrunken and wrinkled.

(iii) Nymph

The first-stage nymph is about 2 mm long, milky white and relatively plumper and less active than older stages. It is quite bare of scales and bristles. After three moults, a covering of scales appears and after the fourth moult the first pair of styles appears; the second pair are visible after eight or nine moults.

(iv) Adult

The stage at which reproduction begins is not known; probably it is about the tenth instar.

The mature insect has a body length of about 12 mm ($\frac{1}{2}$ in.). Its silvery appearance is due to the covering of tiny spade-shaped scales. In this connection,

Robert Hooke compares it with the iridescence of pearls and similar effects depending on numerous 'very thin shells or laminated orbiculations'.

As already mentioned, the body has a carrot-like outline. The head bears long whip-like antennae composed of numerous small joints and a pair of compound eyes, each with only twelve facets. The mouthparts are primitive and not unlike those of the cockroach. The mandibles are used for biting off small particles or for scraping away at surfaces, such as paper.

Silverfish live to a large extent on carbohydrates such as starch and dextrin, and are said to be able to digest certain forms of cellulose. Some forms of paper (chemically pulped) are apparently more digestible than others (mechanically pulped). In addition to carbohydrates, they welcome small amounts of protein in the form of portions of dead insects and the sizes, gums and glues found on wallpapers, book bindings, etc. They may bite very small holes in fabrics (cotton, linen, artificial silk) but cannot subsist on such a diet.

The internal digestive system is similar in plan to that of the cockroach and includes a sack-like crop, a toothed 'gizzard' and a midgut bearing blind food pouches.

The three thoracic segments are rather similar; each bears a simple pair of legs. The insect normally progresses by a series of short rapid runs interrupted by sudden pauses. The feet end in claws which enable it to run up rough materials such as paper or plaster, but it cannot climb vertical polished surfaces. In addition to the styles on the abdominal segments, the females bear ovipositors, made up of two pairs of processes originating below the eighth and ninth abdominal segments. The male has a pair of genital processes and a small median penis on the ninth segment. Copulation has not been observed but it is known that the sperm are transferred in a tiny bag, the 'spermatophore'.

(v) Habits of adults and nymphs

Silverfish are nocturnal in habit and are rarely seen in the daytime when they hide in various crevices. They may be found in various parts of the house but more especially in bathrooms and sculleries, possibly because of their need for moisture. They commonly hide behind skirting boards, under loose wallpaper, etc, during the day and emerge to forage for food at night. If they are surprised by a light during the night, they may remain motionless for a moment and then run rapidly to a harbourage. Owing to their inability to climb up polished surfaces, they are sometimes trapped in wash basins, in sinks or in china or glass utensils in kitchen cupboards.

Silverfish are probably transported from one house to another in bales of furniture or packages of food and other goods.

(c) Quantitative bionomics: ecology

(i) Temperature

The silverfish has been less extensively studied than the firebrat (see page 332); but it is clear that its temperature preferences are lower. Thus eggs will hatch at 22° C, the lowest temperature tested; but they fail to hatch at 37° C or above. The upper lethal temperatures are also lower (judging by experiments on a foreign related species *Ctenolepisma longicaudata*), which died after $\frac{1}{2}$ h at 45° C; 1 h at 44° C or 15 h at 41·5° C [354].

Over the normal biological range, incubation takes 43 days at 22° C and 19 days at 32° C. Maturation requires 90–120 days at 27° C; at this temperature, the instar length increases from 2 to about 20 days (by the tenth instar) and 40 days by the fifteenth instar (and to the end of the life). The adults live about $3\frac{1}{2}$ years at 27° C; 2 years at 29° C and $1\frac{1}{2}$ years at 32° C.

(ii) Humidity

Eggs of *Lepisma* will only hatch at humidities above 50 per cent R.H. at 22° C (71° F) and above 75 per cent R.H. at 29 and 32° C (84 and 90° F). The nymphs will live and mature above 60 per cent R.H. but do not do well below 75 per cent R.H.; the optimum is 90 per cent R.H. [564].

(d) Importance

Both types of bristletail may cause damage to paper (especially wallpaper) and fabric, although in Britain they are seldom numerous enough to cause extensive trouble on this account. Occasionally silverfish will eat away the paste sticking wallpaper to the wall and they may destroy pastes binding insulating wrappings to pipes. Sometimes, too, they cause damage to ancient books or documents or other valuable articles in museums and similar collections.

In most well-kept houses they are regarded as an unpleasant nuisance, especially as their quick darting movements are liable to produce uneasy sensations.

(e) Control [646]

Bristletails are not particularly easy to kill by insecticides if they are living under their optimum conditions. The adults are very long-lived and continue reproducing over a long period; so that if any escape destruction, the infestation will gradually increase again.

It is often possible to get rid of silverfish by rendering conditions unfavourable to them. Damp rooms or houses should be thoroughly heated and aired. If the air is kept warm and dry, this pest cannot survive for very long.

Among modern insecticides which have been recommended are 0·5 per cent of either *gamma* HCH, propoxur or chlorpyrifos as dust or residual sprays; and for dusting inaccessible spaces, silica aerogel. Where insecticidal use is not desirable, traps may be simply made of small glass ointment jars. The outside is covered with adhesive tape, to allow the insects to crawl up exploring, and if they fall in, they cannot climb out.

13.1.2 BOOKLICE (PSOCOPTERA)

(a) Distinctive features

The Psocoptera form a small and rather degenerate group of insects related to the primitive order Isoptera (termites or white ants). On anatomical grounds, it is also believed that the parasitic Mallophaga (bird lice) are descended from early nest-living Psocoptera.

Booklice are small or minute insects with soft yellowish or greyish bodies. The more typical forms have four delicate wings, though they do not readily fly; but an evolutionary tendency has resulted in the loss of wings in many species. Another feature of the order is a tendency to suppress the male sex. In some species, dwarf males occur; in others males are rare or perhaps completely absent. In such cases the females reproduce without fertilization.

Some 800–900 Psocoptera are known, the vast majority of them living out of doors. The genera common in houses may be distinguished by the following simple key:

1. No wing rudiments (Fig. 13.2)	*Liposcelis*
Wing flaps present	(2)
2. Pale yellow or whitish in colour	(3)
Dark brown or black in colour	*Lepinotus*
3. Rows of dark spots on the front margins of some abdominal segments. An active insect (Fig. 13.2)	*Trogium*
Such spots absent. Sluggish	*Psyllipsocus*

(b) Life history [69]

Some of the earlier studies on booklice are somewhat unreliable because the environmental conditions were not carefully controlled (or even recorded in some cases) and these insects are particularly sensitive to such factors. Extensive controlled investigations have been made in recent years on *Liposcelis granicola* and many of the following biological notes are taken from the results. Unfortunately, however, it seems that this species is not a typical British insect, being adapted to a rather warmer environment.

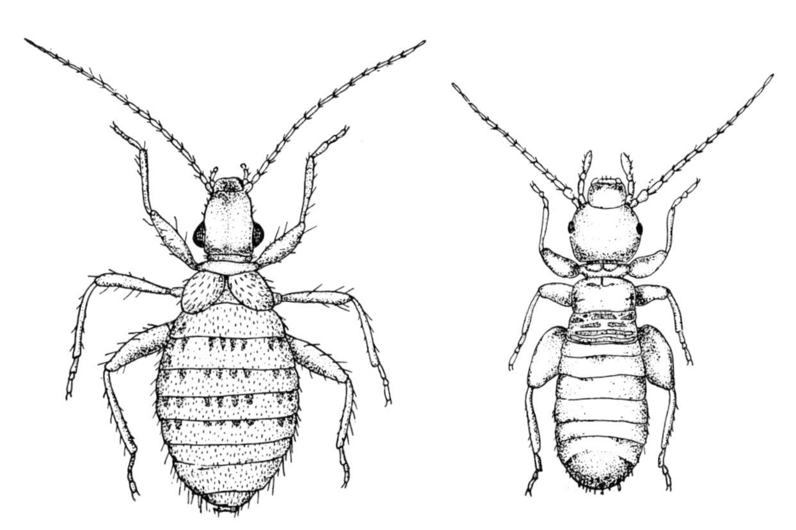

Figure 13.2 Psocoptera (booklice). Left: *Trogium pulsatoritum*. After Brit. Museum (Nat. Hist.) [66].× 30. Right: *Liposcelis granicola*. (After Broadhead and Hobby [69].) × 40.

(i) Oviposition

The female lays her eggs separately (one or two per day), extruding them by internal hydraulic pressure. The egg is sticky when laid and adheres to the substrate. Sometimes the females cover their eggs after laying them with fragments of food or rubbish.

Some of the outdoor species of the group lay small batches of eggs at intervals and cover them with a small silken web. (This production of silk by an adult insect is most unusual; the faculty is usually confined to larvae.)

(ii) Egg

The eggs of booklice, like those of all small animals, are very large in comparison to the parent. *Liposcelis granicola* lays an egg about a third the length of its own body. The egg is smooth and has a bluish-pearly lustre.

(iii) Nymph

The first-stage nymph breaks its way out of the egg shell with the help of a special file-like organ. The young nymph is seen to resemble the adult in general shape, though differing in some proportions and being, of course, much more delicate and paler in colour.

There are four nymphal stages in *Liposcelis granicola*. Other Psocoptera have between three and eight nymphal moults. The habits of the nymphs, for the most part, resemble those of the adults.

(iv) Adult

The heads of booklice are relatively large and bear a pair of rather long thread-like antennae, which are waved from side to side as they walk about. The compound eyes are poorly developed, those of *Liposcelis granicola*, for example, have only seven facets.

The mouthparts are of the biting type suitable for nibbling off small fragments of food. Booklice can apparently discriminate between small particles of suitable and unsuitable food. Thus they seem to be able to pick out fragments of dried yeast from a yeast–starch mixture, provided that the constituents are not too finely ground.

In laboratory experiments, dried brewer's yeast was found to be the most suitable food and the insects flourished on it. Wholemeal flour was much less satisfactory but was greatly improved if allowed to become mouldy. These insects, like many others, evince a vital dependence on vitamins of the B group. However, although they demand a rich diet (e.g. yeast or dried egg) in order to proliferate, individuals can subsist for long periods on a meagre diet, such as pure starch, with little or no growth reproduction.

The thorax bears the walking legs, the last pair, in some species (e.g. *Liposcelis granicola*) having swollen femora, as in leaping insects. However, they have never been seen to jump, although they can run about actively. On encountering an unexpected obstacle, they usually retreat rapidly backwards for a short distance, pause, and then move forward again.

Adults of *Liposcelis* are like nymphs in being completely devoid of wings. Other common genera, however, develop small flap-like wing rudiments, like adult bed bugs.

The abdomen, as usual, contains the sexual organs. *L. granicola* is parthenogenetic and therefore no mating has been observed. In a related species the following procedure has been recorded. The sexes were apparently unaware of each other until at a distance of 2–3 mm apart. The male then became very excited and vibrated his antennae at intervals towards the female who sometimes responded in the same way. Finally, he ran forward, over the female's body, retreated so as to bring his abdomen underneath hers and copulation was then effected. The union lasted from 10 min to over an hour, in different cases. During this period, the female walked about, dragging the smaller male behind her.

An interesting habit of one or two species of booklice (e.g. *Trogium pulsatorium* and *Lepinotus* sp.) which is possibly connected with mating, is the production of tapping sounds by beating the abdomen rhythmically against the substrate (cf. death-watch beetle, page 428). Although this habit was noted as long ago as the 17th century, it is not very easily detected. The sound produced is rather faint on most surfaces, being loudest when the insect is resting on paper.

(c) Quantitative data [69]

At 25° C and 75 per cent R.H., the incubation period of *L. granicola* is 11 days and the nymphal period about 15 days. On a good diet (yeast) the females laid up to three eggs per day at first, but gradually declined to a rate of one per week. The adult life on this diet was 6 months, during which time a total of about 200 eggs was produced in 9 months. On a good diet the adult life was shorter, though more eggs were laid.

Accounts of earlier authors for other species give smaller numbers of eggs (20–60); but conditions may not have been optimum in their experiments.

The species *L. granicola* is rather susceptible to low temperature and dies after 3 h at 0° C. On the other hand, it is rather resistant to high temperature and requires 42·5° C at 75 per cent R.H. and 100 per cent R.H. or 40·5° C at 30 per cent R.H. to kill it in a 24-h exposure.

The lower limits of humidity which will permit development are 55 per cent R.H. at 25° C and 65 per cent R.H. at 35° C.

The short life cycle, long adult life and large numbers of eggs produced by *L. granicola*, suggest that it might have a great potentiality for proliferation. Laboratory studies indicate, however, that a great falling off of fertility is caused by overcrowding and that dense populations (as with mites) do not occur.

(d) Importance

The natural habitats of the booklice include crevices in tree-trunks, under bark, or weathered fences and walls (especially among lichen or moss) and in birds' nests. They live on fragments of animal and vegetable matter, particularly on fungi and lichens.

Inside houses, these little insects may be found in many situations, running over walls, shelves or furniture. As the common name implies, they are often to be found crawling over the backings or between the pages of books. They apparently feed on mildews and moulds, scarcely visible to the human eye, which tend to form on wallpaper, bookbindings, leather and upholstered furniture and various foodstuffs. They are therefore especially prevalent under damp conditions which favour the growth of minute moulds and mildews. For example, they are often observed in new houses owing to the moisture drying out of the plaster.

Since booklice feed mainly on superficial moulds and fungi, they seldom cause any serious damage and they are virtually harmless in small numbers. Occasionally, when favourable conditions give rise to large infestations, they may cause losses to articles readily deteriorated by minor injuries (e.g. valuable furs and fabrics or insect collections and herbaria). Moreover, the appearance of these little insects in any numbers, crawling over furniture, is distasteful to many people; and though they are quite harmless to man, they have sometimes given rise to alarmed complaints, and even litigation, from confusion with true lice.

(e) Control

Booklice are not easy to eradicate by direct attack. Insecticidal treatments with no residual effect (such as fumigation) are of little value, since reinfestation from natural sources is very probable, where conditions are suitable. The use of powder insecticides is only slightly more effective.

The simplest and most certain method of eradicating booklice from a dwelling house is to ensure that it is so dry that the minute moulds and fungi, which serve as their food, cannot grow. If ordinary domestic heating is not effective, the house should be examined for structural defects which may be responsible for local damp patches. If it is difficult or impossible to obtain absolutely dry conditions, the moulds may be destroyed by the use of a fungicide. A solution of 2 per cent formaldehyde in clear industrial spirit, applied as a spray, makes a suitable treatment.

13.1.3 PLASTER BEETLES [67, 274]

(a) Distinctive features

The name 'plaster beetles' is sometimes given to tiny beetles belonging to the familes Lathridiidae and Cryptophagidae. They are not only moderately closely related, but have in common the habit of feeding on moulds and fungi in both the larval and adult stage [272, 322].

Adults of the two families can be most conveniently separated by the number of tarsal segments: tarsi of Cryptophagidae are 5-segmented, except for males of some species which have 4 segments; tarsi of Lathridiidae are 3-segmented, except for males of some species with 2 segments on the first pair of legs. Some of the common species are as follows.

(i) Lathridiidae

Macrogramme filum
This is one of the commonest in European dwellings and it also occurs in North America. The brownish beetles are 1·2–1·6 mm long and have 2-segmented antennal clubs.

Lathridius minutus (Fig. 8.5, page 342)
Adults are 1·2–2·4 mm long; pale reddish brown to black; when black, antennae and legs are reddish brown. The mature larva is about 2·2 mm long, whitish, with moderately sparse outstanding hairs. Cosmopolitan.

Aridius nodifer
Adults about 2 mm long; rather similar to *L. minutus*, but with more distinct

ridges on the elytra; dark brown, but legs paler. Larva whitish with numerous long recurved hairs. Cosmopolitan.

Thes bergrothi
Adults rather similar to above, 1·8–2·2 mm long; reddish brown in colour. European.

(ii) Cryptophagidae

Cryptophagus acutangulus (Fig. 8.5, page 342)
Adult about 1·9–3·0 mm, dark brown. Mature larva, 2·8–3 mm, yellowish white, rather sparsely covered with erect hairs; two horn-like projections at hind end. Cosmopolitan.

(b) Life history

Several species have been reared in the laboratory in petri dishes, on fungal cultures. The eggs are laid singly among the fungal hyphae. The larvae, like the adults, feed on the conidia and the hyphae. Pupation occurs in crevices, with no special pupal cell.

(i) Quantitative bionomics (days)

Lathridius minutus: E, 5–6; L, 12–17; P, 6–7; total, 24–30. *Aridius nodifer*: E, 5–7; L, 20–28; P, 3–4; total, 27–32. (All at 17–18° C.)

Macrogramme arga: E, 4·8; L, 12·3; P, 7·8; total, 24·9. *Adistemia watsoni*: E, 5·5; L, 13·3; P, 9·6; total, 28·0. (Averages at 23° C.)

(c) Prevalence and importance

These small beetles are widely distributed and thrive in moist, secluded places, where moulds and fungi can grow. They are common in the warm damp foundations of straw and hay ricks and they can occur in large numbers in debris and litter in warehouses (see page 348). In dwellings, they proliferate in cellars and damp store rooms on cheese, jam or fabrics which have become mildewed.

A situation in which plaster beetles are liable to become a nuisance is in recently built or reconditioned houses in which the plaster has not completely dried. On these slightly damp walls, especially if they have been papered, small growths of fungi in the form of greyish or whitish patches may form, and these provide food for the beetles, which crawl all over the walls.

(d) Prevention and control

To avoid annoyance from these beetles, every effort should be made to hasten the drying of newly plastered rooms. Every trace of mustiness should be removed by keeping the rooms warm and well aired; if necessary oil stoves should be employed. It is advisable to refrain from papering walls until the plaster has dried thoroughly.

Where the beetles are present, they can be killed by the usual contact insecticides in powder form or as aerosol sprays. The latter should be applied with care to avoid staining the walls.

13.1.4 SPIDERS

(a) Distinctive features

The distinctive features of spiders in general have been dealt with in Chapter 8 (page 465). This section deals with some harmless but sometimes disagreeable domestic spiders, mainly those in Britain.

(i) Recognition of domestic species (Fig. 13.3)

About a score of different spiders may be encountered indoors or in outhouses and stables; of these, about half a dozen are fairly common. It is not feasible to give a scientific key for their identification, but they can probably be distinguished by the following notes.

1. Very lanky legs, nearly 5 times as long as the body. Pale brown in colour. Body grows to 10 mm. 'Scaffolding type' web.
Pholcus phalangoides (Pholcidae)

2. Moderately lanky legs, nearly twice as long as the body. Mottled colouring. 'Sheet-type' web. *Tegenaria* spp. (Agelinidae)
(*a*) Rather pale and not very distinctly marked. Opisthosoma yellowish brown with dark mottling. Body grows to 11 mm (leg length about 25 mm).
T. domestica
(*b*) Opisthosoma with reddish-brown longitudinal band between yellowish spots with dark borders. Body grows to 20 mm (leg length about 43 mm).
T. parietina
(*c*) Opisthosoma with khaki longitudinal stripe with dark lateral markings. Body grows to 19 mm (leg length about 26 mm). *T. atrica*

3. Rather short legs, about as long as the body.
(*a*) Tiny, pink in colour. Body about 2 mm (leg length 2·5 mm). No web.
Oonops domesticus (Oonopidae)
(*b*) Moderate size, with a shiny abdomen suffused with brown markings. Body about 7 mm (leg length 7 mm). 'Scaffolding type' web.
Steatodes bipunctata (Theridiidae)
(*c*) Moderate size, glossy, mouse grey. Body about 11 mm (leg length 10 mm). No web. *Herpyllus blackwalli* (Gnaphosidae)

(b) Life history [62, 63, 176]

Female spiders lay their eggs on a saucer of silk, which is then covered by more silk to form an egg sac. This may be suspended on a silk thread away from harm (e.g. *Tegenaria domestica*) or partly camouflaged with debris (*T. atrica*), or the silk covering may be no more than a few random threads which enables the female to carry the egg bundle about (*Pholcus phalangoides*). In large spiders, dozens or even hundreds of eggs may be laid in a batch (*Steatodes* lays 100–150); but small ones lay few but often. The tiny *Oonops*, for example, lays two at a time.

The young spiders hatch from the eggs devoid of hair, spines or pigmentation and unable to feed or spin. After a few days, during which it continues to consume egg yolk, it moults into a very small but typical spider. Generally, however, it remains inside the egg sac for days, weeks or even months, according to season and weather. Finally, however, the time comes for the young spiders to disperse and seek food.

The habits of the young and growing spiders are not greatly different from those of the adults. All are carnivorous and prey mainly on insects. [Some large

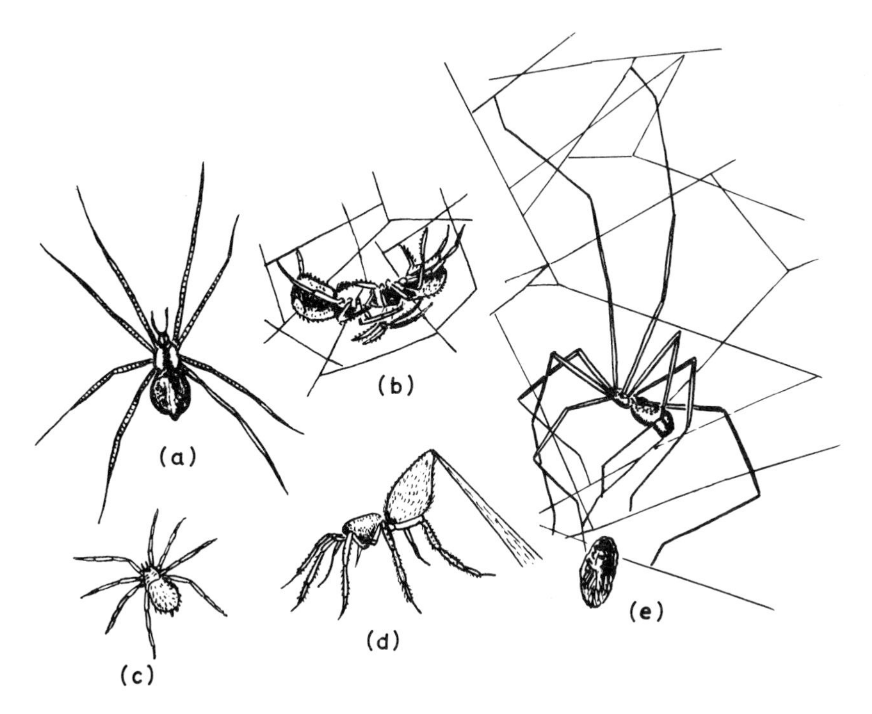

Figure 13.3 Some common domestic spiders. (*a*) *Tegenaria parietina*; (*b*) two *Steatodes bipunctata* mating; (*c*) *Oonops domesticus*; (*d*) *Herpyllus blackwalli* raising a band of silk as protection from an adversary in the rear; (*e*) *Pholcus phalangoides* wrapping an insect in silk. Magnifications: (*a*) × 0·6; (*b*) × 1·6; (*c*) × 4; (*d*) × 1·6; (*e*) × 1. (After Bristowe [62].)

spiders (*Pholcus*, *Tegenaria*) will also feed on woodlice and they not infrequently prey on these in cellars.] Some forms wander about and either catch their prey by stealth (*Oonops*) or by speed (*Herpyllus*). These do not build webs, though they form small silken cells to rest in when not hunting.

A variety of different types of snare are constructed of silk by the web-builders. Sheet webs are perhaps the best known. Those of *Tegenaria* include a tubular retreat for the spider for use in emergency. *Tegenaria* and its relatives characteristically move about on the upper surface of the web, while other spiders cling to the underside. Scaffolding webs are irregular meshes of threads spun in various directions. Some of the threads bear drops of sticky secretion to trap blundering insects.

As spiders grow, like insects they have to moult. The number of moults varies, size being the important factor. A very tiny spider may need only three moults to become adult as compared with ten in a very large species. Females usually moult once or twice more than males before maturity, and a few long-lived spiders (*Tegenaria*) moult once a year in the adult stage.

Mating in spiders is of considerable interest. The male must signal his presence (by visual or tactile signals) to prevent the female confusing him with a possible prey. Copulation is most unusual. The male spins a tiny web and extrudes sperm fluid on it. He then takes this up in special hollow organs at the tips of his pedipalps. On successfully grappling with a female, he introduces these palpal organs (either simultaneously or one at a time) into the vagina under her opisthosoma. The female stores sperm in a spermatheca and can utilize it for fertilizing several subsequent egg batches.

(c) Special peculiarities of domestic spiders

(i) Tegenaria spp.

The *Teganaria* spider, though harmless, arouse distinct uneasiness from their rapid movements and general creepiness. A *T. atrica* female can cover 330 times her body length in 10 s (say 60 cm s^{-1}).

T. domestica is almost world-wide in distribution and is very closely restricted to human dwellings.

T. parietina (sometimes called the Cardinal spider, from a legend connected with Cardinal Wolsey) is restricted to buildings in south-east England.

T. atrica is much less restricted to buildings than the other two species, but tends to enter dwellings in cool weather in the autumn.

(ii) Pholcus phalangoides

This spider occurs only in houses in the south of England and Ireland, where the average temperature exceeds 10° C throughout the year. It spends much time motionless on its almost invisible scaffolding web. When an insect is caught in it the spider uses a long leg to wrap silk threads round the prey, without the necessity of approaching closely. In winter in cold situations, this lanky spider adopts an odd rigid position and hibernates until warm weather returns.

(iii) *Oonops domesticus*

A nocturnal hunter, it normally moves slowly and deliberately, though it can run rapidly forward or backward. Insects of suitable size are often found by touch and appear to be hypnotized by gentle stroking before they are attacked and gripped with tarsal claws.

(iv) *Steatodes bipunctata*

This spider commonly builds its scaffolding-type web beside windows in outhouses, attics and unused rooms, and may be found in cellars. It is rare or absent from northern Britain.

(v) *Herpyllus blackwalli*

Widely distributed in Britain (and Europe), this fierce hunting spider trusts to rapidity of action. An ability to survive for months without water suits it for the arid environment of a human dwelling. It hunts by night, hiding behind pictures or in wall crevices during the day.

(d) Importance

Spiders are the dominant group of their class and surpass all other arachnids in number and variety of species, in their complexity of habits and in range of distribution. Nevertheless, they are only about as numerous as one of the smaller insect orders; about 550 British species are known.

Spiders have attracted comparatively little attention from man. A practical reason for this is their universally carnivorous diet which causes nearly all to be innocuous or distinctly beneficial to man as predators of insects. A very small number are harmful as, for example, the extremely poisonous *Latrodectus* spp. (especially *L. mactans* the black widow). The bites of the well-known *Tarantula* of southern Italy are much less serious, despite the legend of the tarantella dance prescribed as a cure. Unpleasant spider bites do occur there, however, due to *Latrodectus tredecemguttatus* (see page 466).

From time to time, very large hairy spiders, with a crab-like stance, are discovered in hands of bananas. The most common, in bananas imported from the West Indies, is *Heteropoda venatoria*, common in many tropical countries. Despite its fierce appearance it is not dangerous and it is not unwelcome in many tropical homes for its destruction of cockroaches. Since about 1948 other large banana spiders from West Africa have been encountered in Britain at Covent

Garden. These have conspicuous black spots on the abdomen and rings on the legs and belong to the genus *Torania*.

A few species of spiders habitually live indoors in Britain and some others are more or less frequent invaders in autumn and winter. The majority of them, however, like most insects, find the interior of houses too dry for them. They thrive best in damp and neglected rooms; whereas they are rare in well-kept houses, particularly if they are centrally heated. Apart from the direct adverse effect of very dry air on spiders (especially the egg stage), it results in the presence of fewer insects on which to feed.

There are a few minor objections to those spiders which may be encountered indoors. These, especially the large ones, are liable to arouse feelings of disgust or even dismay. Also, old discarded and dusty webs are unsightly and their presence is regarded as evidence of bad housekeeping.

13.1.5 THE FURNITURE MITE (GLYCIPHAGUS DOMESTICUS) [65]

(a) Distinctive characters

Glyciphagus domesticus is a fairly typical acarid mite (see page 368, Fig. 8.8) with a rather round body covered with long bristles, which are seen to be feathered, on high magnification. It is a widely distributed species, which may occur in large numbers on dried plant or animal remains in houses and stables.

(b) Life history [284, 293]

There is an egg, larva and first nymphal ('protonymph') stage. About half the protonymphs then pass to a second nymphal ('deutonymph') stage which eventually moults to produce the adult. The other protonymphs remain inside the cast skin and change into a rather amorphous oval mass; this is a 'hypopal' stage (see page 369) which is resistant to adverse conditions of draught and often remains resting for as much as 6 months. Finally, it produces an active deutonymph which moults to give a normal adult.

(i) Speed of development

Under optimum conditions (23–25° C and 80–90 per cent R.H.) the direct life cycle takes about 22 days. The hypopal forms greatly prolong this, however.

(c) Importance

Glyciphagus domesticus can occur on various foodstuffs, such as flour, sugar and cheese. In addition it may proliferate on furniture stuffed with vegetable fibres and, under rather damp conditions, may cause a nuisance in dwellings, especially in little-used rooms, and sometimes called the 'house mite'. In former years, it was very prevalent on furniture stuffed with green Algerian fibre.

(d) Control

The active stages of the furniture mite are very sensitive to desiccation. Accordingly a thorough warming, airing and general drying out of infested rooms soon curtails the nuisance.

13.2 GARDEN INVADERS

13.2.1 SPRINGTAILS (COLLEMBOLA)

(a) Distinctive features

Collembola are small, fragile insects, rarely exceeding 5 mm in length. Many of them have an abdominal appendage which enables them to leap a few inches through the air, and this is the reason for the common name of the group.

Collembola are evidently rather primitive insects, for they show no trace of metamorphosis and they are assigned to the archaic wingless sub-order Apterygota. Yet in several ways they are rather specialized and they have evidently evolved further, in their own way, than other wingless groups. Thus, the antennae are usually reduced to 4 segments; the mouthparts are deeply retracted into the head; and the abdomen is composed of only 6 segments, less than in any other kind of insect.

There are two main groups; the Arthropleona, with the body more or less cylindrical and the abdomen clearly segmented; and the Symphypleona, with the abdomen nearly globular and without clear segmentation.

(b) Life history and appearance

The eggs of Collembola are smooth and spherical, usually laid in small clusters. The young forms resemble their parents in appearance and during development merely increase in size and pigmentation, with little change in structure. They continue to moult throughout life, apparently at irregular intervals. The appearance of some typical, well grown specimens can be seen in Fig. 13.4. The colouring may range from dull blue-black to greenish yellow or a variety of brighter colours. The body is usually covered with hairs or scales.

Compound eyes are absent and there are only simple ocelli, often surrounded by a pigmented area. The mouthparts are adapted for biting in most species though they are withdrawn into the head. The food usually consists of fragments of live (or more often dead) plant tissue. A few prey on other small insects.

The unusual, characteristic organs of the group occur on the abdomen. The springing organ is a forked tail-like appendage carried under the fourth abdominal segment. Normally it is folded forward and kept in place by another retaining appendage under the third segment. When this releases the spring, it flies backward and projects the insect through the air.

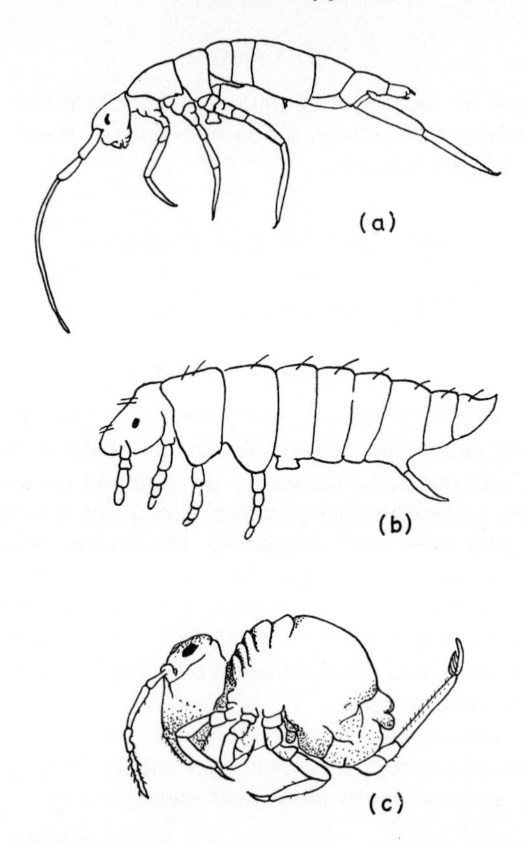

Figure 13.4 Collembola (springtails). (*a*) *Tomocerus plumbeus*; (*b*) *Hypogastura purpurescens*; (*c*) *Sminthurides aquaticus*. (*a*), (*c*) after Willem [617]; (*b*) after Strebel [558]. (*a*) × 7½; (*b*) × 35; (*c*) × 40.

The sexes are separate, though very similar in appearance, and copulation takes place before egg production.

(c) Habits and physiology [558]

Most Collembola live in crevices and crannies, under debris or in the soil. The principal reason for their retiring life is their need of the very high atmospheric humidity which is to be found in such microclimates. Many species depend on cutaneous respiration and these are probably even more sensitive to desiccation than those with a tracheal system.

The domestic Collembola (i.e. sometimes occurring in houses) such as *Hypogastrura purpurescens*, thrive between temperatures of 3° C and 15° C. They are resistant to low temperature and cold death only occurs at − 5 to − 15° C. Other species are also very cold-resisting and many occur in various situations in the Arctic Circle.

Collembola have senses which detect heat and cold and various smells and tastes. They retreat from the light mainly as an escape reaction (when disturbed by mechanical shock, etc).

(d) Quantitative data [558]

Eggs of *H. purpurescens* hatch in 4–15 days. The first moult occurs in about a week and other moults occur at irregular intervals. Females mature in 6–7 weeks. Total life 7–8 months.

Tomocerus minor moults at intervals of 1–4 weeks and lives about a year.

Sminthurinus niger eggs hatch in 10–14 days and females mature in 2–3 weeks.

Hypogastrura viatica observed throughout life at temperatures of 5–15° C, moulted on the average 9 times and lived an average of 78 days.

(e) Importance [413]

Collembola are very widespread and numerous in nature. They are found in the soil, in decaying vegetable matter (dead leaves, rotting wood), among herbage, under the bark of trees, etc. Nearly all live harmless scavenging lives, scarcely observed by man. A few species (e.g. *Hypogastrura viatica* and *Tomocerus minor*) are very common in the gravel of percolating filters at sewage works, where they perform a useful function in feeding on the biological film and preventing clogging, especially in winter, when other insects in the filters are quiescent (see page 387).

For the purposes of this chapter, we are interested in a few species which occasionally intrude into dwelling houses. Species of *Hypogastrura*, *Lepidocrytus*, *Seira*, *Tomocerus*, etc, have been reported as occurring in houses. Some forms breed in cellars; others are common in flower-pots of plants kept indoors. They also occur in bathrooms and in the neighbourhood of sinks and drains. Though they are quite harmless to man, they are sometimes accused of causing bites, probably on account of their jumping habits which recall the behaviour of fleas.

(f) Control

The sensitivity of Collembola to desiccation usually renders it an easy matter to eradicate them from dwelling houses by drying and airing affected rooms. An American pest-control operator described an infestation of *Seira nigromaculata* in decayed insulation of a refrigeration plant. This was eradicated by injecting the insulating material at 2 m (6 ft) intervals with a 5 per cent solution of rotenone in methyl formate.

13.2.2 CRICKETS

(a) Distinctive features

The crickets are distantly related to the cockroaches; both were formerly included in the order Orthoptera, which is now restricted to the jumping types, of which there are three groups:

(a) 'Short horned' grasshoppers with short antennae and the females with short ovipositors. This family includes true locusts.

(b) 'Long horned' grasshoppers with long, thread-like antennae, often as long as the body, and the females with long ovipositors.

(c) Crickets with antennae and ovipositors usually long. Their wings are carried flat on the back and folded down over the sides, whereas the grasshoppers hold them obliquely, like the sides of a roof.

True crickets belong to the family Gryllidae and there are three British species in the sub-family Gryllinae: *Acheta domestica*, the house cricket, *Gryllus campestris*, the field cricket and *Nemobius sylvestris*, the ground cricket. Of these, only *Acheta domesticus* is liable to invade houses, which it also does in North America. In that region, however, other members of the genus *Acheta* sometimes breed in vast numbers and invade urban areas [296]. These have all generally been described as *Acheta (Gryllus) assimilis*; but the type specimen of this species is lost and the description is vague. More recent studies have suggested that there are about 25 species in *Acheta* in the U.S.A., one or more of which have been responsible for the invasions [4]. There are at least 5 eastern species, *A. pennsylvanicus* being most common. These 'field crickets' may be distinguished from the house cricket, as follows.

General colour, straw or dull brown-yellow; head and pro-notum, dark brown mottled with pale areas. Head with an irregular transverse band between the eyes, with narrower light bars before and behind this band Fig. 13.5. *Acheta domesticus*

General colour varying from solid black to dark brown, tending to reddish paler specimens, rather than straw coloured. Head and pronotum black with or without pale markings. Head without transverse bands. *Acheta* spp. (North America)

Other families related to the Gryllidae include species which may be troublesome in gardens and occasionally enter houses, generally in basements. The Gryllotalpidae, or mole crickets, can be distinguished by their spade-like fore legs, furnished with strong teeth for digging. *Gryllotalpa gryllotalpa* is rare in Britain but common in continental Europe. *Scapteriscus acletus* and *S. vicinus*, the southern mole crickets, occur in the southern states of the U.S.A. [621]. A primitive family in the same group is the Gryllacridae or camel crickets, mainly tropical, but with a few species in North America (e.g. *Ceuthophilus californicus* and *C. pallidus*, which are dark brown insects, wingless and with enormous

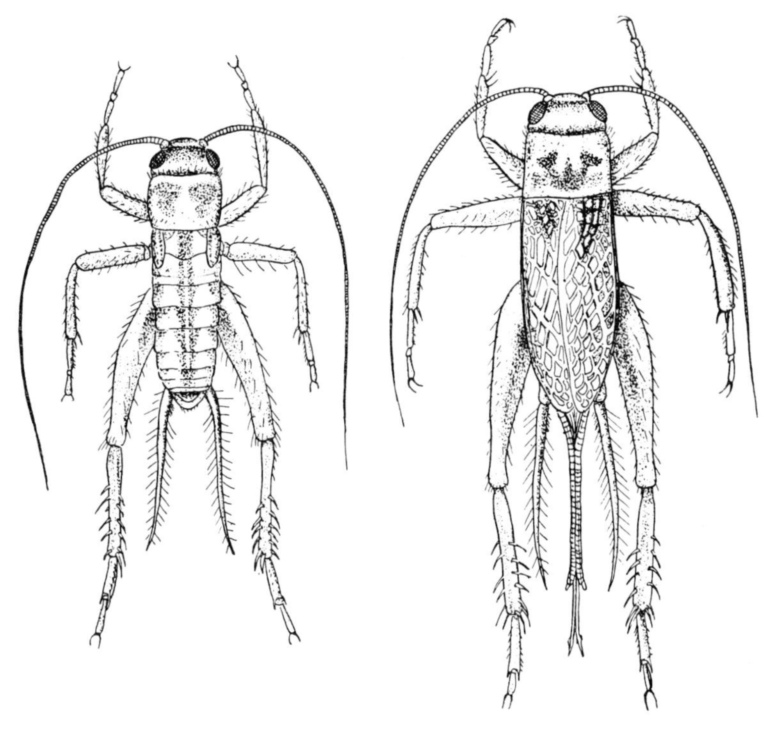

Figure 13.5 The house cricket, *Acheta domesticus*. Left: male nymph: × 6. Right: adult female. × 5. (After Anon. [66a].)

jumping legs) [290]. Another primitive family is the Stenopelmatidae, of which the Jerusalem cricket (*Stenopelmatus fuscus*) occurs in the western U.S.A. This rather large species (2–5 cm) is also wingless, with a big head [26].

(b) Life history of Acheta domesticus [318]

(i) Oviposition

Observations on crickets in captivity suggests that the females usually dig out a pit about $1\frac{1}{2}$ cm deep in soft earth before laying their eggs. The pit is dug by scratching with the front legs, large particles being removed in the mouthparts. Then the abdomen is inserted in the hole and the ovipositor plunged into the bottom. Under laboratory conditions, eggs may also be laid in various moist situations as under damp paper or in moist foods.

(ii) Eggs

The eggs are banana-shaped, smooth and white. They are about 2·4 mm long and

0·3 mm wide. They are rather sensitive to desiccation and on long exposure to dry air they shrivel and collapse. On the other hand, they are liable to be attacked by moulds in air saturated with water vapour.

(iii) Nymphal stages

The first-stage nymph is recognizably like its parents, but differs in many proportions; for example, the head is relatively larger and the hind (jumping) legs relatively smaller.

Development is slow and has not been studied in detail. There is an indefinite number of moults; the maximum recorded is 11; but 7–9 are more common. The characters of the adult appear gradually and by the fourth stage the rudimentary ovipositors of the females can already be distinguished.

(iv) Adults

The head bears a pair of long whip-like antennae and fairly well-developed compound eyes. The mouthparts are of the primitive biting type, not unlike those of cockroaches.

Crickets are omnivorous but appear to prefer soft foods such as raw or cooked vegetables and fruit, bread, dough and similar substances especially if damp and decaying; also cooked or raw meat (including dead or even live insects). They seem to dislike hard-baked flour products and do not eat unhusked cereals. This is apparently not due to inability to bite hard things, for they can bite through the chitin of insects and nibble leather and even wood. Indeed, crickets have a curious habit of biting holes in various fabrics (wood, cotton, artificial silk, etc) which do not serve them as food.

Attempts to rear crickets in captivity are handicapped by a strong tendency towards cannibalism. Crickets with any injury to the legs or other organs are soon set upon and eaten by their companions. This practice is most common under dry conditions and with a lack of moist food.

The thorax of the cricket bears well-developed legs, of which the last pair are enlarged in the form characteristic of jumping insects. The powerful leaping muscles are in the swollen 'femur' segment, and the long narrow 'tibia', which applies the leverage, lies folded forward beneath it in repose. This segment bears two rows of pig-like spines on the hind (under) surface, which gives purchase on rough ground.

Running and not jumping is the normal mode of progression of crickets. Usually the insects move about in short runs (in which the older nymphs cover about 5 cm) interrupted by short pauses of $\frac{1}{2}$–1 s. This mode of progress (which resembles that of the housefly) may require about 20 s to cover 50 cm (20 in). The younger nymphs can climb slowly up vertical glass surfaces but this power is lost by the older stages which can only climb up rough materials.

Jumping is resorted to as an escape reaction when danger threatens. Often from 5 to 15 jumps follow one after another, but the insect gradually shows fatigue and the distance covered gradually declines. The older nymphs can cover about 15–20 cm reaching a height of 6–8 cm in their first leaps, but this falls off to about 2–3 cm; after about 24 jumps they are unable to leap further. As the insect travels through the air, the body swings round through an angle of 100° to 170° so that the next jump follows at an acute angle to the last one. This serves to confuse a pursuer.

The thorax of the cricket also bears the two pairs of wings which are folded over the body in repose. These wings are not used for flight, nor apparently even for gliding. The forewings are stiffer than the hind pair and have a certain protective function. Contrary to the usual rule among Orthoptera, the right wings overlap the left.

The wings of the two sexes are easily distinguished even when closed, for those of the males are modified for producing the characteristic chirruping. Each forewing bears a file-like serrated edge which rests against a ridge or scraper on the other wing. During the chirruping the two forewings are raised to an angle of 45° with the abdomen and are moved laterally in and out, so as to rub the scrapers over the files. Other parts of the wing are developed as vibratory 'tympana' to amplify the sound.

The chirruping is associated with sexual activity and it has been demonstrated that female crickets can be enticed by the sound of males relayed over the telephone. The sound is perceived by auditory organs on the front legs.

The abdomen of the cricket terminates in a pair of prominent cerci which are able to detect air vibrations over a great range of frequencies. In the female there is also present a long needle-like ovipositor.

The mode of copulation has been noted by several observers. The male chirrups and waves his antennae. Later he displays a curious movement, vibrating his body from side to side or back and forth. Later the chirruping changes in character and at this point the female usually mounts his back. If this does not happen, he frequently moves backwards towards her, pushing his abdomen underneath hers. Copulation lasts about 2 min, in which time the male transfers the sperm in a tiny bulb (known as a 'spermatophore'). This remains attached to the female for about an hour after mating and finally falls off. Sometimes the female detaches the spermatophore and eats it. The male secretes a new spermatophore which may take between $\frac{1}{2}$ and 1 h. After this, the insects are able to copulate again although this may not happen for several hours or a day or so.

(c) Quantitative bionomics [87, 205, 318, 388]

In laboratory colonies, with a favourable diet, preoviposition periods of 10 days at 28° C or 5 days at 35° C were observed. The average numbers of eggs laid were 728 and 1060, respectively, over a 5-week period; and the females survived for a further 2–3 weeks without oviposition.

Breeding in houses or warm fermenting dumps may be continuous rather than seasonal; all stages may be found in spring and autumn. The warm original home of crickets is evident in their preference for warmth expressed, in their choice of harbourages. They are found indoors in bakeries, kitchens and boiler houses (among coke, etc) as well as under the hearths of primitive dwellings. They are not active at temperatures below 20° C. Nevertheless, they display fair resistance to cold. Nymphs and adults survived an exposure overnight to temperatures ranging between −4° C and −8° C.

The resistance to starvation is not very great, at least in the younger stages. Second-stage nymphs died, on the average, after about a week without food at room temperature (and in less if also deprived of water). A few, however, survived nearly 3 weeks.

(i) Speed of development

At 23° C: E, 46–51; N, 81–238. At 26° C: E, 23; N, 108–115. At 35° C: E, 12; N, 27–35. At 40° C: E, 13; N, 35–41.

(d) Importance

The house cricket originated in hot Palaerctic deserts (Persia and Sahara) and is ill-adapted to the European winter. In Britain and similar northern latitudes, crickets live out of doors in the summer but tend to seek shelter in the autumn; many of them tend to invade houses. It seems that in former times it was very common for a few crickets to spend the winter in warm crannies of the kitchen. The chirruping noises (made by the adult males) were not regarded as particularly unpleasant; rather the reverse.

Whether modern nerves are less robust, or whatever the reason, the noises of crickets are usually found highly objectionable today. However, infestations in small dwelling houses are less frequent, possibly because of improved construction. In recent years, trouble from crickets is almost always associated with refuse dumps. The large bulk of rubbish offers a very favourable environment for crickets. In addition to shelter and scraps of food, the dump provides warmth from fermentation and putrefaction. Crickets may survive in crevices in large rubbish dumps throughout the winter, but in the autumn, large numbers are liable to migrate and invade houses in the vicinity. These hordes of migrants are almost reminiscent of locust swarms.

Whatever the opinion about the presence of a few crickets in a house may be, there can be no doubt about the objectionable nature of these autumnal mass invasions. Not only do the creatures damage foodstuffs, but they are very prone to bite large holes out of many fabrics and the noise of their concerted chirruping is intolerable. Indeed, it was officially recorded at an inquest at Alderney Edge on 12 January 1934, that a Sanitary Inspector of Hale U.D.C. had been driven by crickets to desperation and suicide.

(e) Control

(i) In houses

Control of crickets inside houses may be done by powder insecticides or by sprays, as recommended for cockroaches (see page 319).

(ii) On refuse dumps [100]

The liability to nuisance from crickets on a refuse dump can be greatly reduced by controlled tipping properly conducted (see page 381). A badly kept dump, with a large exposure of recently tipped rubbish, provides breeding grounds for hordes of crickets. Empty food tins offer shelter and scraps of food, so that they ought to be crushed and covered up as soon as possible.

Where outbreaks of crickets occur in spite of reasonably well applied tipping methods, it may be necessary to cover up as much as possible of the dump with a 15 cm (6-in.) layer of earth, fine ashes, soot or lime. Coarse clinker is useless; the crickets thrive under it.

If covering is difficult or impossible, it may be necessary to resort to insecticides. Probably the simplest effective treatment is to make one or two weekly applications of a powder insecticide with the aid of a rotary blower. Dusts containing either 0·6 per cent *gamma* HCH or 5 per cent DDT should be used at the rate of 1 cwt $acre^{-1}$.

13.2.3 EARWIGS (DERMAPTERA)

(a) Distinctive features

The earwigs form a small order of insects (about 1000 species) of a very characteristic appearance, exemplified by the well-known common European earwig. In Britain there are only 5 native and 4 or 5 introduced or casual species. Some important species in the northern zone are as follows.

(i) Forficulidae

Forficula auricularia (the common European earwig) (Fig. 13.6)
Adult females are about 16 mm long, with straight forceps, curved only at the tip. Males are dimorphic. Some are only slightly larger than the females, but with sickle-shaped forceps about 3·5 mm long; others are much larger, with long sickle-shaped forceps 7·5 mm long. This species also occurs in North America.

(ii) Labidae

Labia minor
Colour; yellowish tawny. Lengths: body, 5–5·5 mm; forceps, 1·5–2·5 mm.

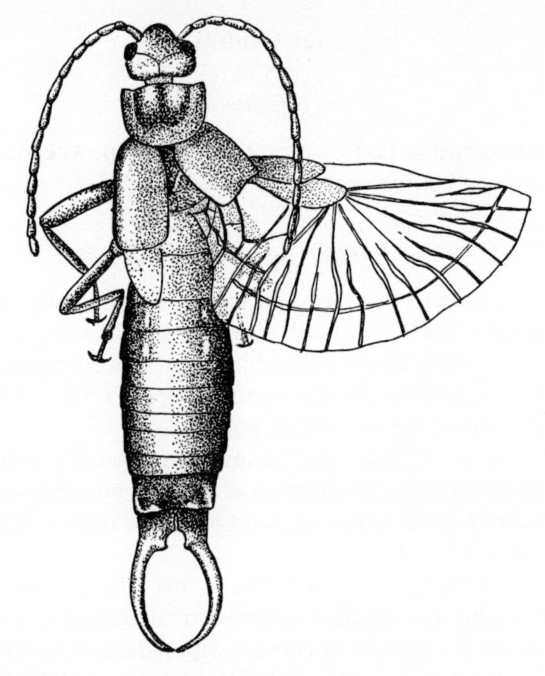

Figure 13.6 Forficula auricularia (the European earwig). (After Chopard [111].) × 5.

(iii) Labiduridae

Labidura riparia (the striped earwig)
Colour, pale brown with distinct dark markings, to chestnut brown with less distinct markings (the markings differ in three French races) about 2–2·5 cm long. The females with forceps like *F. auricularia*; males dimorphic, but with forceps only slightly curved.

This species occurs round the coastal regions of France, mainly in the south; and it is also widely distributed in the southern U.S.A.

(iv) Carcinophoridae

Euborellia annulipes (the ringlegged earwig)
A wingless species, dark brown to black, with a yellowish underside. About 12–15 mm long. Forceps of male short and somewhat twisted; female forceps normal but rather stumpy.

Widely distributed in the U.S.A. and probably indigenous.

E. cincticollis
A closely similar African species, introduced into California and spread through the state. Usually wingless but with a few winged or brachyterous specimens.

The different genera may be distinguished as follows.

1. Antennae darkbrown or black with two white segments.
(13 and 14, *E. annulipes*; 15 and 16, *E. cincticollis*) *Euborellia*
Antennae uniformly dark. (2)
2. Eleven abdominal tergites plus a telson. Forceps not sickle-shaped. *Labidura*
Ten abdominal segments. Forceps sickle-shaped in male. (3)
3. 2nd tarsal segment bilobed. *Forficula*
2nd tarsal segment normal *Labia*

(b) Life history [133, 363]

(i) Oviposition

The eggs are deposited in small covered cells in the upper 2 in. of soil, in early spring. A batch of about 30 is laid by each female and the mother remains with the eggs and tends them. If the female is removed from the eggs (except in the last few days of incubation) they usually become either mouldy or desiccated and die.

(ii) Egg

The eggs of *Forficula auricularia* are oval and white, about 1 mm × $1\frac{1}{4}$ mm in size. The young first-stage nymphs remain with their mother who appears to brood over them like a hen with chicks, as described by De Geer. However, it is possible that the earwig's maternal solicitude has sometimes been exaggerated. After a few days, the young disperse and live alone.

(iii) Nymph

There are normally four nymphal stages, though as many as six have been observed. The nymphs resemble the adults in most particulars, although, of course, the wings are not fully developed until after the last moult. There are fewer segments in the antennae; 8 in the first instar, increasing gradually to 14 in the adult of *F. auricularia.*

(iv) Adult

The general appearance of the adults must be well known to everybody. The mouthparts are formed for biting and resemble those of the order Orthoptera to which the earwigs are related. As already mentioned, earwigs are both carnivorous and vegetarian, the relative importance of the two habits being uncertain.

The head carries compound eyes of moderate size (500–1000 facets); but the insect has rather poor visual capacity. Thus, an object which a honey bee could resolve into 64 points of light would appear as a single spot to the earwig.

Earwigs are nocturnal in habits and spend the daytime resting in dark crevices, preferably away from the ground. The reflexes responsible for this behaviour are as follows: they avoid light and are attracted by dark objects; they prefer to walk upwards and tend to remain at rest with as many parts of the body as possible in contact with solid objects.

The thorax bears the scale-like rudiments of the first pair of wings, underneath which (and protected by them) lie the second pair, folded up in a complex way. The common earwig very rarely flies; it relies on walking and is also widely distributed while hiding in horticultural stock and other articles moved about by man. About half the known species of earwigs in temperate zones are similarly averse to flying. The most likely one to fly is the small, dark *Labia minor* which may occasionally be seen on the wing, sometimes in large numbers.

On the end of the abdomen are the characteristic forcep-like cerci, curved and sickle-shaped in the males, straight in the females. The primary function of these forceps is uncertain. Once or twice they have been observed to impale the insect's prey and sometimes (in *Labia minor*) they are curved over the body to help release or secure the wings. Possibly their function is to intimidate larger animals by their vicious appearance.

Striped earwigs (*Labidura riparia*) live in burrows, under vegetable debris in sandy soil. The young nymphs remain in the maternal burrow until after their first moult and then disperse. These earwigs are predaceous on small arthropods especially (in French coastal areas, amphipods).

The ringlegged earwig (*Euborellia annulipes*) is omnivorous in habit.

(c) Quantitative data

(i) Forficula auricularia

Studies on the European earwig in the neighbourhood of Washington D.C. indicate an incubation period of 73 days in winter, and 20 days in spring. The nymphal period occupied about 50 days in the laboratory and 70 days or more out of doors.

In Britain, the sexes mate in the autumn and hibernate together in cells in the soil, which the males leave first in early spring. The eggs laid about this time hatch about April and the new adults appear towards the end of June. Some females rear a partial second brood [133].

(ii) Euborellia annulipes [48]

There are five nymphal stages and the females outnumber the males by 4:1. These females lay up to 4 batches of eggs after mating. (At 20–29° C) E, 6–7; N, $10+9{\cdot}5+12+14+18=63$ days.

(iii) Labidura riparia [105, 114]

The females lay 3 or 4 batches of eggs and tend them until hatching. (At 20–25° C) E, 8–10; (At 26° C) E, 7; total, 56 days.

(d) Importance

As household pests, earwigs are merely intruders from the garden which do not breed indoors. Sometimes, however, they enter houses in large numbers and become very objectionable. Low-built country houses are most prone to these invasions, particularly if they are surrounded by herbage or covered with creeper. New housing estates are often troubled by plagues of earwigs, perhaps consisting of those dispossessed of their natural habitat by the building operations.

Earwigs are a minor horticultural pest. They sometimes damage beautiful flowers by eating holes in the petals; but, on the other hand, they are beneficial to the gardener by preying on other insects.

(e) Control

Insecticides used indoors are merely palliatives of domestic invasion of earwigs. The insects should be attacked in the garden by baiting (see below) if very numerous, but often it may be sufficient to cut away vegetation or creepers from around the windows of affected rooms. Additional protection may be obtained by smearing a band of creosote, as a repellent, along the base of the outside wall.

Large infestations, such as in new building estates, may be attacked by a poison bait made of a fluorine compound with bran and fish oil or molasses. Crumbs of this mixture should be sprinkled in likely places and covered by boards or tiles. These covers will attract the earwigs and at the same time prevent the poison bait being eaten by domestic animals.

Modern contact insecticides, like HCH or propoxur, may be used around houses; but in one case, this was counterproductive, because some predatory ants were killed more readily than the earwig in question (*Labidura riparia*) [229].

13.2.4 GROUND BEETLES (CARABIDAE)

The ground beetles belong to an enormous family of rather primitive beetles, mainly occurring in the soil, under stones, in rotting wood, moss, etc. They are mainly fairly large, black beetles (though a few are metallic or violet coloured) with some similarity to the Tenebrionidae (cf. *Tenebrio molitor*, Fig. 8.4 page 340). Most of them are predaceous, both as larvae and as adults.

A few species are quite often encountered in barns, outhouses, granaries and even in house basements under rubbish, preying on other beetles and their larvae. Among the species that have been found indoors in numbers are, *Harpalus rufipes* (black, about 10 mm long), *Pristonychus terricola* (bluish black, about 15 mm long) and *Sphodrus leucopthalmus* (black, about 22 mm long).

Many Carabidae occur in North America and several species have been recorded as nuisance invaders of dwellings. *Stenolophus* (*Agonoderus*) *lineola* is common throughout the U.S.A. and *Agonum maculicolle*, the tule beetle, is troublesome in central California [463]. *Nomius pygmaeus*, the stink beetle, occasionally invades houses in the western states. It is particularly unpleasant because of its offensive smell, which contaminates surfaces on which it has crawled [274].

These do no harm; in fact they are beneficial to the extent that they prey on other insects. However, their presence in large numbers is generally regarded as disagreeable and is certainly a sign that thorough cleansing measures are desirable to eliminate the debris under which they shelter and other insects on which they feed.

13.2.5 BAG WORM MOTHS (PSYCHIDAE)

The family Psychidae is widely distributed, but there are few representatives in Britain. The sexes are very different, the males being fairly normal moths, though the wings are only thinly clothed with hairs and scales; but the females are wingless and more or less degenerate. The larvae collect pieces of leaf or other debris to form little cases, which they carry about with them. In many genera, the adult females remain inside these bags and are sought out by the males, who mate with them from outside.

Some of these bag worm moth larvae occasionally desert their host plants and crawl up walls of buildings, sometimes entering rooms. The sight of numerous small creatures moving about in their little cases may cause mild consternation and lead to a call for identification.

The trouble is seldom serious enough to call for drastic control measures.

13.2.6 THE CLOVER MITE (BRYOBIA PRAETIOSA) [214, 415]

(a) Distinctive features

The clover mite is related to the red spider mite, a common pest of fruit trees. The species *Bryobia praetiosa* belongs to a group of related, very similar, plant feeding mites, some of them being orchard pests. The clover mite, however, is mainly annoying from its habit of invading buildings in large numbers.

The body of the adult mite is about 0·7–0·9 mm long and shaped like a pie in a pie-dish (Fig. 13.7).

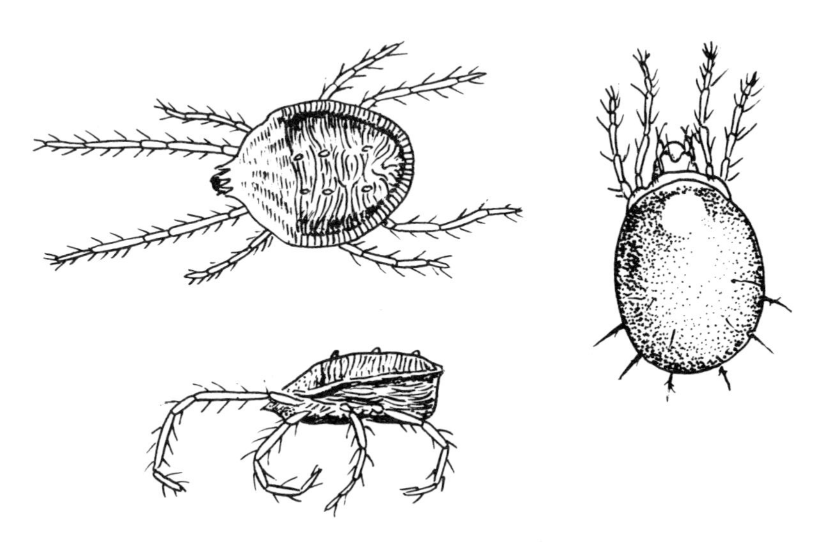

Figure 13.7 Garden mites which may invade houses. Left: *Bryobia praetiosa* × 35 (after Vitzthum [589]). Right: *Trichoribates trimaculatus* × 40 (after Hughes [292]).

(b) Life history

Eggs are laid in a dry situation, such as under bark of trees or in crevices in walls or window frames of buildings. The larvae emerge and make their way down to the grass to feed; but they do not travel far. Subsequently, they return to shelter (often near to the crevice where the eggs were laid) and moult to the first nymphal stage. The nymph wanders down to grass level to feed and finally returns to shelter to moult again. The second nymphal stage repeats this performance and finally the adult female emerges. Males are very rare indeed and most females reproduce parthenogenetically. The females return to the shelter to lay eggs, sometimes more than once. It is thus evident that throughout development there is continual wandering from the crevice shelters down to ground level and back. The eggs tend to be laid amid clusters of cast skins.

(i) Life cycle

Eggs cease to hatch somewhere between 7° C and 2° C. There is also an upper limit about 30° C. It appears that the period of maximum activity is late spring, with a dormant period in very hot summers and renewed activity in autumn, when the mites seek shelter for hibernation. (They overwinter in both the egg and adult stage.) Observations on related races, in Europe, suggest that there may be about five generations in a season.

(c) Importance

There are scattered records of clover mites causing annoyance by entering houses dating back to 1863. The trouble has become much more prevalent, however, with the great increase in house building following the Second World War. The mites are encouraged by planting of lawns in the vicinity of the new buildings.

While the mites are not actually harmful, their invasion of living-rooms in large numbers is very disagreeable and their crushed bodies may stain the walls.

(d) Control [171, 214]

Since mites do not travel far from their shelters, they can be severely discouraged by removal of a layer of turf from the base of infested walls (or trees). An 18-in. or 2-ft gap should be sufficient, and it need not be left as bare earth as most herbaceous plants do not encourage the mite. The removal of a turf layer should be supplemented by spraying infested walls with an acaricide. A Ministry leaflet recommends the following [415]. 'Where vegetation is dense, malathion should be used at about 0·1 per cent . . . as a drenching spray of 4·9 litre/100 m^2 (10 gall per 1000 ft^2). On short grass, paths, walls, etc, the concentration should be increased five times to 0·5 per cent and the spray applied at 1–2 litre/m^{-2} (2–4 gal per 1000 ft^2).

A wettable powder formulation is preferable for walls. Walls should be sprayed to a height of about 1·5 m (5 ft), with special attention to the underside of window sills, and the adjacent turf to a distance of about 1·8 m (6 ft) from the building. Sprays should be applied in mid-September as a prevention, otherwise in April or May when movement of mites between the building and the turf is evident.'

For mites indoors, pyrethrins in an aerosol dispenser make a convenient treatment; but much can be done with a vacuum cleaner.

13.2.7 BEETLE MITES (ORIBATEI)

The Oribatei form a large group of the sub-order Sarcoptiformes, comprising many families of free-living mites. In the adult stage, they are characteristically covered by hard, dark armour, which gives rise to the common name. (A typical beetle mite is shown in Fig. 13.7). The three nymphal stages have soft leathery integuments and are very sensitive to desiccation. Being restricted to humid environments, they are common in soil, humus and vegetable debris, which they apparently feed on and break down into fine particles. A few species have been found to act as intermediate hosts of anoplocephalid tapeworms (e.g. of sheep).

Normally these mites are harmless and unnoticed; but occasionally they proliferate in vast numbers in vegetation adjacent to houses which the wandering adults may invade. This gives rise to minor concern and requests for identification of these curious little creatures.

Special control measures are seldom necessary, apart from clearing away the vegetable debris which is sheltering the mite population.

13.2.8 WOODLICE (CRUSTACEA: ISOPODA)

(a) Distinctive features

Very few of the enormous class Crustacea have succeeded in colonizing the land. About 35 species of one group, the Isopoda, live on land in Britain and 9 or 10 are of minor economic importance as horticultural pests. A few genera are quite common and well known; for examples, *Oniscus*, *Porcellio* and *Armadillium* (Fig. 13.8). These are known as woodlice in Britain and as pillbugs or sowbugs in U.S.A. Pillbug is a descriptive name for *Porcellio* or *Oniscus* which are able to roll themselves up into a tight, pill-like ball when disturbed.

Three of the more common British species are illustrated in Fig. 13.8. The genera may be distinguished by the following key.

1. Uropods (terminal appendages) not extending beyond final abdominal segment. *Armadillium*
Uropods distinct and projecting at end of abdomen. (2)
2. Flagellum of antenna (i.e. portion beyond sharp bend) with 2 segments. Top of head with 2 distinct lateral lobes. *Porcellio*
Flagellum of antenna with 3 segments. Top of head without pronounced lobes. *Oniscus*

(b) Life history [120, 252, 286]

The following details were noted in studies of *Armadillium vulgare*; but other woodlice are not greatly different. The females carry their eggs in a brood pouch until they hatch. From 50 to 150 young have been counted in a single brood. Incubation lasts 50–100 days. The first stage young are white, about 2 mm long, and have only 6 pairs of legs. This stage lasts about a day, after which there is a moult to the second stage, also with 6 pairs of legs. The third and subsequent stages, which have 7 pairs of legs, last 2–3 weeks each. They begin to reproduce when half-grown (about 7 mm long). The final size attained is about 15 mm in length, after about a year.

In England, there is usually one generation a year; but in France at the latitude of Toulouse, there may be three.

(c) Importance

These land-living Crustacea feed on decaying vegetable or animal matter and occasionally they eat and damage growing plants. They feed mainly at night. During the day they are often found sheltering under boards, flower-pots and so forth in neglected corners of the garden or greenhouse. Sometimes, especially in

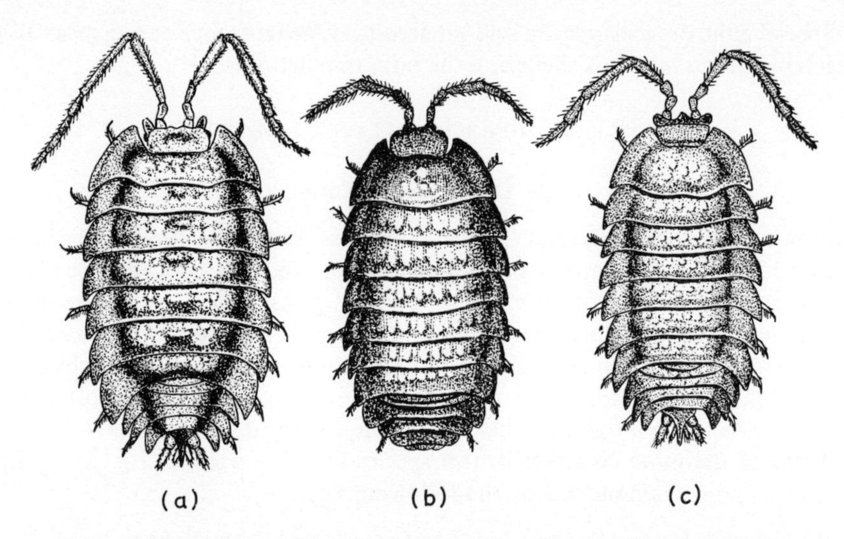

Figure 13.8 Isopoda (woodlice). (*a*) *Oniscus asellus*; (*b*) *Armadillium vulgare*; (*c*) *Porcellio scaber*. (After Collinge[120].) All × 3.

autumn or winter, they will invade basements or other damp rooms of houses, where their presence is objectionable. This happens when they are present in large numbers in the garden or in a greenhouse or conservatory adjacent to the house.

(d) Control

Woodlice which have invaded a room may be simply eradicated by ordinary cleansing measures; they do not, of course, breed indoors. To prevent further invasions, the numbers present in the adjoining garden or greenhouse must be reduced. To do this, all rubbish and such articles as boards, boxes and flower-pots must be cleared up. Then poison traps can be laid by scattering calcium arsenate at the rate of about $\frac{1}{2}$ oz to 10 yd^2 on moist soil and covering with pieces of board or tile. The woodlice which congregate under these covers for shelter will be killed.

13.3 OUTDOOR SWARMS

13.3.1 THE SEAWEED FLY (COELOPA FRIGIDA) [163, 456]

(a) Distinctive features

Certain genera of flies have adopted the habit of breeding in seaweed cast up on the shore. They include a muscid, *Fucellia*; a borborid, *Thoracochaeta*; and a dryomyzid, *Heleomyza*. But, most prevalent are members of the Coelopidae, especially *Coelopa frigida*. This species breeds along most of the European coastline from Biscay to the Arctic and also in eastern and western America and the Far East at similar latitudes.

(i) Common species of Coelopidae

Coelopa frigida is dull black or dark brown in colour with a flattened shape giving a 'squashed' appearance. Head and legs are very bristly, but the thorax has long bristles at the sides only. The top of the thorax bears tiny bristles, randomly distributed except for three distinct longitudinal rows. The wings are clear, except for a brownish tinge. Size: very variable (3–11 mm).

C. pilipes resembles *C. frigida*; but bristles of head, legs and tip of the abdomen are replaced by a furry pile of fine hairs. *Orygma luctuosa* and *Oedoparea buccata* are much less bristly, and the latter is mainly orange in colour. *Malacomyia sciomyzina* is less than 4 mm long, has a yellow head and legs and is not outstandingly bristly.

(b) Life history [80, 163]

Females lay batches of about 80 eggs. In laboratory culture, 3–5 batches are produced. Eggs are laid among rotting seaweed, especially heaps of *Laminaria* or *Fucus*. The larvae which emerge resemble fly maggots; they feed on the slimy decomposition products of the weed. The larvae are never found in the drier surface layer of wrack, but always a few inches down in the moist, decomposing and sometimes pulp-like mass below. It is notable that the larvae are either present in very large numbers (over 600 have been found per litre) or they are absent altogether; they do not survive at all well in small numbers. The most prolific sites are large banks of wrack, often formed at the foot of cliffs or boulders, where it piles up because it cannot be scattered or pushed further back. Narrow 'strings' of wrack along the tide mark tend to dry up and do not decompose so well. There are three larval stages and the pupae are formed in clusters, in the puparia characteristic of the higher flies.

The adult flies keep close to the surface of the seaweed in cold winter conditions; in warmer weather they may fly about in columns a foot or two above the surface. Sometimes they travel away from the shore and, on occasions, can travel long distances inland; but they probably do not breed anywhere except in rotting seaweed.

Adults of *C. frigida* have the curious character of being very readily attracted by the odours of trichlorethylene or carbon tetrachloride; this is not shown by other coelopids, except perhaps *Malacomyia*.

(i) Speed of development

An old record of natural development gives a total of about 6 weeks (E, 2·5; L, 28; P, 14). Under laboratory conditions, at 25° C, development takes 12 days.

(ii) Seasonal abundance

Seaweed flies tend to be most numerous in mild autumns. During the summer, less wrack is thrown up and decomposition is too rapid. During very cold winters the development is retarded and the adults do not fly far from the seaweed.

(c) Prevalence and importance

In most years, seaweed flies are no more than a very minor nuisance; but occasionally, circumstances lead to excessive breeding. The causes responsible probably include excessive accumulations of seaweed during a long mild autumn. Severe outbreaks along the south coast were experienced in 1950. In this season, the flies penetrated inland as far as London (some were even taken in Oxford).

Large clouds of flies tend to annoy visitors at seaside resorts, which is bad for the tourist trade. Apart from being a nuisance on the beaches by day, the flies are attracted by the lights of cafés and places of amusement by night. Also, they have the curious habit of congregating in vast numbers at dry-cleaning establishments and garages, which make use of trichlorethylene, causing surprise and some consternation.

(d) Control [163]

The most efficient and simplest method of preventing a seaweed fly nuisance is to destroy the breeding ground. This can be accomplished by pushing the piles of seaweed down below high-water mark, so that the sea will disperse them. (Small bulldozers may be necessary for this task.) The larvae will be either eaten by birds or drowned, after being washed from the wrack by the incoming tide.

In cases where an acute nuisance problem exists, recourse may be made to insecticides. The following compounds have been used in Britain. Dimethoate at 0·05 g m^{-2}, pirimiphos-methyl at 0·1 g m^{-2}, or *gamma* HCH at 0·2 g m^{-2}.

13.3.2 ST MARK'S FLIES; FEVER FLIES (BIBIONIDAE)

Flies of this family are mainly black, hairy flies, fairly robust, the females of some species having clouded wings. Those of the male are clear, except for a dark mark on the front edge; in both sexes the hind wing veins tend to be atrophied. The antennae are many-jointed (Nematocera) but quite short and arise from in front of the eyes, just above the mouth.

Two common species are: *Dilophus febrilis* (sometimes called, without justification, the fever fly) and *Bibio marci* (St Mark's fly, probably because it often begins to emerge on St Mark's Day, 25 April) [122, 160].

D. febrilis is about 6 mm long and can be distinguished by a circlet of spines on the tip of each fore tibia and by a double collar of spines on the front part of the thorax. *B. marci* is 10–12 mm long, the biggest British bibionid. It has a single large spine on each front tibia.

In North America, the most widespread species of this group is *Bibio albipennis*. Other species are *Dilophus occipitalis* and *D. orbatus* [156]. Another prevalent form is *Plecia nearctica*, commonly described as the lovebug or two-headed bug, from its habit of flying about *in copula* [264]. These occur throughout the Gulf states. When in large numbers, they are troublesome to motorists, from clogging the cooling radiator and obscuring the windscreen.

Large numbers of bibionids swarm in some rural districts in early spring. The females mainly crawl over grass stems and other vegetation, while the males hover, rather sluggishly, overhead. Apart from the minor annoyance caused by the adult swarms, they are believed to be beneficial in pollenating fruit trees. The larvae, however, live in the earth, feeding on plant roots, and are sometimes agricultural pests.

CHAPTER FOURTEEN

Appendices

14.1 CHEMICAL APPENDIX

The following index of insecticides used to control arthropod pests of public health importance gives common names approved by the British Standards Institution. There are also some other common names and trade names, which are denoted by inverted commas. The physical data are from Martin and Worthing [378], with the following abbreviations:

m.p., melting point; b.p., boiling point; v.p., vapour pressure; D, density; sol. H_2O, solubility in water at 20–30° C; SOS, readily soluble in organic (aromatic) solvents; LD50, etc, for acute toxicity.

'Abate' See temephos.
'Actelic' See pirimiphos-methyl.
Allethrin (±)-3-allyl-2-methyl-4-oxycyclopent-2-enyl (±)-*cis-trans*-chrysanthemate (Fig. 4.1); b.p. 160° C; LD50: rat-oral 920 mg kg^{-1}.
'Altosid' See methoprene.
Amitraz *N*, *N*-di-(2, 4-xylyliminomethyl) methylamine *o,p* $(Me)_2C_6H_4$–N: CH–N(Me)CH:N–$C_6H_4(Me)_2$*o,p* Acaricide; m.p. 87° C; sol. H_2O 1 mg l^{-1}; SOS; LD50: rat-oral 800 mg kg^{-1}, dermal 1600 mg kg^{-1}.
'Baygon' See propoxur.
'Baytex' See fenthion.
Bendiocarb 2, 2-dimethyl-1, 3-benzodioxol-4-yl *N*-methyl carbamate (Fig. 4.2); m.p. 130° C; v.p. 5×10^{-6} Torr at 25° C; sol. H_2O 40 ppm; LD50: rat-oral 35–100 mg kg^{-1}, dermal 1000 mg kg^{-1}.
Benzyl benzoate Acaricide (Fig. 4.2); insol. H_2O; SOS; b.p. *c.* 320° C.
BHC See HCH.
Bromophos 4-bromo-2, 5-dichlorophenyl dimethyl phosphorothionate (Fig. 4.2); m.p. 53° C; v.p. $1{\cdot}3 \times 10^{-4}$ Torr at 20° C; sol. H_2O 40 ppm; LD50: rat-oral 4000–7000 mg kg^{-1}.
Bufencarb 3-(1-methylbutyl) phenylmethyl carbamate and 3-(1-ethylpropyl) phenylmethyl carbamate (Fig. 4.2); b.p. 125° C at 0·04 Torr; v.p. 3×10^{-5} Torr at 30° C; sol. H_2O 50 mg l^{-1}; SOS; LD50: rat-oral 87 mg kg^{-1}, rabbit-dermal 680 mg kg^{-1}.
Butacarb 3, 5-di-*tert*-butylphenyl methyl carbamate (Fig. 4.2); m.p. 103° C; v.p. 2×10^{-6} Torr at 20° C; sol. H_2O 15 mg l^{-1}; LD50: rat-oral 1800 mg kg^{-1}.
Camphechlor (Fig. 4.2); yellow wax, softening at 70–90° C; v.p. 0·2–0·4 Torr at

25° C; D 1·65; sol. H_2O 3 ppm; SOS; LD50: rat-oral 85 mg kg^{-1}, dermal 1075 mg kg^{-1}.

Carbaryl 1-naphthyl methyl carbamate (Fig. 4.2); m.p. 142° C; v.p. < 0·005 Torr at 26° C; D 1·23; sol. H_2O 40 ppm; SOS; LD50: rat-oral 850 mg kg^{-1}, dermal > 4000 mg kg^{-1}.

Chlorbenside 4-chlorobenzyl-4-chlorophenyl sulphide; *p*-Cl–C_6H_4–SCH_2–C_6H_4*p*–Cl Acaricide; m.p. 72° C; v.p. $2{\cdot}6 \times 10^{-6}$ Torr at 20° C; LD50: rat-oral > 10 000 mg kg^{-1}.

Chlordane 1, 2, 4, 5, 6, 7, 8, 8-octachloro-3a, 4, 7, 7a-tetrahydro-4, 7-methanoindene (Fig. 4.2); Tech. product a viscous liquid; D 1·6; v.p. 1×10^{-5} Torr at 25° C; LD50: rat-oral 457–590 mg kg^{-1}.

Chlorfenethol 1, 1-bis(4-chlorophenyl) ethanol; *p*-Cl–C_6H_4–C(Me)(OH)–C_6H_4*p*–Cl Acaricide; m.p. 70° C; insol. H_2O; SOS; LD50; rat-oral 930–1400 mg kg^{-1}.

Chlorfensulphide 4–chlorophenyl-2, 4, 5-trichlorophenylazo sulphide; *o,p,m*–$Cl_3C_6H_4$–N: NS–C_6H_4*p*-Cl Acaricide; m.p. 124° C; insol. H_2O; LD50: mouse-oral > 3000 mg kg^{-1}.

Chlorpyrifos diethyl-3,5,6-trichloropyridyl phosphorothioate (Fig. 4.2); m.p. 43° C; v.p. $1{\cdot}9 \times 10^{-5}$ Torr at 25° C; sol. H_2O 2 ppm; LD50: rat-oral 163–135 mg kg^{-1}, rabbit-oral > 1000 mg kg^{-1}, dermal 2000 mg kg^{-1}.

Crotamiton *N*-crotonyl-*N*-ethyl-O-toluidine (Fig. 4.2); sol. H_2O 0·25 %; SOS; Acaricide.

DDT *p,p'*-dichloro-diphenyl trichlorethane (Fig. 4.2); m.p. 109° C; v.p. $1{\cdot}9 \times 10^{-7}$ Torr at 20° C; LD50: rat-oral 180 mg kg^{-1}, dermal 2500 mg kg^{-1}.

DDVP See dichlorvos.

'Delnav' See dioxathion.

Diazinon diethyl-2-isopropyl-6-methyl-4-pyrimidinyl phosphorothionate (Fig. 4.2); b.p. 84° C at 0·0002 Torr; v.p. $1{\cdot}4 \times 10^{-4}$ at 20° C; D 1·12; sol H_2O 0·004 %; SOS; LD50: rat-oral 150–600 mg kg^{-1}, dermal 455–900 mg kg^{-1}.

Dibutyl succinate BuO–C(O)–CH_2–CH_2C(O)–OBu repellent; m.p. −29° C; b.p. 108° C at 4 Torr; insol. H_2O; SOS; LD50: rat-oral *c.* 8000 mg kg^{-1}.

Dicapthon *o*-(2-chloro-4-nitrophenyl)*o,o*-dimethyl phosphorothioate (Fig. 4.2); m.p. 52° C; insol. H_2O; SOS; LD50: rat-oral 330–400 mg kg^{-1}, guinea pig-dermal > 2000 mg kg^{-1}.

Dichlorvos 2, 2-dichlorovinyl dimethyl phosphate (Fig. 4.2); b.p. 35° C at 0·05 Torr; v.p. $1{\cdot}2 \times 10^{-2}$ Torr at 20° C; D 1·4; sol. H_2O 1 %, kerosene 2·5 %; SOS; LD50: rat-oral 56–80 mg kg^{-1}, dermal 75–107 mg kg^{-1}.

Dicofol 2, 2, 2-trichloro-1, 1-di(4-chlorophenyl) ethanol *p*–Cl C_6H_4–C(OH)(CCl_3)–C_6H_4–*p*–Cl Acaricide; m.p. 79° C; insol. H_2O; SOS; LD50: rat-oral 684–809 mg kg^{-1}, rabbit-dermal 1870 mg kg^{-1}.

Dicophane B.P. See DDT.

Dieldrin (HEOD) 1, 2, 3, 4, 10, 10-hexachloro-6, 7-epoxy-1, 4, 4a, 5, 6, 7, 8, 8a-octahydro-exo-1, 4-endo-5, 8-dimethanonaphthalene (Fig. 4.2); m.p. 176° C;

v.p. $1{\cdot}8 \times 10^{-7}$ Torr at 20° C; sol. H_2O 19 ppm; SOS; LD50: rat-oral 45 mg kg^{-1}, dermal 60 mg kg^{-1}.

Diflubenzuron 1-(4-chlorophenyl)-3-(2, 6-dichlorobenzoyl) urea (Fig. 4.3); m.p. 210–230° C, pure 231° C; sol. H_2O 0·2 ppm; LD50: rat-oral *c*. 4000 mg kg^{-1}, dermal-rabbit 2000 mg kg^{-1}.

'Dilan' mixture of 1, 1-bis(4-chlorophenyl)-2-nitropropane and 1, 1-bis(4-chlorophenyl)-2-nitrobutane; brownish solid liquid above 65° C; D 1·28; LD50: rat-oral 480–8000 mg kg^{-1}.

Dimethoate dimethyl-S-(*N*-methylcarbamoylmethyl) phosphorothiolothionate (Fig. 4.2); m.p. 52° C; v.p. $8{\cdot}5 \times 10^{-6}$ Torr at 25° C; D 1·28; sol. H_2O 2·5 g/100 ml; SOS; LD50: rat-oral 500–600 mg kg^{-1}.

Dimethyl thianthrene Acaricide; oily substance; insol. H_2O; SOS.

'Dimilin' See diflubenzuron.

Dioxacarb 2-(1, 3-dioxolan-2-yl)-phenyl *N*-methyl carbamate (Fig. 4.2); m.p. 115° C; v.p. 3×10^{-7} Torr at 20° C; sol. H_2O 6000 ppm; SOS; LD50: rat-oral 80–150 mg kg^{-1}, rabbit-dermal 1950 mg kg^{-1}.

Dioxathion 1, 4-dioxan-2, 3-ylidine bis(*o*, *o*-diethyl phosphorothionate) (Fig. 4.2); liquid; D 1·26; insol. H_2O; SOS; LD50: rat-oral 23–43 mg kg^{-1}, dermal 63–235 mg kg^{-1}.

'Dipterex' See trichlorphon.

Dipropyl isocinchomerate PrOC(O)–(C_5NH_3)–C(O)OPr Repellent; b.p. 150° C at 1 Torr; insol. H_2O; SOS; LD50: rat-oral 5000–6000 mg kg^{-1}.

'Dursban' See chlorpyrifos.

'Eurax' See crotamiton.

Fenchlorphos dimethyl-2, 4, 5-trichlorophenyl phosphorothionate (Fig. 4.2); m.p. 40° C; v.p. 8×10^{-4} Torr at 25° C; sol. H_2O 40 ppm; SOS; LD50: rat-oral 1750 mg kg^{-1}, dermal 2000 mg kg^{-1}.

Fenitrothion 3-methyl-4-nitrophenyl phosphorothionate (Fig. 4.2); b.p. 140–145 at 0·1 Torr; v.p. 6×10^{-6} Torr at 20° C; D 1·32; insol H_2O; SOS; LD50: rat-oral 250–500 mg kg^{-1}, mouse-dermal > 3000 mg kg^{-1}.

Fenthion dimethyl-3-methyl-4-methylthiophenyl phosphorothionate (Fig. 4.2); b.p. 87° C at 0·01 Torr; v.p. 3×10^{-5} Torr at 20° C; D 1·25; sol. H_2O 55 ppm; SOS; LD50: rat-oral 190–615 mg kg^{-1}, dermal 400 mg kg^{-1}.

'Gammexane' See HCH.

'Gardona' See tetrachlorvinphos.

HCH *gamma* isomer γ-1, 2, 3, 4, 5, 6-hexachlorcyclohexane (Fig. 4.2); m.p. 113° C; v.p. $9{\cdot}4 \times 10^{-6}$ Torr at 20° C; sol. H_2O 10 ppm; SOS; LD50: rat-oral 90 mg kg^{-1}, dermal 500 mg kg^{-1}.

Iodophenphos *o*-2, 5 dichloro-4-iodophenyl-*o*, *o*-dimethyl phosphorothionate (Fig. 4.2); m.p. 76° C; v.p. 8×10^{-7} Torr at 20° C; sol. H_2O 2 ppm; LD50: rat-oral 2100 mg kg^{-1}, dermal > 2000 mg kg^{-1}.

Jodophenphos See iodophenphos.

'Kelthane' See dicofol.

'Korlan' See fenchlorphos.

'Landrin' mixture of 3, 4, 5-trimethyl-phenyl methyl carbamate and 2, 3, 5-trimethyl-phenyl methyl carbamate (Fig. 4.2).

Leptophos *o*-(2, 5-dichloro-4-bromophenyl)-*o*-methyl-phenyl phosphorothionate C_6H_5–P(S) (*o*Me)–O–C_6H_2Cl-*o*, *m*Br–*p*; m.p. 70° C; D 1·53; sol. H_2O 2·4 ppm; LD50: rat-oral 50 mg kg^{-1}, dermal > 8000 mg kg^{-1}.

Lindane See HCH

Malathion S-[1, 2-di(ethoxycarbonyl) ethyl] dimethyl phosphorothionate (Fig. 4.2); m.p. 2·9° C; b.p. 157° C at 0·7 Torr; v.p. 4×10^{-5} Torr at 30° C; D 1·23; sol. H_2O 145 ppm; LD50: rat-oral 2800 mg kg^{-1}, rabbit-dermal 4100 mg kg^{-1}.

Methiocarb 4-methylthio-3, 5-xylyl-*N*-methyl carbamate (Fig. 4.2); m.p. 121° C; pract. insol. H_2O; SOS; LD50: rat-oral 100 mg kg^{-1}, dermal 350 mg kg^{-1}.

Methoprene isopropyl (2E, 4E)-11-methoxy-3, 7, 11-trimethyl-2, 4-dodecadienoate (Fig. 4.3); b.p. 100° C at 0·05 Torr; v.p. $2{\cdot}4 \times 10^{-5}$ Torr at 25° C; sol. H_2O 1·4 ppm; SOS; LD50: rat-oral > 34 000 mg kg^{-1}.

Methoxychlor 1, 1, 1-trichloro-2, 2-di(4-methoxyphenyl) ethane (Fig. 4.2); m.p. 89° C; pract. insol. H_2O; SOS; LD50: rat-oral 6000 mg kg^{-1}.

'Mitigal' See dimethyl thianthrene.

'Mobam' benzo (*b*) thien-4-yl methyl carbamate (Fig. 4.2).

'Nuvanol N' See iodophenphos.

Parathion *o*, *o*-diethyl *o*-*p*-nitrophenyl phosphorothioate, (Fig. 4.2); b.p. 162° C at 0·6 Torr; v.p. $3{\cdot}8 \times 10^{-5}$ Torr at 20° C; D 1·27; sol. H_2O 24 ppm; LD50: rat-oral 4–13 mg kg^{-1}, dermal 7–21 mg kg^{-1}.

Perfenate 2, 2, 2-trichloro-1-(3, 4-dichlorophenyl) ethyl acetate (Fig. 4.2); m.p. 85° C; v.p. $1{\cdot}5 \times 10^{-6}$ Torr at 20° C; sol. H_2O 50 ppm; SOS; LD50: rat-oral > 10 000 mg kg^{-1}.

Permethrin 3-phenoxybenzyl-(±)-*cis-trans*-3-(2, 2-dichlorovinyl)-2, 2-dimethyl-cyclopropane carboxylate (Fig. 4.1); m.p. 39° C; b.p. 200° C at 0·01 Torr; sol. H_2O 2 mg l^{-1}; SOS; LD50: rat-oral 1300 mg kg^{-1}.

Phenothrin 3-phenoxybenzyl-(+)-*cis-trans*-chrysanthemate (Fig. 4.1); v.p. 1·64 Torr at 200° C; sol. H_2O 2 ppm; SOS; LD50: rat-oral > 10 000 mg kg^{-1}, dermal > 5000 mg kg^{-1}.

Phenthoate S-α-ethoxycarbonylbenzyl dimethyl phosphorothiolothionate (Fig. 4.2); m.p. 17·5° C; D 1·2; sol. H_2O 11 ppm; LD50: rat-oral 300–400 mg kg^{-1}, dermal > 4800 mg kg^{-1}.

Phoxim α-cyanobenzylidine-amino diethyl phosphorothionate (Fig. 4.2); m.p. 6° C; D 1·2; sol. H_2O 7 ppm; SOS; LD50: rat-oral 2000 mg kg^{-1} dermal 1000 mg kg^{-1}.

Pirimicarb 2-dimethylamino-5, 6-dimethylpyrimidin-4-yl dimethyl carbamate; m.p. 91° C; v.p. 3×10^{-5} Torr at 30° C; sol. H_2O 0·27 g/100 ml; SOS; LD50: rat-oral 147 mg kg^{-1}.

Pirimiphos-methyl 2-diethylamino-6-pyrimidin-4-yl dimethyl phophorothionate (Fig. 4.2); D 1·6; v.p. 1×10^{-4} Torr at 30° C; sol. H_2O 5 ppm; LD50: rat-oral 1180–2050 mg kg^{-1}, rabbit-dermal > 2000 mg kg^{-1}.

Propoxur 2-isopropoxyphenyl *N*-methyl carbamate (Fig. 4.2); m.p. 87° C; v.p.

10^{-2} Torr at 120° C; SOS; LD50: rat-oral 90–128 mg kg^{-1}, dermal 800–1000 mg kg^{-1}.

Resmethrin 5-benzyl-3-furylmethyl-(±)-*cis-trans*-chrysanthemate (Fig. 4.1); m.p. 48° C; b.p. 180° C at 0·01 Torr; LD50: rat-oral *c*. 2000 mg kg^{-1}

'Rogor' See dimethoate.

'Ronnel' See fenchlorphos.

'Sevin' See carbaryl.

'Sumithion' See fenitrothion.

'Sumithrin' See phenothrin.

Temephos *o*, *o*, *o*′, *o*′-tetramethyl-*o*, *o*′-thiodi-*p*-phenylene phosphorothioate (Fig. 4.2); m.p. 30° C; D 1·59; insol. H_2O; LD50: rat-oral 8600–13 000 mg kg^{-1}, rabbit-dermal 1930 mg kg^{-1}.

Tetrachlorvinphos *cis*-chloro-1-(2, 4, 5,-trichlorophenyl) vinyl dimethyl phosphate (Fig. 4.2); m.p. 97° C; v.p. $1{\cdot}5 \times 10^{-6}$ Torr at 25° C; sol. H_2O 11 ppm; LD50: rat-oral 4000–5000 mg kg^{-1}.

Tetradifon 2, 4, 4′, 5-tetrachlorodiphenyl sulphone *o,p,m*–$Cl_3C_6H_2$–$S(O_2)$–C_6H_4Cl–*p* Acaricide; m.p. 149° C; v.p. $2{\cdot}4 \times 10^{-10}$ Torr at 20° C; sol. H_2O 0·02 g/100 ml; LD50: rat-oral 14 700 mg kg^{-1}.

Tetraethyl thiuram monosulphide Acaricide; m.p. *c*. 33° C; SOS.

Tetramethrin 3, 4, 5, 6-tetrahydrophthalimidomethyl-(±)-*cis-trans*-chrysanthemate (Fig. 4.1); m.p. 65–80° C; b.p. 190° C at 0·1 Torr; v.p. $3{\cdot}5 \times 10^{-8}$ Torr at 20° C; D 1·11; SOS; LD50: rat-oral 20 000 mg kg^{-1}.

Thiabendazole 2-(4′-thiazolyl) benzimidazole acaricide; m.p. 304° C; sol; H_2O 50 ppm; LD50: oral-rat 3330 mg kg^{-1}.

Trichlorphon dimethyl-2, 2, 2-trichloro-1-hydroxyethyl phosphate (Fig. 4.2); m.p. 84° C; v.p. $7{\cdot}8 \times 10^{-6}$ Torr at 20° C; D 1·7; sol. H_2O 15·4 g/100 ml; SOS; LD50: rat-oral 560–630 mg kg^{-1}, dermal 2000 mg kg^{-1}.

'Trolene' See fenchlorphos.

14.2 BIOLOGICAL APPENDIX

14.2.1 GENERAL

Key 1a to principal orders by characters of nymphs and adults

1. One or two pairs of wings present. (9)

No wings visible (i.e. either *Apterygota*, or young nymphs or degenerate groups). (2)

2. Tip of abdomen bearing springing organ. Small, declicate insects with 6-segmented abdomen. *Collembola*

Tip of abdomen bearing cerci. (3)

Tip of abdomen without special appendages. (4)

3. Cerci thread-like, often with similar thread-like tail between them. Small delicate carrot-shaped insects. *Thysanura*

Cerci forcep-like, typical 'earwig' shape. *Dermaptera*
Cerci variable. Usually robust insects with spined legs.
Dictyoptera (Blattidae)

4. Small insects (less than a $\frac{1}{4}$ in.) with rather soft pale (yellowish) cuticles. (5)
Larger insects or with fairly hard dark (brown) cuticles. (6)

5. Antennae short, never more than 5 segments. Parasitic on warm-blooded animals.
Biting mouthparts. *Mallophaga*
Sucking mouthparts (withdrawn inside head). *Siphunculata*
Antennae with more than 9 (usually 13–50) segments. Not parasitic.
Psocoptera

6. Mouthparts adapted for biting (tooth-like mandibles). (7)
Mouthparts adapted for piercing and sucking. (8)

7. Pronounced waist between thorax and abdomen. Antennae 'elbowed'. Ant-like. *Hymenoptera* (Formicidae)

8. Antennae short and stout lying in groove. Body laterally compressed. Flea-like. *Siphonaptera*
Antennae slender. Bug-like. *Hemiptera* (Cimicidae)

9. One pair of wings only. Hind pair reduced to small rods. *Diptera*
Two pairs of wings present. (10)

10. Wings similar in texture. (11)
Front wings stiffer (usually leathery or horny) than hind pair and covering them when at rest. (12)

11. Wings opaque, covered with minute scales. *Lepidoptera*
Wings transparent, hind pair linked to front ones by tiny hooks.
Hymenoptera

12. Front wings horny or leathery, devoid of 'veins'. (13)
Front wings leathery, showing some 'veins'. (14)

13. Front wings always very short. Forcep-like cerci present on abdomen.
Dermaptera
Front wings usually completely covering abdomen. No cerci present.
Coleoptera

14. Biting mouthparts. (15)
Piercing mouthparts carried in rostrum pointing backwards under head.
Hemiptera

15. Hind legs enlarged and adapted for jumping. *Orthoptera*
All legs similar. *Dictyoptera* (Blattidae)

Key 1b to principal orders by larval characters

1. Thoracic legs present. (2)
No thoracic legs. (4)

2. 'Prolegs' absent from abdomen (simple larva or grub type). *Coleoptera*
'Prolegs' present on abdomen (caterpillar type). (3)
3. Five pairs of 'prolegs' (or less), the first pair on the 3rd or subsequent segment of the abdomen. *Lepidoptera*
Six to eight pairs of 'prolegs', the first pair on the 2nd abdominal segment. *Hymenoptera* (Sawflies)
4. Head distinct (grub type). (5)
Head reduced, capable of retraction into thorax (maggot type). (7)
5. Abdominal spiracles all similar. (6)
Last pair of abdominal spiracles larger than the rest, often borne on long air tubes. Head moderately armoured, similar in colour to body. *Diptera*
6. Head darker in colour than body and well armoured. Sluggish larvae. *Coleoptera*
Head similar in colour to body (whitish). Small active larvae, with thrusting processes at tip of abdomen. *Aphaniptera*
7. Usually several abdominal spiracles present. Opposite mandibles present. Found in little cells of earth, paper or wax constructed by parents. Often communal insects. *Hymenoptera*
Usually only one pair of large spiracles situated at tip of abdomen. Mouthparts reduced to non-opposable vertical black hooks. Found in decaying organic matter. *Diptera*

Key 1c to main orders by pupal characters

1. Pupa inside a brown puparium formed of the last larval skin. Pupa with one pair of wing rudiments. *Higher Diptera*
No puparium. Pupa either free or inside a cocoon. (2)
2. Limbs glued to the body. *Lepidoptera* (almost all)
Limbs free. (3)
3. No wing buds. (4)
Wing buds present. (5)
4. Wasp-waist evident. Pupae in cells in colonies (ant-hills). *Hymenoptera* (part) (ants)
No wasp-waist. Pupae in small irregular cocoons incorporating rubbish. *Siphonaptera*
5. One pair of wing buds only. *Diptera* (part)
Two pairs of wing buds. (6)
6. Front (upper) pair unveined (future elytra). *Coleoptera*
Front (upper) pair usually much larger than hind pair. Veined. Wasp-waist often present. *Hymenoptera* (part)

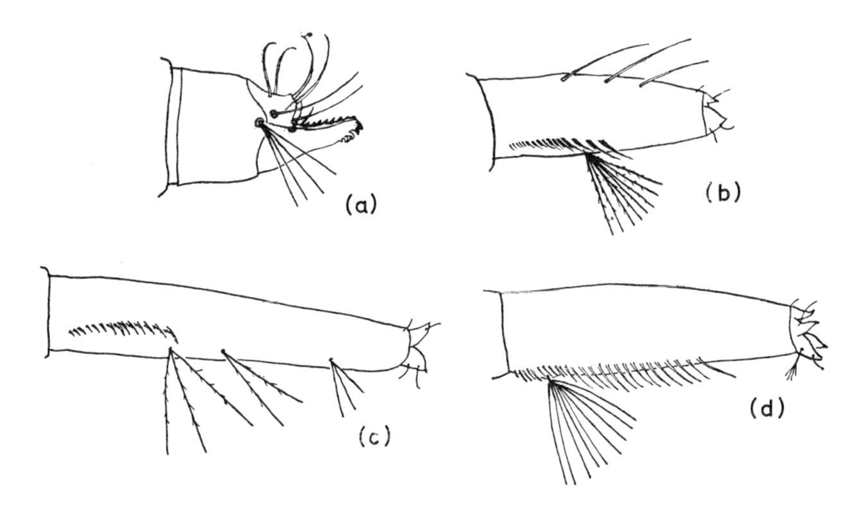

Figure 14.1 Siphons of larvae of British culicine mosquitoes. (*a*) *Mansonia richiardii*; (*b*) *Aedes rusticus*; (*c*) *Culex pipiens*; (*d*) *Culiseta annulata*. (After Marshall [377].)

14.2.2 MOSQUITOES OF THE NORTH TEMPERATE ZONE

Key 2a to main branches of the family Culicidae (adult females)

1. No scales on the wings (though small hairs may be present). Wing venation similar to Fig. 5.5d, page 170. Mouthparts small, unfitted for bloodsucking. (*Dixinae*)
Scales at least on hind border of wing. (2)
2. Few if any scales on the wing (i.e. hind margin only). Mouthparts small, unfitted for piercing. (*Chaoborinae*)
Scales on wing as well as along hind margin. Mouthparts long and, in female, adapted for piercing. (*Culicinae*) (Mosquitoes)

Key 2b to main branches of the family Culicidae (larvae)

1. Thorax narrow, distinctly segmented (Fig. 5.5d). (*Dixinae*)
Thorax markedly broader than abdomen without distinct segmentation. (2)
2. Antennae prehensile with long and strong apical spines. No bushy 'mouth brush'. (*Chaoborinae*)
Antennae not prehensile. Mouth brush present. *Culicinae* (Mosquitoes)

Key 2c to main genera of Culicidae (adult females)

1. Palpi as long as proboscis (Fig. 5.2). Abdomen without scales. *Anopheles*
Palpi shorter than proboscis (Fig. 5.2, page 138). Abdomen with scales. (2)

2. Abdomen long-tapering and pointed. Claws on fore and mid feet toothed. *Aedes*
Abdomen slightly tapering and blunt. Claws all simple. (3)
3. Tarsi without pale rings. *Culex*
Tarsi with pale rings. (4)
4. White rings on 5th tarsal segments of all legs and median band on first tarsal segments. Wings never spotted. *Mansonia*
Either white rings on 5th tarsal segments *or* median band on 1st tarsal segments (never both). Common species with spotted wings. *Culiseta*

Key 2d to main genera of Culicidae (larvae) (Fig. 14.1)

1. Siphon modified for piercing submerged stems of aquatic plants (Fig. 14.1) *Mansonia*
Siphon normal in shape. (2)
2. Siphon with 4 (occasionally 3 or 5) hair tufts. *Culex*
Siphon with only 1 hair tuft. (3)
3. Hair tuft near centre of siphon. *Aedes*
Hair tuft near base of siphon. *Culiseta*

Key 2e to some important European mosquitoes (adult females) (based on Edwards *et al.* [162]) (Fig. 14.2)

(i) Culicines

1. Abdomen long-tapering and pointed; claws on fore and mid feet toothed. *Aedes* (10)
Abdomen slightly-tapering and blunt; claws all simple. (2)
2. Tarsi without pale rings. (3)
Tarsi with pale rings. (4)
3. Dorsal abdominal pale stripes constricted laterally, and sometimes centrally, so as to appear convex or bilobed; patches of dark scales on pale underside of abdomen on median line and usually also on sides; knee-spots distinct and white. General colouring dark brown and white. *Culex pipiens*
Dorsal abdominal pale stripes parallel, unconstricted; dark scales rare on underside, or at least restricted to centre line; knee-spots absent; general colouring, yellowish brown and dirty white. *Culex molestus*
4. White rings on 5th tarsal segments of all legs and median band on first tarsal segment; wings unspotted; wing scales all very broad. *Mansonia richardii*
Either white rings on 5th tarsal segments *or* median band on 1st tarsal segments (not both). (5)
5. Wings with dark spots; large species. (6)
Wings without dark spots. (8)

6. Pale rings near tip of each femur and in middle of 1st segment of each leg. (7)
Such rings absent. *Culiseta alaskaensis*
7. Abdomen with conspicuous white bands. *Culiseta annulata*
Abdomen not distinctly banded; suffused with yellow. *Culiseta subochroea*
8. Pale rings on the tarsi. (9)
No pale rings on the tarsi. (14)
9. Rings extending on both sides of the joints. (10)
Rings confined to the base of each segment. (11)
10. Thorax fawn-coloured with two narrow white lines. *Aedes caspïus*
Thorax with dark brown stripe, on each side of which is a broad ashy-white area. *Aedes dorsalis*
11. Abdomen mainly yellow, unbanded. *Aedes flavescens*
Abdomen distinctly banded. (12)
12. Thorax yellowish towards the sides. *Aedes annulipes*
Thorax darker. (13)
13. Pale bands of abdomen of uniform width. *Aedes cantans*
Pale bands of abdomen constricted in the middle. *Aedes vexans*
14. Abdomen unbanded. (15)
Abdomen with transverse pale bands. (16)
15. Thorax ornate; knees silvery. *Aedes geniculatus*
Thorax reddish-brown; knees dark. *Aedes cinereus*
16. Abdomen with a faint median longitudinal pale stripe, at least towards the tip. *Aedes rusticus*
Abdomen without a trace of such a stripe. (17)
17. Abdominal bands (at least the last few) narrowed in the middle, their hind margins ∧-shaped. *Aedes punctor*
Abdominal bands not narrowed in the middle. (18)
18. Fore- and mid-femora conspicuously sprinkled with pale scales in front. (19)
Fore- and mid-femora with only a few pale scales in front. (20)
19. Dark parts of abdomen with scattered pale scales. *Aedes detritus*
Dark parts of abdomen without scattered pale scales. *Aedes leucomelas*
20. Hind tibia with a white stripe on outer side. *Aedes sticticus*
Hind tibia without such a stripe. *Aedes communis*

(ii) Anophelines

1. Costa with pale spots alternating with dark areas; in Mediterranean region. (2)
Costa entirely dark. (3)
2. Costa with two pale spots. *Anopheles hyrcanus psuedopictus*
Costa with four pale spots. *Anopheles superpictus*

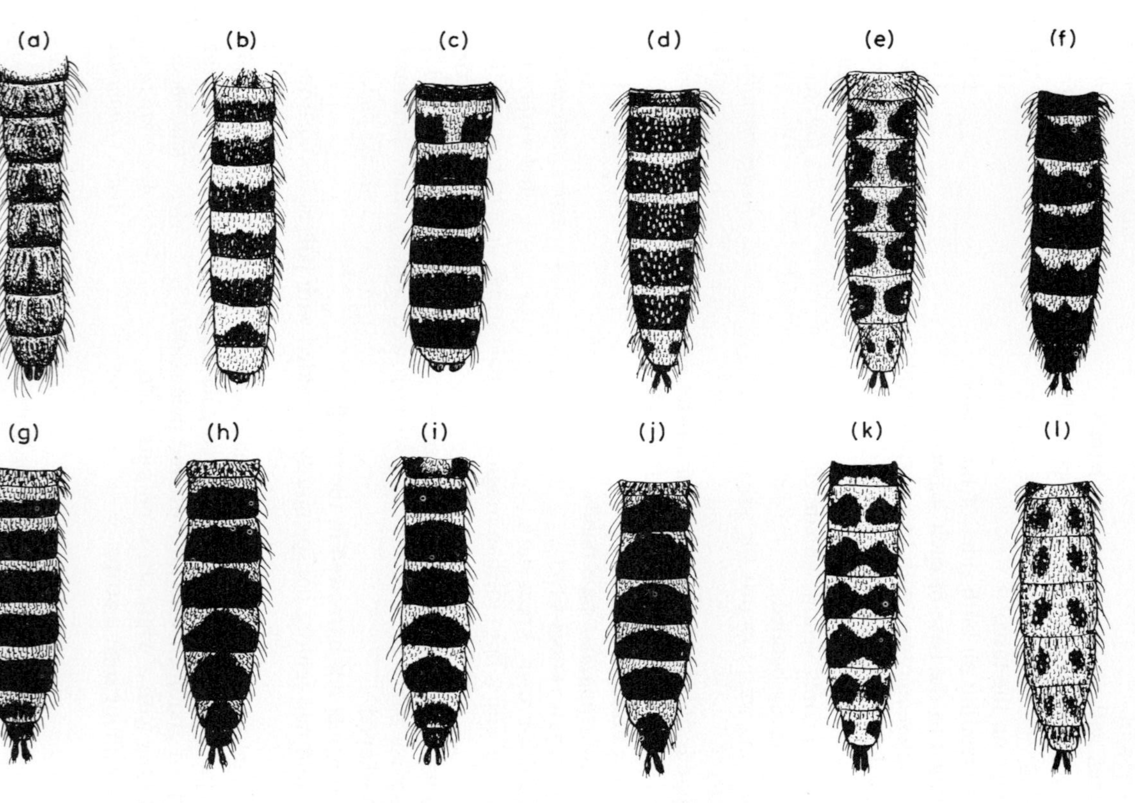

Figure 14.2 Dorsal views of the abdomens of some European mosquitoes. (*a*) *Anopheles maculipennis*; (*b*) *Culex pipiens*; (*c*) *Culiseta annulata*; (*d*) *Aedes detritus*; (*e*) *Ae. caspius*; (*f*) *Ae. vexans*; (*g*) *Ae. intrudens*; (*h*) *Ae. sticticus*; (*i*) *Ae. punctor*; (*j*) *Ae. dianteus*; (*k*) *Ae. rusticus*; (*l*) *Ae. dorsalis*. (After Coe *et al.* [117] and Mohrig [430].)

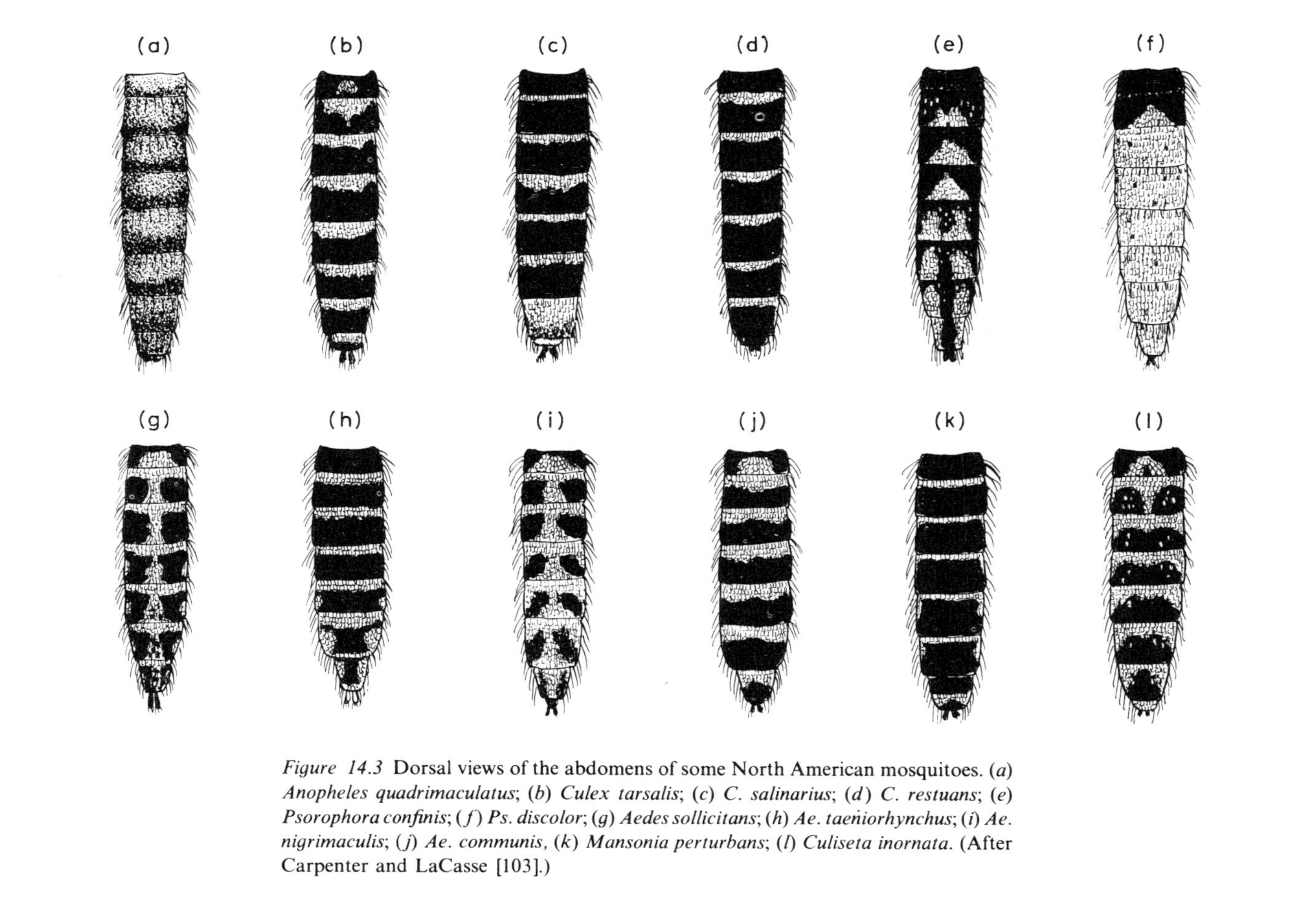

Figure 14.3 Dorsal views of the abdomens of some North American mosquitoes. (*a*) *Anopheles quadrimaculatus*; (*b*) *Culex tarsalis*; (*c*) *C. salinarius*; (*d*) *C. restuans*; (*e*) *Psorophora confinis*; (*f*) *Ps. discolor*; (*g*) *Aedes sollicitans*; (*h*) *Ae. taeniorhynchus*; (*i*) *Ae. nigrimaculis*; (*j*) *Ae. communis*, (*k*) *Mansonia perturbans*; (*l*) *Culiseta inornata*. (After Carpenter and LaCasse [103].)

3. Wing scales aggregated in places giving a naked eye appearance of spots.
Anopheles maculipennis complex (see page 146)
No such spots. (4)
4. Head with a frontal tuft of white scales. (5)
No such white tuft. *Anopheles algeriensis*
5. Body brown; wing about 5·5–6 mm. *Anopheles claviger*
Body blackish; wing about 5 mm. *Anopheles plumbeus*

Key 2f to some important North American mosquitoes (adult females) (based on Tinker and Stojanovich [580]) (Fig. 14.3)

(i) Culicines

1. Legs with white bands. (2)
Legs unbanded, entirely dark, or only last segments white. (10)
2. Proboscis unbanded. (3)
Proboscis with white band. (6)
3. Tarsal segments with white bands at basal ends only. (4)
Tarsal segments with white bands at both ends. (5)
4. Thorax with lyre-shaped marking of silvery scales; silver scales on clypeus and palpi. *Aedes aegypti*
Thorax without lyre-shaped marking; clypeus and palpi dark. Abdomen with pale bands on anteriors of segments 3–6, notched in the middle.
Aedes vexans
5. Abdomen pointed; segment 7 of abdomen narrowed, segment 8 much narrowed and retractile; mesonotum with a broad patch of pale scales on each side. *Aedes atropalpus*
Abdomen blunt; segment 7 not narrowed, segment 8 not retractile; mesonotum with 4 fine longitudinal white lines; mixed dark and light scales on wings.
Orthopodomyia sp.
6. Abdomen blunt at the tip. (7)
Abdomen pointed at the tip. (8)
7. White stripe on side of femur; no pale band near apex of tibia; no pale ring at middle of 1st segment of hind tarsus. *Culex tarsalis*
No stripe on side of femur; pale band present near apex of tibia; pale ring present at middle of 1st segment of hind tarsus. *Mansonia perturbans*
8. Pale ring at middle of 1st segment of hind tarsus; mixed dark and light scales on wings. (9)
No pale ring at middle of 1st segment of hind tarsus; only dark scales on wings; legs uniformly dark; upper surface of abdomen with only pale basal cross bands.
Aedes taeniorhynchus
9. Pale band near apex of femur; white spots on side of tibia; upper surface of abdomen with cross bands only. *Psorophora confinnis*

No pale band on femur or tibia; no spots on side of tibia; upper surface of abdomen with long longitudinal stripe as well as cross bands.
Aedes sollicitans

10. Abdomen pointed; segment 7 of abdomen narrowed; segment 8 much narrowed and retractile; mesonotum with a broad patch of pale scales on each side. *Aedes triseriatus*

Abdomen blunt; segment 7 of abdomen not narrowed; segment 8 not retractile; mesonotum without patch of pale scales. (11)

11. Base of sub-costa without a tuft of hairs on underside of wing; spiracular bristles absent; proboscis short and straight. Genus *Culex* (12)

Base of sub-costa with a tuft of hairs on underside of wing; spiracular bristles present; proboscis long and recurved. Genus *Culiseta* (16)

12. Abdomen with white scales at apex of segments. *Culex territans*

Abdomen with white scales at bases of segments. (13)

13. Abdominal segments with distinct basal bands or lateral spots of white scales. (14)

Abdominal segments with narrow dingy white basal bands. (15)

14. Abdomen with rounded pale bands; mesonotum without pale spots, covered with relatively coarse brownish, greyish or silvery scales.
Culex pipiens quinquefasciatus (*fatigans*)

Abdomen with pale bands almost straight; mesonotum with two or more pale spots (sometimes only one or two scales) against a background of fine coppery scales. *Culex restuans*

15. Pleuron with several bands of broad scales, each group usually comprising more than 6 scales; 7th and 8th abdominal segments almost entirely covered with dingy yellow scales. *Culex salinarius*

Pleuron with few or no scales (when present, rarely more than 5 or 6 in a single group); 7th and 8th abdominal segments banded. *Culex nigripalpus*

16. With broad lightly scaled wings; legs and wings sprinkled with white scales; a large species. *Culiseta inornata*

Wings and legs entirely dark scaled; a small dark species. *Culiseta melanura*

(*ii*) *Anophelines*

1. Wings with definite areas of white or yellowish scales. (2)

Wings without definite areas of white or yellowish scales. (4)

2. Palpi entirely dark; costa with two white spots. *Anopheles punctipennis*

Palpi marked with white; costa with one or two spots. (3)

3. Costa with white spot at tip of wing only. *Anopheles crucians*

Costa with two white spots. *Anopheles pseudopunctipennis*

4. Wings spotted; thoracic bristles not very long; mesothorax dull in rubbed specimens; medium size. *Anopheles quadrimaculatus*

Wings not spotted; thoracic bristles very long; mesothorax shiny in rubbed specimens; small species. *Anopheles barberi*

14.2.3 HOUSEFLIES AND BLOWFLIES

Key 3a to some common flies of public health importance

This key only includes a very restricted selection of Diptera and can only be used to assist in identification of (a) flies liable to occur inside houses; (b) bloodsucking flies and others which may sometimes be confused with them.

1. Abnormal, flattened flies, with legs projecting laterally, giving a crab-like appearance. All bloodsucking parasites of animals. (Some wingless, or with stunted wings) (Fig. 5.6d, e, page 178). *Pupipara*
Normal flies, gnats, etc. (2)
2. Antennae with at least 8, nearly similar segments. *Nematocera* (4)
Antennae with less than 6 similar segments. (3)
3. Large, robust flies, with the eyes banded or spotted in life. Bare not bristly, often with a pattern on body or wings or both. *Tabanidae* (Horseflies; clegs)
Small flies; or if large and robust, then very bristly. (10)
4. Back of the thorax bearing a prominent V-shaped groove. Usually rather large with long legs. *Tipulidae* (Daddy longlegs)
Without V-shaped groove. (5)
5. Ocelli present. Moderate-sized (7 mm long) gnat-like flies with mottled wings (Fig. 6.2a, page 212) *Anisopus* (Window gnat)
Ocelli absent. (6)
6. Costal vein (along anterior edge of wing) running round beyond tip of the wing. Scales or hairs often present on the wing. (7)
Costal vein ending before the tip of the wing which does not bear hairs or scales. (8)
7. Wings almond-shaped, hairy, carried at a roof-like angle in repose (Fig. 6.2b). *Psychodidae* (Moth flies)
Wings oblong with rounded tip, usually bearing scales, carried flat on top of each other in repose. *Culicidae* (Mosquitoes) (see pages 521–572)
8. Antennae shorter than the thorax, consisting of closely-set ring-like segments never plumose. Body thick-set with legs strong, not long and thin (Fig. 5.5). *Simuliidae* (Blackflies)
Antennae longer than the thorax and usually hairy or bushy. (9)
9. Back of the thorax with a longitudinal groove. Legs thin, forelegs often especially long. Mouthparts incapable of bloodsucking (Fig. 5.5e). *Chironomidae*
Back of the thorax without a longitudinal groove. Mouthparts capable of bloodsucking. Mostly minute insects with dappled wings (Fig.5.5b) *Ceratopogonidae*
10. Rather large robust flies or with distinct metallic green or blue coloration.

(Mainly 10 mm long or more with a wing-span of 18 mm or more)
Calliphoridae (Blowflies) (11)
Moderate-sized flies (mostly 5–8 mm long with a wing-span about 10–18 mm) rather similar to the common housefly in appearance. (14)
Very small flies (not more than 4 mm long with 8 mm wing-span). (17)

11. Large hairy fly, with thorax bearing longitudinal black and grey strips; abdomen chequered grey and black. (Length about 13 mm; wing-span about 22 mm) (Fig. 6.1c, page 193). *Sarcophaga*
Thorax and abdomen more uniform. Bluish or green metallic tinge. (12)

12. Highly metallic greenish-blue or greenish-yellow flies (Fig. 6.1e). *Lucilia*
(*Dasyphora cyanella* is similar, but has two longitudinal marks on the back of the thorax.)
Blue-grey or blue-black flies not highly metallic. (13)

13. Blue with much grey dusting overlying the blue. Flies producing loud buzzing sound in flight (Fig. 6.1a). *Calliphora* (Bluebottles)
Midnight blue, almost black blowflies, smaller than the common bluebottles.
Phormia

14. Piercing mouthparts in the form of a forward-projecting proboscis visible from above (Fig. 5.1c, page 136). Fly rather resembling an ordinary housefly in other respects (Fig. 6.1h). *Stomoxys calcitrans* (Stable fly)
Sucking proboscis not projecting in front of head. (15)

15. Fourth wing vein very sharply bent forward so that it nearly joins the third near the wing edge. (16)
Fourth wing vein gently curving forward. Legs partly brown; tip of scutellum reddish (about 8 mm long with 16 mm wing-span) (Fig. 6.1g).
Muscina stabulans
Fourth wing vein parallel to or diverging from the third. Small flies (about 6 mm long with 12 mm wing-span) (Fig. 6.1d). probably *Fannia*

16. Wings carried flat one over the other on the back when at rest. Thorax bearing golden-yellow bristles (Fig. 6.1f). *Pollenia rudis* (Cluster fly)
Wings carried at a diverging angle when at rest. Thoracic bristles black (Fig. 6.1b). *Musca*

17. Very small flies (about 2 mm long with 5 mm wing-span). Yellowish or brownish in colour. (18)
Shining black flies, rather larger than above. (19)

18. Bristle on the antenna feathered. Attracted by ripe fruit, etc. (Fig. 6.2e, f, g).
Drosophila (Fruit flies)
(*D. funebris* has the back of the thorax a shining reddish brown; while *D. repleta* has a dull grey-brown thorax with each bristle arising from a darker spot on it.)
Bristle on antenna plain. Lemon-yellow flies with black markings. Sometimes hibernate indoors in large numbers (Fig. 6.2c, page 212). *Thaumatomyia notata*

19. Anterior veins of wing much thicker than others. (Rarely seen; but larvae and pupae occur in milk curds) (Fig. 6.2h). *Phoridae*

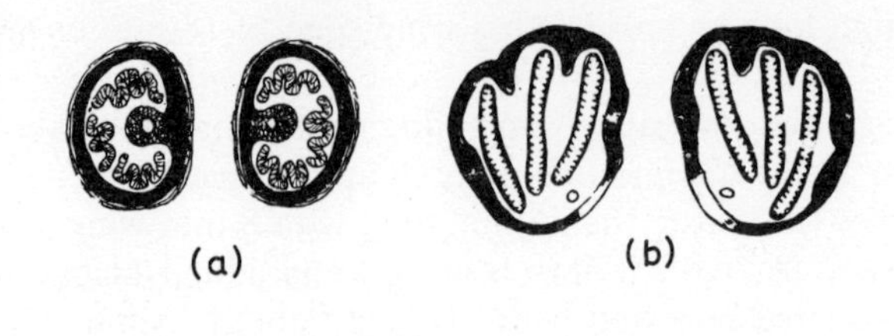

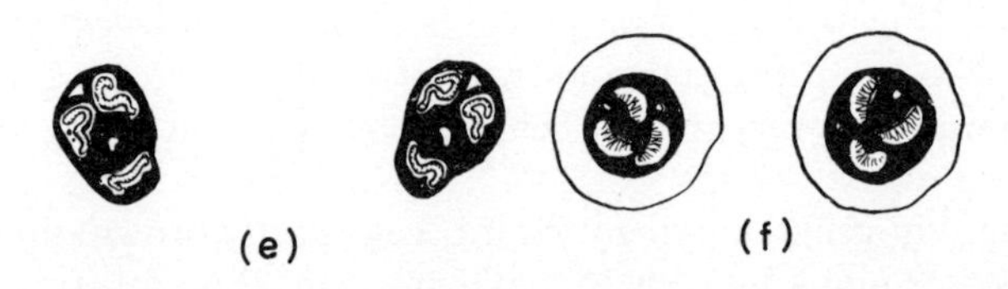

Figure 14.4 Spiracles of larvae of certain *Muscidae* and *Caliphoridae*. (*a*) *Musca domestica*; (*b*) *Sarcophaga* sp.; (*c*) *Calliphora vicina*; (*d*) *Lucilia sericata*; (*e*) *Stomoxys calcitrans*; (*f*) *Muscina stabulans*. Not all to same scale. (After Smart [542].)

Anterior waist between thorax and abdomen. Wings often with a black spot at tip. *Sepsis*
More robust, without pronounced waist. (Larvae in cheese on bacon) (Fig. 6.2d). *Piophila*

Key 3b to larvae of Diptera liable to cause myiasis (after Smart [542])

1. Larvae with a distinct head. (2)
Headless maggots. (3)
2. Nearly cylindrical, naked (Fig. 6.3a). *Anisopus*
Spindle-shaped, body surface bearing hairs (Fig. 6.3b). *Psychoda*
3. Somewhat flattened, bearing projecting tail-like processes on the dorsal side. (4)
Without tail-like processes on dorsal side. (5)
4. Processes simple (Fig. 6.3d, page 214). *Fannia canicularis*

Processes pinnate (Fig. 6.3d, page 214). *Fannia scalaris*

5. Small maggots, not exceeding 10 mm in length when fully grown; body rather cylindrical with the thickest part about the middle; posterior end not truncate; posterior spiracles small and raised above the general surface. (*Acalypterates*) (6)

Larger maggots up to 18 mm in length; body definitely tapering from behind forward, the thickest part being just before the sharply truncate hind end; posterior spiracles flush with the surface or sunk in a fold not obviously small. (*Calypterates*) (7)

6. Stouter larvae; posterior spiracles raised on 2 flask-like papillae which are fused together basally; ventro-lateral processes small, rather flap-like (Fig. 6.3). *Drosophila*

Thinner larvae; posterior spiracles raised on 2 cones; ventro-lateral projections finger-like. *Sepsis*

Thinner larvae; posterior spiracles raised on 2 cones on the dorsal surface of which there is a small vertical papilla; ventro-lateral processes tapering and slightly curved upward; in life, the mature larvae jump about actively (Fig. 6.3e). *Piophila casei*

7. Posterior spiracles sunken and often completely hidden by the overlapping of the short fleshy processes that surround the truncated area bearing the spiracles; the slits in the spiracles nearly vertical (Fig. 14.4d). *Sarcophaga*

Posterior spiracles not so sunken; slits on the spiracles straight but not vertical. *Calliphora* or *Lucilia* (Differences shown in Fig. 14.4)

Posterior spiracles not sunken; slits bent or sinuous. *Musca domestica*, *Stomoxys calcitrans* or *Muscina stabulans* (Differences shown in Fig. 14.4)

14.2.4 PARASITES

Key 4a to genera of some common fleas in buildings (Fig. 14.5)

1. Neither oral nor thoracic combs present. (2)
Thoracic comb present; genal combs absent. (3)
Both thoracic and genal combs present. (4)
2. Row of bristles along back of the head. *Xenopsylla*
Only one bristle (on each side) at the back of the head. *Pulex*
3. Thoracic comb with more than 24 teeth. *Ceratophyllus*
Thoracic comb with fewer than 24 teeth. *Nosopsyllus*
4. Eye absent. *Leptopsylla*
Eye present. (5)
5. Genal comb of 5 rounded spines arranged vertically. *Spilopsyllus*
Genal comb of 8 sharp spines arranged horizontally. *Ctenocephalides*

(Note: Of these genera *Pulex, Xenopsylla, Ctenocephalides* and *Spilopsyllus* belong to the family Pulicidae; *Ceratophyllus* and *Nosopsyllus* belong to the family Ceratophyllidae; and *Leptopsylla* belong to the family Leptopsyllidae.)

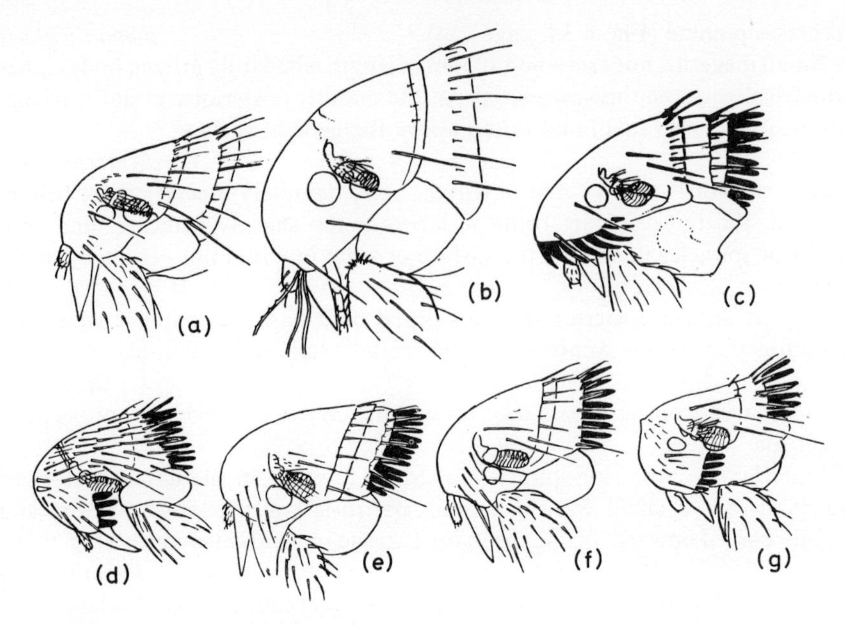

Figure 14.5 Heads of common fleas. (*a*) *Xenopsylla cheopis*; (*b*) *Pulex irritans*; (*c*) *Ctenocephalides felis*; (*d*) *Leptosylla segnis*; (*e*) *Ceratophyllus gallinae*; (*f*) *Nosopsyllus fasciatus*; (*g*) *Spilopsyllus cuniculi*. All females except (*g*). (After Smart [542].)

Key 4b to genera of Ixodidae (based on that of Arthur [17])
(See Fig. 7.9, page **293**)

1. Anal groove surrounding the anus in front. *Ixodes*
Anal groove, either distinct behind the anus, or faint and indistinct. (2)
2. Without eyes. *Haemaphysalis*
With eyes. (3)
3. Palps as broad, or broader than long. (4)
Palps longer than broad. (6)
4. Large ticks, usually ornate. Base of the capitulum rectangular. With festoons. *Dermacentor*
Small ticks, Not ornate. Base of the capitulum usually hexagonal. (5)
5. Anal grooves distinct. With festoons. *Rhipicephalus*
Anal grooves absent in females; weakly present in males. No festoons. *Boophilus*
6. Scutum not ornate. Second palpal segment at least twice as long as the third. *Hyalomma*
Scutum ornate. Second palpal segment less than twice as long as the third. *Amblyomma*

14.2.5 STORED PRODUCT PESTS

COCKROACHES

Key 5a to species of cockroaches liable to occur indoors (based on that given by Cornwell [124])

1. Small (15 mm or less, including forewings). (2)
Large (Over 15 mm long). (3)
2. Yellowish brown, with two longitudinal marks on the pronotum. Adults about 10–15 mm long. *Blattella germanica*
Pronotum with transparent lateral edges, remainder dark brown, though paler at centre. Chestnut markings on the wings giving the appearance of wide bands. (10–15 mm long). *Supella longipalpa*
3. Moderately large (15–55 mm long). (4)
Very large (over 55 mm long). *Blaberus* spp.
4. Forewings noticeably shorter than the abdomen. (5)
Forewings reaching tip of abdomen, or longer. (7)
5. Forewings covering thorax and extending over part of the abdomen. (6)
Forewings reduced to short lobes, not covering the thorax. Dark reddish brown or black. (Female) *Blatta orientalis*
6. Larger (25–29 mm long). Ashy coloured, with a pattern resembling a lobster on the pronotum. *Nauphoeta cinerea*
Smaller (20–24 mm long). Dark reddish brown to black. (Male) *Blatta orientalis*
7. Pronotum 6–7 mm wide, with hind margins sinuate. Medium size (18–24 mm long). Shining brown to black. *Pycnoscelus surinamensis*
Pronotum more than 6–7 mm wide. Larger (over 28 mm). (8)
8. Posterior half of forewings distinctly mottled, with two pronounced dark lines in the basal area. Large (40–50 mm). *Leucophaea maderae*
9. Base of forewings with pale streak on outer edges. Pronotum strikingly marked with a dark central area and pale edging. (30–35 mm long). *Periplaneta australasiae*
Base of forewings without pale streak on outer edges. Pronotum of uniform colour; edging only moderately conspicuous. (10)
10. Pronotum uniformly dark in colour. Body very dark brown to black. (31–35 mm long). *Periplaneta fuliginosa*
Pronotum not entirely uniform in colour, but with pale edging, moderately conspicuous. (11)
11. Cerci tapering, last segments twice as long as wide. Shiny red-brown, with pale yellow edging on pronotum. (28–44 mm). *Periplaneta americana*
Cerci spindle shaped, last segment less than twice as long as wide. (31–37 mm). *Periplaneta brunnea*

ANTS

Key 5b to sub-families of Formicidae important in the North Temperate Zone [148]

1. Pedicel distinctly two-jointed. Myrmecinae
(*Monomorium,*[1] *Solenopsis,*[1, 2] *Pheidole,*[1] *Tetramorium*[1] *Pogonomyrmex,*[2] *Crematogaster*[1])
Pedicel not two-jointed. (2)
2. Gaster (abdomen behind pedicel) constricted between first two segments. Ponerinae
(*Hypoponera*[1])
Gaster not constricted. (3)
3. Gaster with 5 visible segments from above; anal aperture circular and surrounded with a fringe. Formicinae
(*Formica,*[2] *Lasius,*[1] *Prenolepis*[1])
Gaster with only 4 segments visible from above; anal aperture slit-like, not surrounded by bristles. Dolichoderinae
(*Iridomyrmex,*[1] *Tapinoma*[1])

[1] Genera with house-invading species.
[2] Genera with severe stinging species.

A key to some important house-invading species follows.

Key 5c to some ants liable to invade houses in North America (based partly on Eckert and Mallis [159]) (see Fig. 8.3, page 324)

1. One segment in the pedicel. (2)
Two segments in the pedicel. (7)
2. Anal opening terminal, circular and surrounded by a fringe of hairs. (3)
Anal opening not terminal, but ventral, slit-shaped and not surrounded by a fringe of hairs. (4)
3. Thorax very narrow; abdomen triangular, workers about 2·3–2·5 mm long. *Prenolepis imparis*
Thorax broader; abdomen oval-oblong, workers about 3·4–5 mm long. *Lasius niger*
4. Last segment of thorax with a pyramid-like projection; workers, 1·5–2 mm; reddish black with black abdomen. *Dorymyrmex pyramicus*
Last segment of thorax without pyramid-like projection. (5)
5. Petiole difficult to see, lying somewhat in horizontal position and close to overhanging abdomen. Dark brown.
Workers emit strong odour when crushed. *Tapinoma sessile*
Petiole erect and visible. Brown, thorax and legs lighter. No odour on being crushed. *Iridomyrmex humilis*
6. Very small ants. Workers less than 1·75 mm long. (7)

Workers larger than 1·75 mm. (8)

7. Antenna with 2-segmented club. Light yellow. Workers 1·0–1·5 mm. *Solenopsis molesta*

Antenna with 3-segmented club. Reddish-sandy colour. Workers about 2 mm. *Monomorium pharaonis*

8. Pedicel inserted dorsally into abdomen, which is heart-shaped, flattened ventrally and coming to a sharp point. *Crematogaster* spp.

Pedicel inserted terminally into abdomen. (9)

9. Antenna with three-jointed club. Head with deep furrow on vertex. *Pheidole* spp.

Antenna without 3-jointed club. Head without furrow. Pair of spines on last segment of thorax. Blackish-brown, with pale legs. Workers 2–3 mm. *Tetramorium caespitum*

N.B. The *pedicel* is the portion of the body between thorax and abdomen. It may consist of one joint, the *petiole*, or a *petiole* and *post-petiole*.

MOTHS

Key 5d to some common moth pests of stored products (from Hinton and Corbet [276])

N.B. The various wing patterns are shown in Fig. 14.6

1. Hindwing with fringe short, the hairs not nearly half the breadth of the wing. *Pyralidina* (2)

Hindwing with fringe long, the hairs at least as long as half the breadth of the wing. *Tinaeina* (8)

2. Labial palps prominent and curved upwards. (3)

Labial palps inconspicuous in males of *Corcyra cephalonica* and *Paralipsa gularis* but conspicuous in remainder, where they are straight in front of the head or slightly curved downwards. (6)

3. Uppersides forewing pale ochreous buff, with basal and apical areas purple-brown, the pale centres divided from the darker areas by narrow white lines. Uppersides hindwing smoky-black, crossed by 2 narrow whitish lines. *Pyralis farinalis* (The 'meal moth')

Uppersides forewing dull greyish brown with markings obscure; 2 bands (light or dark) can be distinguished across the wing.
Uppersides hindwing greyish white or greyish buff and unmarked. (4)
(N.B. The three following species have rather similar wing patterns and are difficult or impossible to distinguish in rubbed specimens.)

4. Uppersides forewing with the outer band well defined, rather sinuate, pale and bordered on each side by a narrow dark line, the dark bordering being more intense anteriorally. *Ephestia elutella* (The 'warehouse moth')

Uppersides with the outer band very obscure and not well defined. (5)

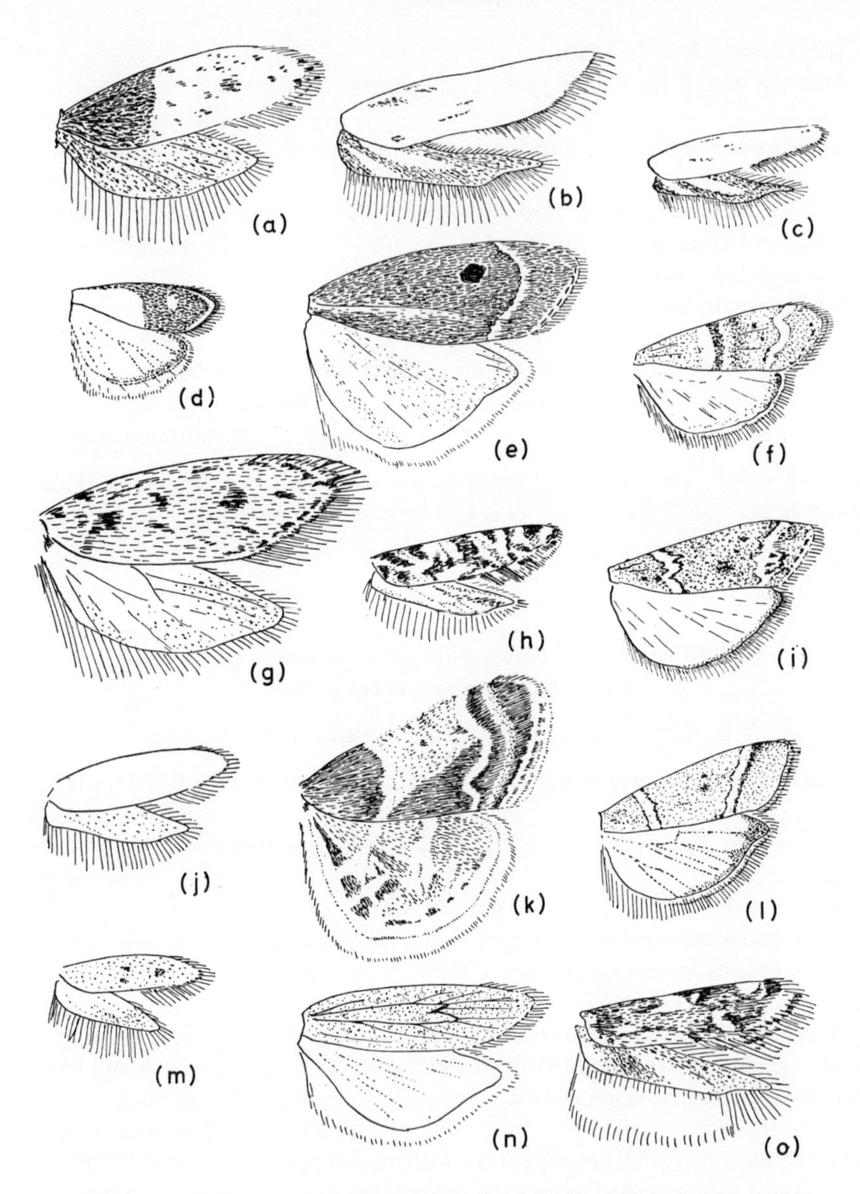

Figure 14.6 Wings of moths which attack stored food or fabrics. (*a*) *Trichophaga tapetzella*; (*b*) *Sitotroga cerealella* (female); (*c*) the same (male); (*d*) *Plodia interpunctella*; (*e*) *Paralipsa gularis* (female); (*f*) *Ephestia cautella*; (*g*) *Hofmannophila pseudospretella*; (*h*) *Nemapogon granella*; (*i*) *Ephestia kühniella*; (*j*) *Tineola bisselliella*; (*k*) *Pyralis farinalis*; (*l*) *Ephestia elutella*; (*m*) *Tinea pellionella*; (*n*) *Corcyra cephalonica* (female); (*o*) *Endrosis sarcitrella*. (All × 3 approximately.) (After Hinton and Corbet [276].)

5. Upperside forewing with the inner band dark and straight (at right angles to front margin), rather broad and continuous and with a broad pale band along its inner edge. *Ephestia cautella* (The 'tropical warehouse moth')
Upperside forewing with the inner band oblique, rather irregular and consisting of dark streaks or spots, and without a pale band along its inner edge.
Ephestia kuhniella (The 'Mediterranean flour moth')

6. Upperside forewing with the basal third pale yellowish buff, this pale area separated from the outer reddish-brown area by a dark-brown line. Head without a projecting tuft of scales. *Plodia interpunctella* ('Indian-meal moth')
Upperside forewing uniformly coloured pale buff-brown. Head with a projecting tuft of scales. Labial palps inconspicuous in males, long and prominent in females. (7)

7. Upperside forewing without spots but with the veins slightly darkened. Upperside hindwing darker in male than in female. *Corcyra cephalonica*
Upperside forewing with a black spot at or beyond the centre of the wing. Male has a reddish-yellow streak in centre of forewing, absent in female.
Paralipsa gularis

8. Labial palps long, sickle-shaped, sharply pointed and curved upwards. Head smooth. (9)
Labial palps shorter, rather blunt, straight or nearly straight, projecting in front of the head or sloping downwards. Head roughly haired. (11)

9. Hindwing apex very elongated, sharply pointed and needle-like. Upperside forewing pale ochreous buff and, usually, a black dot can be seen beyond centre of the wing; upperside hindwing with a whitish stripe, running from wing base to beyond centre for about $\frac{2}{3}$ the length of the wing.
Sitotroga cerealella (The 'Angoumois grain moth')
Hindwing apex may be pointed but not needle-like. Upperside hindwing without yellowish stripe. (10)

10. Head and at least the front of thorax conspicuously white. Upperside forewing shining buff, speckled with dark brown and usually 2 or 3 blackish spots. *Endrosis sarcitrella* (The 'white shouldered house moth')
Head and thorax brown, upperside forewing buff-brown to dark buff-brown, with 3 more or less distinct brown spots.
Hofmannophila (= *Borkhausenia*) *pseudospretella* (The 'brown house moth')

11. Upperside forewing with the basal third deep chocolate brown, contrasting sharply with the whitish outer two-thirds, and with wing apex slightly darkened.
Trichophaga tapetzella (The 'tapestry moth')
Upperside forewing more or less uniformly coloured. (12)

12. Upperside forewing pale ochreous buff, entirely unmarked and without dark dusting. *Tineola bisselliella* (The 'common clothes moth')
Upperside forewing with dark markings and some dark dusting. (13)

13. Upperside forewing with a prominent chain of dark reddish-brown conjoined spots, running obliquely across centre of wing; ground colour buff with dark dusting and other irregular dark-brown markings.

Tinea granella (The 'corn moth')

Upperside forewing uniformly pale shining buff or brown with only 3 faint, dark spots; dark dusting very slight.

Tinea pellionella (The 'case-bearing clothes moth')

Key 5e to the more common lepidopterous larvae infesting stored products

The nomenclature used here for the setae of the thorax and abdomen is that elaborated by Fracker and used by Hinton. It is illustrated in Fig. 14.7. Each seta is indicated by a Greek letter, as follows:

α ALPHA	η ETA	π PI
β BETA	θ THETA	ρ RHO
γ GAMMA	κ KAPPA	σ SIGMA
δ DELTA	μ MU	τ TAU
ε EPSILON	ν NU	ι IOTA

Not all of these setae are required for the purposes of the abridged key given below, but they are included in the diagram for completeness. Some of the setae are always associated with others in definite groups, thus: KAPPA group ($\theta+\kappa+\eta$) or PI group ($\nu+\pi$ on thorax, $\nu+\pi+\iota$ on abdomen). KAPPA and ETA vary considerably in position relation to each other; therefore, to avoid confusion I have followed Hinton in calling the larger seta KAPPA and the smaller ETA in all cases. When THETA is present, it is always smaller than ETA and usually nearer to the spiracle.

Note. It is useful to remember that any larva (excluding first stage) found on stored food which has a large sclerotized ring enclosing a membranous area at the base of RHO on the mesothorax, combined with white or nearly white cuticles and dark conspicuous oval plates at the base of abdominal setae, is a species of *Ephestia*.

The following three keys are after Hinton and Corbet [276], nos. 6b and c being modified by Mr G. Talbot.

1. Prolegs short, narrow and often indistinct with only 2 crotchets each. Body 7 mm long or less and always inside a grain (except first stage) (Fig. 8.7).

(Gelechidae) *Sitotroga cerealella*

Prolegs distinct and well developed with many crotchets. (2)

2. Two setae in KAPPA group of prothorax. *Pyralidae* (3)

Three setae in KAPPA group of prothorax. *Tinaeina* (9)

3. KAPPA group of prothorax in a vertical line. First abdominal segment without a sclerotized ring enclosing the base of RHO. (5)

KAPPA group of prothorax in nearly horizontal line. First abdominal segment with a sclerotized ring enclosing base of RHO. Spiracles with black or heavily marked rims. *Gallerinae* (4)

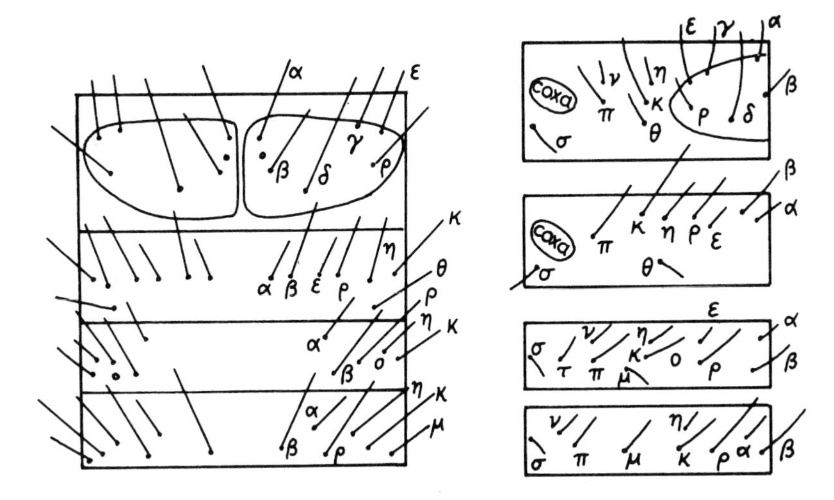

Figure 14.7 Chaetotaxy diagram for larvae of stored product moths. (After Hinton [273].) The segments illustrated are (top to bottom) 1st and 2nd thoracic and`1st and 9th abdominal.

4. Abdominal spiracles with posterior part of rim distinctly thicker than anterior part. Cuticle of abdomen white. Dorsal setae on abdomen without small oval basal plates. *Corcyra cephalonica*

Abdominal spiracles with posterior part of rim only very slightly thicker than anterior part. Cuticle of abdomen greyish white. Dorsal setae on abdomen (segments 1–8) arise from small oval plates. *Paralipsa gularis*

5. Mesothorax with a sclerotized ring enclosing a membranous area at the base of RHO. *Phycitinae* (6)

Mesothorax without a sclerotized ring enclosing a membranous area at the base of RHO (Pyralidinae). Head with only 4 distinct ocelli on each side. *Pyralis farinalis*

6. First 8 abdominal segments with only RHO of the 8th segment arising from sclerotized plate. (Fig. 8.7b, page 355). *Plodia interpunctella*

First 9 abdominal segments with all dorsal setae (except, sometimes, EPSILON) arising from sclerotized plates. (7)

7. Eighth abdominal segment with seta EPSILON separated from spiracle by less than the diameter of the latter. Spiracle as broad or nearly as broad as membranous part enclosed by sclerite of RHO. *Ephestia cautella*

Eighth abdominal segment with EPSILON separated from spiracle by considerably more than the diameter of the latter. (8)

8. Eighth abdominal segment with spiracles not more than $\frac{2}{3}$ as broad as membranous part enclosed by sclerite of RHO. Prothorax with diameter of spiracle much less than distance between KAPPA group setae. Outer tooth of mandible forming part of outer margin of mandible (Fig. 8.7a). *Ephestia elutella*

Eighth abdominal segment with spiracle as broad or broader than membranous part enclosed by sclerite of RHO. Spiracle of prothorax with diameter as great or greater than distance between KAPPA group setae. Outer tooth of mandible, when seen from inner ventral view, does not form part of outer margin of mandible, but is displaced mesally so that this margin is part of second or larger tooth. *Ephestia kuhniella*

9. KAPPA and ETA of abdominal segments 1–8 widely separated and in a more or less horizontal line. KAPPA group on prothorax not in a straight line. *Tinaeidae* (11)

KAPPA and ETA of abdominal segments 1–8 close together and in a more or less vertical line. KAPPA group on prothorax in a straight line. *Oecophoridae* (10)

10. Head with 4 ocelli on each side. Spiracle of 8th segment sometimes round, but usually with vertical diameter distinctly greater. Labium without a basal depression (Fig. 8.7c). *Hofmannophila pseudospretella*

Head with 2 ocelli on each side. Spiracle of 8th segment always round. Labium with a large basal pit. *Endrosis sarcitrella*

11. Head without ocelli. PI setae of meso- and meta-thorax in a fully oblique or nearly horizontal line. PI group of 9th abdominal segment absent. EPSILON of first 8 abdominal segments dorsal and considerably anterior to spiracle. Spiracles of 7th abdominal segment approximately as large as 8th (Fig. 8.7d, page 355). *Tineola bisselliella*

Ocelli present. (12)

12. Head with 6 ocelli on each side. Spiracles of 7th abdominal segment about $\frac{3}{4}$ as large as those of 8th. Without a case. *Tinea granella*

Head with 1 distinct ocellus on each side. Spiracles of 7th abdominal segment only $\frac{1}{2}$ to $\frac{2}{3}$ as large as those of 8th. Always in case shaped like a pillow. *Tinaea pellionella*

BEETLES

Key 5f to adults of beetles commonly occurring as pests of stored food with a few species which attack fabrics and wood (from Hinton and Corbet [276]) (Figs. 8.4–8.6, pages 340–345)

1. Antennae about $\frac{1}{2}$ as long as body, borne on prominent tubercle and capable of being reflexed backwards. All tibiae with 2 spurs (Cerambycidae). Domestic wood-borer. Adult 10–20 mm (Fig. 11.2a, page 431). *Hylotrupes*

Not with this combination of characters. (2)

2. Mouthparts at the end of a prominent snout-like projection between the eyes (Curculionidae). (3)

Head without distinct snout. (5)

3. Elytra not quite covering top of abdomen. Pests of food (Fig. 8.6a, b) *Sitophilus*

Elytra covering tip of abdomen. Larvae wood-borers. (4)

4. Elytra with apical sides dilated and bent upwards. Head deeply constricted just behind eyes (Fig. 11.2c, page 431). *Euophryum*
Elytra with apical sides not flexed up. Head only slightly constricted behind eyes. *Pentarthrum*
5. Elytra leaving at least 1 dorsal abdominal segment exposed. (3)
Elytra completely covering the back. (5)
6. Antennae with a distinct 3-segmented club. Elytra not striate (Fig. 8.5f). *Carpophilus* (Nitidulidae)
Antennae not clubbed or only indistinctly so. Elytra deeply striate. (7)
7. Each eye with a deep indentation extending back from the base of the antennae (Fig. 8.6d, e). *Bruchidae* 'Pea or bean weevils'
Eyes without indentation (Antheribidae) (Fig. 8.6c). *Araecerus fasciculatus* 'Coffee-bean weevil'
8. Elytra with a blue or blue-green metallic lustre on at least apical four-fifths. Cleridae (*Necrobia*)
Elytra never blue or blue green. (9)
9. Prothorax with 6 large acute teeth on each side; upper surface with 3 longitudinal ridges. 2·5–3·5 mm. (Dark brown, rather flat) (Fig. 8.4b). *Oryzaephilus* (Silvanidae)
Prothorax with 2 large teeth on each side, or 1, or none. Back of prothorax with 2 ridges or none. (10)
10. Small flat beetles (1·3–5 mm) with 2 longitudinal ridges on back of prothorax, parallel and near to sides. Antennae long and without club (Fig. 8.4a). *Cryptolestes* (Cucujidae)
Prothorax without longitudinal ridges. (11)
11. Hood-shaped prothorax covered with tubercles, rather coarse, especially in front. (Cylindrical and brown; antennae with a large loose 3-segmented club) (Fig. 8.5h). *Rhizopertha dominica* (Bostrychidae) 'Lesser grain borer'
Prothorax with a few, very fine, tubercles, or none. (12)
12. Prothorax with front corners distinctly toothed or with a very much thickened, nearly smooth, oval, disc-like area, and with a distinct tooth-like projection on the middle of each side (Fig. 8.5a). *Cryptophagus* (Cryptophagidae)
Prothorax without front corners toothed or thickened. (13)
13. Antennae usually more than half as long as the body, none of the segments being broader than the second. Prothorax constricted at the rear to form a kind of waist (except *Gibbium*). Tarsi 5-segmented. *Ptinidae* (11) (Spider beetles)
Antennae less than half the body length and apical segments broader than the second, forming a club. (16)
14. Elytra bare and shiny (Fig. 8.6h). *Gibbium*
Elytra striated and hairy. (15)
15. Breadth of elevated part of head between bases of antennae equal to a fourth or less of the first antennal segment (Fig. 8.6f). *Ptinus*
Breadth of the elevated part of the head between the bases of the antennae equal

to more than half of the 1st antennal segment (Fig. 8.6g). *Niptus*

16. Antennae with large, compact, very distinct, 2-segmented club, 3–5 mm. (Narrow, parallel, brown beetles similar in appearance to *Tribolium* but differing in having all tarsi, instead of only the hind tarsi, 4-segmented and the antennal club with 2 instead of 3 or more segments) (Fig. 11.1g, i, page 425).
(Lyctidae) *Lyctus* ('Powder post beetle')

Antennae with club of 3 or more segments (except some species of *Anthrenus*, but the latter are very convex, nearly round beetles covered with white and brown or black scales) or without a distinct club. (17)

17. Dorsal surface without hairs or scales visible with a hand lens. (18)

Dorsal surface with scales or hairs (visible, at least, if the specimen is viewed against the light with a hand lens), × 15. (26)

18. Distinctly flattened beetles, 5–11 mm. Prothorax distinctly separated from the base of the elytra. Tarsi apparently 4-segmented (Fig. 8.5e).
(Ostomidae) *Tenebroides* ('Cadelle')

Distinctly convex beetles. Base of thorax touching base of elytra. (19)

19. Tarsi all 3-segmented. Sides of the prothorax with a flattened, leaf-like edge and the middle slightly depressed. 1·2–2·4 mm (Fig. 8.5c).
Lathridiidae (*Enicmus*, *Lathridius*)

Front and middle tarsi 5-segmented, hind tarsi 4-segmented; prothorax smoothly convex without leaf-like edge, 4 mm or more. *Tenebrionidae* (20)

20. Large (20–24 mm) black beetle. *Blaps mucronata* ('Churchyard beetle')

Less than 18 mm long. (21)

21. Brown or black beetle about $\frac{1}{2}$ in. (14–18 mm) long (Fig. 8.4e).
Tenebrio ('Mealworm beetle')

Less than 7 mm long. (22)

22. Side margins of the head projecting into and nearly dividing the eyes. (23)

Eyes not divided and more or less round (Fig. 8.4f). *Palorus*

23. Body 4·5–7 mm long, rather broad and black or dark brown in colour (Fig. 8.4g). *Alphitobius*

Body usually less than 4·5 mm long, narrow, sub-cylindrical and reddish brown in colour. (Except *Tribolium destructor* which is 5–6 mm long and black, but of typical *Tribolium* shape). (24)

24. Antennae shorter than head with a very distinct, compact, 5-segmented club. Hind tarsi with basal segment not as long as the combined length of the two following. 2·5–3 mm (Fig. 8.4h).
Latheticus oryzae ('Long-headed flour beetle')

Antenna distinctly longer than head, with a compact 3-segmented club, or a loose 4-segmented club, or gradually broadening from the base with no distinct club. Hind tarsi with basal joint as long as combined length of two following. (25)

25. Elytra with fine longitudinal ridges, at least at the sides. Males without distinct tubercles on the head and without large, upwardly curved teeth on mandibles (Fig. 8.4d). *Tribolium*

Elytra without any longitudinal ridges. Males with 2 prominent tubercles on middle of head and each mandible with a large conspicuous tooth curved upwards (Fig. 8.4e). *Gnathocerus*

26. Antennae rest in cavities on the prothorax visible from the front. Body tortoise-shaped, conspicuously clothed in alternating patches of white and blackish or brownish scales. 2–4 mm. (Dermestidae) *Anthrenus*

Antennae not received in cavities on the prothorax visible from in front. Body oblong-oval and never with scales (though sometimes with black and white hairs). (27)

27. Head without a median ocellus. Elytra usually striate. (28)

Head with a median ocellus. (29)

28. Rather large beetles (5·5–10 mm) densely covered with hairs. (Dermestidae) (Fig. 10.2a, b, page 410). *Dermestes*

Smaller beetles (2–4 mm). (27)

29. Antennal club 3-segmented. First segment of hind tarsus half as long as second (4–5 mm) (Fig. 10.2c). *Attagenus*

Antennal club at least 4-segmented. First segment of hind tarsus much longer than second ($1\frac{1}{2}$–3 mm) (Fig. 8.5d). *Trogoderma*

30. Tarsi all 4-segmented except the front tarsi of males which are 3-segmented. Antennal club 3- or 4-segmented (Fig. 8.5b). *Mycetophagus*, *Typhaea* (Mycetophagidae)

Tarsi all 5-segmented. Antennal club 3- or 8-segmented and always much longer than remainder of antenna. Small brown beetles with the head held under the prothorax and not visible from above. Anobiidae (31)

31. Antenna with segments 4–10 serrate. Elytra not striate (Fig. 8.5g). *Lasioderma serricorne* ('Cigarette beetle')

Antenna with large loose 3-segmented club. Elytra distinctly striate. (32)

32. Prothorax with basal middle part, when seen from the side, very strongly humped. 3–5 mm (Fig. 11.1a, page 425). *Anobium punctatum* ('Common furniture beetle')

Prothorax with basal middle part not humped. 2–3 mm (Fig. 8.5i). *Stegobium paniceum* ('Bread beetle')

14.2.6 FABRIC PESTS

Key 6a to the larvae of the clothes moths and house moths

1. Eighth abdominal segment with the first 2 setae below the spiracle close together. Head with 2 or 4 ocelli on each side. (2)

Eighth abdominal segment with the first 2 setae below the spiracle very widely separated. Head with not more than 1 ocellus on each side. (3)

2. Head with 4 ocelli on each side. Labium without a pit or sclerotized ring (Fig. 8.7c, page 355). *Hofmannophila pseudospretella*

Head with only 2 ocelli on each side. Labium with a large but thin sclerotized ring and often also with a distinct pit. *Endrosis sarcitrella*

3. Larva in a flattened, portable and very definitely formed fusiform case. Prothorax nearly black. *Tinaea pellionella*

Larva free or in a web or silken gallery but never in a well-formed portable case. Prothorax yellowish or reddish, never blackish. (4)

4. Head without ocelli or at any rate ocellar lenses. Spiracles of 8th abdominal segment more or less equal in size to those of 7th (Fig. 8.7d). *Tineola bisselliella*

Head with a distinct ocellus on each side. Spiracles of 8th abdominal segment twice or nearly twice as large as those of seventh. *Trichophaga tapetzella*

Key 6b to the larvae of five genera of dermestid beetles (after Hinton) [274]

1. Abdomen with the 10th segment large and forming a distinctly sclerotized ring in which sternum and tergum are fused together as in the preceding segments; 9th segment with 2-dorsal conical processes which are usually large and conspicuous (Fig. 10.2e, page 410). *Dermestes*

Abdomen with the 10th segment never forming a sclerotized ring, much reduced and entirely membranous or with only the ventral part sclerotized; 9th segment without dorsal conical processes. (2)

2. Dorsal surface without spear-headed, segmented hairs. Maxilla with palp 4-segmented. (3)

Dorsal surface with spear-headed, segmented hairs. Maxilla with palp 3-segmented. (4)

3. Abdomen with a caudal brush of extremely long, slender hairs (Fig. 10.2f). *Attagenus*

Abdomen without a caudal brush of long slender hairs. Thorax and abdomen with marginal setae of tergites club-shaped. *Thylodrias*

4. Abdomen with caudal tufts of spear-headed hairs arising from an entirely membranous recess in the outer hind margins of tergites v, vi and vii. No tuft of these hairs on segment viii. Heads of the spear-headed hairs with 5 struts (Fig. 10.2g). *Anthrenus*

Abdomen with tufts of spear-headed hairs arising from hard portions of the outer hind margins of tergites v, vi and vii; additional tuft present on viii. Heads of the spear-headed hairs with 6 struts (Fig. 8.5d, page 342) *Trogoderma*

14.2.7 WOOD-BORING PESTS

BEETLES

Key 7a to the larvae of some genera of domestic wood-boring beetles

1. Body straight, reaching maximum length of 25–30 mm. (6)

Body distinctly curved, reaching maximum length of 10 mm or less (2)
2. Legs absent. Curculionidae (*Euophryum* or *Pentarthrum*)
Legs present. (3)
3. Legs 3-jointed; 8th abdominal spiracle larger than the others and rather prominent. *Lyctus*
Legs 5-jointed; 8th abdominal spiracle not noticeably larger than the others. (4)
4. Spinules present on the hypopleural folds of the first 6 abdominal segments. (The hypopleural folds lie on the ventro-lateral edges of the body). *Ptilinus*
Spinules not present on abdominal segments, as above. (5)
5. Spinules absent from 9th abdominal segment (6 mm or less). *Anobium*
Spinules present on 9th abdominal segment (9 mm or less). *Xestobium*
6. Head and legs well developed. Pronounced tubercles on 3rd and 4th abdominal segments. *Nacerdes*
Head reduced, legs exceedingly small. No unusual tubercles on 3rd and 4th abdominal segments. *Hylotrupes*

TERMITES

Key 7b to termites in the Northern Temperate Zone

(i) Winged adults of three important families

1. Fore area of the wing with three or more prominent veins beyond the basal wing scale. No fontanelle. (2)
Fore area of the wing with only two prominent veins. Fontanelle present, though perhaps inconspicuous. *Rhinotermitidae*
2. No ocelli. One or more spines on the shaft of the tibia, as well as at the tip. Antennae usually more than 22 segments. *Termopsidae*
Ocelli present. No spines on the shaft of the tibia, only at the tip. Antennae usually with more than 22 segments. *Kalotermitidae*

(ii) Soldiers of three families

1. Eyes present. *Termopsidae*
Eyes absent. (2)
2. Fontanelle present. *Rhinotermitidae*
Fontanelle absent. *Kalotermitidae*

References

1. Abbott, W. S. and Billings, S. C. (1935). (Repellents for clothes moths), *J. Econ. Entom.* **28**, 493.
2. Adam, N. K. (1944). (Spreading pressure of oil films), *Bull. entom. Res.* **36**, 269.
3. Ahmed, S. M. (1976). (Baits for cockroaches), *Internat. Pest Control* **18**(4), 4.
4. Alexander, R. D. (1957). (Systematics of field crickets), *Ann. ent. Amer.* **50**, 584.
5. Allen, J. R. and West, A. S. (1964). (Reaction to mosquito bites), *Proc. 1st Internat. Congr. Parasitol.*, Rome.
6. Anderson, J. F. and Kneen, F. R. (1969). (Flooding against *Chrysops*), *Mosquito News* **29**, 239.
7. Anon. (1945). (Hot air disinfestor), *Industrial Gas Times*, June.
8. Anon. (1957). (Cockroaches as scavengers), *Pest Control* **25**(2), 26.
9. Anon. (1959–63). (Cockroach prevalence in Denmark), *Ann. Rpts. Pest Infestation Lab.*, Copenhagen.
10. Anon. (1971). (Blackflies in the Americas), WHO/VBC/71. 283.
11. Anon. (1973). (Incidence of Gonorrhoea in various countries), *Postgrad. Med. J.* (June suppl.), 7–11.
12. Anon. (1974). (Pets, parasites and petulent patients), *Lancet* February, 157.
13. Anon. (1974). (Health of the Schoolchild), *Dept. Educat. Sci. Rpt.*
14. Anon. (1976). (Head louse incidence in U.S.A.), *Infect. Diseas.* **6**, 21.
15. Anon. (1978). (Fly resistance; use of IDI), *Ann. Rpt. Pest Infest. Lab.* Copenhagen. pp. 46–47.
16. Arevad, K. (1965). (Fly resting sites), *Ent. exp. et Applic.* **8**, 175.
17. Arthur, D. R. (1962). *Ticks and Disease*, Pergamon, Oxford. 445 pp.
18. Axtell, R. C. (1961). (Mite predators of fly maggots), *Farm Research* **27**, 4.
19. Back, E. A. (1935). (Clothes moths and their control), *U.S. Dept. Agric. Bull. no.* 1353.
20. Back, E. A. (1939). (Centipedes and Millipedes), *U.S. Dept. Agric. Leaflet* no. 192.
21. Back, E. A. and Cotton, R. T. (1926). (Low temperature against clothes moth), *Refridg. Engin.* **13**(12), 365.

21a. Back, E. A. & Cotton, R. T. (1922). (Grain pests). *U.S. Dep. Agr. Farmer's Bull.* no. 1260.

22. Back E. A. and Cotton, R. T. (1934). (Low temperature against clothes moth) *Refrigeration World* **59**, 26.
23. Bacot, A. (1914). (Flea bionomics), *J. Hyg. (Plague supplement)* **13**, 447.
24. Bailey, S. F. *et al.* (1965). (Dispersion of *Culex tarsalis*), *Hilgardia* **37**, 73.
25. Baker, W. C. *et al.* (1947). (Impregnation against flies), *Publ. Hlth. Repts.* Washington **62**, 597.
26. Baker, N. W. (1971). (Jerusalem crickets), *Pacific Discovery* **24**(2), 12.
27. Baker, L. F. and Southern, N. D. (1977). (Trap for cockroaches), *Internat. Pest Control* **19**(4), 8.
28. Banks, N. and Snyder, T. E. (1920). (New World Termites), *Bull. U.S. nat. Museum* **108**, 1.
29. Barlow, F. and Hadaway, A. B. (1955). (Insecticidal lacquer), *Trop. Disease Bull.* **52**, 307.
30. Barlow, F. and Hadaway, A. B. (1975). (Potency of Pyrethroids), *PANS* **21**, 233.
31. Barnes, W. *et al.* (1967). (Brickettes containing larvicide), *Mosquito News* **27**, 488.
32. Baronov, N. (1935). (Blackflies from the Danube), *Vet. Arhiv.* **5**, 58, 97.
33. Bartley, W. and Mellanby, K. (1944). (Scabies mite populations), *Parasitol.* **35**, 207.
34. Bar-Zeev, M. and Gothilf, S. (1972). (Flea repellents), *J. med. Entom.* **9**, 215.
35. Bar-Zeev, M. and Gothilf, S. (1973). (Tick repellents), *J. med. Entom.* **10**, 71.
36. Beadles, M. L. *et al.* (1975). (Hornfly control; Hawaii), *J. econ. Entom.* **68**, 781.
37. Beard, R. L. (1963). (Bee venom), *Ann. Rev. Entom.* **8**, 1.

38. Beard, R. L. (1974). (Block baits for termites), *Conn. agric. exp. Sta. Bull.* no. 748.
39. Baerden, R. H. and Steelman, C. D. (1971). (Larvicide "Flit MLO"), *J. econ. Entom.* **64**, 469.
40. Beatson, S. H. (1972). (Pharaoh's ant as vectors), *Lancet*, 19 February, 425.
41. Becker, G. (1940). (Biology of *Anobium punctatum*), *Z. Pflkr.* **50**, 159.
42. Becker, G. (1943). (Biology of *Anobium punctatum*), *Z. ang. Ent.* **30**, 104.
43. Benjamini, E. *et al.* (1960). (Reactions to insect bites) *Exp. Parasitol.* **10**, 214.
44. Benjamini, E. *et al.* (1963). (Flea bite reactions), *Exp. Parasit.* **14**, 75.
45. Berthold, R. and Wilson, B. R. (1967). (Resting sites of cockroaches), *Ann. ent. Soc. Amer.* **60**, 347.
46. Bettini, S. (1977). (Personal communication), Instituto Superiore di Sanita, Rome.
47. Bettini, S. and Finizio, E. (1968). (Biting midges in Tuscany), *Riv. Parasitol.* **29**, 33.
48. Bharadwaj, R. K. (1966). (Bionomics of *Euborellia*, an earwig), *Ann. ent. Soc. Amer.* **59**, 441.
49. Biagi, F. and Delgado-y-Gamica, R. (1974). (Therapy for scabies), *Internat. J. Dermatol.* **13**, 102.
50. Bills, G. T. (1973). (Fly predators in poultry litter), *British poultry Sci.* **14**, 209.
51. Bishopp, F. C. (1939). (The stable fly), *U.S. Dep. agric. Farmers Bull.* no. 1097.
52. Black, R. J. (1964). (Solid waste from towns), *California Vector Views* **11**, 51.
53. Bletchley, J. R. (1953, 1959). (Wood condition and beetle attack), *Ann. appl. Biol.* **40**, 218; *Brit. Wood Protect. Assn. Convention*, 1.
54. Bletchley, J. R. (1961). (Wood condition and beetle attack), *Ann. appl. Biol.* **49**, 362.
55. Blommers, L. (1978). (Personal communication), *Inst. voor Tropische Geneeskunde*, Leiden.
56. Bohart, R. M. and Bechtel, R. C. (1957). (Californian wasps), *Bull. Californian Insect Survey* **4**, 73.
57. Bogdandy, S. (1927). (Bean leaf trap for bed bugs), *Naturwiss.* **15**, 474.
58. Boles, H. P. *et al.* (1974). (Dichlorvos against dermestids) *J. econ. Entom.* **67**, 308.
59. Borkovec, A. B. (1962). (Chemosterilants), *Science* **137**, 1034.
59a. Bovingdon, H. H. S. (1931). (Pests of tobacco). *Tobacco*, London, Aug. 1st.
59b. Bovingdon, H. H. S. (1933). (Pests of tobacco). *Empire Marketing Bd. EMB 67* H.M.S.O.
60. Boyd, M. F. (ed.) (1949). *Malariology*, Vol. II, Collins, Philadelphia, London. 695 pp.
61. Bremer, H. G. (1964). (Pyrethrum smoke), *J. econ. Entom.* **57**, 62.
62. Bristowe, W. S. (1939). *The Comity of Spiders*, Ray Society, London. 560 pp.
63. Bristowe, W. S. (1958). *The World of Spiders*, Collins, London, 304 pp.
64. Britten, E. B. (1961). (Wood boring beetles), *Brit. Mus. (Nat. Hist) Econ. Series* no. 11a.
65. British Museum (Nat. Hist). (1930). (Furniture mites), *Econ. leaflet* no. 2.
65a. British Museum (Nat. Hist.) (1929). (Bristletails). *Econ. leaflet* no. 3.
66. British Museum (Nat. Hist.). (1945). (Booklice), *Econ. leaflet* no. 4.
66a. British Museum (Nat. Hist.) (1940). (Crickets). *Econ. leaflets* no. 5.
67. British Museum (Nat. Hist). (1967). (Plaster Beetles), *Econ. leaflet* no. 6.
68. British museum (Nat. Hist). (1973). (Bed bugs), *Econ. Series* no. 5.
69. Broadhead, E. & Hobby, B. M. (1945). (Booklice), *Discovery* **6**, 143.
70. Brown, A. W. A. (1951). (Mosquito attractants), *Bull. ent. Res.* **42**, 57.
71. Brown, A. W. A. (1961). (Attractants for houseflies), *J. econ. Entom.* **54**, 670.
72. Brown, A. W. A. and Morrison, P. E. (1955). (Air spray against horseflies), *J. econ. Entom.* **48**, 125.
73. Brown, W. B. (1959). (Van fumigation), *Sanitarian*, **67**, 401.
74. Brown, W. B. and Heseltine, H. K. (1964). (Fumigants), *Internat. Pest Control* **6**.
75. Bruce-Chwatt, L. J. and Abela-Hyzeler, P. (1975). (Malaria in England), *Health Trends* **7**, 18.
76. Bry, R. E. *et al.* (1960). (Solid fumigants for dermestids), *J. econ. Entom.* **53**, 966.
77. Bry, R. E. *et al.* (1972). (Paper cases against clothes moth), *J. econ. Entom.* **65**, 1735.
78. Bry, R. E. *et al.* (1973). (Impregnation against dermestids), *J. econ. Entom.* **66**, 546.
79. Burke, H. E. *et al.* (1922). (Lead cable beetle), *U.S. Dept. agric. Bull.* no. 1107.
80. Burnet, B. and Thompson, V. (1960). (Biology of seaweed flies), *Proc. R. entom. Soc. Lond.* **39**, 15.

81. Burns, E. C. *et al.* (1961). (*Bacillus thuringiensis*), *J. Econ. Entom.* **54**, 913.
82. Busvine, J. R. (1942). (Relative toxicity: fumigants), *Nature* **150**, 208.
83. Busvine, J. R. (1944). (Hot air disinfestor), *Bull. entom. Res.* **35**, 115.
84. Busvine, J. R. (1946). (Inheritance of colour; lice), *Proc. R. entom. Soc. Lond.* (*A*) **21**, 98.
85. Busvine, J. R. (1948). (Head and Body lice), *Parasitol.* **39**, 1.
86. Busvine, J. R. (1950). (Eradicating mosquitoes), *Discovery*, March, 85.
87. Busvine, J. R. (1955). (Bionomics of crickets), *Proc. R. entom. Soc. Lond.*, **39**, 15.
88. Busvine, J. R. (1955). (Queries to advisory entomologist), *Mon. Bull. Ministry Hlth. & PHLS* **14**, 178.
89. Busvine, J. R. (1958). (Insecticide resistant bed bugs), *Bull. Wld. Hlth. Org.* **19**, 1041.
90. Busvine, J. R. (1961). (Insecticide resistance in England), *Sanitarian*, December, 192.
91. Busvine, J. R. (1965). (Wire screen mesh for houseflies), *J. Hyg. Cambridge* **63**, 305.
92. Busvine, J. R. (1966). (Fumigants for lice), *J. Hyg. Cambridge* **64**, 45.
93. Busvine, J. R. (1971). (Insecticide resistance), *PANS* **17**, 135.
94. Busvine, J. R. (1976). *Insects Hygiene and History*, Athlone, London. 262 pp.
95. Busvine, J. R. (1978). (Head and Body lice), *Systematic Entom.* **3**, 1.
96. Busvine, J. R. and Vasuvat, C. (1966). (Bin fumigation for lice), *J. Hyg. Cambridge* **64**, 45.
97. Busvine, J. R. *et al.* (1950–51). (Aircraft disinfestation), *Mon. Bull. Ministry Hlth & PHLS* **9**, 80; **10**, 30.
98. Buxton, P. A. (1940). (Lethal high temperature: lice), *Br. med. J.* no. 4, 130.
99. Buxton, P. A. (1947). *The Louse*, Arnold, London.
100. Caesar, L. and Dunstan, A. (1939). (Control of crickets), *Rep. entom. Soc. Ontario* **69**, 101.
101. Campbell, E. and Black, R. J. (1960). (Fly maggots in refuse), *Californian Vector Views* **7**, 9.
102. Carlson, D. A. *et al.* (1971). (Sex attractant; flies), *Science* **174**, 76.
103. Carpenter, S. J. and LaCasse, W. J. (1955). (*North American mosquitoes*), *University of California Press*, Berkley. 360 pp.
104. Casida, J. E. (1970). (Synergists), *J. agric. Fd. Chem.* **18**, 753.
105. Caussanel, C. (1970). (Bionomics of *Labidura*; earwig), *Ann. Soc. Entom. France* **6**, 589.
106. Cavill, G. W. and Robertson, P. L. (1965). (Ant stings), *Science* **149**, 1337.
107. Cawley, B. M. *et al.* (1974). (Aircraft disinfestation: pyrethroids), *Bull. Wld. Hlth. Org.* **51**, 537.
108. Cepalak, J. *et al.* (1976). (Horseflies in Czechoslovakia), Abstr. in *Rev. appl. Entom.* (*B*) **64**, no. 909.
109. Chadwick, P. R. (1976). (Pyrethroids against cockroaches), *Internat. Pest Control* **18**(1), 15.
110. Cheema, P. S. (1956). (Biology of *Tinea pellionella*), *Bull entom Res.* **47**, 167.
110a. Chittenden, F. H. (1897). (Grain pests). *U.S. Dep. Agric. Farmer's Bull.* no. 45, 24 pp.
111. Chopard, L. (1922). (Earwigs etc), *Faune de France*; Vol 3. *Orthoptères et Dermetères.* Paris. pp. 212.
112. Chung, K. H. *et al.* (1975). (Sterile male release: *Stomoxys*), *Korean J. Plant Prot.* **12**, 41.
113. Clements, B. W. *et al.* (1975). (ULV aerosols against *Stomoxys*), Abstr. in *Rev. appl. Entom.* (*B*) **63**, no. 3299.
114. Clements, R. H. and Kerr, S. H. (1969). (Earwigs in Florida), *Pest Control* **37**, 26.
115. Cloudsley-Thompson, J. L. (1958). *Spiders, Scorpions, Centipedes and Mites*, Pergamon, London, New York. 228 pp.
116. Coe, R. L. (1943). (*Drosophila repleta* in England), *Entom. mon. Mag.* **79**, 204.
117. Coe, R. L. *et al.* (1950). *Handbooks for Identification of British Insects.* Vol. 9, (2) *Diptera. R. entom. Soc., London.*
118. Cole, J. H. (1962). (Biology; house moths), *Bull. entom. Res.* **53**, 83.
119. Cole, J. H. and Whitfield, F. G. S. (1962). (Testing moth proofing), *J. Textile Res.* **53**, 236.
120. Collinge, W. E. (1914). (Biology of woodlice), *J. Board Agric.* **21**, 206.
121. Colwell, A. E. and Shorey, H. H. (1975). (Fly mating behaviour), *Ann. entom. Soc. Amer.* **68**, 152.
122. Colyer, C. N. and Hammond, C. O. (1951). *Flies of the British Isles*, Warne, London.
123. Conway, J. A. (1972). (Control of *Stomoxys*; barn spraying), *Internat. Pest Control* **14**, 11.

124. Cornwell, P. B. (1968). *The Cockroach*, Vol. I, Hutchinson, London. 391 pp.
125. Cornwell, P. B. (1974). (Incidence of bugs and fleas in Britain), *Internat. Pest Control* **16**(4), 17.
126. Cornwell, P. B. (1976). *The Cockroach*, Vol. II, Assoc. Business Programmes, London.
127. Cornwell, P. B. and Bull, J. O. (1960). (Radiation for killing insects), *J. Sci. Fd. Agric.* **11**, 754.
128. Cotton, R. T. (1941). *Pests of Stored Grain*, Burgess, Minneapolis.
129. Cragg, J. B. (1955). (Life history; blowflies), *Ann. appl. Biol.* **42**, 197.
130. Craig, G. B. (1963). (Genetic control of Insects), *Bull. Wld. Hlth. Org.* **29** (*Suppl*) 89.
131. Cramer, R. D. (1976). (Controlled release insecticides), *Internat. Pest Control* **18**(2), 4; **18**(3), 4.
132. Creighton, W. S. (1950). (*Ants of North America*), *Harvard Bull. Mus. Comp. Zool.* **104**, 585 pp.
133. Crumb, S. E. *et al.* (1941). (Biology of earwigs), *U.S. Dept. Agric. Tech. Bull.* no. 766.
134. Culpepper, G. H. (1946). (Bionomics; mass-reared lice), *J. econ. Entom.* **39**, 472.
135. Cymorek, S. (1964). (Biology of wood-boring beetles), *Z. angew. Ent.* **55**, 84.
136. Danish Pest Control Laboratory (1977). *Ann. Rpt* for 1976, Copenhagen.
137. Davey, T. H. and Gordon, R. M. (1938). (Mosquito excluding screen), *Ann. trop. Med. Parasitol.* **32**, 413.
138. David, W. A. L. (1944). (Box or bin fumigation), *Bull. entom. Res.* **35**, 79, 101.
139. David, W. A. L. (1946). (Aerosol behaviour), *Bull. entom. Res.* **37**, 1.
140. Deoras, P. J. and Joshee, A. K. (1961). (Insect bite mechanism), *Current Science* **30**, 465.
141. Derbeneva-Ukhova, P. A. (1934). (Bionomics of blowflies; in Russian), *Med. Parasitol.* **3**, 403.
142. Derbeneva-Ukhova, P. A. (1937). (Fly breeding material; in Russian), *Med. Parasitol.* **6**, 3.
143. Derbeneva-Ukhova, P. A. (1940). (Fly breeding materials; in Russian), *Med. Parasitol.* **9**, 323, 521.
144. Detinova, T. S. (1962). (Age grading: Diptera), *Wld. Hlth. Org. Monogr. Series* no. 47.
145. Dmitriev, G. A. (1978). (Control of ticks), *Internat. Pest Control* **20**(5), 10.
146. Doane, R. W. *et al.* (1936). *Forest Insects*, McGraw Hill, New York. 463 pp.
147. Donaldson, R. J. (1976). (Head louse prevalence in Britain), *R. Soc. Hlth. J.* **96**, 55.
148. Donnisthorpe, H. S. J. (1927). *British Ants*, Routledge, London. 436 pp.
149. Dorman, S. C. *et al.* (1938). (Bionomics; blowflies), *J. econ. Entom.* **31**, 44.
150. Doszhanov, T. N. and Nurtazin, A. T. (1967). (Control of sheep ked), Abstr. in *Rev. appl. Entom.* (*B*) **64**, no. 2071.
151. Dow, R. P. (1955). (Cockroach habitats; Texas), *J. econ. Entom.* **48**, 106.
152. Duda, O. (1924). (Fruit flies), *Ent. Medd.* **14**, 164.
153. Dyte, C. E. (1960). (Insecticidal lacquers), *Pest Technology* February, 98.
154. Dyte, C. E. (1960). (Electrical insecticide vaporizers), *Food Trade Rev.* **33**(3), 35.
155. Ebeling, W. (1961). (Dehydrating dusts) *Hilgardia* **30**, 531.
156. Ebeling, W. (1964). (Flies invading houses), *P.C.O. News* **24**(4), 23.
157. Ebeling, W. (1966). (Termites; damage estimate), *University of California Div. agric Sci Man.* 38.
158. Ebeling, W. (1975). *Urban Entomology*, University of California Press, Berkley. 695 pp.
159. Eckert, J. E. and Mallis, A. (1937). (Californian ants), *University of California agric. Exp. Sta. Circular* no. 342, 37 pp.
160. Edwards, F. W. (1925). (Bibionidae), *Ann. appl. Biol.* **12**, 263.
161. Edwards, J. P. and Clarke, B. (1978). (Eradication of *Monomorium* by methoprene), *Internat. Pest Control* **20**(1), 5.
162. Edwards, F. W., Oldroyd, H. and Smart, J. (1939). *British Bloodsucking Flies*, British Mus. (Nat. Hist). 156 pp.
163. Egglishaw, H. J. (1960). (Seaweed flies), *Pest Technology* **3**, 212.
164. Eichler, D. A. and Nelson, G. S. (1971). (*Simulium* and cattle disease), *J. Helminth* **45**, 245.
165. Eichler, W. and Lutgens, R. (1962). (Entomophobia), *Ang. Parasitol.* **3**, 79.
166. Elliott, R. (1955). (Larvicide brickettes), *Trans. R. Soc. trop. Med. Hyg.* **49**, 528.
167. Elliott, M. (1976). (Properties of pyrethroids), *Environm. Hlth. Perspectives* **14**, 3.
168. Elliott, M. *et al.* (1973). (Photostable pyrethroid), *Nature* **246**, 169.
169. Elliott, M. *et al.* (1974). (New powerful pyrethroid), *Nature* **248**, 710.
170. El-Sawaf, S. K. (1956). (Bionomics; Bruchids), *Bull. Soc. entom. d'Egypte* **40**, 49.

171. English, L. L. and Snetsinger, R. (1957). (Control of *Bryobia*), *J. econ. Entom.* **50**, 135.
172. Entente Interdepartementale pour la Demoustication du Littoral Mediterraneen (1977). (Personal communication).
173. Epstein, E. and Orkin, M. (1977). (Scabies; clinical aspects), in *Scabies and Pediculosis*, Lippincott, Philadelphia.
174. Evans, B. R. and Porter, J. E. (1965). (Cockroaches on ships), *J. econ. Entom.* **58**, 479.
175. Evans, B. R. *et al.* (1963). (Insects on aircraft), *Mosquito News* **23**, 9.
176. Ewing, H. E. (1918). (Biology of spiders), *Proc. Iowa Acad. Sci.* **25**, 177.
177. Eyles, D. E. and Cox, W. W. (1943). (Mosquito abundance), *J. nat. Malaria Soc.* **2**, 71.
178. Finnegan, S. (1945). (Acari as scrub typhus vectors), *Brit. Mus. (Nat. Hist.) Econ. Series* no. 16.
179. Ferriere, C. (1941). (Enemies of clothes moths), *Mitt. Schweiz entom. Ges.* **18**, 374.
180. Ferris, G. (1951). *The Sucking Lice*, Mem. Pacific Coast Entom. Soc., I.
181. Fisher, I. and Morton, R. S. (1970). (Crab lice in Britain), *Brit. J. venereal Dis.* **46**, 362.
182. Fisher, R. C. (1937–41). (Biology of *Xestobium rufo-villosum*), *Ann. appl. Biol.* **24**, 600; **25**, 155; **27**, 545.
183. Flemmings, M. B. and Ludwig, D. (1964). (Bionomics of lice in culture), *Ann. entom. Soc. Amer.* **57**, 560.
184. Flintoff, F. (1950). (Municipal cleansing practice), *Contractors' Record & municipal Engng.*, London.
185. Forest Products Research Laboratory (1956). (Ambrosia beetles), Leaflet no. 50.
186. Forest Products Research Laboratory (1959). (*Anobium punctatum*), Leaflet no. 8.
187. Forest Products Research Laboratory (1963). (*Hylotrypes bajulus*), Leaflet no. 14.
188. Forest Products Research Laboratory (1963). (*Lyctus* spp.) Leaflet no. 3.
189. Forest Products Research Laboratory (1963). (*Xestobium rufo-villosum*), Leaflet no. 4.
190. Fredeen, F. J. (1964). (Rearing blackflies), *Canad. J. Zool.* **42**, 527.
191. Freeman, J. A. (1964). (Damage by food pests), *Ann. appl. Biol.* **53**, 200.
192. Freeman, J. A. (1973). (Grain infestation in international trade), in *Grain Storage*, Ch. 5. Westport, Connecticut, U.S.A.
193. Freeman, J. A. and Turtle, E. E. (1947). *Insect Pests of Food*, H.M.S.O., London.
194. Frey, W. (1962). (*Sitophilus zea-mais*), *Nachr. Deutsch Pflandienst.* **14**, 145.
195. Frontali, N. and Grasso, A. (1964). (*Latrodectus* venom), *Arch. Biochem. Biophysics* **106**, 213.
196. Gahan, C. J. and Laing, F. (1932). (Wood-boring beetles), *Brit. Mus. (Nat. Hist.) Econ Ser.* no. 11.
197. Gahan, J. B. (1957). (Mosquito larvicides), *Mosquito News* **17**, 198.
198. Gaon, J. D. and Murray, E. S. (1966). (Typhus in Yugoslavia), *Bull. Wld. Hlth. Org.* **35**, 133.
199. Geiger, W. B. *et al.* (1942). (Mothproofing), *Textile Res.* **13**, 21.
200. Georghiou, G. P. *et al.* (1971). (Fly resistance in California), *Bull. Wld. Hlth. Org.* **45**, 43.
201. Georghiou, G. P. *et al.* (1972). (Insecticide tests; *Leptoconops*), *Mosquito News* **32**, 205.
202. Georghiou, G. P. *et al.* (1975). (Mosquito resistance in California), *Proc. 43rd ann. Conf. Californian Mosq. Control Assn*, 41.
203. Gerberg, E. (1977). (Personal communication), Insect Control and Research Inc., Baltimore.
204. Gerberg, E. (1957). (New World Lyctidae), *U.S. Dept Agric. Tech Bull* no. 1157.
205. Ghouri, A. S. and McFarlane, J. (1956). (Bionomics of crickets), *Canad. Entom.* **90**, 156.
206. Gillies, P. A. *et al.* (1971). (Aerial application of larvicide), *Mosquito News* **31**, 528.
207. Gless, E. *et al.* (1958). (Control of chicken lice), *J. econ. Entom.* **51**, 229.
208. Goodwin-Bailey, K. F. *et al.* (1957). (Aerosol test method), *Ann. appl. Biol.* **45**, 347.
209. Gordon, R. M. and Crewe, W. (1948). (Insect bite mechanism), *Ann. trop. Med. Parasit*, **42**, 334.
210. Gordon, R. M. and Lumsden, R. W. (1938). (Insect bite mechanism), *Ann. trop, Med. Parasit.* **33**, 259.
211. Gosswald, K. (1932). (Ants), *Z. hyg. Zool.* **30**, 202.
212. Gotaas, H. (1956). (Composting), *Wld. Hlth. Org. Monogr. Ser.* no. 31.
213. Gould, G. E. and Deay, H. O. (1940). (Biology of Cockroaches), *Purdue University agr. exp. Sta. Bull.* no. 451.

214. Gradidge, J. M. (1958). (The clover mite), *Sanitarian* **66**, 480.
215. Gradidge, J. M. (1963). (Control of *Fannia*), *Sanitarian* **71**, 400.
216. Graham-Smith, G. S. (1913). *Flies and Disease*, Cambridge University Press.
217. Gratz, N. G. (1977). (Head lice; incidence), in *Scabies and Pediculosis*, Lippincott, Philadelphia.
218. Green, A. A. (1951). (Bionomics of blowflies), *Ann. appl Biol.* **38**, 475.
219. Green, A. A. (1951–4). (Flies at slaughter houses), *Ann. appl Biol.* **38**, 475; **40**, 705; **41**, 165.
220. Green, A. A. (1952). (Control of Blowflies), *J. R. sanit. Inst.* **72**, 621.
221. Green, A. A. (1963). (Fly control), *Proc. 1st Brit. Pest Control Conf.*, Oxford.
222. Greenberg, B. (1959). (Blowflies as vectors), *Am. J. trop. Med. Hyg.* **8**, 618.
223. Greenberg, B. (1960). (Blowflies as vectors), *Ann. entom. Soc. Amer.* **53**, 125.
224. Greenberg, B. (1962). (Bacterial survival in flies), *J. Insect Path.* **4**, 216.
225. Greenberg, B. (1963). (Blowflies as vectors), *Am. J. Hyg.* **77**, 177.
226. Greenberg, B. (1971, 1973). *Flies and Disease* Vols. I and II, Princeton University Press. 856 and 447 pp.
227. Greenberg, B. *et al.* (1970). (Disease transmission by flies), *Infection and Immunity* **2**, 800.
228. Griswold, G. H. and Crowell, M. (1936). (Humidity and clothes moths), *Ecology* **17**, 241.
229. Gross, H. R. and Spink, W. T. (1969). (Control of earwigs), *J. econ. Entom.* **62**, 686.
230. Groth, V. (1974). (Screens to protect cattle from *Stomoxys*), Abstr. in *Rev. appl. Entom.* (*B*) **62**, no. 1908.
231. Grout, R. A. (ed) (1949). *Hive and Honey Bee*, Hamilton (Illinois), Dendent. 652 pp.
232. Grundemann, A. W. (1947). (Biology; *Triatoma sanguisuga*), *J. Kansas entom. Soc.* **20**, 77.
233. Gulati, P. V. *et al.* (1977). (Scabies in India), *Internat. J. Dermat.* **4**, 281.
234. Gunn, D. L. (1935). (Ecology of cockroaches), *J. exp. Biol.* **12**, 185.
235. Hackett, L. W. (1937). *Malaria in Europe*, Oxford University Press. 336 pp.
236. Haines, T. W. and Palmer, E. C. (1955). (Cockroach habitats; Georgia), *Ann. trop. Med. Hyg.* **4**, 1131.
237. Hair, J. A. and Howell, D. E. (1970). (Biology & control; *Amblyomma americanum*), *Oklahoma agric. expt. Sta. Bull.* B-679.
238. Hallas, T. *et al.* (1977). (House pests in Denmark; trends), *Ent. Medd.* **45**, 77.
239. Hamer, O. (1944). *Flies associated with cattle*, Isogtrykken, Copenhagen.
240. Hammad, S. M. (1955). (*Pentarthrum huttoni*), *Proc. R. entom Soc.* (*A*) **30**, 33
241. Hansens, E. J. (1951). (DDT against *Stomoxys*), *J. econ. Entom.* **44**, 482.
242. Harker, J. E. (1956). (Diurnal rhythm: *Periplaneta*), *J. expt. Biol.* **37**, 154.
243. Harris, E. C. (1963). (Control of woodworm), *Brit. woodpreserving Soc. ann. Convention.*
244. Harris, E. G. *et al.* (1976). (Fly control in Malta), *PANS* **22**, 207, 215.
245. Harris, R. L. *et al.* (1973). (Juvenoids for controlling biting flies), *J. econ. Entom.* **66**, 1099.
246. Harris, W. V. (1954). (Termites in Europe), *Entom. mon. Mag.* **90**, 194.
247. Harris, W. V. (1961). *Termites*, Longman, London. 187 pp.
248. Hartley, R. S. *et al.* (1943). (Moth proofing), *J. Soc. Dyers Colourists* **59**, 266.
249. Hase, A. (1917). (Bed bug behaviour), *Monogr. angew. Ent.* no. 1, Berlin.
250. Hase, A. (1929). (Reaction to bug bites), *Munch. med. Wochenschr.* **76**, 202.
251. Hase, A. (1932, 1934). (Mothproofing), *Anzeig. Schadlsk.* **7**, 73; **9**, 35, 85.
252. Hatchett, S. P. (1947). (Biology of woodlice), *Ecol. Monogr.* **17**, 47.
253. Hawkes, H. A. (1951). (Biology of window gnat), *Ann. appl. Biol.* **38**, 592.
254. Haydak, M. H. (1953). (Cockroach protein requirements), *Ann. entom. Soc. Amer.* **46**, 547.
255. Hays, S. B. and Arrant, F. S. (1960). (Poison baits), *J. econ. Entom.* **53**, 188.
256. Hebra, F. (1844). (Scabies), *Med Jahrb. Ostrereich Stat.* **46**, 280; **47**, 44.
257. Hecht. O. (1933). (Insect bite reactions), *Z. Haut u. Geschlkr.* **44**, 241.
258. Hecht, O. (1970). (Light reactions of flies), *Bull. ent. Soc. Amer.* **16**, 94.
259. Heileson, B. (1946). *Studies on Acarus scabiei*, Rosenkilde, Copenhagen. 370 pp.
260. Heller-Haupt, A. and Busvine, J. R. (1974). (Acaricides for house dust mites), *J. med. Entom.* **11**, 551.
261. Herms, W. B. (1961). *Medical Entomology*, (5th edn.). Macmillan, New York. 616 pp.

262. Hernandez-Perez, E. (1976). (Scabies therapy), *Arch. Dermatol.* **112**, 1400.
263. Herrick, G. W. and Griswold, G. H. (1932). (Fumigants for clothes moth), *J. econ. Entom.* **25**, 243.
264. Hetrick, N. E. (1970). (Biology of the love-bug), *Florida Entom.* **53**, 23.
265. Hewitt, C. G. *et al.* (1912). (Flies and disease), *Rpts. local Gov. Bd. Hlth. med. subjects (N.S.)* no. 66.
266. Hewitt, M. *et al.* (1971). (Human infestation from pet animals), *Brit. J. Dermat.* **85**, 215.
267. Hewitt, M. *et al.* (1973). (Mites and dermatitis), *Brit. J. Dermat.* 89, 401.
268. Hickin, N. E. (1971). *Termites; a world problem*, Hutchinson, London. 232 pp.
269. Hickin, N. E. (1975). *The insect factor in wood decay*, 3rd edn, Hutchinson, London. 383 pp.
270. Hilger, C. (1899). (Fleas in German dwellings), cited by Peus, F. (q. v.).
271. Hill, M. A. (1946). (Biting midges), *Discovery* **7**, 200.
272. Hinton, H. E. (1941). (Systematics; Lathridiidae), *Bull. entom. Res.* **32**, 191.
273. Hinton, H. E. (1943). (Stored food moth larvae), *Bull. entom. Res.* **34**, 163.
274. Hinton, H. E. (1945). *Beetles associated with stored Products*, Brit. Mus. (Nat. Hist.). 443 pp.
275. Hinton, H. E. (1956). (Characters of house moths), *Bull. entom. Res.* **47**, 251.
276. Hinton, H. E. and Corbett, A. S. (1963). (Common insect pests of stored food), *Brit. Mus. (Nat. Hist.) Econ. Ser.* no. 15.
277. Hinton, H. E. and McKenny-Hughes, A. W. (1947). (*Euophryum confine*) *Mon. Bull. Min. Hlth. & PHLS* **6**, 173.
278. Hirst, S. (1920). (Biting spiders & centipedes), *Brit. Mus. (Nat. Hist.) Econ. Ser.* no. 6.
279. Hirst, S. (1922). (Injurious mites), *Brit. Mus. (Nat. Hist.) Econ. Ser* no. 13.
280. Hoechst Chemical Co. (1975). (ULV application), *Booklet.*
281. Holan, G. (1971). (Biodegradable DDT analogues), *Nature*, **232**, 644.
282. Hopkins, L. (1955). (Polistine wasps), *J. econ. Entom.* **48**, 161.
283. Hopper, J. M. (1971). (Head lice in British Columbia), *Canad. J. Publ. Hlth.* **62**, 154.
284. Hora, A. M. (1934). (*Glyciphagus domesticus*; biology), *Ann. appl. Biol.* **21**, 483.
285. Horsfall, W. R. (1955). *Mosquitoes. bionomics and relation to disease*, Ronald, New York. 723 pp.
286. Howard, H. W. (1940). (Biology of woodlice), *J. Genetics* **40**, 83.
287. Howe, R. W. (1957). (*Lasioderma* bionomics), *Bull. entom. Res.* **48**, 9.
288. Howe, R. W. (1978). (Principles of food storage), *Outlook on Agriculture* **9**, 198.
289. Howe, R. W. and Burgess, H. D. (1953). (*Ptinus* bionomics), *Bull. entom. Res.* **44**, 461.
290. Hubble, T. H. (1936). (Systematics; camel crickets), *Univ. Florida Publ. biol. Sci.* **2**, 551.
291. Hudson, B. W. *et al.* (1960). (Flea bite reactions), *Expt. Parasitol.* **9**, 18, 264.
292. Hughes, A. M. (1961). *The Mites of stored Food, Min. Agric. Fd. Fish. Tech. Bull.* no. 9.
293. Hughes, A. M. and Hughes, T. E. (1958). (*Glyciphagus domesticus*; biology), *Proc. Zoo. Soc. London* **108B**, 715.
294. Hull, G. and Davidson, R. H. (1958). (Biology of *Supella longipalpis*), *J. econ. Entom.* **51**, 608.
295. Husieny, T. M. el *et al.* (1976). (Pyrethroids against flies in Arabia), *PANS* **22**, 317.
296. Hutchins, R. E. and Langston, J. M. (1953). (Field crickets invade towns in U.S.A.), *J. econ. Entom.* **46**, 169.
297. Hyde, M. B. (1962). (Asphyxiation against grain pests), *Ann. appl. Biol.* **50**, 362.
298. Jackson, W. B. and Maier, P. P. (1961). (Movement of cockroaches in sewers), *Ohio J. Sci.* **61**, 220.
299. James, M. T. (1947). (Myiasis in America), *U.S. Dep. Agric. Misc. Publ.* no. 631.
300. Jamnback, H. and Collins, D. E. (1955). (Control of blackflies), *New York State Mus. Bull* no. 350.
301. Janisch, E. (1935). (Bed bugs; harmful high temperature), *Z. Parasitenk.* **7**, 408.
302. Jensen, K. (1933). (Wood beetle control; by heat), *Mitt. Ges. Vorratschutz* **9**, 15.
303. Johnson, C. G. (1941). (Bed bug bionomics), *J. Hyg.* Cambridge **41**, 345.
304. Johnson, C. G. and Mellanby, K. (1942). (Itch mite populations), *Parasitol.* **34**, 285.
305. Johnson, D. R. (1973). (Recent malaria in U.S.A.), *Mosquito News* **33**, 341.

306. Jones, B. B. and Owen, F. (1934). *The Scientific Aspects of controlled Tipping*, City of Manchester.
307. Juranek, D. D. (1977). (Head lice in U.S.A.), in *Scabies and Pediculosis*, Lippincott, Philadelphia. 203 pp.
308. Juranek, D. D. and Schultz, M. G. (1977). (Scabies epidemiology), in *Scabies and Pediculosis* Lippincott, Philadelphia.
309. Kashef, H. (1955). (*Stegobium* bionomics), *Ann. Soc. Ent. France* **124**, 5.
310. Kaston, B. J. (1973). *How to know the Spiders*, Brown, Dubuque, Iowa, 289 pp.
311. Keiding, J. (1962). (Thiourea for maggot control), *11th Internat. Congr. Entom. Vienna*, **2**, 618.
312. Keiding, J. (1965). (Resting sites of flies), *Riv. di Parasit.* **26**, 45.
313. Keiding, J. (1974). (Fly resting sites), in *Control of Arthropods of Medical & Veterinary Importance*, Plenum, New York. 138 pp.
314. Keiding, J. (1976). (Pyrethrin resistance in flies), *Pestic. Sci.* **7**, 283.
315. Keilin, D. (1916). (*Pollenia rudis* biology), *Bull. Sci. France et Belge* **49**, 15.
316. Kelsey, J. M. *et al.* (1945). (*Anobium punctatum* biology), *N.Z. J. Sci. Technol.* **27**, 59.
317. Kemper, H. (1936). (Bed bugs; biology and control), *Klientiere & Pelztiere* **12**(3), 107.
318. Kemper, H. (1937). (Cricket biology), *Z. hyg. Zool. Schdlbkf.* **29**, 69.
318a. Kemper, H. (1938). (Stored food pests). *Z. hyg. Zool. Schädl.bk.* **30**, 34.
319. Kemper, H. (1939). *Die Nahrungs & Genussmittels Schadlinge & ihre Bekampfung*, Schops, Leipzig. 270 pp.
320. Kemper, H. (1958). (Urticating hairs on larvae), *Proc. 10th Internat. Congr. Entom.*, Montreal. Vol. 3, p. 719.
321. Kennedy, J. S. (1939). (Mosquito flight), *Proc. zool. Soc. Lond.* **109**, 221.
322. Kerr, T. W. and McLean, D. (1956). (Lathridiidae; biology and control), *J. econ. Entom.* **49**, 269.
323. Kettle, D. S. (1962). (Biting midge bionomics), *Ann. Rev. Entom.* **7**, 401.
324. Kilpatrick, J. W. *et al.* (1962). (Fly bait dispenser), *J. econ. Entom.* **55**, 951.
325. Kitchener, J. A. *et al.* (1944). (Dehydrating dusts), *Trans. Farad. Soc.* **40**, 10.
326. Knipling, E. F. (1960). (Sterile male control), *J. econ. Entom.* **53**, 415.
327. Knülle, W. (1963). (*Acarus siro*), *Z. angew. Entom.* **51**, 300.
328. Kobayashi, H. and Mizushima, H. (1937). (Fly biology), *Keijo J. Med.* **8**, 19.
329. Koller, R. (1937). (*Piophila casei* biology), *Z. hyg. Zool. Schdlbkf.* **29**, 104.
330. Kovchazov, G. (1978). (Personal communication), Inst. Hygiene and Epidemiology, Varna, Bulgaria.
331. Kozhanechikov, I. V. (1946). (Fly biology), *C.R. Acad. Sci. USSR* **51**, 241.
332. Kuhlow, F. (1977). (Personal communication), Bernhard-Nocht Inst. Schiffs- und Tropenkrankheit, Hamburg.
333. Kutz, F. W. (1974). (Electric mosquito repelling device), *Mosquito News* **34**, 369.
334. Kuzina, O. S. (1938, 1940). (Fly breeding sites; in Russian), *Med Parasit.* **7**, 244; **9**, 340.
335. LaBreque, G. C. (1963). (Chemosterilants against flies), *J. econ. Entom.* **56**, 150.
336. LaBreque, G. C. *et al.* (1975). (*Sterile male release for Stomoxys*?), U.S. Dep. Agric. ARS Gainesville Fla. 449.
337. Laing, F. (1932). (Housemoth biology), *Entom. mon. Mag.* **68**, 77.
338. Laing, F. (1946). (Cockroach), *Brit. Mus. (Nat. Hist.) Econ. Ser.* no. 12.
339. Lamizana, M. T. and Mouchet, J. (1976). (Head lice in France), *Med. et Maladies Infect.* **6**, 48.
340. Larson, E. B. and Thomsen, M. (1940). (*Stomoxys* biology), *Vidensk. Medd. Dansk naturhist. Foren.* **104**, 1.
341. Latta, R. and Yeomans, A. H. (1943). (Fumigation bags), *J. econ. Entom.* **36**, 402.
342. Laviorpierre, M. M. *et al.* (1959). (Insect bite mechanism), *Ann. trop. Med. Parasitol.* **53**, 235, 347.
343. Lee, C. W. (1973). (Blackfly larvae control), *PANS* **19**, 190.
344. Leeson, H. S. (1932). (Flea bionomics), *Parasitol.* **24**, 196.
345. Legner, E. F. and Olton, G. S. (1968). (Fly larvae; parasites), *Ann. entom. Soc. Amer.* **61**, 1306.

346. Legner, E. F. *et al.* (1967). (Fly parasites), *Ann. entom. Soc. Amer.* **60**, 462.
347. Legner, E. F. *et al.* (1975). (Biological control of fly maggots), *Californian Agric.* **29**, 8.
348. Leikine, L. I. (1942). (Fly breeding sites; in Russian), *Med. Parasit.* **11**, 82.
349. Levi, H. W. (1959). (*Latrodectus* biology), *Trans. Amer. Micro. Soc.* **78**, 7.
350. Lidror, R. and Lifshitz, Y. (1965). (Head lice in Israel), *Publ. Hlth. (Israel)* **8**, 1.
351. Liljedahl, L. A. *et al.* (1976). (Aircraft disinfection tests), *Bull. Wld. Hlth. Org.* **54**, 391.
352. Lindberg, U. H. and Ulff, B. (1968). (Cinchonic repellents), *Acta Pharm. Sueica* **5**, 44.
353. Lindquist, A. W. *et al.* (1944). (Systemic insecticides in mammals), *J. econ. Entom.* **37**, 128.
354. Lindsay, E. (1939). (Firebrat biology), *Proc. R. Soc. Victoria* 52, 35.
355. Linley, J. R. and Davies, J. B. (1971). (Biting midge biology and control), *J. econ. Entom.* **64**, 264.
356. Lloyd, L. (1943, 1945). (Sewage flies), *Ann. appl. Biol.* **30**, 47, 358; *J. Inst. Sewage Purific.* **1**, 119.
357. Loeb, J. *et al.* (1917). (*Drosophila* bionomics), *J. biol. Chem.* **32**, 105.
358. Logan, J. A. (1953). *The Sardinian Project*, Johns Hopkins Press, Baltimore. 415 pp.
359. Luke, J. E. and Snetzinger, R. (1972). (Firebrat control), *J. econ. Entom.* **65**, 917.
360. MacConnell, J. G. *et al.* (1970). (*Solenopsis* ant venom), *Science* **168**, 840.
361. Macdonald, G. (1946). (Mosquito larvicides), *Trop. Disease Bull.* **43**, 885.
362. Maciver, D. R. (1963). (Mosquito coils), *Pyrethrum Post* **7**(2), 22.
363. Mackie, D. B. *et al.* (1942). (Earwig biology), *Bull. Californian Dept. Agric.* **31**, 110.
364. MacLeod, J. and Craufurd-Benson, H. J. (1941). (Louse incidence), *Parasitol.* **33**, 278.
365. MacLeod, J. and Donelly, J. (1961). (Sterile male release; blowflies), *Entom. exp. & appl.* **4**, 101.
366. Madeock, E. C. (1949). (Head lice in England), *Mon. Bull. Min. Hlth. & PHLS* **8**, 26.
367. Magnan, A. (1934). *La Vol des Insectes*, Herman, Paris.
368. Mahdi, *et al.* (1976). (Pyrethroid against flies; Cairo), *PANS* **22**, 443.
369. Majori, G. *et al.* (1971). (*Leptoconops* habitat; Tuscany), *Riv. di Parasitol.* **32**, 277.
370. Makara, G. (1964). (Personal Communication), Parasitology Laboratory, Budapest.
371. Mallis, A. (1944). *Handbook of Pest Control*, McNair-Dorland, New York. 554 pp.
372. Mallis, A. *et al.* (1961). (Cockroach dispersion), *Pest Control* **29**(6), 32.
373. Mansour, K. *et al.* (1934). (Constituents of wood), *Biol. Rev.* **9**, 363.
374. Marchall, T. K. (1957). (Treatment of stings), *Practitioner* **178**, 712.
375. Markin, G. P. (1970). (*Iridomyrmex* biology), *Ann. entom. Soc. Amer.* **63**, 1238.
376. Markina, V. V. *et al.* (1971). (*Simulium* repellent; in Russian), Abstr. in *Rev. appl. Entom. (B)* **61**, no. 302.
377. Marshall, J. F. (1938). *The British Mosquitoes*, Brit. Mus. (Nat. Hist). 341 pp.
378. Martin, H. and Worthing, C. R. (1977). *Pesticide Manual*, 5th edn, Brit. Crop Prot. Council. 593 pp.
379. Mason, B. *et al.* (1977). (Pyrethroid against flies; Monrovia), *Internat. Pest Control* **19**, 8.
380. Mathews, G. A. (1975). (Droplet size determination), *PANS* **21**, 213, 343.
381. Mathis, W. *et al.* (1969). (Flies in rural areas), *J. econ. Entom.* **62**, 1288.
382. Mathis, W. *et al.* (1970). (Air barrier to flies), *J. econ. Entom.* **63**, 29.
383. Mathlein, R. (1967). (Repellents for food pests), *Swedish nat. Inst. Plant Prot. Control* **13**, 112.
384. Matsumura, F. (1975). *Toxicology of Insecticides*, Plenum New York, London. 503 pp.
385. Maunder, J. W. (1971). (Malathion for head lice), *Commun. Med.* **126**, 145.
386. McClure, H. F. (1935). (Bite of *Melanolestes), Entom. News* **46**, 138.
387. McCoy, C. E. (1958). (Control of fly breeding), *J. econ. Entom.* **51**, 411.
388. McFarlane, J. E. (1959). (Cricket bionomics), *Canad. Entom.* **37**, 913.
389. McGovern, T. P. *et al.* (1974). (Cockroach repellents), *J. econ. Entom.* **67**, 639.
390. McKiel, J. A. and West, A. C. (1961). (Insect bite reaction), *Pediatric Clinic N. Amer.* **8**, 795.
391. McLintock, J. and Depner, K. R. (1954)). (Horn fly biology), *Canad. Entom.* **86**, 20.
392. McMullen, A. I. and Hill, M. N. (1977). (Lethecin film drowns pupae), *Nature* **267**, 244.
393. Medical Research Council (1942). (Bed bugs and slums), *Rpt. Cmttee on Bed bug Infest.*, HMSO, London. 64 pp.
394. Meifert, D. W. *et al.* (1978). (Lethal trap for *Stomoxys*), *J. econ. Entom.* **71**, 290.

395. Mellanby, K. (1932). (Low temperature & *Xenopsylla*), *J. exp. Biol.* **9**, 222.
396. Mellanby, K. (1935). (Lethal temperatures; *Cimex*), *Parasitology* **27**, 111.
397. Mellanby, K. (1939). (Bed bug behaviour), *Parasitol.* **31**, 200.
398. Mellanby, K. (1939). (Low temperature; *Blatta*), *Proc. R. Soc.* (*B*) **127**, 473.
399. Mellanby, K. (1941). (Scabies incidence), *Medical Officer* **66**, 141.
400. Mellanby, K. (1941). (Head louse incidence), *Medical Officer* **65**, 39.
401. Mellanby, K. (1942). (Itch mite survival), *Bull. entom. Res.* **33**, 267.
402. Mellanby, K. (1943). *Scabies*, Oxford University Press. 81 pp.
403. Mellanby, K. (1944). (Scabies transmission), *Parasitol.* **35**, 197.
404. Mellanby, K. (1946). (Mosquito bite reaction), *Nature* **158**, 554.
405. Mellanby, K. *et al.* (1942). (Lethal temperature; itch mite), *Bull. entom. Res.* **33**, 267.
406. Melvin, R. (1934). (Fly development), *Ann. entom. Soc. Amer.* **27**, 406.
407. Metcalf, R. L. (1955). *Organic Insecticides*, Interscience, New York. 392 pp.
408. Metcalf, R. L. *et al.* (1971). (Biodegradable DDT analogues), *Bull. Wld. Hlth. Org.* **44**, 363.
409. Micks, D. W. *et al.* (1974). (Larvicide tests: Flit MLO), *Am. J. trop. Med. Hyg.* **23**, 270.
410. Miller, L. A. (1951). (Canadian tabanid biology), *Canad. J. Zool.* **29**, 240.
411. Miller, R. W. and Pickins, L. G. (1975). (Horn fly control), *J. med. Entom.* **12**, 141.
412. Miller, R. W. and Pickins, L. G. (1975). (Horn fly control), *J. econ. Entom.* **68**, 810.
413. Mills, H. B. (1930). (Collembola), *Proc. Iowa Acad. Sci.* **37**, 389.
414. Milo, A. B. de and Borkovec, A. B. (1974). (New chemosterilant), *J. econ. Entom.* **67**, 457.
415. Ministry of Agriculture, Fisheries and Food (1974). (*Bryobia* mites), *Advis. leaflet* no. 305.
416. Ministry of Agriculture, Fisheries and Food (1976). (Cockroaches), *Advis. leaflet* no. 383.
417. Ministry of Agriculture, Fisheries and Food (1976). (Wasps), *Advis. leaflet* no. 451.
418. Ministry of Agriculture, Fisheries and Food (1976). (Grain beetle), *Advis. leaflet* no. 492.
419. Ministry of Agriculture, Fisheries and Food (1977). (Bruchids), *Advis. leaflet* no. 126.
420. Ministry of Agriculture, Fisheries and Food (1977). (Ants indoors), *Advis. leaflet* no. 366.
421. Ministry of Agriculture, Fisheries and Food (1977). (Pests in food stores), *Advis. leaflet* no. 483.
422. Ministry of Agriculture, Fisheries and Food (1977). (Grain weevils), *Advis. leaflet* no. 219.
423. Ministry of Agriculture, Fisheries and Food (1977). (Pests in farm stores), *Advis. leaflet* no. 368.
424. Ministry of Agriculture, Fisheries and Food (1977). (Mites in food), *Advis. leaflet* no. 489.
425. Ministry of Agriculture, Fisheries and Food (1978). (Warehouse moth), *Advis. leaflet* no. 546.
426. Ministry of Agriculture, Fisheries and Food (1978). (Bacon pests), *Advis. leaflet* no. 373.
427. Ministry of Health (1962). *Control of Mosquito Nuisance in Great Britain*, Memo 238/med, H.M.S.O., London. 38 pp.
428. Ministry of Housing and Local Government (1961). *Pollution of Water by tipping Waste*, H.M.S.O., London.
429. Missiroli, A. (1948). (Houseflies and disease), *Internat. Congr. trop. Med. & Malaria*, Washington.
430. Mohrig, W. (1969). (German Culicidae), *Parasitol. Schiftereihe* **18**, 260.
431. Monro, H. A. (1960). (Fumigants), *Ann. Conf. Pub. Hlth. Inspectors*, Scarborough.
432. Mooser, H. *et al.* (1956). (Larvicidal oils), *Schweiz Z. allg. Path. Bakt.* **19**, 552.
433. Morgan, N. O. and Retzer, H. J. (1974). (CO_2-operated dust gun), *J. econ. Entom.* **67**, 563.
434. Morlan, H. B. *et al.* (1962). (Sterile male control; mosquitoes), *Mosquito News* **22**, 295.
435. Mouchet, J. (1977). (Personal communication), ORSTOM Bondy, France.
436. Mourier, H. (1964). (Behaviour of houseflies), *Vidensk. Medd. Dansk naturh. Foren.* **127**, 181.
437. Mourier, H. (1965). (Behaviour of houseflies), *Vidensk. Medd. Dansk naturh. Foren.* **128**, 221.
438. Mourier, H. and Hannine, S. ben (1969). (Housefly parasites), *Vidensk. Dansk naturh. Foren.* **132**, 211.
439. Mourier, H. (1978). (Personal communication), *Statens Skadedyrlaboratorium*, Lyngby, Denmark.
440. Mulla, M. S. (1963). (Eyefly repellents), *J. econ. Entom.* **56**, 753.
441. Mulla, M. S. *et al.* (1973). (Bait for eyeflies), *J. econ. Entom.* **66**, 1094.
442. Mulla, M. S. *et al.* (1977). (Eyefly attractants), *J. econ. Entom.* **70**, 644.

443. Nagel, I. (1921). (Clothes moth biology), *Z. angew. Entom.* **7**, 164.
444. Nagyar, K. and Makara, G. (1974). (Control of cattle flies), *Abstr. in Rev. appl. Entom.* (*B*) **64**, no. 239.
445. Natvig, L. R. (1948). (Scandinavian Culicidae), *Norsk Entom. Tidsskrift Suppl.* I, 567 pp., Oslo.
446. Newell, W. (1909). (*Iridomyrmex* biology), *J. econ. Entom.* **2**, 174.
447. Nicoli, R. M. and Sautet, J. (1955). (Louse resistance; France), *Inst. nat. d'Hygiene Monogr.* no. 8, Paris.
448. Nielson, B. O. (1963). (Biting midges; Denmark), Naturhist. Mus. Aarhus. 46 pp.
449. Nieschutz, O. (1935). (*Fannia* bionomics), *Zool. Anz.* **110**, 225.
450. Nuorteva, P. (1959). (Blowflies as vectors), *Ann. entom. Fennica* **25**, 1.
451. Nuttall, G. H. (1918). (*Pthirus* biology), *Parasitol.* **10**, 383.
452. Nuttall, G. H. *et al.* (1908). *Ticks; a Monograph of the Ixodidae*, Cambridge University Press. 550 pp.
453. O'Brien, R. D. (1967). *Insecticides*, Academic Press, New York, London.
454. Oda, T. (1966). (Fly dispersal), *Endem. Disease Bull Nagasaki Univ.* **8**, 136.
455. Oldroyd, H. (1946). (Horseflies), *Discovery* **7**, 329.
456. Oldroyd, H. (1954). (Seaweed flies), *Discovery* **15**, 198.
457. Oldroyd, H. (1964). *Natural History of Flies*, Weidenfeld, London. 324 pp.
458. Olsen, R. F. (1964). (Chagas' Disease in U.S.A.), cited by Gerberg (q.v.)
459. Orkin, M. *et al.* (ed) (1977). *Scabies and Pediculosis*, Lippincott, Philadelphia. 203 pp.
460. Page, Ab. and Lubatti, O. F. (1963). (Fumigants), *Ann. Rev. Entom.* **8**, 339.
461. Palicka, P. and Merka, V. (1971). (Scabies in Czechoslovakia), *J. Hyg. Epidem. Microb. Immun.* **15**, 457.
462. Pan American Health Organization (1973). *Control of Lice and Louse-borne Diseases*, Sci. Publ. no. 263.
463. Papp, C. S. (1960). (N. American beetles), *Entom. News* **71**, 69.
464. Parkin, E. A. (1933). (Wood-boring beetles), *Bull. Entom. Res.* **24**, 33.
465. Parkin, E. A. (1934). (*Lyctus bruneus* biology), *Ann. appl. Biol.* **21**, 495.
466. Parkin, E. A. (1943). (Water content of wood and beetle attack), *Ann. appl. Biol.* **30**, 136.
467. Parkin, E. A. (1960). (Damage by clothes moth), *New Scientist* **7**, 1284.
468. Parkin, E. A. (1963). (Pest loss in stored food), *Proc. internat. Seed testing Ass.* **28**, 893.
469. Parkin, E. A. and Woodroffe, G. E. (1961). (*Dermestes* control), *Sanitarian* **69**, 297.
470. Parr, H. C. (1959). (*Stomoxys* control), *Nature* **184**, 829.
471. Parr, H. C. (1962). (*Stomoxys* biology), *Bull. entom. Res.* **53**, 437.
472. Parrish, H. M. (1959). (Danger of wasp and bee stings), *Arch. internat. med.* **104**, 198.
472a. Patton, W. S. (1931). *Insects, Ticks, Mites and venonous Animals.* Grubb, Croydon, 740 pp.
473. Patyk, S. (1976). (Sheep ked control), Abstr. in *Rev. appl. Entom.* (*B*) **66**, no. 353.
474. Pavan, M. (1952). (*Iridomyrmex* venom), *Proc. internat. Cong. Entom. Amsterdam*, **1**, 321.
475. Payot, F. (1918). (*Pthirus* biology), *Bull. Soc. Vaud Sci. nat.* **53**, 127.
476. Peacock, A. D. (1916). (Lice populations), *J. R. Army med. Corps* **27**, 31.
477. Peacock, A. D. (1949–51). (*Monomorium* biology), *Entom. mon. Mag.* **85**, 265; **86**, 129, 171, 294; **87**, 185.
478. Peck, S. M. *et al.* (1943). (Attempted immunisation; scabies), *J. Amer. med. Assn.* **123**, 821.
479. Perfil'ev, P. P. (1966). *Sandflies of USSR*, Acad. Sci. USSR Zool Inst. transl. Jerusalem Israel Prog. Sci. Press. 363 pp.
480. Petersen-Braun, M. (1977). (*Monomorium* biology), *Proc. 8th internat. Cong. Union Study Soc. Insects.*
481. Peus, F. (1937). (*Thaumatomyia* bionomics), *Z. hyg. Zool.* **29**, 207.
482. Peus, F. (1938). *Die Flöhe*, Hygienisch Zool. Monogr. Scops, Leipzig. 106 pp.
483. Pfadt, R. E. *et al.* (1975). (Sheep ked control), *J. econ. Entom.* **68**, 468.
484. Phillip, C. B. (1931). (Horseflies of Minnesota), *Minnesota agric. exp. Sta. Bull.* **80**, 10.
485. Pimentel, D. *et al.* (1951). (Impregnated paper against flies), *Soap* **27**(1), 102.
486. Pippin, W. F. (1970). (*Triatoma* biology), *J. med. Entom.* **7**, 30.

487. Planet, L. (1905). *Araignées*, Derolles, Paris. 341 pp.
488. Poincelot, R. P. (1972). (Composting), *Connecticut agr. exp. Sta. Bull* no. 727.
489. Pomerentzev, B. I. (1950). *Fauna of USSR IV (2) Ixodidae*, Trans. Amer. Inst. Biol. Soc., Washington. 199 pp.
490. Pratt, H. D. and Littig, K. S. (1962). (Important ticks), *U.S. Publ. Hlth. Service Publ.* no. 772, 42 pp.
491. Prevett, P. F. (1959). (*Sitophilus zea-mais*), *Bull. entom. Res.* **50**, 697.
492. Quarterman, K. D. *et al.* (1949). (Fly breeding in Florida), *Amer. J. trop. Med.* **29**, 973.
493. Quate, L. W. (1955). (Sandflies of U.S.A.), *Univ. California Pub. Entom.* **10**, 103.
494. Raastal, J. E. (1974). (Blackflies in Norway), *Proc. 3rd intern. Congr. Parasitol. Munich* **2**, 918.
495. Rageau, J. and Mouchet, J. (1967). (Biting arthropods; Camargue), *Cahiers ORSTROM* **5**, 263.
496. Rasmussen, S. (1967). (Fumigation for wood beetles), *Materialen & Organismen* **2**, 65.
497. Readio, P. A. (1927). (N. American Reduviidae), *Univ. Kansas Sci. Bull.* **17**, 5.
498. Raybould, J. N. (1966). (Fly traps for sampling), *J. econ. Entom.* **59**, 639.
499. Reeves, E. L. and Garcia, C. (1969). (Sticky seeds to control larvae), *Mosquito News* **29**, 601.
500. Reichmuth, W. (1936). (Silverfish biology), *Z. Hyg. Zool Schdlsbk.* **28**, 65.
501. Reynolds, D. G. and Vidot, A. (1978). (Biting midge control; Seychelles), *PANS* **24**, 19.
501a. Richards, O. W. & Herford, G. V. (1930). (Stored food pests in London), *Ann. appl. Biol.* **17**, 367.
502. Richardson, H. (1940). (Millbank disinfestor), *J. R. Army med. Corps* **74**, 121.
503. Richter, I. *et al.* (1976). (Muscalure tests), *Experientia* **32**, 186.
504. Rickett, F. E. *et al.* (1972). (Pyrethrum dermatitis), *Pestic. Sci.* **3**, 57.
505. Rioux, J. A. and Golvan, Y. J. (1969). (*Leishmaniasis in S. France*), Inst. nat. Santé & Recherche Méd., Paris.
506. Ritter, F. J. *et al.* (1977). (Pheromones of *Monomorium*), in *Crop Protection Agents* (edited by McFarlane), J. E., Academic Press, London.
507. Rivnay, E. (1932). (Bed bug behaviour), *Parasitol.* **24**, 121.
508. Rivosecchi, L. (1962). (Sterile male release; houseflies), *Riv. di Malariol* **23**, 71.
509. Roberts, F. H. (1931). (*Piophila* bionomics) *Queensland agric. J.* **35**, 227.
510. Rodriguez, J. G. *et al.* (1970). (Mites control fly larvae), *J. med. Entom.* **7**, 335.
511. Roth, M. and Willis, E. R. (1952). (Cockroach behaviour), *J. exp. Zool* **119**, 483.
512. Roth, M. and Willis, E. R. (1954). (Cockroach reproduction), *Smithsonian Misc. Collect.* **12**, 1.
513. Roth, M. and Willis, E. R. (1957). Medical Importance of Cockroaches, *Smithsonian Misc. Collect.* **134**(10). 1.
514. Roubaud, E. (1936). (Biothermal maggot control), *Q. Bull. Hlth. Org. L. of Nations.*
515. Russel, F. E. *et al.* (1972). (Spider bite treatment), *in Current Therapy*, (edited Conn), A., Saunders, Philadelphia.
516. Rust, M. K. and Reierson, D. A. (1978). (Cockroach contact poisons), *J. econ. Entom.* **71**, 704.
517. Ryckman, R. E. (1962). (*Triatoma* biology), *Univ. California Publ. Entom.* **27**, 93.
518. Ryckman, R. E. and Ryckman, A. E. (1967). (N. American Reduviidae), *J. med. Entom.* **4**, 326.
519. Sacca, G. (1964). (Housefly varieties), *Ann. Rev. Entom.* **9**, 341.
520. Sacca, *et al.* (1968). (Fly dispersal), *Archiv. Soc. med-chirug. di Messina* **12**(3), 1.
521. Saxena, B. P. *et al.* (1977). (New chemosterilant), *Nature* **270**, 512.
522. Schlusche, M. (1936). (Hymenopterous stings), *Zool Jahrber.* (*Anat.*) **61**, 77.
523. Schoof, H. F. *et al.* (1954). (Fly breeding in towns), *J. econ. Entom.* **47**, 245.
524. Schoof, H. F. *et al.* (1961). (Dichlorvos disinsection of aircraft), *Bull. Wld. Hlth. Org.* **24**, 623.
525. Schreck, L. E. *et al.* (1975). (*Stomoxys* traps), *J. med. Entom.* **12**, 338.
526. Schroeter, A. (1977). (Scabies; a venerial disease), *in Scabies and Pediculosis*, Lippincott, Philadelphia. 203 pp.
527. Schrut, A. H. and Waldron, W. G. (1963). (Entomophobia), *J. Am. med. Assn.* **186**, 429.
528. Schulz, N. (1925). (Moth digestion of keratin), *Biochem. Z.* **154**, 124.
529. Scott, H. G. (1964). (Larvae with urticating hairs), *Pest Control* **32**(9), 24.
530. Séguy, E. (1944). *Faune de France* no. 43, (Parasitic insects), Paris.

531. Senevet, G. (1937). *Faune de France* no. 32, (Ixodidae), Paris.
532. Sgrowaska, P. (1977). (Personal communication), Nat. Inst. Hygiene, Warsaw.
533. Shawarby, A. (1953). (Fleas in Egypt), *Bull. entom. Res.* **44**, 377.
534. Sherman M. *et al.* (1962, 1967, 1971). (Feeding insecticide to chickens to kill flies), *J. econ. Entom.* **55**, 990; **60**, 1395; **64**, 1150.
535. Schwardt, H. A. (1936). (Horseflies of Arkansas), *Univ. Arkansas agric. expt. Sta. Bull.* no. 332.
536. Skierska, B. (1976). (Polish biting midges), abstr. in *Rev. appl. Entom.* (*B*) **64**, no. 817.
537. Simmons, S. W. *et al.* (1942). (*Stomoxys* control), *J. econ. Entom.* **35**, 589.
538. Simmons, S. W. *et al.* (1944). (Larvicides for flies), *J. econ. Entom.* **37**, 135.
539. Sitar, A. (1977). (Personal communication), Inst. Hlth. Prot. of Serbia, Belgrade, Yugoslavia.
540. Siverly, R. E. and Schoof, H. F. (1955). (Fly breeding in towns), *Ann. entom. Soc. Amer.* **48**, 258.
541. Smart, J. (1934). (Blackfly biology), *Proc. R. physic. Soc. Edin.* **22**, 217.
542. Smart, J. (1948). *Insects of Medical Importance*, Brit. Mus. (Nat. Hist.).
543. Smith, C. N. (1963). (Chemosterilants), *Bull. Wld. Hlth. Org. (Supp)* **29**, 99.
544. Smith, C. N. *et al.* (1946). (*Dermacentor* biology), *U.S. Dept. Agric. Tech. Bull.* no. 905. 74 pp.
545. Smith, C. N. *et al.* (1964). (Chemosterilants), *Ann. Rev. Entom.* **9**, 239.
546. Smith, C. W. and Micks, D. W. (1968). (*Loxosceles* venom), *Am. J. trop. Med. Hyg.* **17**, 651.
547. Smith, E. L. (1970). (Californian Thysanura), *Pan-Pacific Entom.* **46**, 212.
548. Smith, M. R. (1965). (American ants), *U.S. Dept. Agric. Tech. Bull.* 1326.
549. Smith, L. M. and Lowe, H. (1948). (Californian biting midges), *Hilgardia* **18**, 167.
550. Solomon, M. E. (1961–62). (Food mites), *Sanitarian* **69**, 291; *Ann. appl. Biol.* **50**, 178.
551. Somme, L. (1959). (*Stomoxys* control), *Norsk entom. Tijskr.* **12**, 113.
551a. Southcott, R. V. (1978). (Lepidopterism, esp. Australia), *Rec. Adelaide Children's Hospital* **2**, 87.
552. Staal, G. B. (1975). (Hormones as insecticides), *Ann. Rev. Entom.* **20**, 417.
553. Stannke, H. L. (1948). (Scorpions), *Arizona Sta. Coll. Bull.* no. 71. 61 pp.
554. Steiner, P. (1937). (*Hylotrupes* biology), *Z. angew. Entom.* **23**, 531.
555. Step, E. (1932). *Ants, Bees and Wasps*, Warne, London. 238 pp.
556. Steve, P. C. (1960). (*Fannia* control), *J. econ. Entom.* **53**, 999.
557. StGeorge, R. A. *et al.* (1950). (Old house borer), *Pest Control* **25**(2), 29.
558. Strebel, O. (1932). (Springtails), *Z. morph. Oekol. Tiere* **25**, 31.
559. Stumper, R. (1960). (Ant stings), *Naturwiss.* **47**, 457.
560. Sturtevent, A. H. (1918). (Drosophila as a vector), *J. Parasitol.* **5**, 84.
561. Sullivan, W. N. *et al.* (1962, 1964). (Aircraft disinsection), *Bull. Wld.Hlth. Org.* **27**, 263; **30**, 113.
562. Sullivan, W. N. *et al.* (1978). (Water-based aerosol for disinsection), *Bull. Wld. Hlth. Org.* **56**, 129.
563. Sweetman, H. L. (1934). (Silverfish biology), *Bull. Brooklyn entom. Soc.* **29**, 158.
564. Sweetman, H. L. (1938). (Silverfish biology), *Ecol. Monogr.* **8**, 285.
565. Sweetman, H. L. (1952). (Firebrat bionomics), *Trans. 9th inter. Cong. Entom.* **1**, 411.
566. Sweetman, H. L. and Kulash, W. M. (1944). (*Ctenolepisma*), *J. econ. Entom.* **37**, 444.
567. Swingle, L. D. (1913). (Sheep ked biology), *Univ. Wyoming agric. exp. Sta. Bull.* no. 99.
568. Szabo, J. B. (1964). (Blackflies in Hungary), *Opusc. Zool. Budapest* **5**, 113.
569. Szabo, J. B. (1971). (Blackflies in Hungary), *Parasitol. Hung.* **4**, 169.
570. Taha, A. M. (1963). (Fly dispersal), *J. Egypt. Pub. Hlth. Assn.* **38**, 143.
571. Talens, A. F. de *et al.* (1970). (Landing of houseflies), *J. exp. Biol.* **52**, 233.
572. Tashiro, H. and Schwardt, H. (1949). (Horseflies of U.S.A.), *J. econ. Entom.* **42**, 269.
573. Teschner, D. (1958). (Flies and human faeces), *Z. angew. Entom.* **45**, 153.
574. Theodor, O. (1935). (Sandfly bite reactions), *Z. morph. Oekol. Tiere* **19**, 678.
575. Thomsen, M. (1934). (Control of flies in manure), *Q. Bull. Hlth. Org. L. of Nations* **3**, 304.
576. Thomsen, M. (1938). *Stuefluen og Stikfluen*, Christanson, Copenhagen.
577. Thomsen, M. *et al.* (1936). (Fly breeding sites), *Bull. entom. Res.* **27**, 559.
578. Thornhill, W. E. (1974). (CO_2 pressurised sprayer), *PANS* **20**, 241.

579. Tiflof, V. and Ioff, J. (1932). (Rodent flea biology: in Russian), *Rev. Microbiol.* **11**, 2.
580. Tinker, M. E. and Stojanovich, C. J. (1972). (N. American mosquito key), in *Vector Control in International Health*, Wld. Hlth. Org., Geneva.
581. Titschack, E. (1922, 1936). (Clothesmoth biology), *Z. tech. Biol.* **10**, 1; *Z. wiss. Zool.* **124**, 13; **128**, 509; *Z. ang. Ent.* **23**, 1.
582. Tomlinson, T. G. (1946). (Sewage flies), *Water Poll. Res. tech. Paper* 9.
583. Tomlinson, T. G. and Jenkins, S. H. (1947). (Sewage fly control), *Publ. Wks. Roads Transport Congr.* p. 460.
584. Tomlinson, T. G. and Stride, A. O. (1945). (Sewage flies), *J. Inst. Sewage Purific.* **2**, 140.
585. Trojan, E. (1930). (Sting of hymenoptera), *Z. Morph. Oekol. Tiere* **19**, 678.
586. Tsutsumi, C. (1966, 1968). (Housefly behaviour), *Jap. J. med. Sci. Biol.* **19**, 155; **21**, 195.
587. Tyler, P. S. (1961). (Swarming houseflies), *Sanitarian* **69**, 285.
588. Ussova, Z. V. (1961). (Flies of W. USSR), trans. from Russian by Israel Prog. Sci. Transl. (Simuliidae p. 236 *et seq.*).
588a. Vayssiere, P. & Lepesme, P. (1937). (Bostrychid grain pests), *Agron. colon.* **233**, 129.
589. Vitzhum, H. (1931). *Acarina* (*Handb. d. Zool.*) Gruyter, Berlin. 160 pp.
590. Vladimirova, M. S. *et al.* (1938). (Fly maggot competition; in Russian), *Med. Parasitol.* 7, 755.
591. Voorhorst, R. *et al.* (1969). *House-dust Atopy*, Stafleu, Leiden. 159 pp.
592. Wade, C. F. *et al.* (1961). (Mite predators of flies), *Ann. entom. Soc. Amer.* **54**, 776.
593. Wagner, R. E. (1961). (Wasp control), *J. econ. Entom.* **56**, 628.
594. Wall, W. J. and Swift, A. H. (1954). (Firebrat damage), *J. econ. Entom.* **47**, 187.
595. Wang, J. S. and Brooks, T. S. (1970). (Odorous house ant), *J. econ. Entom.* **63**, 1971.
596. Wassermann, M. *et al.* (1974). (DDT residues in man), *CEC/EPA/WHO Symposium*, Paris.
597. Waterhouse, D. F. (1952). (Keratin digestion by moths), *Austral. J. Sci. Res.* **5**, 143.
598. Waterhouse, D. F. (1957). (Mothproofing agents), *Adv. Pest Control* **2**, 207.
599. Waterhouse D. F. and Norris, K. R. (1965). (Bushfly repellents), *Austral. J. Sci.* **28**, 351.
600. Watkinson, I. A. and Clarke, B. S. (1974). (Insect hormones), *PANS* **19**, 488.
601. Watson, J. A. (1964). (Firebrat biology), *J. Insect Physiol.* **10**, 305.
602. Watt, J. *et al.* (1948). (Flies and disease), *Publ. Hlth. Repts. Wash.* **63**, 1319.
603. Wattal, B. L. and Kalra, N. (1961). (Bug trap), *Indian J. Malariol.* **15**, 157.
604. Weidhaas, D. E. and LaBreque, G. C. (1970). (Housefly populations), *Bull. Wld. Hlth. Org.* **43**, 72.
605. Weidhaas, D. E. *et al.* (1962). (Sterile male release; mosquitoes), *Mosquito News* **22**, 283.
606. Weiser, J. (1963). (Biological control), *Bull. Wld. Hlth. Org.* **29**, (*Suppl*) 107.
607. White, R. E. (1962). (Anobiidae of Ohio), *Ohio Biol Sci. N.S.* **1**(4), 1.
608. Whitfield, F. G. *et al.* (1958). (Clothes moth biology), *Lab. Practice* **7**, 210.
609. Whitsel, R. H. and Schoeppner, R. F. (1966). (*Leptoconops* biology), *California Vector Views* **13**, 17.
610. Wichmand, H. (1953). (Impregnated cardboard for flies), *Nature* **172**, 758.
611. Wiesmann, R. (1962). (Fly senses), *J. Hyg. Epid. Microbiol, Immun.* **6**, 303.
612. Wiesmann, R. (1962). (Fly behaviour), *Mitt. Schweiz entom. Ges.* **35**, 69.
613. Wigglesworth, V. B. (1944). (Dehydrating dusts), *Nature* **153**, 493.
614. Wilhelmi, J. (1919). (Fly breeding choice), *Z. angew. Entom.* **5**, 261.
615. Wilkinson, R. N. *et al.* (1971). (Larvicidal pellets), *J. econ. Entom.* **64**, 11.
616. Wille, J. (1920). (Cockroach biology), *Monog. angew. Ent. (Beih. Z. angew. Ent.)* **7**, 1.
617. Willem, V. (1900). (Thysanura), *Mem. Sav. Etr. Acad. R. Belge* **58**, 1.
618. Williams, R. W. (1951). (*Culicoides* in Alaska), *Ann. entom. Soc. Amer.* **44**, 173.
619. Willis, E. R. and Lewis, N. (1957). (Cockroach resistance to starvation), *J. econ. Entom.* **50**, 438.
620. Willis, E. R. *et al.* (1958). (Cockroach biology), *Ann. entom. Soc. Amer.* **51**, 53.
621. Wisecup, C. B. and Hayslip, N. C. (1953). (Mole cricket control), *U.S. Dept. Agric. Leaflet* no. 237.
622. Wolfe, L. S. and Peterson, D. G. (1958). (Trapping blackfly larvae), *Canad. J. Zool.* **36**, 863.
623. Wolfe, L. S. and Peterson, D. G. (1959). (Blackflies near Quebec), *Canad. J. Zool.* **37**, 137.

624. Woodroffe, G. E. (1951). (House moth biology), *Bull. entom. Res.* **41**, 529.
625. Woodroffe, G. E. (1953). (Insects in bird nests), *Bull. entom. Res.* **44**, 739.
626. Woodroffe, G. E. and Southgate, B. J. (1951). (Pests from bird nests), *Proc. zool. Soc. London* **121**, 55.
627. World Health Organization (1961). (Aircraft disinsection), *11th Rpt. Insecticide Committee Tech. Rpt. Ser.* no. 206.
628. World Health Organization (1963). (Resistance and Vector control), *13th Rpt. Insecticide Committee Tech. Rpt. Ser.* no. 265.
629. World Health Organization (1967). (Aircraft disinsection), *16th Rpt. Insecticide Committee Tech. Rpt. Ser.* no. 356.
630. World Health Organization (1967). (Preventing reintroduction of Malaria), *Rpt. Insecticide Committee Tech. Rpt. Ser.* no. 374. 32 pp.
631. World Health Organization (1970). (Resistance & vector control), *17th Rpt. Insecticide Committee Tech. Rpt. Ser.* no. 443.
632. World Health Organization (1971). (Application of pesticides), *18th Rpt. Insecticide Committee Tech. Rpt. Ser.* no. 465.
633. World Health Organization (1972). *Vector Control in International Health*, Geneva, 144 pp.
634. World Health Organization (1973). (Plague incidence), *WHO Chronicle* **27**, 369; **28**, 71.
635. World Health Organization (1973). *Specifications for Pesticides*, 4th edn. WHO, Geneva. 333 pp.
636. World Health Organization (1974). (Typhus incidence), *WHO Chronicle* **28**, 427.
637. World Health Organization (1974). *Equipment for Vector Control*, 2nd edn. WHO, Geneva. 179 pp.
638. World Health Organization (1975). (Toxicity schedule), *WHO Chronicle* **29**, 397.
639. World Health Organization (1976). (Resistance), *22nd Rpt. Insecticide Committee Tech. Rpt. Ser.* no. 585.
640. Wright, C. G. (1960). (*Lyctus plannicollis* biology), *Ann. entom. Soc. Amer.* **53**, 285.
641. Wright, C. G. (1965). (Cockroach habits), *J. econ. Entom.* **58**, 1032.
642. Wright, J. E. and Spates, G. E. (1972). (Fly control; juvenoids and parasites), *Science* **178**, 1292.
643. Wright, J. E. *et al.* (1976). (Juvenoids for fly control), *J. econ. Entom.* **69**, 79.
644. Wright, R. H. (1962). (Mosquito behaviour), *Canad. Entom.* **94**, 1022.
644a. Yamaguchi, T. *et al.* (1976). Water-based aerosols. *Aerosol Reports* **15**, 255.
645. Zahar, A. R. (1951). (Blackfly biology), *J. Animal Ecol.* **20**, 33.
646. Zeigler, T. W. (1955). (Silverfish control), *Pest Control* **23**(6), 9.
647. Zhdanova, T. G. (1976). (Biting midges of Ukraine; in Russian), abstr. in *Rev. appl. Entom.* (*B*) **64**, no. 2030.
648. Zivkovic, V. (1971). (Blackflies in Yugoslavia), *Acta veterin. Belgrade*, **21**, 225.
649. Zivkovic, V. (1975). (Blackflies from Danube now), *Acta veterin, Belgrade*, **25**, 279.
650. Zivkovic, V. and Burany, B. (1972). (Blackflies in Yugoslavia), *Acta veterin. Belgrade* **22**, 133.
651. Zivkovic, V. and Petrovic, Z. (1976). (Medical entomology in Yugoslavia), *Acta veterin. Belgrade* **26**, 9.
652. Zumpt, Z. (1965). *Myiasis in the Old World*, Butterworth, London. 267 pp.

Index

Numbers in bold type refer to pages with illustrations